U0135116

A+ GRE/GMAT 數學

GRE General Test

任教美國洛杉磯華裔數學老師編撰

本書遵照主辦美國大學研究所入學考試 GRE
以及 GMAT 的教育測試中心（ETS）最新試題指引編撰，
簡單易學！

本書內容包括全部所需數學複習以及八份模擬試題，
附詳細試題解題說明。

★ Amazon、書店、學校　美國暢銷 20 萬冊 ★

作者 / 楊蓉昌

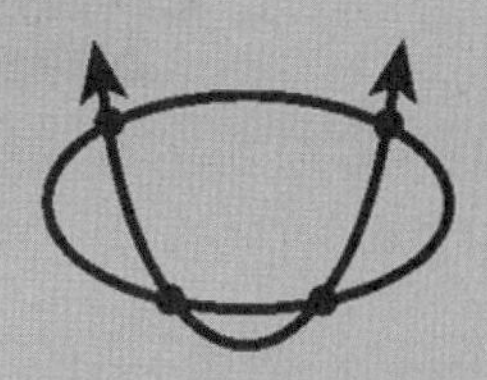

A$^+$ GRE / GMAT數學

作者／楊蓉昌
出版日期／2017年10月
版次／初版一刷
ISBN／978-986-95526-4-6
定價／NT$ 450

出版　城邦印書館股份有限公司
地址　台北市中山區民生東路二段141號B1
電話　02-2500-2605　傳真　02-2500-1994

PREFACE

I am a math teacher in Los Angeles. In order to provide a clear and easy way for my students to learn math in my classes, I designed **notes** which simplify and outline the concepts and skills in course textbooks. The notes focus on the correct and easy steps in the problem-solving process. Each topic is organized in one or two pages of explanation for easy understanding. I wrote this book following the same pattern of problem-solving in math books used by schools across Asia. This pattern has been proven to be excellent for student-learning.

The **GRE program** and tests are designed by the Graduate Record Examination Board and Education Test Service (ETS).
The **GMAT program** and tests are designed by the Graduate Management Admission Council (GMAC) and Educational Test Service (ETS).

This book focuses on the Quantitative Reasoning of GRE Test and the Quantitative Section of GMAT Test. The content is simple and easy to read like a math dictionary for the students to study and review all the math topics and skills necessary to ace the GRE/GMAT Math Tests. This book will help you to be familiar with different types of questions on the tests and facilitate the learning process. If you are not familiar with a specific topic in math, you can easily find it and learn it in this book.

Many GRE/GMAT test-prep books in the market emphasize a variety of test-taking strategies. I believe the best strategy for a student's success is to practice, practice, and more practice. Since GRE/GMAT Math Tests are not tied to specific textbooks or teaching methods, the more practice tests a student does, the better the students can ace the test on the real test day. The practice tests in this book can help you find out which are as you need to work on. The content of this book covers all math you need to review for the GRE/GMAT Math tests.

The goal of this book is to give the students the tools to master the GRE/GMAT Math tests with confidence and optimism. If you are a math professional, you may send your comments or suggestions to us.

Rong Yang
Los Angeles, California

楊蓉昌

Taipei, Taiwan

HOW TO USE THIS BOOK

1. The computer-adaptive software is not presented in this book. A GRE POWERPREP11 software will be sent to you from Educational Test Service (ETS) for tutorial when you register for the GRE revised General Test. The computer software for GMAT tutorial is available for download at no charge at www.mba.com.
2. Questions are sometimes repeated in actual GRE or GMAT Math Tests. Therefore, you should study and get used to all of the practice test questions in this book.
3. Once you have evaluated your performance on the practice tests in this book, you can find what types of preparation you might want to do for the actual test.
4. The **practice tests** for GRE math **subject test** could be downloaded at no charge as a PDF file from www.ets.org/gre. Use this book as your math dictionary when needed help for the actual test.
5. If there are any math areas or topics you cannot find in this book or you want to learn how to use the graphing calculator, check out A-Plus Notes Series:
   "A-Plus Notes for Algebra, A Graphing Calculator Approach"

HOW TO STUDY THE PRACTICE TESTS IN THIS BOOK

1. Work on each practice test to see how well you do, check your answers, and count your score using the answer key at the end of each test.
2. In the practice tests in this book, similar questions of different cases are often put together for you to study them simultaneously. (See Pages 366, 390, 399, 423)
3. There are many overlap in math areas between GRE and GMAT. Work on all of the practice tests in this book.
4. If your answer to a question is incorrect, find out how to reach the right answer by reading the related areas and topics in this book. You can find Table of Contents on Page 5, Areas and Topics Covered each test on Page 10,11, 12, and Index on Page 467 ~ 470.

ATTENTION: MATH TEACHERS AND STUDENTS

This book will make the teachers' job and the students' learning process easier if the students have it on hand in class. This book is a complete set of lecture "notes" for the students. They only need to read it.

Most schools in the United States take back the textbook from the student after the course is completed. This book is simple and easy like a math dictionary for the student to review when needed in the future. You may use it as a reference book together with the textbook, or use it as the textbook in your classes.

You may visit this book and readers' reviews, or write your comments at

www.amazon.com

The Publisher

TABLE Of CONTENTS

It covers a variety of test questions on specific field of mathematics And does not show specific pattern of test questions. The practice test for GRE math subject test could be downloaded at no charge as a PDF file from www.ets.org/gre .
Use this book as your math dictionary when needed help for studying the actual test.

UNDERSTANDING THE GRE TESTS

1. The **GRE** (Graduate Record Examinations) includes **General Test** and **Subject Tests** It is administrated by Graduate Record Examination Board and Educational Testing Service (**ETS**). Most colleges require its applicants to take the General Test, the Subject Test, or both, as a factor to assist the admissions for graduate study.

2. The GRE General Test is needed by almost all graduate programs. It is used to test aptitude instead of specific knowledge of a student. You can search for information including test bulletins, registration forms, and which subject test is needed for colleges that interest you at www.ets.org/gre/revised/bulletinandforms.

3. The GRE General Test is a computer-delivered test given year-round at computer-based test centers in the U.S. and worldwide, and at paper-based test centers in the countries where computer-delivered testing is not available. A GRE POWERPREP practice tests will be sent to you for tutorial when you register for the computer-delivered GRE Test.

4. **Computer-delivered GRE General Test** (3 hours and 45 minutes)
   Analytical Writing (1 section, 2 writing tasks (essays), 30 min. per task)
   Verbal Reasoning (2 sections, 20 questions per section, 30 min. each)
   Quantitative Reasoning (2 sections, 20 questions per section, 35 min. each)
   Unscored section (20 questions, 35 minutes) and Research Section
   a. An unidentified unscored section (verbal or math) may be included in the test and may appear in any order in the test. They are tested for possible use in future tests and will not count toward your score.
   b. An identified research section for research purpose may be included in place of the unscored section. It will always appear at the end of the test and not count toward your score.
   c. An on-screen calculator will be provided for the Quantitative Reasoning sections. It has three memory buttons: memory recall [MR], memory clear [MC], and memory sum [M+].
   d. You may skip questions. The testing software has a "mark and review" feature that allows you to mark questions you want to revisit or change your answers.
   e. Some questions require you to select one or more answer choices, or enter a numeric answer.
   f. you should answer all questions on the Verbal Reasoning and Quantitative Reasoning tests and not to leave any question unanswered. Nothing is subtracted from a score if you answer a question incorrectly.

-----Continued-----

5. **Paper-delivered General Test** (3 hours and 30 minutes)
   Analytical Writing (2 sections, 1 writing task per section, 30 min. per section)
   Verbal Reasoning (2 sections, 25 questions per section, 35 min. per section)
   Quantitative Reasoning (2 sections, 25 questions per section, 40 min. per section)
   a. You will enter all your answers in the test book itself.
   b. A basic calculator on the Quantitative Reasoning will be provided at the test site. You are not allowed to use your own calculator.
   c. You are free to skip questions. You may revisit or change your answer to any question by erasing it completely.
   e. Some questions require you to select one or more answer choices, or enter a numeric answer.
   f. you should answer all questions on the Verbal Reasoning and Quantitative Reasoning tests and not to leave any question unanswered. Nothing is subtracted from a score if you answer a question incorrectly.
6. On the GRE General Test, the analytical writing score is reported on a scale from 0 to 6, in half-point increments. The verbal or quantitative scores are reported separately on a scale from 130 to 170, in 1-point increments. First the raw score is computed. The raw score is the number of questions answered correctly. A raw score is converted to a scaled score. The scaled score and percentile ranking of your score will tell you what percent of test takers scored beneath a given score in order to compare with those of students who have taken different tests at different dates. Small differences in score scale will look like small difference. Bigger differences (10 or 20) in score will look like a big difference.
7. You will receive your scores on the computer-delivered GRE General Test about 10 to 15 days after the test date. You will receive your scores on the paper-delivered GRE General Test within 6 weeks. All scores you earned within 5 years will be reported to the colleges you designate.
8. The Quantitative Reasoning test questions on **GRE General Test** include four types of questions. **Quantitative Comparisons, Multiple-choice questions (Select One Answer Choice), Multiple-choice questions (Select One or More Answer Choices), and Numeric Entry Questions.** It covers the following basic math topics:
   1. Arithmetic 2. Algebra 3. Geometry 4. Data Analysis
9. The **GRE Subject Tests** measure how well you have majored and learned in a specific fields of study in college. The math subject Tests are given at paper-based test centers worldwide three times a year (October, November, and April).
   It covers the areas in **Algebra** (25%), **Additional Topics** (25%), and **Calculus** (50%).
   Additional Topics (Pre-Calculus): Set Theory, Sequences and Series, Probability and Statistics, Permutations and Combinations, Polar and Complex Coordinates, Numerical Analysis, Geometry, Trigonometry, Topology, Logic, Limit and Continuity.

The test consists of approximately 66 multiple- choice questions for 170 minutes.
The total scaled scores is from 200 to 990.

UNDERSTANDING THE GMAT TEST

1. The **GMAT** (Graduate Management Admission Test) is administrated by the Graduate Management Admission Council (**GMAC**) and Educational Testing Service (**ETS**) for Graduate Business Education. It is used as one significant factor among other factors to assist the admissions for graduate study in business and management programs.

2. The GMAT does not measure the skill of any particular subject area of an applicant. It measures overall ability how an applicant is ready for the graduate business and management programs. It enrolls students from different fields in undergraduate study and work experience.

3. The GMAT is needed by almost all business schools. You can search for information including test bulletins, registration forms, or colleges that interest you at Website: www.gmac.com, www.mba.com, or E-mail: gmat@ets.org.

4. The GMAT Test is a computer-adaptive test (**CAT**). It has four-part test consisting of four separately timed sections:
   Analytical Writing : Analysis of an argument (1 essay, 30 minutes)
   Integrated Reasoning : Analyze and make conclusions from multiple sources and formats (12 questions, 30 minutes)
   Verbal (41 multiple-choice questions, 75 minutes)
   Quantitative (Math) (37 multiple-choice questions, 75 minutes)

5. In the **Analytical Writing**, you type your essay using the computer. The cope of your essay and your score report will be sent to the colleges you choose.

6. In the **Integrated Reasoning**, the questions can be verbal, quantitative, or both. You need to make conclusions from graphic, numeric, and verbal information. You have an online calculator in this section.

7. In the **Verbal** and **Quantitative** multiple-choice questions, different students will be given different questions. The computer selects the next question that best reflects your ability level on all previous questions. When you answer the easier questions correctly, you get a chance to answer harder questions for earning a higher score. It does not mean you will get an easier question if you got the previous question wrong. Don't waste time at estimating the difficulty of any question. There are only very few questions that are too easy or too hard for you. You cannot skip or go back to change your answers. If you don't know the answer, you must eliminate as many wrong answers as possible, then select and confirm the best choice.

-----Continued-----

8. The score on the Verbal section, it requires only basic English language. It includes reading comprehension, critical reading, and sentence correction.

9. The score on the Analytical Writing Assessment (**AWA**) is reported on a scale from 0 to 6, in a half-point increment, by the average of scores given by expert readers on your essay. A score of zero is given to essay that is off-topic, or is blank. Then, the average score is rounded.

10. The score on the Integrated Reasoning is determined on a scale of 1 to 8, in a one-point increments. You must answer all parts of a single question to earn credit.

11. The **Total GMAT** scores on the Verbal and Quantitative multiple-choice questions are determined by a complex procedure, based on the number of questions you answer correctly or incorrectly, the level of difficulty and the other statistical properties of the questions. The total scores on 41 verbal questions and 37 quantitative questions are combined to produce a total multiple-choice questions scaled score, on a 200 to 800 scale, in a 10-point increment. The scaled score and percentile ranking in your score report will explain how your score on the test compares with those of students who have taken different tests at different dates.

12. The software for tutorial is available for the GRE computer-adaptive test. You can download it at no charge at www.mba.com.

13. The **GMAT Math Test (Quantitative Section)** includes two types of multiple -choice questions, **Problem Solving Questions** and **Data Sufficiency Questions** . It covers the following basic math knowledge:

1. Arithmetic
2. Algebra
3. Geometry
4. Word Problems

In the Data Sufficient Questions, each answer statement must be able to find only one single value to be sufficient.

Areas and Topics Covered
(GRE revised Math General Test)

The GRE revised Math General Test covers the following areas:

1. Arithmetic **2. Algebra**
3. Geometry **4. Data Analysis**

You are not expected to have studied every topics on the test.

Areas	GRE Topics	pages	GRE Topics	pages
Arithmetic	**1.1** Integers **1.2** Fractions **1.3** Exponents and Roots **1.4** Decimals	15 ~ 20 21 ~ 31 37 ~ 39 21 ~ 35	**1.5** Real Numbers **1.6** Ratio **1.7** Percent	15 25 25 ~ 33
Algebra	**2.1** Operations with Algebraic Expression **2.2** Rules of Exponents **2.3** Solving Linear Equations and Systems **2.4** Solving Quadratic Equat. **2.5** Solving Linear Inequal.	95 ~ 97 101 119 ~ 123 163 133 127	**2.6** Functions **2.7** Applications for Word Problems **2.8** Coordinate Geometry **2.9** Graphs of Functions	109,133,149 171 ~ 185 111 ~ 115 251 ~ 295 109 ~ 114 135, 149 151 ~ 157
Geometry	**3.1** Lines and Angles **3.2** Polygons **3.3** Triangles	214 235 221~ 234	**3.4** Quadrilaterals **3.5** Circles **3.6** Three-Dimensional Figures	235 239 281~293
Data Analysis	**4.1** Graphing Methods for describing data **4.2** Numerical Methods for Describing Data **4.3** Counting Methods	57 ~ 60 57 ~ 60 61 ~ 76	**4.4** Probability **4.5** Distributions of Data, Random Variables, and Probability Distributions **4.6** Data Interpretation	67 ~ 72 57 ~ 76 57 ~ 76

Source: (ETS) The Official Guide to the GRE revised General Test.

Areas and Topics Covered
(GMAT Math Test)

The GMAT Math Test covers the following areas:

1. Arithmetic **2. Algebra**
3. Geometry **4. Word Problems**

You are not expected to have studied every topics on the test.

Areas	GMAT Topics	pages	GMAT Topics	pages
Arithmetic	**1.** Integers	15, 17	**7.** Powers and Roots	37 ~ 39
	2. Fractions	21, 31	**8.** Descriptive Statistics	57~72
	3. Decimals	21, 29	**9.** Sets	61
	4. Real Numbers	15	**10.** Counting Methods	62, 73
	5. Ratio and Proportion	25	**11.** Discrete Probability	67
	6. Percents	25 ~ 33		
Algebra	**1.** Algebraic Expressions	91 ~ 97	**6.** Solving Quadratic Equations	133
	2. Equations	109	**7.** Exponents and Radicals	101
	3. Solving Linear Equations	119 ~ 123	**8.** Solving Inequalities	127~129
	4. Solving Two Linear (System) of Equations	163	**9.** Solving Absolute-Value Equation and Inequality	131
	5. Solving Polynomial Equations by Factoring	139 ~ 143	**10.** Functions	109
Geometry	**1.** Lines	214	**6.** Triangles	221~234
	2. Intersecting Lines and Angles	219	**7.** Quadrilaterals	235
			8. Circles	255
	3. Perpendicular Lines	115, 219	**9.** Rectangular Solids and Cylinders	281~293
	4. Parallel Lines	115, 219	**10.** Coordinate Systems in Geometry	111~115 251~295
	5. Polygons (Convex)	235		
Word Problems	**1.** Rate Problems	25	**6.** Profit	33~35
	2. Work Problems	171~185	**7.** Sets	61
	3. Mixed Problems	171~185	**8.** Geometry Problems	213~293
	4. Interest Problems	35	**9.** Measurement Problems	297
	5. Discount	34	**10.** Data Interpretation	57~60 71~72

Source: Graduate Management Admission Council (GMAC) GMAT Official Guide

Areas and Topics Covered

(GRE Math Subject Test)

The GRE Math Subject Test covers the following areas:

1. Algebra **2. Additional Topics** (Pre-Calculus) **3. Calculus**

You are not expected to have studied every topics on the test. Calculus is a special topic and is not shown in this book. You need to take the two-year calculus course during college years.

Areas	Topics	pages
Algebra	Polynomials Linear Transformations System of Linear Equations Matrices,Determinants,Cramer's Rule Vectors Number Theory Eigenvalues and Eigenvectors Abstract Algebra	139 ~ 149 151 ~ 153 , 275 ~ 278 163 ~ 166 77 ~ 80 207 ~ 214 49 ~ 55 89 ~ 90 not covered
Additional Topics (Pre-Calculus)	Set Theory Sequences and Series Probability and Statistics Permutations and Combinations Polar and Complex Coordinates Numerical Analysis Geometry Trigonometry Topology Logic Limit and Continuity	61 65 ~ 70 , 81 ~ 82 57 ~ 72 73 ~ 76 301 ~ 308 not covered 215 ~ 297 315 ~ 348 not covered 63 ~ 64 , 83 ~ 88 211 ~ 214
Calculus	Differentiability and Integrability Differential and Integral Calculus of one and several variables connected with Geometry, Trigonometry, and Differential Equations	You need to study the two-year calculus course during college years.

Source: Educational Testing Service (ETS) GRE Subject Test Official Prep Book

Review the Topics and Related Skills

(GRE/GMAT Math Tests)

You are not expected to have studied every topic on the tests.

Chapter 1: Arithmetic

GRE Subject Test Only

Chapter 2: ALGEBRA

GRE Subject Test Only

Chapter 3: GEOMETRY

Chapter 4: TRIGONOMETRY
(GRE Subject Test Only)

Chapter One : Arithmetic

1-1 Real Numbers

In mathematics, the following numbers are called **real numbers**:

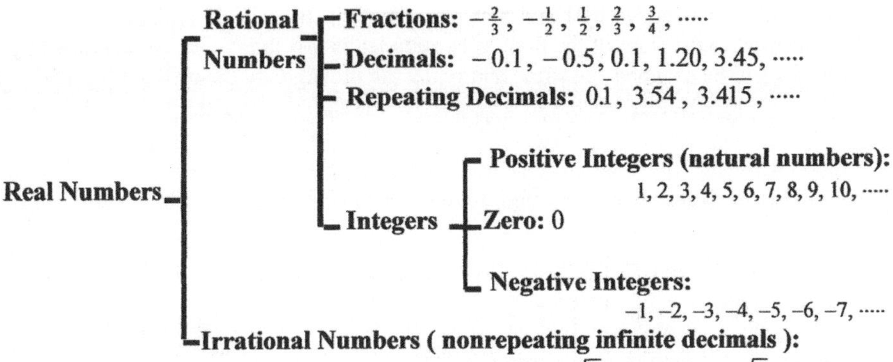

Definition of Real Numbers: They are numbers which can be located on the number line.

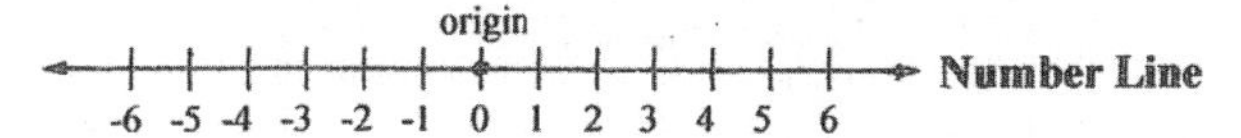

Real numbers include **rational numbers** and **irrational numbers**.

Rational Numbers: They are numbers which can be written as the ratio of two integers.

$0.1 = \frac{1}{10}$, $0.5 = \frac{5}{10} = \frac{1}{2}$, $1.2 = \frac{12}{10} = \frac{6}{5}$, $0.\bar{1} = \frac{1}{9}$, $0.\overline{54} = \frac{54}{99} = \frac{6}{11}$.

Every rational number can be expressed as either a terminating decimal or as a repeating decimal.

Irrational Numbers: They are numbers which cannot be written as the ratio of two integers. They are nonrepeating, and nonterminating (infinite) decimals.

Integers: They are positive integers, negative integers, and zero.
Zero is an integer, neither positive nor negative.

Positive Integers: They are numbers which we use to count objects. (1, 2, 3, 4, 5, 6, ·····)
They are also called **natural numbers** or **counting numbers**.

Whole Numbers: They are 0 and positive integers. (0, 1, 2, 3, 4, 5, 6, ·····)

Prime Number: It is a whole number other than 0 and 1 which is divisible only by 1 and itself. (2, 3, 5, 7, 11, 13, 17, 19, 23, 29, 31, 37, 41, ·····). 2 is the only even prime.

Composite Number: It is a whole number other than 0 and 1 which is not a prime number. (4, 6, 8, 9, 10, 12, 14, 15, 16, 18, 20, 21, 22, 24, ·····)

Rule of Rounding : When an exact computation is not needed for the answer of a problem, we use the **rules of rounding** to estimate numbers. We can round each number to the nearest 10, 100, 1000, ····. It depends on how accurate the answer we need. The symbol "≈" means "approximately equal to".

1. If the digit to the right of the digit to be rounded is 5 or more, we **round up** by increasing the digit to be rounded by **1** and replacing the digits to the right with zeros.
2. If the digit to the right of the digit to be rounded is less than 5, we **round down** by leaving the digit to be rounded **the same** and replacing the digits to the right with zeros.

$$928{,}426 \approx 928{,}430 \approx 928{,}400 \approx 928{,}000 \approx 930{,}000.$$

Place –Value of Numbers: Every digit that make up a number has a **place-value** name.

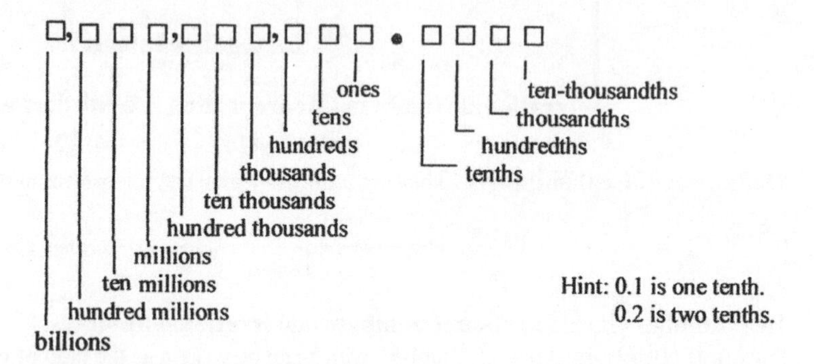

The short word name of 21, 234, 567, 809.1234 is:

" 21 billion, 234 million, 567 thousand, 809 and 1 thousand 234 ten-thousandths."

Read or write in **the number-word name**:

"twenty-one billion, two hundred thirty-four million, five hundred sixty-seven thousand, eight hundred nine and one thousand two hundred thirty-four ten-thousandths."

To write a number in words, the comma "," can be omitted.
To write a number in words, we must use a hyphen "-" between 20 and 100.
Never use the word "and" when we write the whole-number part.
To write the decimal point, we write the word "**and**".

1-2 Integers and Absolute Value

Rules for **adding** any two integers a and b :

1) The numbers to be added are called the **addends**. The answer is called the **sum**.

2) **Commutative** and **associative** hold for addition.

Commutative: $8+4=4+8$. Associative: $(8+4)+2=8+(4+2)$.

3) No sign on a number means it is positive. $1=+1$, $2=+2$.

4) If a and b are both positive, then the sum is positive. $2+3=5$.

5) If a and b are both negative, then the sum is negative. $-2+(-3)=-5$.

6) **Absolute Value**: The positive of any number "a". It is denoted by $|a|$.

$$|3|=3,\quad |-3|=3,\quad |-2|=2,\quad -|-3|=-3$$

If a is positive and b is negative and $|a|>|b|$, then $a+b$ is positive.

$3+(-2)=1$. (3 has the larger absolute value, the sum is positive.)

If a is positive and b is negative and $|a|<|b|$, then $a+b$ is negative.

$2+(-3)=-1$. (-3 has the larger absolute value, the sum is negative.)

7) **The Property of Opposites**: $a+(-a)=0$. $3+(-3)=0$.

8) **The Property of Additive Identity**: $a+0=a$. $3+0=3$.

9) Addition of numbers can be written in **horizontal form** or **vertical form**.

10) To add numbers in vertical form, we line up like values.

You may consider a positive number as **"receiving"**, a negative number as **"giving away"**. The sum is **"the net result"**.

$-5+3 = -2$ → Give away 5 and receive 3, -2 is the net result.

$-3+5 = 2$ → Give away 3 and receive 5, 2 is the net result.

$(-3)+(-5) = -8$ → Give away 3 and give away 5, -8 is the net result.

Examples

1. $8+4=12$, $4+8=12$, $-8+(-4)=-12$, $8+(-4)=4$, $4+(-8)=-4$.

2. $8+(-8)=0$, $-8+8=0$, $8+0=8$, $0+8=8$, $0+(-8)=-8$.

3. $|18|=18$, $|-18|=18$, $-|-18|=-18$. **4.** $|15+(-24)|+6=|-9|+6=9+6=15$.

Evaluate each expression if $a=2$, $b=-3$, $c=-4$.

5. $a+b=2+(-3)=-1$. **6.** $a+|b|+(-|c|)=2+|-3|+(-|-4|)=2+3+(-4)=1$.

7. John weighed 126 pounds. He gains 25 pounds this month. How much does John weigh this month ?

Solution: $126+25=151$ pounds.

8. Jorge lost 10 pounds in January, lost 9 pounds in February, and lost 4 pounds in March. What was his total weight loss for these three months ?

Solution: $(-10)+(-9)+(-4)=-23$ pounds. He lost 23 pounds.

Rules for **subtracting** any two integers a and b :

1. The number to be subtracted is called the **subtrahend**. The number we subtract from in a subtraction is called the **minuend**. The answer is called the **difference**.

Minuend – Subtrahend = Difference.

$8-2=6$.

2) **Commutative** and **associative** do not hold for subtraction.

$a-b \neq b-a$. **Examples:** $8-4 \neq 4-8$.

$(a-b)-c \neq a-(b-c)$. $(8-4)-2 \neq 8-(4-2)$.

3) Subtraction is the **inverse** of addition. To subtract a number, we add its **opposite**.

$-(-a)=a$, $a-b=a+(-b)$, $a-(-b)=a+b$.

Examples: $-(-8)=8$, $8-2=8+(-2)$, $8-(-2)=8+2$.

4) $+(-a)$ and $-(+a)$ are usually simplified to "$-a$". $+(-8)=-8$, $-(+8)=-8$.
$a+(-b)$ is usually simplified to "$a-b$". $8+(-4)=8-4$.

5) $-(a+b)=-a-b$. $-(a-b)=-a+b$. A negative sign in front of a parenthesis means " the **opposite** of the expression in the parenthesis ". $-(a+b-3)=-a-b+3$.

6) If the signs are the **same**, we combine the numbers and keep the sign.

$3+2=5$; $-3-2=-5$.

If the signs are different, we take the **difference** and use the sign of the number having the larger absolute value. $3-1=2$; $-3+2=-1$.

Examples

1. $8+(-4)=8-4=4$.
$4+(-8)=4-8=-4$.
$-4-8=-12$.
$8-(-4)=8+4=12$.
$-8-(-4)=-8+4=-4$.
$-8-8=-16$.

2. $-56-(-24)=-56+24=-32$.
$56-(-24)=56+24=80$.

3. $-15-12=-27$.
$-(15-12)=-15+12=-3$.
$-(15+12)=-15-12=-27$.

4. Evaluate each expression if $x=-4$, $y=3$, and $z=-2$.

a) $x-y-z=-4-3-(-2)=-7+2=-5$.

b) $-x+|z|-|x|=-(-4)+|-2|-|-4|=4+2-4=6-4=2$.

5. John weighed 196 pounds. He lost 25 pounds. How much does John weigh then ?
Solution: $196-25=171$ pounds.

6. Find the distance between two points –5 and 7 on the number line.
Solution: $d=|7-(-5)|=|7+5|=|12|=12$. Or $d=|-5-7|=|-12|=12$.

7. What is the distance between 125 feet (below sea level) and 102 feet (below sea level) ?
Solution: $d=|-125-(-102)|=|-125+102|=|-23|=23$ feet.

Rules for **multiplying** any two integers a and b :

1. The number to be multiplied is called the **multiplicand**. The multiplicand is multiplied by the **multiplier**. The answer is called the **product**.
   The multiplicand and the multiplier are called the **factors** of the product.
   Multiplicand × multiplier = product
   $2 \times 4 = 8 \rightarrow$ 2 and 4 are the **factors** of product 8.
2) Different ways to indicate multiplication: $2 \times 4 = 2 \cdot 4 = 2(4) = (2)4 = (2)(4) = 8$.
3) If a and b have the same sign, then the product is positive.
   $(a)(b) = ab$, $(-a)(-b) = ab$. **Examples**: $(8)(4) = 32$, $(-8)(-4) = 32$.
   If a and b have opposite signs, then the product is negative.
   $(a)(-b) = -ab$, $(-a)(b) = -ab$. **Examples**: $(8)(-4) = -32$, $(-8)(4) = -32$.
4) **Distributive property** : $a(b + c) = ab + ac$, $a(b - c) = ab - ac$.
5) **Multiplication** is **commutative** and **associative** : $ab = ba$, $(ab)c = a(bc)$.
6) The product of any integer and -1 is the additive inverse of the integer.
   $(-1)\, a = -a$, $(-1)(-1) = -(-1) = 1$.
   $(-a)(-b) = (-1 \cdot a)(-1 \cdot b) = (-1)(-1)(ab) = 1(ab) = ab$.
7) $a \cdot 1 = 1 \cdot a = a$ and $a \cdot 0 = 0 \cdot a = 0$.

Examples

1. $(5)(4) = 20$, $(-5)(-4) = 20$.
   $(5)(-4) = -20$, $(-5)(4) = -20$.
2. $-8(0) + 8(0) = 0 + 0 = 0$.
   $-8 \cdot 0 \cdot 6 = 0$.
3. $4(-1)(-2) = -4(-2) = 8$
4. $-2 \cdot -3 \cdot -4 = 6 \cdot -4 = -24$.
5. $4(-2)(-3)(-4) = -8(-3)(-4)$
   $= 24(-4) = -96$.
6. $-2(3-4) = -2(3) - 2(-4) = -6 + 8 = 2$.
   Or $-2(3-4) = -2(-1) = 2$.
7. Harvey sold 110 books last month. Each book earned a \$3 profit. How much was the profit last month ?
   Solution: \$3 × 110 = \$ 330.
8. There are 12 eggs in one box. You bought 5 boxes. How many eggs did you buy ?
   Solution: $12 \times 5 = 60$ eggs.
9. A number is the product of 12 and 8, minus 15. What is the number ?
   Solution: $12 \times 8 - 15 = 96 - 15 = 81$.
10. Light travels at the rate of about 300 million meters per second. It takes 8 minutes and 20 seconds for the light from the Sun to reach the Earth. Find the distance from the Sun to the Earth.
    Solution: $8 \times 60 + 20 = 500$ seconds.
    $d = 300$ million $\times 500 = 150{,}000$ million $= 150$ billion meters.
11. If your heart beats 75 times per minute, how many times does your heart beat in one hour ?
    Solution: $75 \times 60 = 4{,}500$ beats.

Rules for **dividing** any two integers a and b :

1. The number to be divided is called the **dividend.** The dividend is divided by the **divisor**. The answer is called the **quotient**.

Dividend ÷ divisor = quotient

$8 \div 4 = 2 \rightarrow 2$ is the quotient of $(8 \div 4)$.

2) Different ways to indicate division: $8 \div 4 = 4\overline{)8} = \frac{8}{4} = 8/4 = 8 \cdot \frac{1}{4} = \frac{1}{4} \cdot 8 = 2$.

3) If a and b have the same sign, then the quotient is positive.

If $b \neq 0$, $\frac{a}{b} = \frac{a}{b}$ and $\frac{-a}{-b} = \frac{a}{b}$. **Examples:** $\frac{8}{4} = 2$, $\frac{-8}{-2} = 4$.

If a and b have opposite signs, then the quotient is negative.

If $b \neq 0$, $\frac{a}{-b} = -\frac{a}{b}$ and $\frac{-a}{b} = -\frac{a}{b}$. **Examples:** $\frac{8}{-2} = -4$, $\frac{-8}{2} = -4$.

4) Distributive property : If $c \neq 0$, $\frac{a+b}{c} = \frac{a}{c} + \frac{b}{c}$ and $\frac{a-b}{c} = \frac{a}{c} - \frac{b}{c}$.

5) Division is not **commutative** and **associative** :

$\frac{a}{b} \neq \frac{b}{a}$; $a \div b \div c \neq a \div (b \div c)$. **Examples:** $\frac{8}{4} \neq \frac{4}{8}$, $8 \div 4 \div 2 \neq 8 \div (4 \div 2)$.

6) $\frac{0}{a} = 0$; $\frac{a}{0}$ is **undefined** (it means " **Division by zero is not allowed.** ").

Examples

1. $8 \div 2 = 4$. **2.** $-8 \div (-2) = 4$. **3.** $8 \div (-2) = -4$. **4.** $-8 \div 2 = -4$.

5. $\frac{0}{8} = 0$, $\frac{8}{0}$ is undefined. **6.** $8 \div (-4) \cdot 10 = -2 \cdot 10 = -20$.

7. You need 60 eggs. Each box contains 12 eggs. How many boxes should you buy ?
Solution: $60 \div 12 = 5$ boxes.

8. Roger worked 40 hours last week. He earned $240. How much does he earn per hour ?
Solution: $240 \div 40 = \$6$.

9. 900 yearbooks are packaged in 60 cartons. How many yearbooks are in each carton ?
Solution: $900 \div 60 = 15$ yearbooks.

10. In a fund-raising program, a class has planned to sell 720 Christmas cards. There are 12 volunteers to sell the cards. How many cards must each student sell ?
Solution: $720 \div 12 = 60$ cards.

11. A player scored a total of 1,692 points in 47 games. What is his average points per game ?
Solution: $1{,}692 \div 47 = 36$ points.

12. If your heart beats 25 times in 20 seconds, how many times does your heart beat in one minute ?
Solution: $(60 \div 20) \times 25 = 3 \times 25 = 75$ times. Or $25 \times (60 \div 20) = 25 \times 3 = 75$ times.

1-3 Decimals and Fractions

A decimal has a **whole-number** part and a **decimal** part separated by a **decimal point**. To compare decimals, we write one decimal under the other and line up the decimal points and like place-values. Starting from left to right, compare the first digits that are **not alike**. Add zeros after the last digit when necessary.

4.125	3.4**6**2	5.430**4**
2.346	3.4**7**3	5.430**0**

Compare the digits 4 and 2, 4 > 2, we have 4.125 > 2.346. –4.125 < –2.346.
Compare the digits 6 and 7, 6 < 7, we have 3.462 < 3.473. –3.462 > –3.473.
Compare the digits 4 and 0, 4 > 0, we have 5.4304 > 5.43. –5.4304 < –5.43.
Every integer can be written as an equal decimal by adding one or more zeros following the decimal point. 1 = 1.0 = 1.00 = 1.000 = ···· ; 10 = 10.0 = 10.00 = 10.000 = ····
Decimals include:

1. **Terminating** (or **finite**) **decimal.** It ends in a given place. 0.1, 0.3, 2.54.
2. **Nonterminating** (or **infinite**) **decimal.** It does not end in any given place.
   3.1415····. 1.4142····. 1.7320····. 2.71828····.
3. **Repeating decimal.** It is a nonterminating decimal that repeats a pattern.
   To write a repeating decimal, we place a bar above the digit that repeats.
   $0.111\cdots = 0.\overline{1}$. $0.333\cdots = 0.\overline{3}$. $2.125454\cdots = 2.12\overline{54}$.

Rules to **add** or **subtract** with decimals:

1. We write one decimal under the other and line up the decimal points.
   This automatically line up the place-values.
2. Add zeros after the last digit when necessary.
3. Add or subtract as with whole numbers.

To add or subtract decimals and whole numbers, we rewrite each whole number as an equal decimal with a decimal point and zeros after the last digit.

33.050	33.050	12.00	34.00
+ 24.573	– 24.573	+ 34.45	– 12.83
57.623	**8.477**	**46.45**	**21.17**

Rules to **multiply** with decimals:

1. We multiply as with two whole numbers.
2. Count the total number of decimal places in both numbers being multiplied.
3. Put the same number of decimal places in the product.
4. Add zeros in the product when necessary.

To multiply a decimal by a power of 10 (10, 100, 1000,·····), we move the decimal point one decimal place to the right for each zero in the power of 10.

Rules to **divide** with decimals:

1. We move both decimal points to the right by the same number of places in order to get a whole-number divisor.
2. Place a decimal point in the quotient directly above the decimal point in the dividend.
3. Add zeros after the last digit on the dividend when necessary.
4. Divide as with whole numbers.
5. Round the quotient as needed.

To divide a decimal by a whole-number divisor, we start from step 2.

Examples

1. $5.16 \div 12 = ?$

$$\begin{array}{r} 0.43 \\ 12\overline{)5.16} \\ \underline{4\,8} \\ 36 \\ \underline{36} \\ 0 \end{array}$$

$5.16 \div 12 = 0.43.$

2. $-5.16 \div 1.2 = ?$

$$\begin{array}{r} 4.3 \\ 1.2\overline{)5.1\,6} \\ \underline{4\,8} \\ 3\,6 \\ \underline{3\,6} \\ 0 \end{array}$$

$-5.16 \div 1.2 = -4.3.$

3. $5.16 \div 0.0012 = ?$

$$\begin{array}{r} 4300. \\ 0.0012\overline{)5.1600.} \\ \underline{4\,8} \\ 36 \\ \underline{36} \\ 0 \end{array}$$

$5.16 \div 0.0012 = 4300.$

4. Find $58 \div 2.4$ and round the solution to the nearest tenth.

Solution:

$$\begin{array}{r} 24.16 \\ 2.4\overline{)58.00} \\ \underline{48} \\ 10\,0 \\ \underline{9\,6} \\ 4\,0 \\ \underline{2\,4} \\ 1\,60 \\ \underline{1\,44} \\ 16 \end{array}$$

$58 \div 2.4 \approx 24.2.$

5. Find $56.4 \div (-2.43)$ and round the solution to the nearest hundredth.

Solution:

$$\begin{array}{r} 23.209 \\ 2.43\overline{)56.40.000} \\ \underline{48\,6} \\ 7\,80 \\ \underline{7\,29} \\ 51\,0 \\ \underline{48\,6} \\ 2\,400 \\ \underline{2\,187} \\ 213 \end{array}$$

$56.4 \div (-2.43) \approx -23.21.$

6. Find the average (mean) of the numbers 31.6, 24.7, 56 and 34.2.

Solution: $\dfrac{31.6+24.7+56+34.2}{4} = \dfrac{146.5}{4} = 36.625$.

Fractions are commonly used in our daily life.

1. A fraction is used to represent each part of a whole number that is divided into smaller and equal parts.
   Suppose we divide the number 1 into 5 equal parts. Each part is represented by the fraction $\frac{1}{5}$. Two parts are represented by the fraction $\frac{2}{5}$, and so on.
   If we divide \$1 into 4 equal parts. Each part is $\frac{1}{4}$ of \$1, or a quarter (25 cents).
2. A fraction is also used to represent a division of two whole numbers.
   2 and 5 are whole numbers. 2 divided by 5 is written as $\frac{2}{5}$ or $2/5$.

In the fraction (division) $\frac{2}{5}$, 2 is called the **numerator**, 5 is called the **denominator**, and the horizontal bar is called the **fraction bar**.

In a fraction, the denominator cannot be 0. We call a fraction with zero denominator as "**undefined**". $\frac{2}{0}$ is undefined (it means that "division by 0 is not allowed.").

Proper fraction: The numerator is less than the denominator. Such as $\frac{1}{2}$, $\frac{2}{3}$, $\frac{4}{5}$, ...

Improper fraction: The numerator is greater than or equal to the denominator.
Such as $\frac{2}{1}$, $\frac{3}{2}$, $\frac{5}{4}$, $\frac{1}{1}$, $\frac{2}{2}$, $\frac{5}{1}$, ...

Mixed numbers: A whole number plus a fraction. Such as $1\frac{1}{2}$, $1\frac{1}{4}$, $3\frac{2}{5}$, ...

Rules to **add** or **subtract** with fractions:

1. If the fractions have a common denominator, we combine the numerators and write the result with the same denominator.
2. If the fractions have different denominators, we rewrite the fractions with a common denominator before adding and subtracting.
3. Convert mixed numbers to improper fractions before adding and subtracting. Or, use vertical form to add or subtract mixed numbers.

Examples

1. $\frac{3}{5}+\frac{1}{5}=\frac{3+1}{5}=\frac{4}{5}$. **2.** $\frac{3}{5}-\frac{1}{5}=\frac{3-1}{5}=\frac{2}{5}$. **3.** $\frac{1}{2}+\frac{1}{3}=\frac{3}{6}+\frac{2}{6}=\frac{5}{6}$.

4. $2\frac{2}{3}+4\frac{5}{6}=2\frac{4}{6}+4\frac{5}{6}=6\frac{9}{6}=7\frac{3}{6}=7\frac{1}{2}$. **5.** $6\frac{1}{4}-3\frac{3}{5}=\frac{25}{4}-\frac{18}{5}=\frac{125}{20}-\frac{72}{20}=\frac{53}{20}=2\frac{13}{20}$.

6.

$$\begin{array}{rl} & 2\frac{2}{3}=2\frac{4}{6} \\ +) & 4\frac{5}{6}=4\frac{5}{6} \\ \hline & 6\frac{9}{6}=7\frac{3}{6}=7\frac{1}{2}. \end{array}$$

7.

$$\begin{array}{rl} & 6\frac{1}{4}=6\frac{5}{20}=5\frac{25}{20} \\ -) & 3\frac{3}{5} \qquad =3\frac{12}{20} \\ \hline & 2\frac{13}{20}. \end{array}$$

Rules to **multiply** with fractions:

1. To multiply with fractions, we multiply their numerators and multiply their denominators. Then write the answer in simplest form.
2. Write each mixed number as an improper fraction before multiplying.
3. It is easier if we can "simplify first when possible" and then multiply.
4. To multiply fractions and whole numbers, we rename each whole number as an equal fraction (use "1" as its denominator). $2 = \frac{2}{1}$, $3 = \frac{3}{1}$, $4 = \frac{4}{1}$, ...

Examples

1. $\frac{1}{3} \times \frac{2}{3} = \frac{1 \times 2}{3 \times 3} = \frac{2}{9}$.
2. $\frac{2}{\cancel{5}_1} \times \frac{\cancel{10}^2}{17} = \frac{2 \times 2}{1 \times 17} = \frac{4}{17}$. (Simplify first and then multiply.)

   OR: $\frac{2}{5} \times \frac{10}{17} = \frac{2 \times 10}{5 \times 17} = \frac{20}{85} = \frac{4}{17}$. (Multiply first and then simplify.)
3. $-4\frac{2}{3} \times (-7) = \frac{14}{3} \times \frac{7}{1} = \frac{14 \times 7}{3 \times 1} = \frac{98}{3} = 32\frac{2}{3}$
4. $\frac{7}{3} \times (-2\frac{2}{3}) = \frac{7}{3} \times (-\frac{8}{3}) = -\frac{7 \times 8}{3 \times 3} = -\frac{56}{9} = -6\frac{2}{9}$.
5. $2\frac{2}{3} \times 4 \times \frac{2}{7} = \frac{8}{3} \times \frac{4}{1} \times \frac{2}{7} = \frac{8 \times 4 \times 2}{3 \times 1 \times 7} = \frac{64}{21} = 3\frac{1}{21}$.

Rules to **divide** with fractions:

1. To divide with fractions, we **multiply** the reciprocal of the divisor.
2. Multiply as the rules.
3. Division by 0 is undefined (it means "Division by zero is not allowed.").

Dividing by a number gives the same result as multiplying by its reciprocal.

$$15 \div 3 = 5 \rightarrow 15 \times \frac{1}{3} = \frac{15}{1} \times \frac{1}{3} = 5 \ ; \ 15 \div \frac{2}{3} = 15 \times \frac{3}{2} = \frac{15}{1} \times \frac{3}{2} = \frac{45}{2} = 22\frac{1}{2}$$

Examples

1. $\frac{1}{3} \div \frac{2}{3} = \frac{1}{\cancel{3}} \times \frac{\cancel{3}}{2} = \frac{1}{2}$.
2. $-\frac{\frac{4}{5}}{\frac{7}{3}} = -\frac{4}{5} \times \frac{3}{7} = -\frac{12}{35}$.
3. $\frac{7}{3} \div (-2\frac{2}{3}) = \frac{7}{3} \div (-\frac{8}{3}) = -\frac{7}{\cancel{3}} \times \frac{\cancel{3}}{8} = -\frac{7}{8}$.
4. $0 \div \frac{2}{5} = 0$; $\frac{2}{5} \div 0$ is undefined.
5. $2\frac{1}{5} \div 3\frac{3}{4} = \frac{11}{5} \div \frac{15}{4} = \frac{11}{5} \times \frac{4}{15} = \frac{44}{75}$.
6. $\frac{4}{7} \div 5 = \frac{4}{7} \div \frac{5}{1} = \frac{4}{7} \times \frac{1}{5} = \frac{4}{35}$.

1-4 Ratios, Rates, Proportions, and Percents

Ratios

A **ratio** is used to compare two numbers. We write a ratio as a fraction in **lowest terms**. If an improper fraction represents a ratio, we do not change it to a mixed number. Note that the lowest terms of $\frac{18}{10}$ is $\frac{9}{5}$, the simplest form of $\frac{18}{10}$ is $1\frac{4}{5}$.

If there are 15 boys and 25 girls in a class, we can compare the number of boys to the number of girls by writing a ratio (fraction). A comparison of 15 boys to 25 girls is $\frac{15}{25}$. It is read as " 15 to 25 ". A ratio can be written in three ways: 15 to 25, 15:25, or $\frac{15}{25}$.

Just as with fractions, a ratio can be written in lowest terms: $\frac{15}{25}=\frac{3}{5}$. We say that the ratio of the number of the boys to the number of the girls is " 15 to 25 ", or " 3 to 5 ".

A ratio $\frac{15}{25}$ can be also represented to the statement " 15 out of 25 ". It is equivalent to the statement " 3 out of 5 ".

If a ratio is used to compare two measurements, we must use the measurements in the same unit (like measures). (see example 4).

A **rate** is a ratio of two unlike measures. A rate is usually simplified to a **per unit form** , or called a **unit rate**. The rate of 120 miles to 8 gallon is $\frac{120}{8}=15$ miles per gallon.

To find a rate in a problem, it always means to find the unit rate.

Examples

1. Write the ratio " 6 to 9 " as a fraction in lowest terms.

   Solution: $\frac{6}{9}=\frac{2}{3}$.

2. Write the ratio " 18 : 10 " as a fraction in lowest terms.

   Solution: $\frac{18}{10}=\frac{9}{5}$.

3. Write the ratio $\frac{1.8}{1.2}$ in lowest terms.

   Solution: $\frac{1.8}{1.2}=\frac{18}{12}=\frac{3}{2}$.

4. Write the ratio $\frac{2h}{150\,\text{min}}$ in lowest terms.

   Solution: $\frac{2h}{150\,\text{min}}=\frac{120\,\text{min}}{150\,\text{min}}=\frac{4}{5}$.

5. You drive 240 miles in 4 hrs. At that rate, how many miles could you drive in 7 hrs. ?

   Solution: unit rate (average speed) $=\frac{240}{4}=60$ mph (miles per hour).

   $\therefore$ distance $=60\times 7=420$ miles in 7 hrs.

Proportions

A **proportion** is formed by two equivalent ratios. In a proportion, we say that the given two ratios are equal or **proportional**.

$\frac{1}{2} = \frac{2}{4}$ is a proportion.

If two ratios are not equal, we say that the two given ratios are **not proportional**.

$\frac{1}{2}$ and $\frac{3}{4}$ are not equal (not proportional). We can not form them as a proportion.

$$\frac{1}{2} \neq \frac{3}{4}$$

To determine if two given ratios are equal (proportional), we use the following statement:

Property of Proportions: In a proportion, the cross products are equal.

If $\frac{a}{b} = \frac{c}{d}$, then $ad = bc$. (b and d are nonzero numbers)

Example: $\frac{1}{2} = \frac{2}{4}$ is a proportion because $1 \times 4 = 2 \times 2$.

Sometimes one of the 4 terms in a proportion is a variable. An equation is formed.

$$\frac{1}{2} = \frac{n}{4}$$

To find n in this proportion, we use equivalent fractions with a common denominator.

$$\frac{1 \times 2}{2 \times 2} = \frac{n}{4} \rightarrow \frac{2}{4} = \frac{n}{4} \quad \therefore n = 2.$$

We can find n also by **cross-multiplying** in the proportion.

$$\frac{1}{2} = \frac{n}{4} \rightarrow 1 \times 4 = 2 \times n, \quad 4 = 2n, \quad n = \frac{4}{2} = 2.$$

Examples

1. Are the ratios $\frac{5}{8}$ and $\frac{15}{24}$ proportional ?

Solution:

Cross multiply: $5 \times 24 = 8 \times 15$

$(120 = 120)$

They are proportional.

2. Are the ratios $\frac{5}{6}$ and $\frac{7}{8}$ proportional ?

Solution:

Cross multiply: $5 \times 8 \neq 6 \times 7$

$(40 \neq 42)$

They are not proportional.

3. Solve $\frac{n}{24} = \frac{5}{8}$.

Solution:

$n \times 8 = 24 \times 5$

$8n = 120 \quad \therefore n = \frac{120}{8} = 15.$

4. Solve $\frac{3}{5} = \frac{x}{20}$.

Solution:

$3 \times 20 = 5 \times x$

$60 = 5x \quad \therefore x = \frac{60}{5} = 12.$

5. Two pounds of cheese cost \$7. How much do five pounds cost ?

Solution: Write a proportion and use cross-products to solve it.

$$\frac{7}{2} = \frac{x}{5}, \quad 7 \times 5 = 2 \times x, \quad 35 = 2x, \quad \therefore x = \frac{35}{2} = \$17.50.$$

Percents

A **percent** is a ratio of a number to 100. The symbol for percent is " %".

Percent means: **1. per hundred (or hundredths).**

$$20\% = \frac{20}{100}$$

2. out of 100.

20% means " 20 out of 100 ". Therefore, $20\% = \frac{20}{100}$.

Then we have: 20% of 100 is 20. We multiply: $100 \times 20\% = 100 \times \frac{20}{100} = 20$.

20% of 200 is 40. $200 \times 20\% = 200 \times \frac{20}{100} = 40$.

$\frac{20}{100} = \frac{2}{10} = \frac{1}{5}$. " 20 out of 100 " is equivalent to " 2 out of 10 ".

" 20 out of 100 " is equivalent to " 1 out of 5 ".

Therefore, $\frac{20}{100}$, $\frac{2}{10}$, and $\frac{1}{5}$ are all equivalent to 20%.

Examples:

1. Find 20% of 300.

Solution:

$300 \times 20\%$
$= 300 \times \frac{20}{100}$
$= 60$.

2. Find $\frac{1}{5}$ of 300.

Solution:

$300 \times \frac{1}{5} = 60$.

3. What number is 5% Of 40 ?

Solution:

$n = 40 \times 5\%$
$= 40 \times \frac{5}{100}$
$= 2$.

4. What number is 8% of 50 ?

Solution:

$n = 50 \times 8\%$
$= 50 \times \frac{8}{100}$
$= 4$.

5. 60 out of 300 students are boys. What percent of the students is boy ?

Solution: $p = \frac{60}{300} = \frac{20}{100} = 20\%$.

We can also use a proportion to change a fraction to a percent.

6. Write $\frac{1}{5}$ as a percent.

Solution:

$\frac{1}{5} = \frac{n}{100}$
$5n = 100$
$n = \frac{100}{5} = 20$
$\therefore \frac{1}{5} = 20\%$.

7. What percent is 60 out of 300 ?

Solution:

$\frac{60}{300} = \frac{n}{100}$
$300n = 6{,}000$
$n = \frac{6{,}000}{300} = 20$
$\therefore \frac{60}{300} = 20\%$.

Notes

1-5 Percents and Decimals

The following examples show us a general relationship between percents and decimals.

$45\% = \dfrac{45}{100} = 0.45$, $120\% = \dfrac{120}{100} = 1.20$, $25.4\% = \dfrac{25.4}{100} = 0.254$

Therefore, we have the following rules:

A) To change a percent to a decimal, we move the decimal point 2 places to the left and remove the % sign.

Examples: **1.** $20\% = 0.2$; $2\% = 0.02$; $0.2\% = 0.002$; $0.02\% = 0.0002$.

2. $25\% = 0.25$; $2.5\% = 0.025$; $0.25\% = 0.0025$.

3. $125\% = 1.25$; $225\% = 2.25$; $925\% = 9.25$.

4. $25.7\% = 0.257$; $125.7\% = 1.257$; $1325.7\% = 13.257$.

B) To change a decimal to a percent, we move the decimal point 2 places to the right and add the % sign.

Examples: **1.** $0.2 = 20\%$; $0.02 = 2\%$; $0.002 = 0.2\%$; $0.0002 = 0.02\%$.

2. $0.25 = 25\%$; $0.025 = 2.5\%$; $0.0025 = 0.25\%$.

3. $1.25 = 125\%$; $2.25 = 225\%$; $9.25 = 925\%$.

4. $0.257 = 25.7\%$; $1.257 = 125.7\%$; $13.257 = 1325.7\%$

Examples

1. What number is 5% of 40 ?
Solution:
$$\begin{aligned} n &= 40 \times 5\% \\ &= 40 \times 0.05 \\ &= 2. \end{aligned}$$

2. Find 80% of 50.
Solution:
$$\begin{aligned} &50 \times 80\% \\ &= 50 \times 0.8 \\ &= 40. \end{aligned}$$

3. What percent is 15% of 80% ?
Solution:
$$\begin{aligned} &80\% \times 15\% \\ &= 0.80 \times 0.15 \\ &= 0.12 = 12\%. \end{aligned}$$

4. Find the sales tax on \$50 if the tax rate is 8.25 %.
Solution:
$$50 \times 8.25\% = 50 \times 0.0825 = \$4.13.$$

Notes

1-6 Percents and Fractions

We have learned how to use a proportion to change a fraction to a percent. In this section, we introduce an easier way to change a fraction to a percent.

To change a fraction to a percent, we first change it to a decimal and then to a percent.

Examples: 1. $\frac{1}{2}=0.5=50\%$, $\frac{1}{5}=0.2=20\%$, $\frac{2}{5}=0.4=40\%$, $\frac{3}{4}=0.75=75\%$

$\frac{3}{8}=0.375=37.5\%$, $\frac{5}{8}=0.625=62.5\%$

2. $\frac{1}{3}=0.333\cdots=33.\bar{3}\ \%=33.3\ \%$ (**It is the nearest tenth of a percent.**)

$\frac{1}{3}=0.33\frac{1}{3}=33\frac{1}{3}\ \%$ (**It is an exact percent.**)

3. $\frac{3}{7}=0.42\overline{857142}=0.42\frac{857142}{999999}=0.42\frac{857142\div142857}{999999\div142857}=0.42\frac{6}{7}=42\frac{6}{7}\%$

(Hint: To find the GCF = 142857, read "Rollabout Divisions" on Page 54.)

To change a percent to a fraction, we change it to an equivalent fraction with a denominator of 100.

Examples

1. $50\ \%=\frac{50}{100}=\frac{1}{2}$, $20\ \%=\frac{20}{100}=\frac{1}{5}$, $40\ \%=\frac{40}{100}=\frac{2}{5}$, $125\ \%=\frac{125}{100}=1\frac{1}{4}$

$12.5\ \%=\frac{12.5}{100}=\frac{125}{1000}=\frac{1}{8}$, $62.5\ \%=\frac{62.5}{100}=\frac{625}{1000}=\frac{5}{8}$

2. $\frac{1}{2}\%=\frac{1}{2}\times\frac{1}{100}=\frac{1}{200}\cdot 2\frac{1}{2}=\frac{5}{2}\%=\frac{5}{2}\times\frac{1}{100}=\frac{5}{200}=\frac{1}{40}$

$33\frac{1}{3}\%=\frac{100}{3}\%=\frac{100}{3}\times\frac{1}{100}=\frac{1}{3}$

Examples

1. Find $5\frac{1}{3}\%$ of 33.

Solution: $33\times5\frac{1}{3}\%=33\times\frac{16}{3}\%=33\times\frac{16}{3}\times\frac{1}{100}=\frac{528}{300}=1.76$.

2. What percent is 12 of 30 ?

Solution: $\frac{12}{30}=0.4=40\%$.

Notes

1-7 Percent and Applications

In this section, we apply the ideas we have learned about percent to solve real consumer problems.

A) Finding a percent of one number of another, we write a " p " for the percent.

Examples:

1. What percent of 40 is 5 ? (5 is what percent of 40 ?)

Solution: $40 \times p = 5 \quad \therefore p = \dfrac{5}{40} = \dfrac{1}{8} = 0.125 = 12.5\%$.

2. Joe spent $50 and sales tax $4 to buy a pair of shoes. What percent is the sales tax ?

Solution: $50 \times p = 4 \quad \therefore p = \dfrac{4}{50} = \dfrac{2}{25} = 0.08 = 8\%$.

B) Finding a number when both the percent and another number are known, we write a " n " for the number.

Examples:

3. 40 is 5% of what number ? (5% of what number is 40 ?)

Solution:

$$40 = n \times 5\%$$

$$40 = n \times 0.05 \quad \therefore n = \frac{40}{0.05} = 800.$$

4. $5\frac{1}{3}\%$ of what number is 1.76 ?

Solution:

$$n \times 5\tfrac{1}{3}\% = 1.76 \ , \quad n \times \frac{16}{3}\% = 1.76 \ , \quad n \times \frac{16}{300} = 1.76$$

$$\therefore n = 1.76 \times \frac{300}{16} = \frac{528}{16} = 33.$$

5. Joe paid $4 sales tax for a pair of shoes he bought. The sales-tax rate is 8%. What is the price of the shoes ?

Solution:

Let $p =$ the price

$$p \times 8\% = 4 \ , \quad p \times 0.08 = 4$$

$$\therefore p = \frac{4}{0.08} = \$50.$$

C) Finding the percent of increase (or decrease), we write the ratio of the amount of increase (or decrease) to the original amount.

Examples

6. What is the percent of increase from 4 to 6 ?

Solution: percent of increase $= \dfrac{6-4}{4} = \dfrac{1}{2} = 0.5 = 50\%$.

7. What is the percent of decrease from 6 to 4 ?

Solution: percent of decrease $= \dfrac{6-4}{6} = \dfrac{2}{6} = \dfrac{1}{3} = 0.33\frac{1}{3} = 33\frac{1}{3}\%$.

8. The bill of electricity was \$60 last month. It decreased to \$45 this month. Find the percent of decrease.

Solution:

percent of decrease $= \dfrac{60-45}{60} = \dfrac{15}{60} = \dfrac{1}{4} = 0.25 = 25\%$.

9. The population of a city increased from 125,000 to 180,000 in 10 years. Find the percent of increase.

Solution:

percent of increase $= \dfrac{180{,}000-125{,}000}{125{,}000} = \dfrac{55{,}000}{125{,}000} = \dfrac{11}{25} = 0.44 = 44\%$.

D) Discount and Commission

A discount is a decrease in the price of a product. A discount can be expressed as a percent of the original price.

A commission is a payment to the salesman for selling products. A commission can be expressed as a percent of the total sales.

Examples

10. A suit that is regularly \$45.90 is on sale for 30% discount. What is the sale price of the suit ?

Solution: the discount: $\$45.90 \times 30\% = 45.90 \times 0.30 = \$13.77$.

the sale price: $\$45.90 - \$13.77 = \$32.13$.

(Or: $\$45.90 \times 70\% = 45.90 \times 0.70 = \$\,32.13$)

11. Mary sold sport products worth \$32,000 this week. She receives 3% commission on her sales. How much is her commission this week ?

Solution: commission: $\$32{,}000 \times 3\% = 32{,}000 \times 0.03 = \$960$.

E) Simple Interest

Interest is the amount of money charged for the use of borrowed money. Interest is also the amount of money earned for the money deposited in a bank.
The interest rate is usually expressed as a percent of the principal over a period of time. If the interest rate is computed on a yearly basis, it is called a yearly rate or an annual rate. When interest is computed on the principal only (no interest on the interest previously earned), it is called **simple interest**.
The formula to compute simple interest is:

$$\text{Interest} = \text{principal} \times \text{rate} \times \text{time}$$
$$\mathbf{I = p\,r\,t}$$

Examples:

12. Find the simple interest you pay if you borrow \$500 for 3 years at a yearly rate 7%.
Solution: $\mathrm{I = p\,r\,t}$, $\mathrm{I} = \$500 \times 7\% \times 3 = \$500 \times 0.07 \times 3 = \$105$.

13. You have a saving account with \$500. The annual interest rate is 3.5%. How much simple interest will your account earn in 3 months ? How much do you have in your account after 3 months ?
Solution: $\mathrm{I = p\,r\,t}$, $\mathrm{I} = 500 \times 3.5\% \times \dfrac{3}{12} = 500 \times 0.035 \times \dfrac{1}{4} = \$4.38$.
Total $= \$500 + \$4.38 = \$504.38$.

F) Compound Interest

Compound interest is computed on "the principal plus the interest previously earned". The principal increases when the interest is compounded. The compound interest is usually given in daily, monthly, quarterly, semiannually, or yearly.
The formula to compute the compound amount (principal plus compound interest) is:

$$A = p(1+r)^n \,, \quad r: \text{annual interest rate} \quad n: \text{number of years}$$

Examples

14. You have a savings account with \$500. The annual interest rate 7%, compounded yearly. How much do you have in your account after 3 years ?
Solution: $A = p(1+r)^n = 500(1+0.07)^3 = 500(1.07)^3 = 500(1.225043) = \$612.52$.

15. \$500 is deposited in a bank at 6%, compounded monthly. Find the new principal after 3 months ?
Solution: monthly rate $r = 0.06 \div 12 = 0.005$
$A = p(1+r)^n = 500(1+0.005)^3 = 500(1.005)^3 = 500(1.01508) = \$507.54$.

Notes

1-8 Exponents, Roots, and Radicals

Exponent (power): In the expression 5^3, 3 is called the **exponent** (or **power**) of 5. 5 is called the **base.** 5^3 is the exponential (or power) form of $5 \times 5 \times 5$.

The exponent 3 in the expression 5^3 means that there are three **repeated factors** of 5.

Exponential form or **power form** a^n is a short way to write a **multiplication** of n repeated factors of a: $a^n = a \cdot a \cdot a \cdots\cdots a \rightarrow n$ repeated factors of a

$5^1 \rightarrow$ read " five to the first power".

$5^2 \rightarrow$ read " five to the second power " or " five squared ".

$5^3 \rightarrow$ read " five to the third power " or " five cubed ".

$5^n \rightarrow$ read " five to the nth power ".

Any number can be written as " **the number to the first power** ".

$1 = 1^1$, $2 = 2^1$, $(-3) = (-3)^1$, $4 = 4^1$, $-5 = (-5)^1$, $99 = 99^1$, $100 = 100^1$, ...

Examples: Write each product using exponents.

$2 \cdot 2 = 2^2$, $2 \cdot 2 \cdot 2 = 2^3$, $2 \cdot 2 \cdot 2 \cdot 2 = 2^4$, $-2 \cdot -2 \cdot -2 \cdot -2 \cdot -2 = (-2)^5$.

$3 \cdot 3 = 3^2$, $3 \cdot 3 \cdot 3 = 3^3$, 7 squared $= 7^2$, 7 cubed $= 7^3$.

The expression $3a^2$ means that a is squared, but not 3. $3a^2 = 3 \cdot a \cdot a$.

The expression $(3a)^2$ means that 3 and a are both squared. $(3a)^2 = 3a \cdot 3a = 9a^2$.

Notice that: $-3^2 \neq (-3)^2$. $-3^2 = -3 \cdot 3 = -9$, $(-3)^2 = (-3) \cdot (-3) = 9$.

$-2^4 = -2 \cdot 2 \cdot 2 \cdot 2 = -16$, $(-2)^4 = (-2) \cdot (-2) \cdot (-2) \cdot (-2) = 16$.

$(-a)^n$ is positive if n is an even number. $(-1)^2 = 1$, $(-1)^{50} = 1$, $(-1)^{100} = 1$.

$(-a)^n$ is negative if n is an odd number. $(-1)^3 = -1$, $(-1)^{49} = -1$, $(-1)^{99} = -1$.

$(-2)^{19} = -2^{19}$, $(-2)^{20} = 2^{20}$.

General rules of exponents (powers): If a is any nonzero real number, then:

1. $a^m \cdot a^n = a^{m+n}$. **2.** $\dfrac{a^m}{a^n} = a^{m-n}$. **3.** $a^0 = 1$ if $a \neq 0$

Rules: 1. To multiply two powers of the same base, we add the powers.

2. To divide two powers of the same base, we subtract the powers.

3. If a is any nonzero number, then $a^0 = 1$. $1^0 = 1$, $2^0 = 1$, $3^0 = 1$, ...,$100^0 = 1$.

0^0 is undefined (it means that the power "0" does not apply to 0 itself).

Examples:

1. $2^{30} \cdot 2^{10} = 2^{30+10} = 2^{40}$. **2.** $\dfrac{2^{30}}{2^{10}} = 2^{30-10} = 2^{20}$. **3.** $\dfrac{2^{10}}{2^{10}} = 1$, or $\dfrac{2^{10}}{2^{10}} = 2^{10-10} = 2^0 = 1$.

Roots and Radicals

Because $5^2 = 25$, we say that the square root of 25 is 5. We write $\sqrt{25} = 5$.
Because $5^3 = 125$, we say that the cubed root of 125 is 5. We write $\sqrt[3]{125} = 5$.
Because $2^4 = 16$, we say that the 4th root of 16 is 2. We write $\sqrt[4]{16} = 2$.
The symbol $\sqrt[n]{b}$ is called a **radical.** We read $\sqrt[n]{b}$ as " **the** n **th root of** b ".
b is called the **radicand,** n is called the **index.**
Index "2" of the square root is usually omitted from the radical. $\sqrt{b}$ means $\sqrt[2]{b}$.
To find the square root of a given number, we find a number that, when squared, equals the given number. However, $\sqrt[n]{b}$ means only the **principal root** of b.

$\sqrt{4} = \sqrt{2 \cdot 2} = \sqrt{2^2} = 2$. $\qquad \sqrt{9} = \sqrt{3 \cdot 3} = \sqrt{3^2} = 3$.

$\sqrt{4} = \sqrt{(-2)^2} = -2$ is incorrect. $\qquad \sqrt{9} = \sqrt{(-3)^2} = -3$ is incorrect.

$\sqrt{(-2)^2} = \sqrt{4} = 2$. $\qquad \sqrt{(-3)^2} = \sqrt{9} = 3$.

$\sqrt{0.01} = \sqrt{(0.1)(0.1)} = \sqrt{(0.1)^2} = 0.1$. $\qquad \sqrt{1.44} = \sqrt{(1.2)(1.2)} = \sqrt{(1.2)^2} = 1.2$

To find the n th root of a given number, we follow the same idea as the above examples and so on.

$\sqrt[3]{8} = \sqrt[3]{2 \cdot 2 \cdot 2} = \sqrt[3]{2^3} = 2$. $\qquad \sqrt[4]{16} = \sqrt[4]{2 \cdot 2 \cdot 2 \cdot 2} = \sqrt[4]{2^4} = 2$.

$\sqrt[3]{-8} = \sqrt[3]{-2 \cdot -2 \cdot -2} = \sqrt[3]{(-2)^3} = -2$. $\qquad \sqrt[4]{(-2)^4} = \sqrt[4]{16} = 2$, **not** -2.

The square root of a given number that is not a perfect square is an **irrational number** (a nonrepeating infinite decimal). We use a calculator to find its decimal approximation, depending on what decimal place we round to.

$\sqrt{3} \approx 1.732$. $\quad \sqrt{5} \approx 2.236$. $\quad \sqrt{7} \approx 2.646$. $\quad \sqrt{8} \approx 2.828$.

$\sqrt{0.15} \approx 0.387$. $\quad \sqrt{17} \approx 4.123$. $\quad \sqrt{140} \approx 11.832$.

To find the square root of a fraction, we write the square root of the numerator over the square root of the denominator.

$$\sqrt{\frac{4}{9}} = \frac{\sqrt{4}}{\sqrt{9}} = \frac{2}{3}. \quad \text{Or: } \sqrt{\frac{4}{9}} = \sqrt{\frac{2}{3} \cdot \frac{2}{3}} = \frac{2}{3}.$$

$$\sqrt{\frac{12}{35}} = \frac{\sqrt{12}}{\sqrt{35}} \approx \frac{3.464}{5.916} \approx 0.586. \quad \text{Or: } \sqrt{\frac{12}{35}} \approx \sqrt{0.343} \approx 0.586.$$

Laws of Exponents

Laws of Exponents (Rules of Powers)

If a and b are real numbers, m and n are positive integers, $a \neq 0$, $b \neq 0$, then:

1) $a^0 = 1$ **2)** $a^m \cdot a^n = a^{m+n}$ **3)** $\frac{a^m}{a^n} = a^{m-n}$ **4)** $\frac{a^m}{a^n} = \frac{1}{a^{n-m}}$

5) $(a^m)^n = a^{mn}$ **6)** $(ab)^m = a^m b^m$ **7)** $a^{-m} = \frac{1}{a^m}$ **8)** $\frac{1}{a^{-m}} = a^m$

9) $\left(\frac{b}{a}\right)^m = \frac{b^m}{a^m}$ **10)** $\left(\frac{b}{a}\right)^{-m} = \left(\frac{a}{b}\right)^m$ **11)** $a^{\frac{1}{m}} = \sqrt[m]{a}$ **12)** $a^{\frac{n}{m}} = (\sqrt[m]{a})^n$

Examples

1. $1^0 = 1,\ 2^0 = 1, \cdots\cdots, 10^0 = 1$
2. $5^{20} \cdot 5^{12} = 5^{20+12} = 5^{32}$
3. $\frac{5^{20}}{5^{12}} = 5^{20-12} = 5^8$
4. $\frac{5^{20}}{5^{-12}} = 5^{20-(-12)} = 5^{20+12} = 5^{32}$
5. $(5^{20})^{12} = 5^{20\times12} = 5^{240}$
6. $\frac{20^{24}}{4^{10}} = \frac{(4\times5)^{24}}{4^{10}} = \frac{4^{24}\times5^{24}}{4^{10}} = 4^{14}\times5^{24}$
7. $3^{-2} = \frac{1}{3^2} = \frac{1}{9}$
8. $\frac{1}{3^{-2}} = 3^2 = 9$
9. $\left(\frac{4}{5}\right)^2 = \frac{4^2}{5^2} = \frac{16}{25}$ or: $\left(\frac{4}{5}\right)^2 = \frac{4}{5} \div \frac{4}{5} = \frac{16}{25}$
10. $\left(\frac{4}{5}\right)^{-2} = \left(\frac{5}{4}\right)^2 = \frac{25}{16} = 1\frac{9}{16}$
11. $25^{\frac{1}{2}} = \sqrt{25} = 5$
12. $32^{\frac{3}{5}} = (\sqrt[5]{32})^3 = 2^3 = 8$
13. $25^{-\frac{1}{2}} = \frac{1}{25^{\frac{1}{2}}} = \frac{1}{\sqrt{25}} = \frac{1}{5}$
14. $32^{-\frac{3}{5}} = \frac{1}{32^{\frac{3}{5}}} = \frac{1}{(\sqrt[5]{32})^3} = \frac{1}{2^3} = \frac{1}{8}$
15. $9^{2-\pi} \cdot 3^{2+2\pi} = (3^2)^{2-\pi} \cdot 3^{2+2\pi} = 3^{4-2\pi} \cdot 3^{2+2\pi} = 3^{4-2\pi+2+2\pi} = 3^6 = 726$

1-9 Order of Operations

To simplify an expression with numbers, we use the following **order of operations**:

1) Do all operations within the grouping symbols first, start with the innermost grouping symbols.
2) Do all operations with exponents.
3) Do all multiplications and divisions in order from left to right.
4) Do all additions and subtractions in order from left to right.

In mathematics, we have agreed the above order of operations to ensure that there is exactly one correct answer.

It is important that we always multiply and divide before adding and subtracting.

Grouping Symbols: A grouping symbol is used to enclose an expression that should be simplified first. First simplify the expression in the innermost grouping symbol. Then work toward the outmost grouping symbol. Grouping symbols are also called **inclusion symbols**.

Parentheses ( ), Brackets [ ], Braces { },
Fraction Bar ——.

Examples

1. $1+2\cdot 3=1+6=7$. **2.** $(1+2)\cdot 3=3\cdot 3=9$. **3.** $2\cdot 3-4=6-4=2$.

4. $4-6\div 2=4-3=1$ **5.** $(4-6)\div 2=-2\div 2=-1$. **6.** $2(3-4)=2(-1)=-2$.

7. $3+2^3=3+8=11$. **8.** $(3+2)^3=5^3=125$. **9.** $3-3^3=3-27=-24$.

10. $2^3\cdot 3^2=8\cdot 9=72$. **11.** $2^3-3^2=8-9=-1$. **12.** $2^3+3^2=8+9=17$.

13. $1^5+2\cdot 3^4=1+2\cdot 81=1+162=163$. **14.** $12-35\div 7+2=12-5+2=7+2=9$.

15. $7-\dfrac{72-30\div 5}{3}+2=7-\dfrac{72-6}{3}+2=7-\dfrac{66}{3}+2=7-22+2=-15+2=-13$.

16.
$$\begin{aligned}&2+3\{4-[6-2(3-1)]\div 2\}\\&=2+3\{4-[6-2(2)]\div 2\}\\&=2+3\{4-[6-4]\div 2\}\\&=2+3\{4-2\div 2\}\\&=2+3(4-1)\\&=2+3(3)=2+9=11.\end{aligned}$$

17.
$$\begin{aligned}&3-2\{5-4^2\cdot[2-3(1-3)]\}\\&=3-2\{5-16\cdot[2-3(-2)]\}\\&=3-2\{5-16\cdot[2+6]\}\\&=3-2\{5-16\cdot 8\}\\&=3-2(5-128)\\&=3-2(-123)=3+246=249.\end{aligned}$$

18. Joe finished 15 assignments and 3 tests (which he received 2 A's and 1 B) in math class. Each assignment earns him 2 points; for each test he can earn 10 points for an A and 7 points for a B. What is his total point ?
Solution: $15\times 2+10\times 2+7\times 1=30+20+7=57$ points.

1-10 Scientific Notations

To write very large numbers or very small numbers, we use the **scientific notations.**

The distance from the Sun to the Earth = 93 million miles = 9.3×10^7 miles = 1.5×10^{11} meters.

The distance from the Earth to the Moon = 250,000 miles = 2.5×10^5 miles = 4×10^8 meters.

The mass of one atom of hydrogen = 1.674×10^{-24} gram.

Scientific Notation: **(A number)** = $m\times10^n$, $1\le m<10$.

m is a number greater or equal to 1, but less than 10. n is an integer.

The following powers of 10 are useful in scientific notations.

$10=10^1$, $100=10^2$, $1{,}000=10^3$, $10{,}000=10^4$, $100{,}000=10^5$,.......

$0.1=10^{-1}$, $0.01=10^{-2}$, $0.001=10^{-3}$, $0.0001=10^{-4}$, $0.00001=10^{-5}$,.......

To write a number in scientific notation, we move the decimal point to get a number that is at least 1 but less than 10. Count the number of decimal places it moves and use it as the power of 10.

$123{,}000{,}000=1.23\times10^8$. (Moved 8 digits to the left. The power is positive.)

$0.0000000123=1.23\times10^{-8}$. (Moved 8 digits to the right. The power is negative.)

The Milky Way is the home galaxy of our solar system and contains about 200 billion stars ($200\times10^9=2\times10^{11}$ stars). A **light-year** is the distance light travels in one year in a vacuum. A light-year is about 5.9 trillion miles (5.9×10^{12} miles). The nearest star to our Sun, Alpha Centauri, is 4.25 light-years (2.51×10^{13} miles) away. The diameter of our solar system is 1.18×10^{11} miles. The size of the universe is estimated at least 2 million light-years (2×10^6 light-years $=1.18\times10^{19}$ miles) in diameter.

Examples (Write in scientific notations.)

1. $100=1\times100=1\times10^2$.

2. $5{,}000=5\times1{,}000=5\times10^3$.

3. $0.02=2\times0.01=2\times10^{-2}$.

4. $0.005=5\times0.001=5\times10^{-3}$.

5. $12{,}000=1.2\times10{,}000=1.2\times10^4$.

6. $0.00012=1.2\times0.0001=1.2\times10^{-4}$.

7. $2{,}350{,}000=2.35\times10^6$.

8. $0.00000235=2.35\times10^{-6}$.

9. $2\times10^5\times6\times10^7=12\times10^{12}=1.2\times10\times10^{12}=1.2\times10^{13}$.

10. $2.35\times10^2\times5.2\times10^4=(2.35)(5.2)\times10^6=12.22\times10^6=1.222\times10\times10^6=1.222\times10^7$.

11. An atom of oxygen has a mass of 2.658×10^{-23} gram. Write it in decimal numeral.

Solution: $2.658\times10^{-23}=0.00000000000000000000002658$ gram.

12. Light travels at the rate of 186,000 miles per second. It takes 8 minutes and 20 seconds for the light from the Sun to reach the Earth. Find the distance from the Sun to the Earth in scientific notation.

Solution: $d=186{,}000\times(8\times60+20)=93{,}000{,}000=9.3\times10^7$ miles.

Notes

1-11 Scale Factors and Drawings

Scale Drawing: A reduced or enlarged figure that is similar to the actual object.
Scale Factor: A ratio of the measurement of a scale drawing to the actual measurement.
Examples

1. The scale of the drawing on a map is 1 inch = 10 inches. What is the actual length of a line if the length on the Map is 8 inches ?
Solution:

Method 1:
Actual length $= 8 \times 10 = 80$ inches.
Method 2:
Let $x =$ the actual length of the line
$\frac{1}{10} = \frac{8}{x}$ $\therefore x = 80$ inches.

2. The scale of the drawing on a map is $\frac{1}{2}$ inch = 10 inches. What is the actual length of a line if the length on the map is $5\frac{1}{2}$ inches ?
Solution:

Method 1: $10 \div \frac{1}{2} = 20$
The scale is 1 in. = 20 in.
Actual length $= 5\frac{1}{2} \times 20 = 110$ in.
Method 2:
Let $x =$ the actual length of the line
$\frac{1/2}{10} = \frac{5\frac{1}{2}}{x}$ $\therefore \frac{1}{2}x = 10(5\frac{1}{2})$
$\frac{1}{2}x = 55$, $x = 110$ inches.

3. The scale of the drawing on a map is $1\frac{1}{2}$ inches = 18 miles. Los Angeles and San Diego are 10 inches apart on the map. What is the actual distance between the two cities ?
Solution:
$18 \div 1\frac{1}{2} = 12$, the scale is 1 inch = 12 miles
Actual distance $= 12 \times 10 = 120$ miles.

4. The scale of the drawing on a map is 1 inch = 1 foot. How tall would the drawing be if you are 5 feet 8 inches tall ?
Solution: The scale is 1 inch = 12 inches

5 feet 8 inches = 68 inches
$68 \div 12 = 5\frac{2}{3}$ inches. Ans.

5. The measurement on the scale drawing is 4 inches. The actual measurement is 5 yards. What is the scale factor ?
(1 yard = 36 inches)
Solution:

5 yards = 180 inches
Scale factor $= \frac{4}{180} = \frac{1}{45}$.

6. The scale of the drawing of a house map is 1 inch = 20 feet. Find the scale factor ?
Solution: 20 feet = 240 inches

Scale factor $= \frac{1}{240}$.

Scale Factors in similar figures

If two geometric figures are similar (the same shape), the ratios of all corresponding dimensions (length, width, height, perimeter, radius, circumference) are equal. These ratios in lowest terms are called the **scale factors**, or **similarity ratios**.

If the scale factor of two similar figures is $a : b$ (or $\frac{a}{b}$) , we have the following formulas:

The ratio of the **surface areas:** $\dfrac{A_1}{A_2}=\left(\dfrac{a}{b}\right)^2$; The ratio of the **volumes:** $\dfrac{V_1}{V_2}=\left(\dfrac{a}{b}\right)^3$

Examples:

1. The following two rectangular solids are similar,

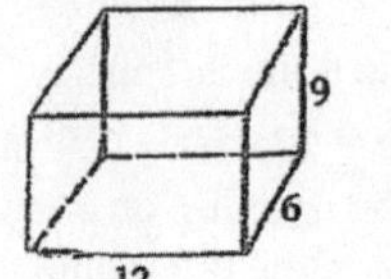

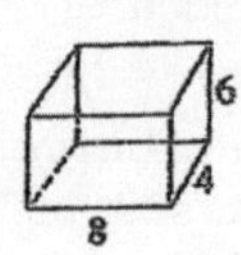

a) what is the ratio of their corresponding sides ?
b) what is the ratio of their surface areas ?
c) what is the ratio of their volumes ?

Solution: a) $\dfrac{a}{b}=\dfrac{6}{4}=\dfrac{9}{6}=\dfrac{12}{8}=\dfrac{3}{2}$. It is the **scale factor**.

b) $\dfrac{A_1}{A_2}=\dfrac{2(12\cdot6+12\cdot9+9\cdot6)}{2(8\cdot4+8\cdot6+6\cdot4)}=\dfrac{2(234)}{2(104)}=\dfrac{468}{208}=\dfrac{9}{4}$. **OR:** $\dfrac{A_1}{A_2}=\left(\dfrac{3}{2}\right)^2=\dfrac{9}{4}$.

c) $\dfrac{V_1}{V_2}=\dfrac{12\cdot6\cdot9}{8\cdot4\cdot6}=\dfrac{648}{192}=\dfrac{27}{8}$. **OR:** $\dfrac{V_1}{V_2}=\left(\dfrac{3}{2}\right)^3=\dfrac{27}{8}$.

2. The following two cylinders are similar,

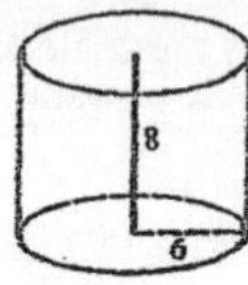

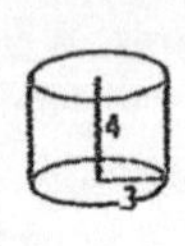

a) what is the ratio of their radii ?
b) what is the ratio of their total surface areas ?
c) what is the ratio of their volumes ?

Solution: a) $\dfrac{r_1}{r_2}=\dfrac{6}{3}=\dfrac{2}{1}$. It is the **scale factor**.

b) $\dfrac{A_1}{A_2}=\dfrac{2(\pi\cdot6^2)+\pi\cdot12\cdot8}{2(\pi\cdot3^2)+\pi\cdot6\cdot4}=\dfrac{168\pi}{42\pi}=\dfrac{4}{1}$. **OR:** $\dfrac{A_1}{A_2}=\left(\dfrac{r_1}{r_2}\right)^2=\left(\dfrac{2}{1}\right)^2=\dfrac{4}{1}$.

c) $\dfrac{V_1}{V_2}=\dfrac{\pi\cdot6^2\cdot8}{\pi\cdot3^2\cdot4}=\dfrac{288\pi}{36\pi}=\dfrac{8}{1}$. **OR:** $\dfrac{V_1}{V_2}=\left(\dfrac{r_1}{r_2}\right)^3=\left(\dfrac{2}{1}\right)^3=\dfrac{8}{1}$.

Examples

3. The following two cones are similar,

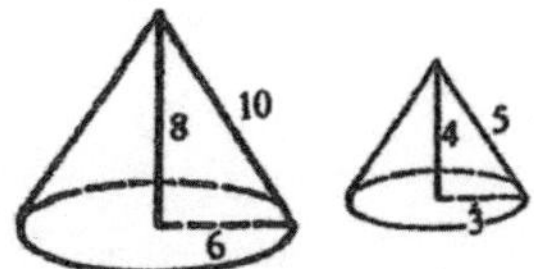

a) what is the ratio of their radii ?
b) what is the ratio of their total surface areas ?
c) what is the ratio of their volumes ?

Solution: a) $\frac{r_1}{r_2}=\frac{6}{3}=\frac{2}{1}$. It is the **scale factor**.

b) $\frac{A_1}{A_2}=\frac{\pi\cdot 6^2+\pi\cdot 6\cdot 10}{\pi\cdot 3^2+\pi\cdot 3\cdot 5}=\frac{96\pi}{24\pi}=\frac{4}{1}$. **OR:** $\frac{A_1}{A_2}=\left(\frac{2}{1}\right)^2=\frac{4}{1}$.

c) $\frac{V_1}{V_2}=\frac{\frac{1}{3}\cdot\pi\cdot 6^2\cdot 8}{\frac{1}{3}\cdot\pi\cdot 3^2\cdot 4}=\frac{96\pi}{12\pi}=\frac{8}{1}$. **OR:** $\frac{V_1}{V_2}=\left(\frac{r_1}{r_2}\right)^3=\left(\frac{2}{1}\right)^3=\frac{8}{1}$.

4. The scale factor of two similar pyramids is 5 : 1. If the surface area of the smaller pyramid is 35 square feet, what is the surface area of the larger pyramid ?
Solution:

Let A_1 = the surface area of the larger pyramid.
A_2 = the surface area of the smaller pyramid.

$\frac{A_1}{A_2}=\left(\frac{5}{1}\right)^2=\frac{25}{1}$ $\therefore A_1=A_2\cdot 25=35\cdot 25=875$ square feet.

5. The scale factor of two similar spheres is 2 : 5. If the volume of the larger sphere is 375 cubic inches, what is the volume of the smaller sphere ?
Solution:

Let V_1 = the volume of the smaller sphere.
V_2 = the volume of the larger sphere.

$\frac{V_1}{V_2}=\left(\frac{2}{5}\right)^3=\frac{8}{125}$ $\therefore V_1=V_2\cdot\frac{8}{125}=375\cdot\frac{8}{125}=24$ cubic inches.

6. If the ratio of the total surface areas of two similar cylinders is 9 : 4,
a) what is the ratio of their radii ? b) what is the ratio of their volumes ?

Solution:

a) $\frac{r_1}{r_2}=\sqrt{\frac{9}{4}}=\frac{3}{2}$. b) $\frac{V_1}{V_2}=\left(\frac{3}{2}\right)^3=\frac{27}{8}$.

Notes

1-12 Repeating Decimals

We can convert a repeating decimal into a common fraction. Most textbooks use the following method to convert a repeating decimal into a common fraction.

Example: Convert $0.5\overline{42}$ into fraction.

Solution:

$$\begin{aligned} \text{Let} \quad N &= 0.5\overline{42} \\ 100N &= 54.2\overline{42} \\ -) \quad N &= \ 0.5\overline{42} \\ \hline 99N &= 53.7 \\ N &= \frac{53.7}{99} = \frac{537}{990} = \frac{179}{330} \end{aligned}$$

$$\therefore\ 0.5\overline{42} = \frac{179}{330}.$$

<u>**Quick Method**</u> **:** For each repeating digit, we write the denominator with a number of 9, and 0 for each nonrepeating digit. Subtract the nonrepeating digits from the numerator.

Examples:

$0.\overline{1} = \frac{1}{9}$, $0.\overline{2} = \frac{2}{9}$, $0.\overline{3} = \frac{3}{9} = \frac{1}{3}$, $0.\overline{9} = \frac{9}{9} = 1$, $0.\overline{12} = \frac{12}{99} = \frac{4}{33}$, $0.\overline{123} = \frac{123}{999} = \frac{41}{333}$

$0.0\overline{1} = \frac{1}{90}$, $0.0\overline{2} = \frac{2}{90} = \frac{1}{45}$, $0.0\overline{3} = \frac{3}{90} = \frac{1}{30}$, $0.0\overline{12} = \frac{12}{990} = \frac{4}{330}$, $0.00\overline{12} = \frac{12}{9900} = \frac{1}{825}$

$0.1\overline{2} = \frac{12-1}{90} = \frac{11}{90}$, $0.1\overline{23} = \frac{123-1}{990} = \frac{122}{990} = \frac{61}{495}$, $0.12\overline{3} = \frac{123-12}{900} = \frac{111}{900} = \frac{37}{300}$

$0.12\overline{345} = \frac{12345-12}{99900} = \frac{12333}{99900} = \frac{4111}{33300}$, $0.0012\overline{3} = \frac{123-12}{90000} = \frac{111}{90000} = \frac{37}{30000}$

$1.\overline{12} = 1\frac{12}{99} = 1\frac{4}{33}$, $0.\overline{73} + 0.\overline{53} = \frac{73}{99} + \frac{53}{99} = \frac{126}{99} = \frac{99}{99} + \frac{27}{99} = 1.\overline{27}$

$0.\overline{314} + 0.\overline{26} = \frac{314314}{999999} - \frac{262626}{999999} = \frac{51688}{999999} = 0.\overline{051688}$, (**Hint:** LCM = 6 digits)

$0.\overline{624} + 0.\overline{1432} = \frac{624624624624}{999999999999} + \frac{143214321432}{999999999999} = \frac{767838946056}{999999999999} = 0.\overline{767838946056}$

Notes

1-13 Factors and Multiples

Before we study the operations with fractions, we need to learn some of the properties of whole numbers. **Number Theory** is the study of the properties of whole numbers. In this chapter, some of the properties of whole numbers are discussed.

Whole numbers: 0 and positive integers. They are 0, 1, 2, 3, 4, 5, 6, 7, 8, and so on.
Divisible: A whole number is said to be divisible by another number if the remainder is 0.
Even numbers: Numbers which are divisible by 2 (such as 2, 4, 6, 8, 10, ·····).
Odd numbers: Numbers which are not divisible by 2 (such as 1, 3, 5, 7, 9, ·····).
0 is considered to be an even number, neither positive nor negative.

Factor: The whole numbers we multiply are called the factors of their product.
Examples: $1 \times 12 = 12$. 1 and 12 are factors of 12.
$2 \times 6 = 12$. 2 and 6 are factors of 12.
A number is said to be **divisible** by each of its factors.
To find the factors of a number, we divide that number by each whole numbers.
If the reminder is 0 (divisible), then the whole number divisor is a factor.
If the reminder is not 0, then the whole number divisor is not a factor.
1 is the smallest factor of every whole number. The only factor of 1 is 1.
A number divided by 0 is undefined. Therefore, 0 is not a factor of any number.
0 divided by any nonzero integer is 0. Therefore, every nonzero integers is a factor of 0.

Multiple: A multiple of a number is the product of that number and any whole number.
Examples: The multiples of 3 are 0, 3, 6, 9, 12, 15, 18, 21, ···· .
The multiples of 4 are 0, 4, 8, 12, 16, 20, 24, 28, ····· .
To find the multiples of a number, we multiply that number by 0, 1, 2, 3, ·····.
Any nonzero integers multiplied by 0 is 0. Therefore, 0 is a multiple of any nonzero integer.

Conventionally, we apply factors and multiples on whole numbers (no negative) only.
(± 3 are the factors of 12 or -12. But, negative numbers are excluded.)

Examples

1. Find all the whole-number Factors of 12.
Solution:
12 is divisible by 1, 2, 3, 4, 6, 12.
The factors are 1, 2, 3, 4, 6,12.

2. Write the first five multiple of 6.
Solution:
Multiplying 6 by 0, 1, 2, 3, 4.
The multiple are 0, 6, 12, 18, 24.

1-14 Divisibility Tests

A whole number is divisible (the remainder is 0) by each of its factors.
To find the factors of a number, sometimes it is helpful to test the divisibility by inspecting the patterns of the digits without actually dividing.

Divisibility Tests: A whole number is:

divisible by **2** : if its last digit is 0, 2, 4, 6, or 8.
divisible by **3** : if the sum of its digits is divisible by 3.
divisible by **4** : if its last two digits are divisible by 4.
divisible by **5** : if its last digit is 0 or 5.
divisible by **6** : if it is divisible by both 2 and 3.
divisible by **8** : if its last three digits are divisible by 8.
divisible by **9** : if the sum of its digits is divisible by 9.
divisible by **10** : if its last digit is 0.
divisible by **11** : if the difference between the sum of the odd digits and the sum of the even digits is 0 or 11.

25894 is divisible by 11 because $(2 + 8 + 4) - (5 + 9) = 0$.
86801 is divisible by 11 because $(8 + 8 + 1) - (6 + 0) = 11$.

Examples

Test each number for divisibility by 2, 3, 4, 5, 6, 8, 9, 10, or 11.

a. 105 *b.* 1248 *c.* 9170 *d.* 78903

Solution:

a. 105 is divisible:
by 3 because the sum of its digits $(1 + 0 + 5 = 6)$ is divisible by 3.
by 5 because its last digit is 5.

b. 1248 is divisible:
by 2 because its last digit is 6.
by 3 because the sum of its digits $(1 + 2 + 4 + 8 = 15)$ is divisible by 3.
by 4 because its last two digits 48 is divisible by 4.
by 6 because it is divisible by both 2 and 3.
by 8 because the last three digits 248 is divisible by 8.

c. 9170 is divisible:
by 2, 5, and 10 because its last digit is 0.

d. 78903 is divisible:
by 3 and 9 because the sum of its digits is divisible by both 3 and 9.
by 11 because $(7 + 9 + 3) - (8 + 0) = 11$.

1-15 Prime Factorization

Prime Number: A whole number other than 0 and 1 which is divisible only by 1 and itself.
Here are the prime numbers: 2, 3, 5, 7, 11, 13, 17, 19, 23, 29, 31, 37,·····.
A prime number has exactly two factors, 1 and itself.

0 is divisible by every nonzero integer (0 has many factors). Therefore, 0 is not a prime number.
1 is divisible only by itself (1 has only one factor, itself). Therefore, 1 is not a prime number (see the other reason at the bottom of this page).
All even numbers greater than 2 are not prime numbers. 2 is the only even prime number.
Therefore, to identify whether or not an odd number is a prime number, we test the divisibility by the prime numbers 2, 3, 5, 7, 11, 13, 17, 19,·····.
Example: 121 is divisible by 11. Therefore, 121 is not a prime number.

Composite Number: A whole number other than 0 and 1 which is not a prime number.
Here are the composite numbers: 4, 6, 8, 9, 10, 12, 14, 15, 16, ·····.
A composite number has more than two factors.
The numbers 0 and 1 are neither prime nor composite.

Prime Factorization: It is to write a composite number as a product of prime numbers.
We can factor a number into prime factors by using either of the following two methods: **Shortcut division** and **Factor tree**.

In prime factorization, we divide the number by a prime number (from small to large) until the bottom row is a prime number. Then, we write the prime factors in order from least to greatest.

Example: Find the prime factorization of 24.
Solution:
Method 1: **Shortcut division** Method 2: **Factor tree**

```
2| 24            24
2| 12          2 × 12
 2| 6            2 × 6
    3              2 × 3
```

Answer: $24 = 2 \times 2 \times 2 \times 3 = 2^3 \times 3$.
(Exponents are often used to express the prime factorization.)

Fundamental Theorem of Arithmetic:
Every composite number can be factored as a product of prime factors in exactly one way, except for the order of the prime factors.

1 is not considered as a prime number to avoid $24 = 1 \times 1 \times 1 \times \cdots \times 2^3 \times 3$.
The Fundamental Theorem of Arithmetic would be false if 1 was defined as a prime number.

1-16 Greatest Common Factor and Least Common Multiple

Greatest Common Factor (GCF): The greatest number that is a factor of two or more given numbers.

Factors of 12: 1, 2, 3, 4, **6**, 12 ; Factors of 18: 1, 2, 3, **6**, 9, 18

The common factors of 12 and 18 are 1, 2, 3, **6**. The GCF of 12 and 18 is **6**.

To find the GCF of two or more given numbers, we write the prime factorizations of the given numbers, and find the **lowest power** of each factor that occurs in **all** of the factorizations. The product of these lowest powers is the GCF.

Example 1: Find the GCF of 12 and 18.

Solution:

$12 = 2^2 \times 3$; $18 = 2 \times 3^2$

$\therefore$ GCF $= 2 \times 3 = 6$.

Example 2: Find the GCF of 54 and 90.

Solution:

$54 = 2 \times 3^3$; $90 = 2 \times 3^2 \times 5$

$\therefore$ GCF $= 2 \times 3^2 = 18$.

Least Common Multiple (LCM): The smallest number that is a multiple of two or more given numbers.

Multiples of 12: 0, 12, 24, **36**, 48, 60, **72**, ···· ; Multiples of 18: 0,18, **36**, 54, **72**, ····

The common multiples of 12 and 18 are **36**, **72**, ····. The LCM of 12 and 18 is **36**.

To find the LCM of two or more given numbers, we write the prime factorizations of the given numbers, and find the **highest power** of each factor that occurs in **any** of the factori zations. The product of these highest powers is the LCM.

Example 3: Find the LCM of 12 and 18.

Solution:

$12 = 2^2 \times 3$; $18 = 2 \times 3^2$

$\therefore$ LCM $= 2^2 \times 3^2 = 36$.

Example 4: Find the LCM of 54 and 90.

Solution:

$54 = 2 \times 3^3$; $90 = 2 \times 3^2 \times 5$

$\therefore$ LCM $= 2 \times 3^3 \times 5 = 270$.

Example 5: Find the GCF and the LCM of $24x^2y^2$ and $84x^3y$.

Solution:

$$24x^2y^2 = 2^3 \cdot 3 \cdot x^2 \cdot y^2 \ ; \ 84x^3y = 2^2 \cdot 3 \cdot 7 \cdot x^3 \cdot y$$

$\therefore$ GCF $= 2^2 \cdot 3 \cdot x^2 \cdot y = 12x^2y$. LCM $= 2^3 \cdot 3 \cdot 7 \cdot x^3 \cdot y^2 = 168x^3y^2$.

The above examples show the following relationship between GCF and LCM.

Relationship between GCF and LCM

Let (a, b) represents the GCF, $[a, b]$ represents the LCM of two numbers a and b.

The product of two numbers a and b is equal to the product of their GCF and LCM.

Formula: $a \times b = (a, b) \times [a, b]$

Example 6. If the product of two numbers is 216 and their GCF is 6, what is their LCM?

Solution: $216 = 6 \times$ LCM $\therefore$ LCM $= 36$.

Example 7. If the product of two numbers is 4860 and their LCM is 270,What is their GCF?

Solution: $4860 = 270 \times$ GCF $\therefore$ GCF $= 18$.

The Shortcut method

1. To find the GCF of two or more given numbers, we choose a prime number (from small to large) that can divide into **all** of the given numbers. Continue to choose the divisors until no divisor can be used. The product of all the divisors is the GCF.
   Or, find the largest number that can divide into all of the given numbers.
2. To find the LCM of two or more given numbers, we choose a prime number (from small to large) that can divide into **any two or more** of the given numbers. Write the quotients and the **remaining** numbers on the next row. Continue to choose the divisors until no divisor can be used. The product of all the divisors and the numbers on the bottom row is the LCM.
3. To find GCF and LCM of two very large numbers, we use the **Rollabout Divisions** to find the largest number that can divide into all of the given numbers.

Example 8: Find the GCF and the LCM of 54 and 90.

Solution:

$$\begin{array}{r|rr} 2 & 54 & 90 \\ \hline 3 & 27 & 45 \\ \hline 3 & 9 & 15 \\ \hline & 3 & 5 \end{array} \qquad \textbf{Or: } \begin{array}{r|rr} 18 & 54 & 90 \\ \hline & 3 & 5 \end{array}$$

$\therefore$ GCF $= 2 \times 3 \times 3 = 18$

LCM $= 2 \times 3 \times 3 \times 3 \times 5 = 270$.

Example 9: Find the GCF and the LCM of 12, 30 and 45.

Solution:

$$\begin{array}{r|rrr} 3 & 12 & 30 & 45 \\ \hline 2 & 4 & 10 & 15 \\ \hline 5 & 2 & 5 & \mathbf{15} \\ \hline & \mathbf{2} & 1 & 3 \end{array} \quad \begin{array}{l} \\ \\ \leftarrow \text{15 is the remaining number.} \\ \leftarrow \text{2 is the remaining number.} \end{array}$$

$\therefore$ GCF = 3. (3 is the only common divisor.)

LCM $= 3 \times 2 \times 5 \times 2 \times 1 \times 3 = 180$.

Example 10: Find the GCF and the LCM of 15, 21 and 35.

Solution:

$$\begin{array}{r|rrr} 3 & 15 & 21 & 35 \\ \hline 5 & 5 & 7 & \mathbf{35} \\ \hline 7 & 1 & \mathbf{7} & 7 \\ \hline & \mathbf{1} & 1 & 1 \end{array} \quad \begin{array}{l} \leftarrow \text{no common divisor except 1.} \\ \leftarrow \text{35 is the remaining number.} \\ \leftarrow \text{7 is the remaining number.} \\ \leftarrow \text{1 is the remaining number.} \end{array}$$

$\therefore$ GCF = 1. (There is no common divisor except 1.)

LCM $= 3 \times 5 \times 7 \times 1 \times 1 \times 1 = 105$.

Example 11: If a and b are integers and $a < b$, GCF = 18, LCM = 270, find a and b.

Solution: (See Example 8)

$270 \div 18 = 15$, 15 is the product of the numbers on the bottom row.

$15 = 1 \times 15$ or $15 = 3 \times 5$

$a = 1 \times 18 = 18$ and $b = 15 \times 18 = 270$. Ans.

Or: $a = 3 \times 18 = 54$ and $b = 5 \times 18 = 90$. Ans.

To find the GCF and the LCM of two large numbers, we use the **Rollabout Divisions** to find the largest number that can divide into all of the given numbers.

The Rollabout Divisions was first invented by Euclid, a famous Greek mathematician who lived about 300 B.C. It was first introduced in U.S.A by Mr. Rong Yang.

Rollabout Divisions

Step 1: Divide the larger number by the smaller number to find the remainder.
Step 2: Divide the smaller number by the remainder in Step 1.
Step 3: Divide the previous remainder by the new remainder.
Step 4: Repeat the Step 3 until no remainder is found.
Step 5: The divisor of the last division is the GCF if the remainder is zero.
Step 6: Find LCM by the same method on page 52.

Example: Find the GCF and the LCM of 3,473 and 1,963.

Solution:

1.
$$1963\overline{)3473}\ \text{quotient } 1,\quad 3473-1963=1510 \text{ remainder}$$

2.
$$1510\overline{)1963}\ \text{quotient } 1,\quad 1963-1510=453 \text{ remainder}$$

3.
$$453\overline{)1510}\ \text{quotient } 3,\quad 1510-1359=151 \text{ remainder}$$

4.
$$151\overline{)453}\ \text{quotient } 3,\quad 453-453=\mathbf{0}$$

We have: GCF =151 Ans.

$$\begin{array}{r|rr} 151 & 3473 & 1963 \\ \hline & 23 & 13 \end{array}$$

$3473 \div 151 = 23$ and $1963 \div 151 = 13$

We have: $\text{LCM} = 151 \times 23 \times 13 = 45149$ Ans.

1-17 Number Properties

Axiom (Postulate): It is a statement that we assume to be true without proof. The following basic axioms (postulates) about the properties of real numbers are used very frequently in the field of mathematics.

For all real numbers *a*, *b*, and *c*, we have :

1. **Reflexive Property**: $a = a$
2. **Symmetric Property**: If $a = b$, then $b = a$.
3. **Transitive Property:** If $a = b$ and $b = c$, then $a = c$.
4. **Substitution Property:** If $a = b$, then a can be replaced by b.
5. **Commutative Property:** $a + b = b + a$ and $ab = ba$
6. **Associative Property:** $(a+b)+c = a+(b+c)$ and $(ab)c = a(bc)$
7. **Distributive Property:** $a(b+c) = ab + ac$ and $(b+c)a = ba + ca = ab + ac$
8. **Multiplicative Property of Zero:** $a \times 0 = 0$ and $0 \times a = 0$
9. **Additive Identity Property:** $a + 0 = a$ and $0 + a = a$
10. **Multiplicative Identity Property:** $a \times 1 = a$ and $1 \times a = a$
11. **Addition Property of Equality:** If $a = b$, then $a + c = b + c$.
12. **Subtraction Property of Equality:** If $a = b$, then $a - c = b - c$.
13. **Multiplication Property of Equality:** If $a = b$, then $ac = bc$.
14. **Division Property of Equality:** If $a = b$ and $c \neq 0$, then $\frac{a}{c} = \frac{b}{c}$.
15. **Additive Inverse (Axiom of Opposites):** $a + (-a) = 0$
16. **Multiplication Inverse (Axiom of Reciprocals):** If $a \neq 0$, then $a \times \frac{1}{a} = 1$.
17. **Property of the Opposite of a Sum:** $-(a+b) = (-a) + (-b)$
18. **Property of the Reciprocal of a Product:** If $a \neq 0$ and $b \neq 0$, then $\dfrac{1}{ab} = \dfrac{1}{a} \times \dfrac{1}{b}$.
19. **Property of Proportions:** If $\dfrac{a}{b} = \dfrac{c}{d}$, then $ad = bc$. (The cross products are equal.)

Notes

1-18 Descriptive Statistics

In order to find a convenient way to describe and compare the information of entire collection of data, we need to organize the data and find a single number that is most significant to represent the data. Then we can analyze and make conclusion about the data. **Descriptive Statistics** is the study of collecting, organizing, and analyzing numerical data.

The **mean, median, mode,** and **range** are important statistical numbers often used in analyzing numerical data. The mean, median, and mode are three kinds of averages.

Mean: It is the sum of the numbers divided by the number of items in the data. It is the arithmetic average and is most commonly used as an average.

Median: It is the middle number in the data when the data are arranged in an increasing order (from least to greatest). For an even number of items, the average of the two numbers in the middle is the median.

Mode: It is the number that appears most often in the data. There may be no mode or more than one mode.

Range: It is the difference between the highest and the lowest numbers.

The mean, median, or mode provides the information about the **central tendency** of the data. Which of these three measures is the best to represent the central tendency depends on the way in which you need to use the data. If the mean is inflated by a few very larger or very small numbers in the data, the median can be used as an average. Mode can be used as an average if we want the most frequent number in the data. The range provides the information about the **spread** of the data.

Examples

1. The heartbeats per minute (pulse rates) for 9 students are listed below.

   66, 71, 72, 69, 70, 71, 72, 85, 72

   Find: **a)** the mean **b)** the median **c)** the mode **d)** the range.

   Solution:

   a) Divide the sum of the numbers by the number of students.

   The mean $= \frac{66+71+72+69+70+71+72+85+72}{9} = \frac{648}{9} = 72$.

   b) Arrange the numbers from least to greatest.

   66, 69, 70, 71, **71**, 72, 72, 72, 85. The median = 71.

   c) 72 appears most often. The mode = 72.

   d) $85 - 66 = 19$ The range = 19.

2. Find the median of the data 15, 19, 6, 17, 12, 12, 17, 11.

   Solution:

   Arrange the data from least to greatest: 6, 11, 12, **12**, **15**, 17, 17, 19

   The median $= \frac{12+15}{2} = 13.5$.

Statistical Graphs

In statistics, tables and graphs are often used to display and represent numerical data. Data represented in table or graph are easily to read and understand.
There are many different kinds of statistical tables and graphs we use in statistics, such as **bar graph, line graph, circle graph, picture graph, Stem-and-Leaf plot, Box-and -Whisker plot**, and others. Three major graphs are shown below. We can estimate data from graphs.

Examples

1. The bar graph below shows the population of a town between 1930 and 2000.
   1. What data unit is used on the horizontal axis ?
   2. What data unit is used on the vertical axis ?
   3. In which year did the population increase the most ?
   4. Estimate the approximate population in 1980.
   5. Find the approximate total increase in population during the years from 1930 to 2000.
   6. In which year did the population stay the same as the previous year.

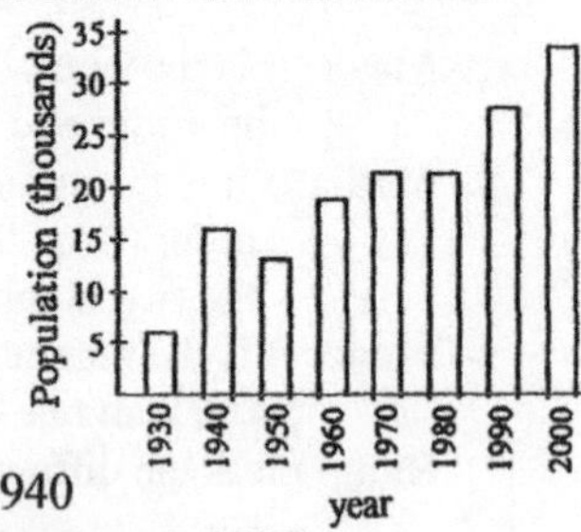

Solution:

1. The year **2.** Thousands of persons **3.** 1940
4. 22,000 **5.** 34,000 – 6,000 = 28,000 persons **6.** 1980

2. Make a line graph to illustrate the given data.

Population of a town (in Thousands)

1930	1940	1950	1960	1970	1980	1990	2000
6	16	13	17	22	22	28	34

Solution:

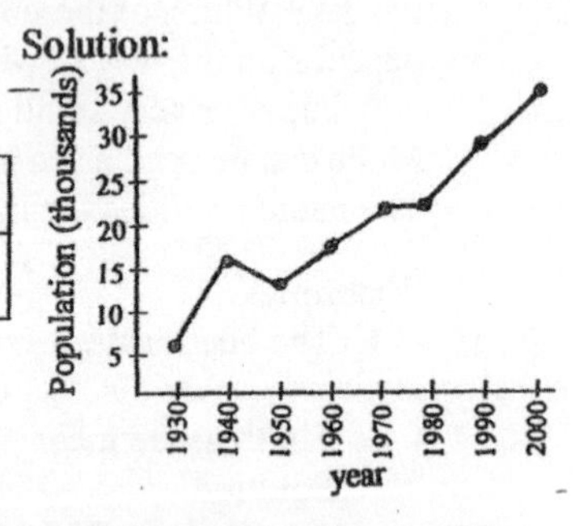

3. The circle graph below shows Ronald's budget to spend his money. His monthly income is \$2,700. Find how much money does he plan to spend on food, clothes, housing, girl friends, savings and others.

Solution:

Food: $\$2,700 \times 25\% = \$675$
Clothes: $\$2,700 \times 8\% = \$216$
Housing: $\$2,700 \times 17\% = \$459$
Girl friends: $\$2,700 \times 3\frac{2}{3}\% = \$99$
Saving: $\$2,700 \times 35\frac{1}{3}\% = \$954$
Others: $\$2,700 \times 11\% = \$297$

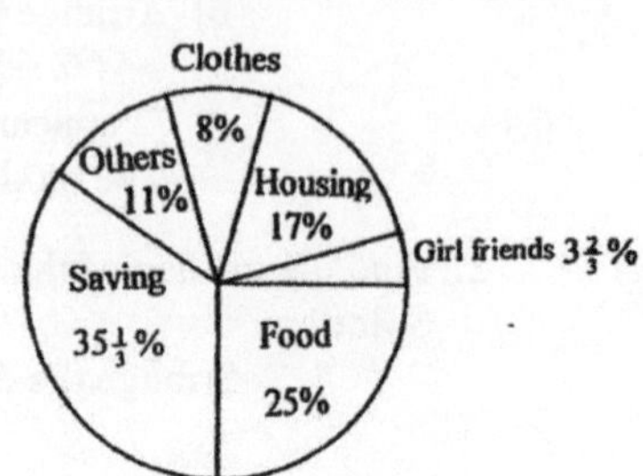

Histogram and Normal Distribution

In statistics, we use a table called **frequency distribution** to show information about the the frequency of occurrences of statistical data. **Histogram** is a statistical graph to describe a frequency distribution.

In a histogram, data are grouped into convenient intervals. A boundary datum in a histogram is included in the interval to its left. If the midpoints at the tops of the bars are connected, a smooth **"bell-shaped"** curve called **normal curve** can result from the plotting of a larger collection of data. The area under the normal curve represents the probability from **normal distribution**. The area under the curve and above the $x-$axis is 1. Therefore, the sum of the probabilities of all possible outcomes is 1 (or 100%).

The probability of each region under a normal curve is shown below. We can find the the probability of a given region under the normal curve.

Frequency Distribution

Test scores	numbers of students
45 ~ 55	3
55 ~ 65	13
65 ~ 75	21
75 ~ 85	27
85 ~ 95	15
95 ~100	7

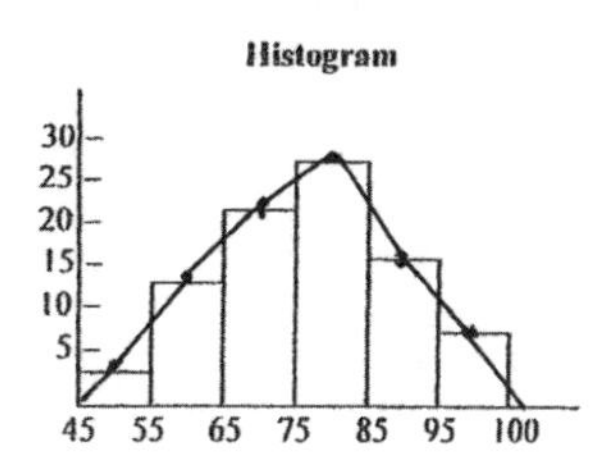

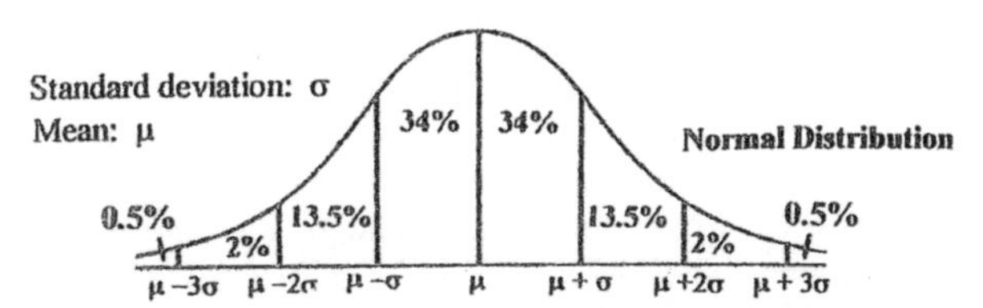

Examples

1. The heights of the students in a high school are normally distributed as shown in the graph. Find the percentage of the students in each the following groups.

   a. Height $>172\,cm$ **b.** Height $<172\,cm$ **c.** Height $>179\,cm$ **d.** Height $<186\,cm$

   Solution:

   a. 50% **b.** 50%

   c. $13.5\% + 2\% + 0.5\% = 16\%$

   d. $50\% + 34\% + 13.5\% = 97.5\%$

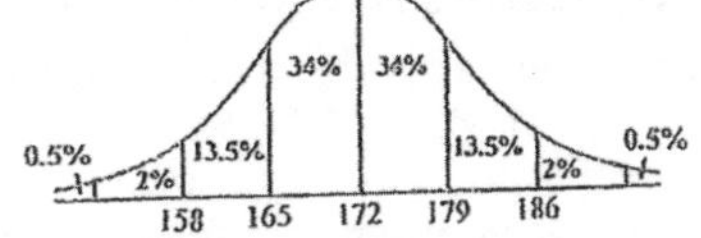

2. The sales of a math book each month are normally distributed as shown in the graph.

   a. What is the probability that a month will sell more than 600 copies in a month ?

   b. What is the probability that a month will sell less than 300 copies in a month ?

   Solution: **a.** $13.5\% + 2\% + 0.5\% = 16\%$

   b. $2\% + 0.5\% = 2.5\%$

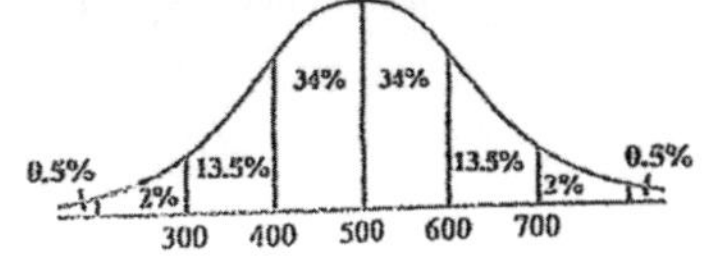

Stem-and-Leaf Plots & Box-and-Whisker Plots

A stem-and-leaf plot is a statistical plot used to organize numerical data. We can use the the stem-and-leave plot to convert data into a bar graph or compare two sets of data.

Examples

1. Draw a stem-and-leaf plot of the data.

13, 4, 23, 34, 40, 52, 55, 52, 9, 38
60, 36, 22, 68, 15, 39, 29, 15, 41

Solution:

Identify the lowest number 4.
Identify the highest number 68.
In the plot, use the tens digits as the stems and the ones digits as the leaves.
Draw a vertical line between the stems and the leaves.
Arrange each row of leaves from smallest to largest.
A legend is used if the decimal point is not included in the stems.

Stems	Leaves
0	4, 9
1	3, 5, 5
2	2, 3, 9
3	4, 6, 8, 9
4	0, 1
5	2, 2, 5
6	0, 8

1 | 3 = 13

2. Draw a stem-and-leaf plot of the data below.

8.9, 4.6, 11.2, 10.2, 8.3, 8.4, 10.4, 8.8
4.2, 8.1, 10.8, 5.7, 10.4, 5.3, 11, 6.4

Solution:

Identify the lowest number 4.2.
Identify the highest number 11.2.
In the plot, use the whole number as the stems and the tenths as the leaves.
Arrange each row of leaves from smallest to largest.

Stems	Leaves
4.	2, 6
5.	3, 7
6.	4
7.	
8.	1, 3, 4, 8, 9
9.	
10.	2, 4, 4, 8
11.	0, 2

A **box-and-whisker plot** is also a statistical plot used to organize numerical data. In the box-and-whisker plot, the data are divided into four sections (2 whiskers and 2 boxes). Each section contains 25% of data. The lines to the left and right of the boxes are the whiskers of the plot. We can use the box-and-whisker plot to identify the three **quartiles** (Q_1, Q_2, Q_3) between the two **outliers** (The minimum and the maximum).
Q_1 is the 1st **(lower) quartile.** Q_2 is the 2nd **quartile** (It is the median.). Q_3 is the 3rd **(upper) quartile. Interquartile Range** = $Q_3 - Q_1$ (It is the spread of the middle half.)

3. Draw a box-and-whisker plot for the test scores of 12 students.

75, 86, 65, 95, **85, 68**, 92, 90, 85, 75, 82, 80

Solution:

Arrange the numbers from least to greatest.
65, 68, 75, 75, 80, **82, 85**, 85, 86, 90, 92, 95

The median = 83.5.
The median of the numbers below 83.5. It is 75.
The median of the numbers above 83.5. It is 88.
The interquartile range = $Q_3 - Q_1 = 88 - 75 = 13$.

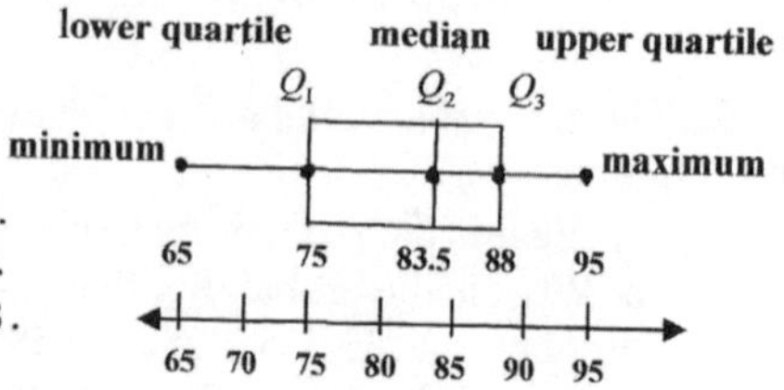

lower quartile Q_1 median Q_2 upper quartile Q_3
minimum maximum
65 75 83.5 88 95
65 70 75 80 85 90 95

1-19 Sets and Venn Diagrams

Set: Any particular group of elements or members. We use braces, { }, to name a set.

U: The **universal** set. It is the set consisting of all the elements.

$A \cap B$: The **intersection** of two sets. The set consists of the elements that are common to both sets A and B.

$A \cup$ B: The **union** of two sets. The set consists of all members of both sets A and B.

ϕ: The **empty** set, or null set. The set consists of no member.
The set $\{0\}$ contains exactly one member "0". The set $\{0\}$ is not an empty set.

$\in$: It indicates " **it is a member of** ". $\notin$: It indicates " **it is not a member of** ".

$A \subset B$ or $A \subseteq B$: If every member in set A is also a member of set B, then A is a **subset** of B, B is a **superset** of A. Every set is the subset of itself. It is agreed that the empty set is also a subset of every set.

If A is a subset of B and $A \neq B$, we write $A \subset B$, then A is a **proper subset** of B. If A is a subset of B and $A = B$, we write $A \subseteq B$.

$A \not\subset B$: Set A is **not a subset** of set B.

A' or A^c: The **complement** of set A. A' is the set consisting of all the elements not in A.

$A \cup A' = U$; $A \cap A' = \phi$

Venn Diagram: A diagram used to indicate relations between sets.

1. $n(A \cup B) = n(A) + n(B) - n(A \cap B)$
2. $n(A \cup B \cup C) = n(A) + n(B) + n(C) - n(A \cap B) - n(B \cap C) - n(A \cap C) + n(A \cap B \cap C)$

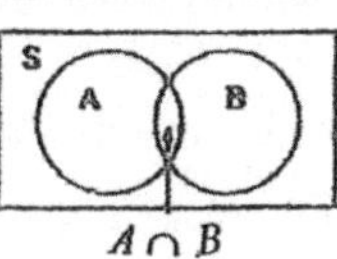

$A \cap B$

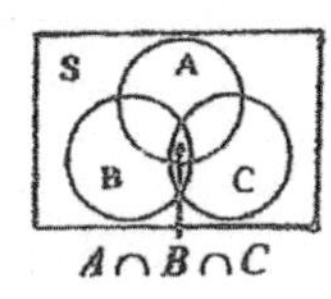

$A \cap B \cap C$

Examples

1. If $A = \{1, 3, 5, 7\}$, $B = \{2, 4, 6, 8\}$, $U = \{1, 2, 3, 4, 5, 6, 7, 8\}$, we have:

$3 \in \{1, 3, 5, 7\}$, $6 \notin \{1, 3, 5, 7\}$, $A \neq B$, $\{1, 3, 5, 7\} = \{1, 5, 7, 3\}$

$A \subset U$, $B \subset U$, $A \not\subset B$, $B \subset B$, $\phi \subset A$, $\phi \subset B$, $A' = B$, $A \cup A' = U$

$A \cup B = U$, $A \cap B = \phi$, $A \cap U = \{1, 3, 5, 7\}$, $U \cap B = \{2, 4, 6, 8\}$

2. In a high school class, there are 25 students who went to see movie A, 18 students went to see move B. These figures include 9 students went to see both movies. How many students are in this class ?

Solution:

$25 + 18 - 9 = 34$ students.

3. In a high school class, there are 17 students in the basketball team, 18 in the volley ball team, 20 in the baseball team. Of these, there are 9 students in both the basket ball and baseball teams, 6 in both volleyball and baseball teams, and 5 in both volleyball and basketball teams. These figures include 2 students who are in all three teams. How many students are in these class ?

Solution:

$17 + 18 + 20 - 5 - 6 - 9 + 2 = 37$ students.

1-20 Counting Principles

1) Principle of Multiplication

If we select one from *m* ways, then we select another one from *n* ways; then all the possible selections are:

$$m \times n \text{ different ways}$$

2) Principle of Addition

If we select one from *m* ways, then **mutually exclusive**, we select another one from *n* ways; then all the possible selections are:

$$m + n \text{ different ways}$$

Examples

1. How many different ways can a 10-question true-false test be answered if each question must be answered ?
   Solution: each question has two selections:
   $2\times2\times2\times2\times2\times2\times2\times2\times2\times2 = 1{,}024$ ways   Or: $2^{10} = 1{,}024$ ways

2. How many license plates of 6 symbols can be made using letters for the first 2 symbols and digits for the remaining symbols ?
   Solution:
   $26 \times 26 \times 10 \times 10 \times 10 \times 10 = 6{,}760{,}000$ license plates

3. How many different selections are possible to assign 4-letter from 26 alphabets ?
   Solution:
   $26\times26\times26\times26 = 456{,}976$ codes

4. How many positive even integers less than 1,000 can be made using the digits 5, 6, 7, and 8 ?
   Solution:
   1-digit even integers: 2
   2-digit even integers: $4 \times 2 = 8$
   3-digit even integers: $4 \times 4 \times 2 = 32$
   Total 42 even integers

5. There are 3 routes from City A to City B, and 4 routes from B to City D. Mutually exclusive, there are 5 routes from City A to City C, and 2 routes from City C to City D. How many different routes can be selected from City A to City D ?
   Solution:
   $3\times4 = 12$ routes
   $5\times2 = 10$ routes
   Total 22 routes

1-21 Logical Reasoning

Logical Reasoning: It is a thinking process to prove a desired conclusion from given facts. Logical Reasoning includes analyzing relevant information, observing patterns, and estimating unknown data to reach a logical conclusion. The logical questions in mathematics may not cover any of the topics in mathematics. We use the following symbols to express a statement. Suppose A is a statement, we use $\sim A$ to represent the negation of A. We read $\sim A$ as "not A" or " it is false that A.".
$A \rightarrow B$ is a conditional statement in the form " if A, then B.".
$B \rightarrow A$ is the **converse** of $A \rightarrow B$. It tells that " If B, then A."
$\sim A \rightarrow \sim$B is the **inverse** of $A \rightarrow B$. It tells that " If not A, then not B.".
$\sim B \rightarrow \sim A$ is the **contraposition** of $A \rightarrow B$. It tells that " If not B, then not A.".
In logical reasoning, we have the following rule:

Rule on Contraposition: $A \rightarrow B$ is logical equivalent to $\sim B \rightarrow \sim A$.

The rule states that given a **If-then** true statement, then there is only one other **If-then** statement (the **contraposition**) is logically true.
The rule also indicates that " If A, then B " does not necessarily mean " If B, then A.", or " If not A, then not B." are true.
A **contradiction** is a false statement because it directly contradicts either the original statement of " If A, then B " with " If A, then $\sim B$ " or " the Rule on Contraposition ".
Examples

1. $A \rightarrow B$: If two angles are vertical angles, then they are congruent. **True**
   $B \rightarrow A$: If two angles are congruent, then they are vertical angles. **False**
   $A \rightarrow \sim B$: If two angles are vertical angles, then they are not congruent. **False**
   $\sim B \rightarrow A$: If two angles are not congruent, then they are vertical angles. **False**
   $\sim A \rightarrow B$: If two angles are not vertical angles, then they are congruent. **False**
   $\sim A \rightarrow \sim B$: If two angles are not vertical angles, then they are not congruent. **False**
   $\sim B \rightarrow \sim A$: If two angles are not congruent, then they are not vertical angles. **True**

2. Given the true statement " If you have measles, then you have a rash.". Write another If-then statement which must be true.
   Solution:
   The contraposition is " If you have no rash, then you have no measles.".
3. Given the true statement " If you have measles, then you have a rash. ".
   Write a If-then statement that cannot be true.
   Solution:
   The contradiction is " If you have measles, then you have no rash. ".
   or " If you have no rash, then you have measles. ".

Most logical questions apply to Rule on Contraposition. However, it may not be easy to prove a statement false. To prove a statement false, we find a contradiction or find one exception. A **counterexample** is an exception that proves a statement false. A single counterexample is sufficient to prove a statement which is false.

Examples

4. Give a counterexample to show that the statement can be false: " $|x+y|=|x|+|y|$ ".

Solution: If $x=4$ and $y=-1$, then $|x+y|=|4+(-1)|=|3|=3$,

and $|x|+|y|=|4|+|-1|=4+1=5$.

We have $|x+y|\neq|x|+|y|$ if x and y are in opposite signs. The statement is false.

5. Give a contradiction to show that the given statement is false:
" If two angles are congruent, then they are vertical angles."
Solution:
A contradiction: "If two angles are not vertical angles, then they are not congruent. "
Hint: It is a contradiction because the contraposition of the given statement is false.
A contraposition must be true if the original statement is true.

6. Give a counterexample to show that the statement is false"
" If two angles are congruent, then they are vertical angles"
Solution:
If two angles are congruent, then they could be two of the three angles in a triangle.

Logical Argument: A logical argument has two given statements (premises) and one conclusion. There are five patterns of logical arguments. Arguments that use these patterns correctly will have a valid conclusion. Arguments that use these patterns incorrectly will have a invalid conclusion.

1. Direct Argument
If p is true, then q is true
P is true.
Therefore, q is true.

2. Indirect Argument
If p is true, then q is true
q is not true.
Therefore, p is not true.

3. Chain Rule
If p is true, then q is true.
If q is true, then r is true.
Therefore, if p, then r.

4. Or Rule
P is true or q is true.
P is not true.
Therefore, q is true.

5. And Rule
p and q not both true.
But q is true.
Therefore, p is not true.

Examples

6. In the logical proof of the statement
" If $x=4$, then $\sqrt{x}$ is an integer. ",
decide whether the following conclusion is true or false.

$\sqrt{x}$ is an integer. Therefore, $x=4$.

Solution:

The conclusion is false.

7. In an indirect proof of the statement
" If $x=4$, then $\sqrt{x}$ is an integer. ",
what is the assumption used to arrive a valid conclusion ?

Solution:

$\sqrt{x}$ is not an integer.
Therefore, $x\neq 4$.

1-22 Sequences and Number Patterns

Patterns appear in nature and our daily life. Studying sequences and number patterns helps us to identify, understand patterns and make predictions.

Sequence: A sequence is a consecutive nature numbers in a certain order.
Each number is called a term of the sequence.

To find the pattern of a sequence, we find the **differences** between any two successive terms.

Examples

1: Find the next five terms of the sequence 2, 4, 6, 8, 10, ·····.
Solution: The terms increase by 2. The differences are constant.
The next five terms are 12, 14, 16, 18, and 20.

2: Find the next five terms of the sequence 106, 95, 84, 73, 62, ····.
Solution: The terms decrease by 11. The differences are constant.
The next five terms are 51, 40, 29, 18, and 7.

3: Find the next five terms of the sequence 1, 3, 6, 10, 15, ·····.
Solution: The differences are not constant.
The next five terms are continued by adding 6, 7, 8, 9, and 10.
The next five terms are 21, 28, 36, 45, and 55.

1 , 3 , 6 , 10 , 15 , **21** , **28** , **36** , **45** , **55**
2 3 4 5 **6** **7** **8** **9** **10** (differences)

4: Find the next five terms of the sequence 2, 6, 12, 20, 30, ·····.
Solution: The differences are not constant.
The next five terms are continued by adding 12, 14, 16, 18, and 20.
The next five terms are 42, 56, 72, 90, and 110.

2 , 6 , 12 , 20 , 30 , **42** , **56** , **72** , **90** , **110**
4 6 8 10 **12** **14** **16** **18** **20** (differences)

5: Each of the five teams plays each of the other teams just once. How many games will there be ? How many games will there be for 8 teams ?
Solution:

Number of teams	1	2	3	4	5	6	7	8
Number of games	0	1	3	6	10	15	21	28

There will be 10 games for 5 teams.
There will be 28 games for 8 teams.

1-23 Arithmetic and Geometric Sequences

Arithmetic Sequence: A sequence which has common different between any two successive terms.

Arithmetic Mean (Average): A single term between two terms of an arithmetic sequences. If p, m, q is an arithmetic sequence, then $m = \frac{p+q}{2}$.

Geometric Sequence: A sequence in which the ratios of consecutive terms are the same.

Geometric Mean: If p, m, q is a geometric sequence, then $m = \pm\sqrt{pq}$.

Formulas

1. The general or nth term of an arithmetic sequence a_1, $a_1 + d$, $a_1 + 2d$, $a_1 + 3d$, ·····, a_n is given by: $a_n = a_1 + (n-1)\,d$

   Where a_1 :1st term, n: position of a_n, d: common difference.

2. The general or nth term of a geometric sequence a_1, $a_1 r$, $a_1 r^2$, $a_1 r^3$, ·····, a_n is given by: $a_n = a_1 r^{n-1}$

   Where a_1: 1st term, n: position of a_n, r: common ratio

Examples

1. Find the 10th term of 25, 33, 41, ·····, a_{10}.

   Solution:

   $$d = 33 - 25 = 8\,,\quad a_{10} = a_1 + (n-1)d = 25 + (10-1)8 = 97$$

2. Find the arithmetic mean between 17 and 53.

   Solution:

   $$m = \frac{p+q}{2} = \frac{17+53}{2} = 35.$$ Hint: the sequence is 17, **35**, 53, 71,·····.

3. Find the 10th term of the geometric sequence 2, 6, 18, ·····.

   Solution: The common ratio $r = \frac{6}{2} = 3$, $a_{10} = a_1 r^{10-1} = 2 \cdot 3^9 = 39{,}366$

4. Find the 8th term of the geometric sequence 2, $\frac{2}{3}$, $\frac{2}{9}$, $\frac{2}{27}$, ·····.

   Solution: The common ratio $r = \frac{2/3}{2} = \frac{2}{3} \cdot \frac{1}{2} = \frac{1}{3}$, $a_8 = a_1 r^{8-1} = 2 \cdot \left(\frac{1}{3}\right)^7 = \frac{2}{3^7} = \frac{2}{2187}$

5. Find the geometric mean of –4 and –9.

   Solution

   $$m = \pm\sqrt{pq} = \pm\sqrt{-4 \cdot -9} = \pm 6$$

   (Hint: The sequence is –4, **6**, –9, ····· or –4, **–6**, –9, ·····.)

1-24 Discrete Probability

In mathematics, we want to find the possibilities of uncertainty. For example, we want to know the chance of winning a lottery, or the chance to have a life of more than 90 years. There are countless of uncertainty in our everyday life. To measure the uncertainties, we use the term "the probability". For example, the probability of tossing a fair coin turned up head is one-half, or 50%.
The main use for the theory of probability is to make decision by performing a survey or experiment that will allow us to study, predict, and make conclusions from data. In the study of probability, the various results in a random experiment are called the **outcomes**. The collection of all possible outcomes is called the **sample space**. Any subset of the sample space is called an **event**.
There are two types of discrete probability, **theoretical probability** and **experimental probability**.

Theoretical Probability: (It is simply called the probability.)

If there are n equally likely outcomes and an event A for which there are k of these outcomes, then the probability that the event A will occur is given by:

$$P(A)=\frac{k}{n}$$

The probability of tossing a fair coin turned up head is $P(H)=\frac{1}{2}$.

The probability of tossing a fair coin turned up tail is $P(T)=\frac{1}{2}$.

The probability of rolling a fair die turned up an even number is $P(E)=\frac{3}{6}=\frac{1}{2}$.

The probability of an event must be between 0 and 1 (or 0% and 100%).

The statement $P(A)=0$ means that event A cannot occur (0%).

The statement $P(A)=1$ means that event A must occur (100%).

If p is the probability of an event, $1-p$ is the probability of an event not occurring.

Experimental Probability:

In real-life situations, sometimes it is not possible to find the theoretical probability of an event. Therefore, we need to find the experimental probability by performing an experiment or a survey. We can use the experimental probability to predict future occurrences of events.

If a sufficient number of trials is conducted n times, and an event A occurs e of these times, then the experimental probability that the event A will occur in another trial is given by: $P(A)=\frac{e}{n}$

A school with 2,000 students has 40 students who are left-handed this year. Then, the probability of a student chosen at random will be left-handed is $\frac{40}{2000}=\frac{1}{50}$. If there is no other information available, we can best estimate that the probability of a student chosen at random next year will be left-handed is $\frac{1}{50}$.

Tree Diagrams

A **tree diagram** is helpful in listing all the possible outcomes (the sample space).
The sample space for tossing two coins is shown below. Use H for heads and T for tails.
There are four possible outcomes (four simple events) {(H, H), (H, T), (T, H), (T, T)}.

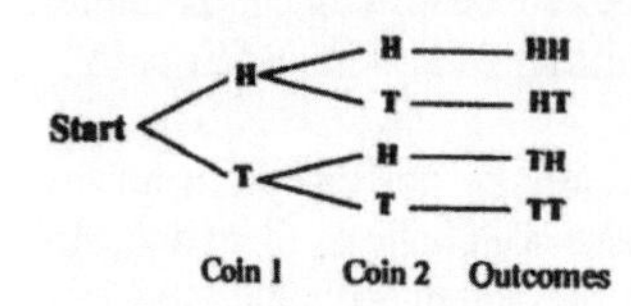

The probability of two heads is $\frac{1}{4}$.
The probability of two tails is $\frac{1}{4}$.
The probability of "one is heads and one is tails" is $\frac{1}{4}+\frac{1}{4}=\frac{1}{2}$.
The probability of "at least one is heads" is $\frac{3}{4}$.

Independent Events: In an experiment, A and B are two independent events if the occurrence of one event does not affect the occurrence of the other.

Dependent Events: In an experiment, A and B are two dependent events if the occurrence of one event affect the occurrence of the other.

Probability of Independent Events:

If A and B are two independent events, then the probability that both A and B occur is

$$P(A \cap B) = P(A) \times P(B). \text{ Or, } P(A \text{ and } B) = P(A) \times P(B).$$

Example 3: A bag contains 5 red balls and 4 white balls. Two balls are drawn at random from the bag. The first ball drawn is put back into the bag before the second ball is drawn. What is the probability that the two balls drawn are both red ?
Solution:

$$P(\text{red}) = \frac{5}{9} \;;\; P(\text{red and red}) = \frac{5}{9} \times \frac{5}{9} = \frac{25}{81}.$$

Probability of Dependent Events:

If A and B are two dependent events, then the probability that both A and B occur is $P(A \cap B) = P(A) \times P(B\,|\,A)$. Or, $P(A \text{ and } B) = P(A) \times P(B\,|\,A)$.
$P(B\,|\,A)$ is called the **conditional probability** of B given that A has occurred.

Example 4: A bag contains 5 red balls and 4 white balls. Two balls are drawn at random from the bag. The first ball drawn is not put back into the bag before the second ball is drawn. What is the probability that the two balls drawn are both red ?
Solution:

$$P(\text{red}) = \frac{5}{9} \;;\; P(\text{red}\,|\,\text{red}) = \frac{4}{8} \;;\; P(\text{red and red}) = \frac{5}{9} \times \frac{4}{8} = \frac{20}{72} = \frac{5}{18}.$$

Mutually Exclusive Events

Mutually exclusive events: Two events which have no elements in common.

The probability of the **union** of any two events having elements in common in a sample space is: (not mutually exclusive events)

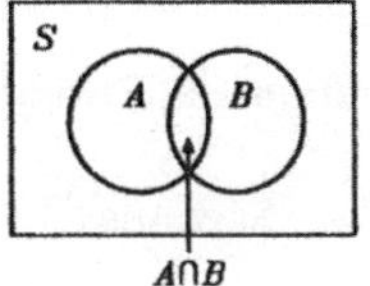

Formula:

$$P(A \cup B) = P(A) + P(B) - P(A \cap B)$$

OR: $P(A \text{ or } B) = P(A) + P(B) - P(A \text{ and } B)$

The probability of the **union of** two **mutually exclusive** events in a sample space is:
(For mutually exclusive events, $P(A \cap B) = 0$)

Formula:

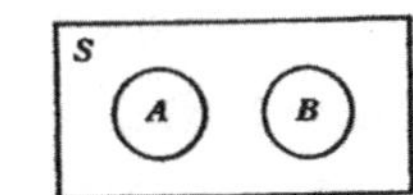

$$P(A \cup B) = P(A) + P(B)$$

OR: $P(A \text{ or } B) = P(A) + P(B)$

Examples

1. A number is drawn at random from $\{1, 2, 3, 4, 5, 6\}$. What is the probability that either a number larger than 3 or an even number will occur ?
   Solution:

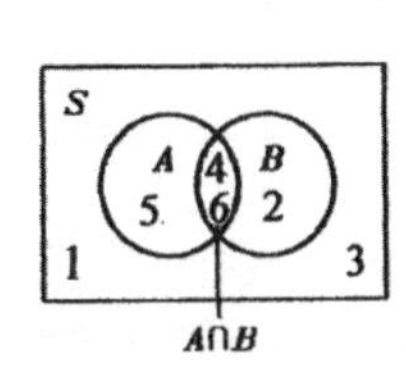

$S = \{1, 2, 3, 4, 5, 6\}$

$A = \{4, 5, 6\}$ $\quad \therefore P(A) = \frac{1}{2}$

$B = \{2, 4, 6\}$ $\quad \therefore P(B) = \frac{1}{2}$

$A \cap B = \{4, 6\}$ $\quad \therefore P(A \cap B) = \frac{2}{6} = \frac{1}{3}$

$A \cup B = \{2, 4, 5, 6\}$

$$P(A \cup B) = P(A) + P(B) - P(A \cap B) = \frac{1}{2} + \frac{1}{2} - \frac{1}{3} = \frac{2}{3}.$$

2. A bag contains 5 red, 4 white and 3 black balls. One ball is drawn at random. What is the probability that it is a red or a white ball ?
   Solution:

(A and B are mutually exclusive events.)

Let A → the event that the ball is red. B → the event that the ball is white.

$$P(A \cup B) = P(A) + P(B) = \frac{5}{12} + \frac{4}{12} = \frac{9}{12} = \frac{3}{4}.$$

3. A bag contains 5 red balls and 4 white balls. Two balls are drawn together at random from the bag. What is the probability that the two balls drawn are of the same color ?
   Solution:

(A and B are mutually exclusive events.)

Let A → the event that two balls are red. B → the event that two balls are white.

$$P(A) = \frac{{}_5C_2}{{}_9C_2} = \frac{10}{36} \; ; \; P(B) = \frac{{}_4C_2}{{}_9C_2} = \frac{6}{36}$$

$$P(A \cup B) = P(A) + P(B) = \frac{10}{36} + \frac{6}{36} = \frac{16}{36} = \frac{4}{9}.$$

Binomial Distribution

Binomial Experiment: It is an experiment which has n independent trials and only two possible outcomes "success and failure". The probability of success of each trial is the same.

Binomial Probability: In a binomial experiment, the probability of x successes in n trials is:

Formula: $P(x) = {}_nC_x P^x (1-P)^{n-x}$ where P is the probability of success on each trial and $(1-P)$ is the probability of failure.

Binomial Distribution: The probabilities of all possible numbers of successes in a binomial experiment .

Suppose we throw a fair coin 10 times. The probability of each throw turned up head is $\frac{1}{2}$.

The probability that none of all 10 throws turned up head is $(\frac{1}{2})^{10} = 0.00098$.

The probability that first throw is head and all the other 9 throws are tails is $(\frac{1}{2})^1(\frac{1}{2})^9 =$ 0.00098. However, the number of combinations of 1 head and 9 tails in 10 throws are ${}_{10}C_1 = 10$. Therefore, the probability of the event that 1 head and 9 tails in 10 throws is:

$$P(1) = {}_{10}C_1(\tfrac{1}{2})^1(\tfrac{1}{2})^9 = 10 \cdot (0.00098) = 0.0098$$

Similarly, The probability of the event that 2 heads and 8 tails is:

$$P(2) = {}_{10}C_2(\tfrac{1}{2})^2(\tfrac{1}{2})^8 = 45 \cdot (0.00098) = 0.0441$$

The probability that the event having each x-value ($x = 0, 1, 2, 3, \cdots, 10$) of throwing a fair coin 10 times can be calculated from:

$P(x) = {}_{10}C_x(\frac{1}{2})^x(\frac{1}{2})^{10-x}$ Therefore: For n throws: $P(x) = {}_nC_x(\frac{1}{2})^x(\frac{1}{2})^{n-x}$

The result of the above probabilities and its histogram of the binomial distribution are shown below:

x	$P(x)$
0	0.00098
1	0.0098
2	0.0441
3	0.1176
4	0.2058
5	0.2470
6	0.2058
7	0.1176
8	0.0441
9	0.0098
10	0.00098

Binomial Distribution

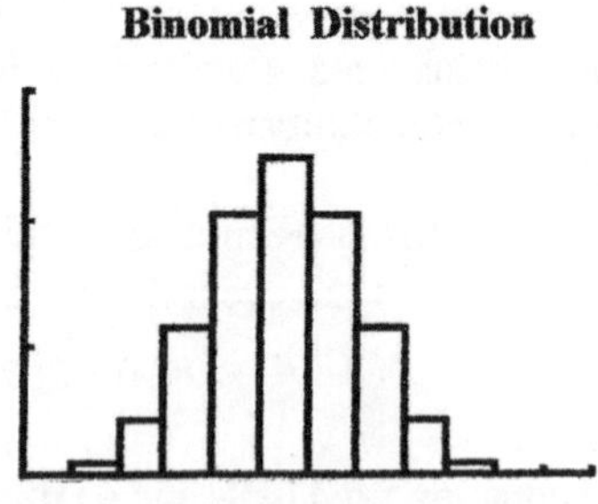

The distribution of the above example is **symmetric** because the probability of success is 0.5. Otherwise, the distribution is **skewed**. When n is large and p is close to 0.5, a binomial distribution is close to a normal distribution.

1-25 Dispersion and Standard Deviation

In real life situation, we want to organize and analyze numerical data and make prediction. In statistics, we use statistical (numerical) values to summarize and compare data. The **mean**, **median**, or **mode** provides the measures of **central tendency** of the data. The **range** provides the measure of **spread** or **dispersion** of the data.

Another measure of spread or dispersion is the **standard deviation** of the data.

To find the standard deviation of the data, we need the measures of mean, deviation, and variation.

Mean is the arithmetic average of the data. It is denoted by M, μ, or $\bar{x}$.
Deviation is the difference of each datum from the mean.
Variation is the dispersion of the data. It is denoted by v.

The variation shows how the data are scattered about the mean. It is computed by squaring each deviation from the mean, adding these squares, and dividing their sum by the number of the entries.

Variation: $v = \dfrac{\sum (x_i - \bar{x})^2}{n}$ where $i = 1, 2, \cdots\cdots, n$

Standard Deviation of a Set of Data (It is denoted by σ)
The standard deviation of a set of data is the square root of the variance:

Standard Deviation: $\sigma = \sqrt{\dfrac{\sum (x_i - \bar{x})^2}{n}} = \sqrt{\dfrac{(x_1 - \bar{x})^2 + (x_2 - \bar{x})^2 + \cdots\cdots + (x_n - \bar{x})^2}{n}}$

where x_1, x_2, $\cdots\cdots$, x_n are each term of the data, and $\bar{x}$ is the mean.

To compare two sets of data, the data set with greater standard deviation are more spread out about the mean.

Example

The scores of 5 games of a high school sport teams are $8, 15, 15, 31, 41$.
Find the standard deviation.
Solution:

The mean $\bar{x} = \dfrac{8 + 15 + 15 + 31 + 41}{5} = 22$

The standard deviation $\sigma = \sqrt{\dfrac{(8-22)^2 + (15-22)^2 + (15-22)^2 + (31-22)^2 + (41-22)^2}{5}}$

$= \sqrt{\dfrac{736}{5}} = \sqrt{147.2} \approx 12.13$.

The Properties of Normal Distribution

In Chapter 5-4, we have learned the concepts of **histogram** and **normal distribution**. The graph of normal distribution is a smooth bell-shaped curve from the plotting of a large collecting of data. The shape of the curve indicates that the frequency of occurrences of data are concentrated around the center (the mean).

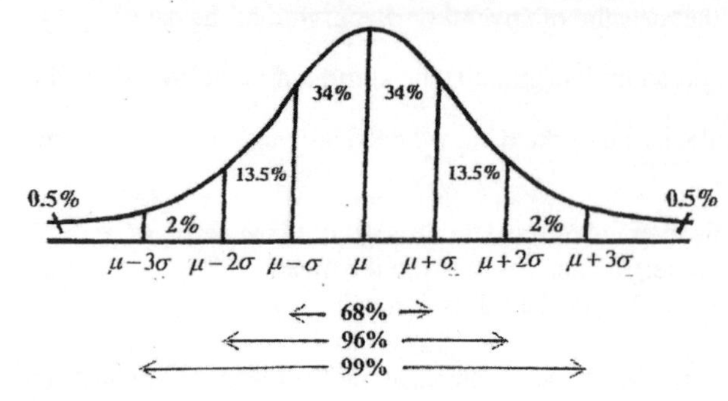

$$y = \frac{1}{\sqrt{2\pi}} e^{\frac{-(x-\mu)^2}{2\sigma^2}}$$

σ : Standard deviation

μ : Mean

e : 2.72828....

Area = 1

Properties of a Normal Distribution

1. The data are symmetrical about the mean.
2. The mean and median are about equal.
3. About **68%** of the values are within one standard deviation from the mean.
4. About **95%** of the values are within two standard deviations from the mean.
5. About **99%** of the values are within three standard deviations from the mean.

Standard Normal Distribution: A normal distribution with $\mu = 0$ and $\sigma = 1$.

Example

The mean of the heights of the 3,000 students in a high school was found to be 172 *cm* with a standard deviation of 7 *cm*.

1. Draw a normal curve showing the heights at one, two, and three standard deviations from the mean.
2. Find the percentage of students whose heights are between 179 and 193 *cm*.
3. How many students whose heights are between 165 and 179 *cm* ?
4. How many students whose heights are less than 158 *cm* ?

Solution:

1. $\mu \pm \sigma = 172 \pm 7 \quad = 179$ and 165

   $\mu \pm 2\sigma = 172 \pm 2(7) = 186$ and 158

   $\mu \pm 3\sigma = 172 \pm 3(7) = 193$ and 151
2. $13.5\% + 2\% = 15.5\%$.
3. $3{,}000 \times 68\% = 2040$ students.
4. $3{,}000 \times 2.5\% = 75$ students.

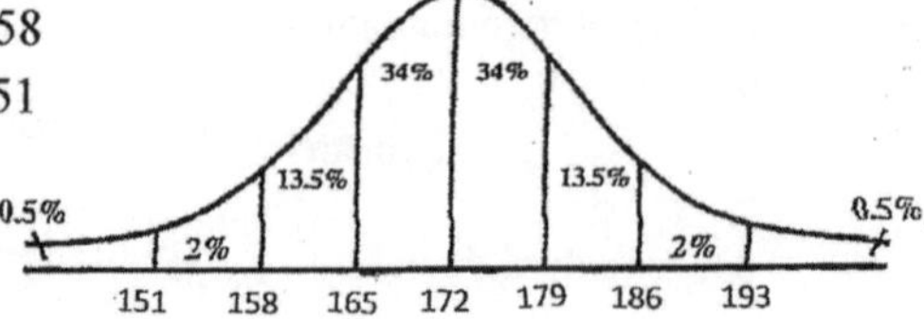

1-26 Permutations and Combinations

Permutation: It is an arrangement of elements without repetitions in a definite order.
Factorial Notation ($n!$): It is the product of consecutive integers beginning with n, decreasing and ending with 1. Read as " n factorial ".

The number of permutations of the 3 letters a, b, c taken 3 letters at a time:
abc, acb, bac, bca, cab, cba → total 6 permutations. $3! = 3\cdot 2\cdot 1 = 6$

The number of permutations of the 4 letters a, b, c, d taken 3 letters at a time:
abc, acb, bac, bca, cab, cba
bcd, bdc, cbd, cdb, dbc,dcb
cda, cad, dca, dac, acd, adc
dab, dba, adb, abd, bda, bad → total 24 permutations. $\frac{4!}{(4-3)!} = \frac{4\cdot 3\cdot 2\cdot 1}{1} = 24$

The number of permutations of n different elements taken n elements at a time is :

$${}_nP_n = n\cdot(n-1)\cdot(n-2)\cdot(n-3)\cdot\cdots\cdot 3\cdot 2\cdot 1$$

Formula: ${}_nP_n = n!$ **Example:** ${}_{10}P_{10} = \frac{10!}{(10-10)!} = \frac{10!}{1!} = 10! = 3,628,800$

The number of permutations of n different elements taken r elements at a time is:

$${}_nP_r = n\cdot(n-1)\cdot(n-2)\cdot(n-3)\cdot\cdots\cdot(n-r+1)$$

Formula: ${}_nP_r = \frac{n!}{(n-r)!}$ **Example:** ${}_{18}P_2 = \frac{18!}{(18-2)!} = \frac{18\cdot 17\cdot \cancel{16!}}{\cancel{16!}} = 306$

The symbols $P(n, \text{r})$ and P_r^n **are often used to denote** ${}_nP_r$**.**

Examples

1. Find the number of permutations of the letters a, b, c taken 3 letters at a time ?
   Solution: ${}_3P_3 = 3! = 3\cdot 2\cdot 1 = 6$ permutations.

2. How many permutations can be made from the letters a, b, c, d taken 4 letters at a time ?
   Solution: ${}_4P_4 = 4! = 4\cdot 3\cdot 2\cdot 1 = 24$ permutations.

3. How many permutations can be made from the letters a, b, c, d taken 3 letters at a time ?
   Solution: ${}_4P_3 = \frac{4!}{(4-3)!} = \frac{4\cdot 3\cdot 2\cdot 1}{1!} = \frac{24}{1} = 24$ permutations.

4. How many arrangements can be selected on a shelf from 10 different books taken 5 books for each arrangements ?
   Solution:

$${}_{10}P_5 = \frac{10!}{(10-5)!} = \frac{10\cdot 9\cdot 8\cdot 7\cdot 6\cdot \cancel{5!}}{\cancel{5!}} = 30,240 \text{ arrangements.}$$

Combinations: It is an arrangement of elements without repetitions and regardless of the order.

The number of combinations of the letters a, b, c taken 3 letters at a time:

abc → 1 combination.

The number of combinations of the letters a, b, c, d, take 3 letters at a time:

abc, bcd, cda, dab → 4 combinations.

Note that the only difference between a permutation and a combination is that we consider "abc, acb, bac, bca, cab, cba" ($3! = 6$) objects as one combination (regardless of the order). Therefore, we divide the formula of ${}_nP_r$ by $r!$ to obtain the formula of ${}_nC_r$.

The number of combinations of n different elements taken r elements at a time:

Formula: ${}_nC_n = 1$ **Example:** ${}_{10}C_{10} = \dfrac{10!}{(10-10)! \cdot 10!} = \dfrac{1}{0!} = \dfrac{1}{1} = 1$. (Hint: $0! = 1$)

The number of combinations of n different elements taken r elements at a time:

Formula: ${}_nC_r = \dfrac{n!}{(n-r)! \cdot r!}$; ${}_nC_r = {}_nC_{n-r}$

Examples: ${}_{18}C_2 = \dfrac{18!}{(18-2)! \cdot 2!} = \dfrac{18!}{16! \cdot 2} = \dfrac{18 \cdot 17 \cdot 16!}{16! \cdot 2} = 153$.

${}_{18}C_{16} = \dfrac{18!}{(18-16)! \cdot 16!} = \dfrac{18!}{2! \cdot 16!} = \dfrac{18 \cdot 17 \cdot 16!}{2! \cdot 16!} = 153$.

The symbols $C(n, r)$, C_r^n and $\binom{n}{r}$ are often used to denote ${}_nC_r$.

Examples

1. Find the number of combinations of the letters a, b, c taken 3 letters at a time ?

   Solution: ${}_3C_3 = \dfrac{3!}{(3-3)! \cdot 3!} = \dfrac{1}{0!} = \dfrac{1}{1} = 1$ combination.

2. How many combinations can be made from the letters a, b, c, d taken 4 letters at a time ?

   Solution: ${}_4C_4 = 1$ combination

3. How many combinations can be made from the letters a, b, c, d taken 3 letters at a time ?

   Solution: ${}_4C_3 = \dfrac{4!}{(4-3)! \cdot 3!} = \dfrac{24}{1! \cdot 6} = 4$ combinations.

4. There are 20 boys in a class. How many different committees of 3 boys can be formed ?

   Solution:

   $${}_{20}C_3 = \frac{20!}{(20-3)! \cdot 3!} = \frac{20 \cdot 19 \cdot 18 \cdot 17!}{17! \cdot 6} = 1{,}140 \text{ committees.}$$

5. There are 20 boys and 15 girls in a class. A committee to be formed consisting of 3 boys and 3 girls. How many different committees are possible ?

   Solution:

   $${}_{20}C_3 \cdot {}_{15}C_3 = \frac{20!}{(20-3)! \cdot 3!} \cdot \frac{15!}{(15-3)! \cdot 3!} = \frac{20 \cdot 19 \cdot 18}{6} \cdot \frac{15 \cdot 14 \cdot 13}{6} = 1140 \cdot 455 = 518{,}700 \text{ committees.}$$

Permutations with Repeating Selections

We have learned to find the number of permutations of n different elements taken "n" or "r" elements at a time without repetitions (each element can be taken only once) . Now, we will find the number of permutations if each element can be taken with repetitions.
The number of permutations of the letters a, b, c taken 3 letters at a time, each letter can be taken repeatedly:

aaa, aab, aac, aba, abb, abc, aca, acb, acc
baa, bab, bac, bba,bbb, bbc,bca, bcb, bcc
caa, cab, cac, cba, cbb, cbc, cca, ccb, ccc $\rightarrow$ 27 permutations. $3^3 = 27$

The number of permutations of n different elements taken r elements at a time and each element can be taken repeatedly:

Formula: $P = n \cdot n \cdot n \cdot n \cdot \cdots = n^r$

Examples

1. Find the number of permutations of the letters a, b, c taken 3 letters at a time, each letter can be taken repeatedly ?
Solution:
$P = n^r = 3^3 = 27$ permutations.

2. How many permutations can be made from the letters a, b, c, d taken 4 letters at a time, each letter can be taken repeatedly ?
Solution:
$P = n^r = 4^4 = 256$ permutations.

3. How many permutations can be made from the letters a, b, c, d taken 3 letters at a time, each letter can be taken repeatedly ?
Solution:
$P = n^r = 4^3 = 64$ permutations.

4. How many permutations can be formed from the letters a, b, c, d, e taken 4 letters at a time, each letter can be taken repeatedly ?
Solution:
$P = n^r = 5^4 = 625$ permutations.

5. How many 4-digit integers can be formed from the digits 1, 2, 3, 4, 5 and repetitions of digits are allowed ?
Solution:
$P = 5^4 = 625$.

6. How many different ways can be chosen to drop 4 letters into 6 different mail boxes ?
Solution: Each letter has 6 choices.
$P = 6 \cdot 6 \cdot 6 \cdot 6 = 6^4 = 1{,}296$ ways.

Hint: 4^6 is incorrect. (4^6 include one same letter to be dropped into 6 boxes.)

7. How many 4-digit integers can be made from the digits 0, 1, 2, 3, 4, 5 and repetitions of digits are allowed ?
Solution.
$P = 5 \cdot 6^3 = 1{,}080$ integers.
(Hint: It cannot begin with 0.)

8. How many 3-digit integers which are larger than 220 can be made using the digits 0,1,2 3,4 and repetitions of digits are allowed ?
Solution:
(2) (2) (1,2,3,4) $\rightarrow 1 \times 1 \times 4 = 4$
(2) (3,4,) (0,1,2,3,4,) $\rightarrow 1 \times 2 \times 5 = 10$
(3,4,) (0,1,2,3,4) (0,1,2,3,4) $\rightarrow 2 \times 5 \times 5 = 50$
$4 + 10 + 50 = 64$.

Permutations with Elements Alike

Now, we learn how to find the number of permutations of n elements that are not all different. The number of permutations of n elements taken n at a time, with n_1 elements alike, n_2 elements alike, n_3 elements alike, ...is:

Formula: $P = \dfrac{n!}{n_1! \cdot n_2! \cdot \cdots \cdot n_k!}$ It is also called **distinquishable permutations.**

Examples

1. How many permutations can be formed from the letters in ACADIA, taken 6 at a time ?
Solution:

$$P = \frac{6!}{3!} = \frac{720}{6} = 120 \text{ permutations.}$$

2. How many permutations can be formed from the letters in ACACIA, taken 6 at at a time ?
Solution:

$$P = \frac{6!}{3! \cdot 2!} = \frac{720}{6 \cdot 2} = 60 \text{ permutation.}$$

3. How many five-digit numerals can be made from 3, 6, 9, and 7 if 3, 6, and 9 can be used once, 7 can be used twice ?
Solution: (3, 6, 9, 7, 7)

$$P = \frac{5!}{2!} = \frac{120}{2} = 60 \text{ numerals.}$$

4. How many different ways to line up 10 balls if 6 are red, 3 are white, and 1 is black ?
Solution:

$$P = \frac{10!}{6! \cdot 3! \cdot 1!} = \frac{10 \cdot 9 \cdot 8 \cdot 7 \cdot \cancel{6!}}{\cancel{6!} \cdot 6 \cdot 1} = 840.$$

5. How many five-digit numerals can be made from 6 and 2 if 6 can be used three times, 2 can be used twice ?
Solution: (6, 6, 6, 2, 2)

$$P = \frac{5!}{3! \cdot 2!} = \frac{120}{6 \cdot 2} = 10 \text{ numerals.}$$

6. How many seven-digit numerals can be made from 2, 5, and 8 if we use 2 and 5 twice, 8 three times ?
Solution: (2, 2, 5, 5, 8, 8, 8)

$$P = \frac{7!}{2! \cdot 2! \cdot 3!} = \frac{5040}{2 \cdot 2 \cdot 6} = 210 \text{ numerals.}$$

7. How many seven-digit numerals can be made from 2, 5, and 8 if we use 2 and 5 three times, 8 once ?
Solution: (2, 2, 2, 5, 5, 5, 8)

$$P = \frac{7!}{3! \cdot 3!} = \frac{5040}{6 \cdot 6} = 140 \text{ numerals.}$$

8. How many short-cut routes can be selected from A to B ?

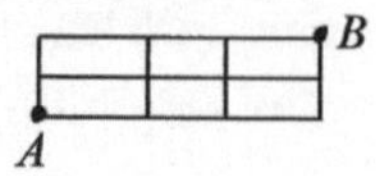

Solution: (→ → → ↑ ↑)

$$P = \frac{5!}{3! \cdot 2!} = \frac{5 \cdot 4 \cdot \cancel{3!}}{\cancel{3!} \cdot 2} = 10 \text{ routes.}$$

1-27 Matrices and Determinants

Matrix: It is a rectangular array of numbers enclosed by brackets.
Element: Each number in a matrix is called the element of the matrix.
Dimensions of a matrix: It is the number of rows and the number of columns in a matrix. The number of rows is given first. A 3×4 matrix is a matrix having 3 rows and 4 columns (read " 3 by 4 matrix ").
Row Matrix: It is a matrix having only one row.
Column Matrix: It is a matrix having only one column.
Square Matrix: It is a matrix having the same numbers in rows and columns.
Zero Matrix: It is a matrix having all elements to be zero.
Identity Matrix: It is a square matrix whose main diagonal has all elements by 1 and all other elements by 0.
Two matrices are equal if and only if they have the same dimensions and the same elements in all corresponding positions.
Matrix addition: We add the corresponding elements of the matrices being added.
Matrix Subtraction: We subtract the corresponding elements of the matrices being subtracted. Or, add the additive inverse of the second matrix.
Addition and subtraction of matrices of different dimensions are not allowed.

Matrix Multiplication: The product (multiplication) of two matrices is the matrix in which each element is the sum of the product of corresponding elements in row of the first matrix and in column of the second matrix. $A_{m\times n} \cdot B_{n\times p} = X_{m\times p}$

The dimension of column in the first matrix must equal the dimension of row in the second matrix. Otherwise, they cannot be multiplied.

Examples

1. Let $A=\begin{bmatrix} 5 & -4 \\ -1 & 2 \end{bmatrix}$, $B=\begin{bmatrix} 3 & 6 \\ 2 & -5 \end{bmatrix}$

   a. $A+B=\begin{bmatrix} 8 & 2 \\ 1 & -3 \end{bmatrix}$ **b.** $A-2B=A+(-2B)=\begin{bmatrix} 5 & -4 \\ -1 & 2 \end{bmatrix}+\begin{bmatrix} -6 & -12 \\ -4 & 10 \end{bmatrix}=\begin{bmatrix} -1 & -16 \\ -5 & 12 \end{bmatrix}$

2. $\begin{bmatrix} 2 & 3 \end{bmatrix}\cdot\begin{bmatrix} 5 \\ 7 \end{bmatrix}=[2\cdot 5+3\cdot 7]=[31]_{1\times 1}$

3. $\begin{bmatrix} 5 \\ 7 \end{bmatrix}_{2\times 1}\cdot\begin{bmatrix} 2 & 3 \end{bmatrix}_{1\times 2}=\begin{bmatrix} 5\cdot 2 & 5\cdot 3 \\ 7\cdot 2 & 7\cdot 3 \end{bmatrix}=\begin{bmatrix} 10 & 15 \\ 14 & 21 \end{bmatrix}_{2\times 2}$

4. $\begin{bmatrix} 3 & -1 \\ 2 & 0 \end{bmatrix}\cdot\begin{bmatrix} -1 \\ 2 \end{bmatrix}_{2\times 1}=\begin{bmatrix} 3(-1)+(-1)\cdot 2 \\ 2(-1)+0\cdot 2 \end{bmatrix}_{2\times 1}=\begin{bmatrix} -5 \\ -2 \end{bmatrix}_{2\times 1}$

5. $\begin{bmatrix} -1 \\ 2 \end{bmatrix}_{2\times 1}\cdot\begin{bmatrix} 3 & -1 \\ 2 & 0 \end{bmatrix}_{2\times 2}$ They cannot be multiplied.

Determinants

The form of a determinant is similar to a matrix. A matrix is formed with brackets enclosing the entries. A **determinant** is formed with vertical bars enclosing the entries. Each entry is called an **element** of the determinant.

The number of elements in any row or column is called the **order** of the determinant.

Determinant of a 2 × 2 matrix A is defined as follows:

$$\det A = \begin{vmatrix} a & b \\ c & d \end{vmatrix} = ad - bc$$

Determinant of a 3 × 3 matrix A is defined as follows:(Three ways)

1. $\det A = \begin{vmatrix} a_1 & b_1 & c_1 \\ a_2 & b_2 & c_2 \\ a_3 & b_3 & c_3 \end{vmatrix} = a_1b_2c_3 + a_2b_3c_1 + a_3b_1c_2 - a_3b_2c_1 - a_1b_3c_2 - a_2b_1c_3$

The products of elements:

1. run down from left to right are positive.
2. run down from right to left are negative.

2. Expansion by its minors across any row

$$\det A = a_1\begin{vmatrix} b_2 & c_2 \\ b_3 & c_3 \end{vmatrix} - b_1\begin{vmatrix} a_2 & c_2 \\ a_3 & c_3 \end{vmatrix} + c_1\begin{vmatrix} a_2 & b_2 \\ a_3 & b_3 \end{vmatrix}$$

3. Copy the first two columns to the right side of the determinant

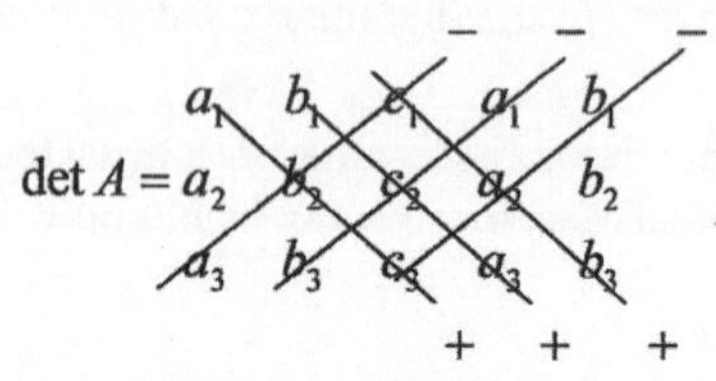

The above formulas are not valid for $n \times n$ if $n \geq 4$.

Examples (Evaluate each determinant)

1. $\begin{vmatrix} 1 & 5 \\ 4 & 8 \end{vmatrix} = 1\cdot 8 - 5\cdot 4 = 8 - 20 = -12$

2. $\begin{vmatrix} -3 & 6 \\ -2 & 4 \end{vmatrix} = -3\cdot 4 - 6\cdot(-2) = -12 + 12 = 0$

3. $\begin{vmatrix} 2 & 1 & -3 \\ 3 & 5 & -4 \\ -4 & 2 & 6 \end{vmatrix} = 60 + (-18) + 16 - 60 - (-16) - 18 = 60 - 18 + 16 - 60 + 16 - 18 = -4$

4. $\begin{vmatrix} \sqrt{2} & -\sqrt{3} \\ \sqrt{15} & \sqrt{6} \end{vmatrix} = \sqrt{2}\cdot\sqrt{6} - (-\sqrt{3}\cdot\sqrt{15}) = \sqrt{12} + \sqrt{45} = 2\sqrt{3} + 3\sqrt{5}$

5. $\begin{vmatrix} \log 5 & \log 2 \\ \log 2 & \log 5 \end{vmatrix} = (\log 5)^2 - (\log 2)^2 = (\log 5 + \log 2)(\log 5 - \log 2) = \log(5\times 2)\cdot\log\frac{5}{2}$

$= \log 10 \cdot \log\frac{5}{2} = 1\cdot\log\frac{5}{2} = \log\frac{5}{2}$

Cramer's Rule

Cramer's Rule gives us an easy method to solve a system of linear equations in *n* variables.

Cramer's Rule: If $D \neq 0$, then the system of *n* linear equations in *n* variables has the following unique solution:

$$x = \frac{D_x}{D}, \quad y = \frac{D_y}{D}, \quad z = \frac{D_z}{D}, \cdots\cdots$$

$D \rightarrow$ The determinant of the coefficient matrix of the variables.
$D_x \rightarrow$ The determinant of replacing the coefficients of x in D by the constants.
$D_y \rightarrow$ The determinant of replacing the coefficients of y in D by the constants.
$D_z \rightarrow$ The determinant of replacing the coefficients of z in D by the constants.

A system of *n* linear equations in *n* variables has a unique solution if and only if $D \neq 0$.

1) If $D \neq 0$, the system of equations has **consistent and independent** equations. The system has a unique solution.

2) If $D = 0$ and at least one of D_x, D_y, D_z, ... $\neq 0$, the system of equations has **inconsistent** equations. There is no solution (they are parallel).

3) If $D = 0$ and all of D_x, D_y, D_z, ... $= 0$, the system of equations has **consistent and dependent** equations. There are unlimited (infinite) number of solutions (they coincide).

To make a wise and efficient choice of solution method, we need to keep each method (substitution, elimination, inverse matrix, Cramer's Rule, and others) in our memory. Inverse matrix and Cramer's Rule are used to solve a system in which the coefficients matrix must be a square. Cramer's Rule is also not efficient if the system has many equations because many determinants needed to be evaluated.
If the coefficient matrix is not invertible (no inverse), the inverse matrix cannot be used to solve a system. The method of row operations is advisable for all kinds of systems. However, a scientific calculator will simply solve a system.

Examples

1. Solve $\begin{cases} 2x - y = 6 \\ 3x + 5y = 22 \end{cases}$

Solution:

$$D = \begin{vmatrix} 2 & -1 \\ 3 & 5 \end{vmatrix} = 10 - (-3) = 13$$

$$D_x = \begin{vmatrix} 6 & -1 \\ 22 & 5 \end{vmatrix} = 30 - (-22) = 52, \quad D_y = \begin{vmatrix} 2 & 6 \\ 3 & 22 \end{vmatrix} = 44 - 18 = 26$$

$$\therefore \; x = \frac{D_x}{D} = \frac{52}{13} = 4, \quad y = \frac{D_y}{D} = \frac{26}{13} = 2.$$

2. Solve $\begin{cases} 3x+2y-2z=11 \\ x+y+2y=-1 \\ 2x-y+2z=-4 \end{cases}$

Solution:

$$D=\begin{vmatrix} 3 & 2 & -2 \\ 1 & 1 & 2 \\ 2 & -1 & 2 \end{vmatrix}=6+2+8-(-4)-(-6)-4=22 \text{ (Consistent and independent)}$$

$$D_x=\begin{vmatrix} 11 & 2 & -2 \\ -1 & 1 & 2 \\ -4 & -1 & 2 \end{vmatrix}=22+(-2)+(-16)-8-(-22)-(-4)=22$$

$$D_y=\begin{vmatrix} 3 & 11 & -2 \\ 1 & -1 & 2 \\ 2 & -4 & 2 \end{vmatrix}=-6+8+44-4-(-24)-22=44$$

$$D_z=\begin{vmatrix} 3 & 2 & 11 \\ 1 & 1 & -1 \\ 2 & -1 & -4 \end{vmatrix}=-12+(-11)+(-4)-22-3-(-8)=-44$$

$$\therefore x=\frac{D_x}{D}=\frac{22}{22}=1,\quad y=\frac{D_y}{D}=\frac{44}{22}=2,\quad z=\frac{D_z}{D}=\frac{-44}{22}=-2$$

---

3. Solve $\begin{cases} x+2y=4 \\ x+2y=-2 \end{cases}$

Solution:

$$D=\begin{vmatrix} 1 & 2 \\ 1 & 2 \end{vmatrix}=2-2=0$$

$$D_x=\begin{vmatrix} 4 & 2 \\ -2 & 2 \end{vmatrix}=8-(-4)=12\neq 0$$

$$D_y=\begin{vmatrix} 1 & 4 \\ 1 & -2 \end{vmatrix}=-2-4=-6\neq 0$$

Inconsistent equations.
They are parallel (no intersection).

Ans: There is no solution.

4. Solve $\begin{cases} x+2y=4 \\ 2x+4y=8 \end{cases}$

Solution:

$$D=\begin{vmatrix} 1 & 2 \\ 2 & 4 \end{vmatrix}=4-4=0$$

$$D_x=\begin{vmatrix} 4 & 2 \\ 8 & 4 \end{vmatrix}=16-16=0$$

$$D_y=\begin{vmatrix} 1 & 4 \\ 2 & 8 \end{vmatrix}=8-8=0$$

Consistent and dependent equations.
(They coincide.)

Ans: There are unlimited number of solutions.

1-28 Arithmetic and Geometric Series

Arithmetic Series: It is the sum of an arithmetic sequence.
Geometric Series: It is the sum of a geometric sequence.
Formulas:

1. The sum of the first n terms of an **arithmetic series** is given by:

$$S_n = a_1 + a_2 + a_3 + \cdots\cdots + a_n = \frac{n}{2}(a_1 + a_n)$$

Where $a_1 \to 1^{st}$ term , $a_n \to$ last term , $n \to n$th terms

2. The sum of the first n terms of a **geometric series** is given by:

$$S_n = a_1 + a_1 r + a_1 r^2 + a_1 r^3 + \cdots\cdots + a_1 r^{n-1} = \frac{a_1(1-r^n)}{1-r} \ , \ r \neq 1$$

Where $a_1 \to$ 1st term , $n \to$ The nth terms , $r \to$ common ratio

(**Hint:** Find n from $a_n = a_1 r^{n-1}$ if n is not given.)

We use the above formula, the Index must begin at $k = 1$.
(see example 5 below)

Examples

1. Find the sum of the arithmetic sequence $1 + 2 + 3 + 4 + \cdots\cdots + 500$.
   Solution: $a_1 = 1$, $a_{500} = 500$, $n = 500$

$$S_{500} = \frac{n}{2}(a_1 + a_{500}) = \frac{500}{2}(1 + 500) = 250(501) = 125{,}250$$

2. Evaluate the series $\sum_{k=1}^{30}(-5k + 10)$.
   Solution: $5, 0, -5, -10, \cdots\cdots$

$$a_1 = 5, \ d = -5, \ a_{30} = -5(30) + 10 = -140$$

$$S_{30} = \tfrac{30}{2}[5 + (-140)] = -2025$$

3. Find the sum of the geometric sequence $1 + 3 + 9 + 27 + \cdots\cdots$ for the first 8 terms.
   Solution:

$$a_1 = 1, n = 8, r = 3, \ S_8 = \frac{a_1(1-r^n)}{1-r} = \frac{1\cdot(1-3^8)}{1-3} = \frac{1-6561}{-2} = \frac{-6560}{-2} = 3{,}280$$

4. $$\sum_{k=1}^{12}\left(\frac{1}{2}\right)^k = \frac{1}{2} + \frac{1}{4} + \frac{1}{8} + \cdots\cdots + \left(\frac{1}{2}\right)^{12} = \frac{\frac{1}{2}[1-(\frac{1}{2})^{12}]}{1-\frac{1}{2}} = 1 - \left(\frac{1}{2}\right)^{12}$$

5. $$\sum_{k=0}^{12}\left(\frac{1}{2}\right)^k = \left(\frac{1}{2}\right)^0 + \sum_{k=1}^{12}\left(\frac{1}{2}\right)^k = 1 + \frac{\frac{1}{2}[1-(\frac{1}{2})^{12}]}{1-\frac{1}{2}} = 1 + \left[1 - \left(\frac{1}{2}\right)^{12}\right] = 2 - \left(\frac{1}{2}\right)^{12}$$

1-29 Infinite Geometric Series

Infinite Geometric Series: It is a geometric series having its terms increases without limit (without bound).

$$a_1 + a_1 r + a_1 r^2 + a_1 r^3 + \cdots\cdots + a_1 r^{n-1} + \cdots\cdots \quad \text{as } n \to \infty.$$

Formula: The sum of a finite geometric series is given by:

$$S_n = \sum_{k=1}^{n} a_1 r^{k-1} = \frac{a_1(1-r^n)}{1-r}$$

Find the formula for the sum of an infinite geometric series.

1) If $|r| \geq 1$, then $|r^n|$ increases without limit as $n \to \infty$, the sum increases without limit as n increases without limit (the series has no sum).

2) If $|r| < 1$, then $|r^n|$ approaches 0 as $n \to \infty$, the sum is a finite number given by:

$$S_\infty = \sum_{k=1}^{\infty} a_1 r^{k-1} = a_1 + a_1 r^2 + a_1 r^3 + \cdots\cdots + a_1 r^{n-1} + \cdots\cdots = \frac{a_1}{1-r}, \quad |r| < 1$$

Examples

1. Find the sum of the infinite geometric series: $9 - 6 + 4 - \frac{8}{3} + \cdots\cdots$

Solution:

$$r = \frac{-6}{9} = -\frac{2}{3}, \quad |r| = \left|-\frac{2}{3}\right| = \frac{2}{3} < 1$$

The series has a sum (a finite number).

$$S_\infty = \frac{a_1}{1-r} = \frac{9}{1-(-\frac{2}{3})} = \frac{9}{5/3} = \frac{27}{5}$$

2. Find the sum of the infinite geometric series: $1 - \sqrt{2} + 2 - 2\sqrt{2} + 4 - 4\sqrt{2} + \cdots$

Solution:

$$r = \frac{-\sqrt{2}}{1} = -\sqrt{2}, \quad |r| = |-\sqrt{2}| = \sqrt{2} = 1.414 > 1$$

The series has no sum. The sum increases without limit as n increases without limit.

3. Find the sum of the infinite geometric series: $1 + \frac{1}{2} + \frac{1}{4} + \frac{1}{8} + \frac{1}{16} + \cdots\cdots$

Solution:

$$r = \frac{1}{2} \div 1 = \frac{1}{2} < 1$$

The series has a sum (a finite number).

$$S_\infty = \frac{a_1}{1-r} = \frac{1}{1-\frac{1}{2}} = 2$$

1-30 Mathematical Logic

Mathematical logic is also called **symbolic logic**. We use symbolic forms and rules of logic to demonstrate a logical statement which can be either true or false, but not both.

Truth Values: T (true) and F (false) are the possible truth values of a sentence.
Statement: It is a sentence said to have a truth value.

$3+4=7$ T ; $6\times 4=25$ F.

Open sentence: A sentence whose truth value cannot be determined until the value of its variable is given.

$x+3=5$ (Its truth value cannot be determined.)
$3x+6=9$ (Its truth value cannot be determined.)

Capital letters, such as A, B, P, Q are used to represent specific statements.
Small letters, such as a, b, p, q are used to represent statements that are not specific.

$\sim p$: It is the negation of p. $\sim p$ has the opposite value of a statement p.
Read "not p." or "it is false that p.".

When p is true, $\sim p$ is false.
When p is false, $\sim p$ is true.

If T (true) and F (false) are the possible truth values of A and B, then we have the following **truth table**:

(1) A: $3+4=7$ (2) B: $3>5$

A	~A
T	F

B	~B
F	T

p is logically equivalent to $\sim(\sim p)$.

p	$\sim p$	$\sim(\sim p)$
T	**F**	**T**
F	**T**	**F**

The truth value of the statement $3+5=8$ is true, and the truth value of its negation $3+5\neq 8$ is false. The truth value of the statement $2\geq 3$ is false, and the truth value of its negation $2<3$ is true.

The negation of the statement " There are 7 days in a week " is " It is not the case that a week has 7 days". The truth value of this statement is true, and its negation is false. The negation of the statement " All polygons have equal sides " is " Some polygons have equal sides ".

A **compound statement** is formed by two simple statements.

The compound statements in logic: There are three major compound statements in logic.

1. $p \wedge q$: It is the statement of a conjunction in the form " p and q ".

$\sim(p \wedge q)$, or $(p \wedge q)'$ represents the negation of a conjunction.

$p \wedge q$ is logically equivalent to $q \wedge p$.

A conjunction is true if both simple statements are true. Otherwise, it is false.

$p: 2+3=5$ T ; $q: 2 \times 4 = 9$ F We have $p \wedge q$ is false.

2. $p \vee q$: It is the statement of a disjunction in the form " p or q "

$\sim(p \vee q)$, or $(p \vee q)'$ represents the negation of a disjunction.

$p \vee q$ is logically equivalent to $q \vee p$.

A disjunction is true if either one or both simple statements are true. Otherwise, it is false. $p: 2+3=6$ F ; $q: 2-1=1$ T We have $p \vee q$ is true.

If p and q represent two simple statements, the truth tables for conjunction and disjunction of these two simple statements are summarized in the following:

p	q	$p \wedge q$
T	T	T
T	F	F
F	T	F
F	F	F

p	q	$p \vee q$
T	T	T
T	F	T
F	T	T
F	F	F

3. $p \to q$: It is the statement of an implication (a conditional statement) in the form " if p, then q." or " if p happens, then q occurs.".

p is the antecedent (or hypothesis) and q is the consequence (or conclusion).

$\sim(p \to q)$, or $(p \to q)'$ represents the negation of $p \to q$.

An implication $p \to q$ is false if p is true and q is false, Otherwise, it is true.

a. If $4 < 8$, then $6 > 2$. The implication is true.

b. If $4 < 8$, then $6 < 2$. The implication is false.

c. If $4 > 8$, then $6 > 2$. The implication is true.

d. If $4 > 8$, then $6 < 2$. The implication is true.

Truth table for $p \to q$:

p	q	$p \to q$
T	T	T
T	F	F
F	T	T
F	F	T

In an implication $p \to q$, a false statement p implies any q:

If the sun rises from the west, then you will not pass the test. T

If the sun rises from the west, then you will pass the test. T

The following two sentences are the same in common words:

Even the sun rises from the west, you will not pass the test.

Only if the sun rises from the west, you will pass the test.

The truth table can be used to find the truth value of any form of compound statements.

1. Find $(\sim p) \vee (\sim q)$.

Solution:

p	q	$\sim p$	$\sim q$	$(\sim p) \vee (\sim q)$
T	T	F	F	F
T	F	F	T	T
F	T	T	F	T
F	F	T	T	T

2. Find $\sim(p \to q)$

Solution:

p	q	$p \to q$	$\sim(p \to q)$
T	T	T	F
T	F	F	T
F	T	T	F
F	F	T	F

Tautology and Contradiction:

$p \vee (-p)$: It is a statement of a tautology in the form "always true".

$p \wedge (-p)$: It is a statement of a contradiction in the form "always false".

p	$\sim p$	$p \vee (\sim p)$	$p \wedge (\sim p)$	$\sim (p \vee (\sim p))$	$\sim (p \wedge (\sim p))$
T	F	T	F	F	T
F	T	T	F	F	T

The negation of a tautology is a contradiction.
The negation of a contradiction is a tautology.

DeMorgan's Law: $\sim (p \wedge q)$ is logically equivalent to $(\sim p) \vee (\sim q)$.
$\sim (p \vee q)$ is logically equivalent to $(\sim p) \wedge (\sim q)$.
We can use truth table to prove the above law.

$p \leftrightarrow q$, or p iff q: The statement of a biconditional (or a double implication) in the form "p if and only if q." or "if p, then q and if q, then p.".

$p \leftrightarrow q$ also means that p is equivalent to q. They have the same truth values.

$p \leftrightarrow q$ is true if p and q are both true or both false.

Converse: $q \to p$ is the converse of $p \to q$.

1. $p \to q$: If two angles whose measures are equal, then they are congruent. T
$q \to p$: If two angles are congruent, then their measures are equal. T

2. $p \to q$: If two angles are vertical angles, then they are congruent. T
$q \to p$: If two angles are congruent, then they are vertical angles. F

When an implication is true, its converse is not necessary true.
When both an implication and its converse are true, it is a biconditional.
$p \leftrightarrow q$: Two angles whose measures are equal if and only if they are congruent.

Inverse: $\sim p \to \sim q$ is the inverse of $p \to q$.

Contrapositive: $\sim q \to \sim p$ is the contrapositive of $p \to q$.

The following truth table shows the truth values of converse, inverse and contrapositive of an implication: (We can verify this truth table by the following examples.)

$p \to q$	$q \to p$	$\sim p \to \sim q$	$\sim q \to \sim p$
T	T	T	T
T	F	F	T
F	**T**	**T**	**F**
F	F	F	F

We have: $p \to q$ is logical equivalent to $\sim q \to \sim p$.

$q \to p$ is logical equivalent to $\sim p \to \sim q$.

Examples:

1. $p \to q$: If two angles whose measures are equal, then they are congruent. T

$q \to p$: If two angles are congruent, then their measures are equal. T

$\sim p \to \sim q$: If two angles whose measures are not equal, then they are not congruent. T

$\sim q \to \sim p$: If two angles are not congruent, then their measures are not equal. T

2. $p \to q$: If two angles are vertical angles, then they are congruent. T

$q \to p$: If two angles are congruent, then they are vertical angles. F

$\sim p \to \sim q$: If two angles are not vertical angles, then they are not congruent. F

$\sim q \to \sim p$: If two angles are not congruent, then they are not vertical angles. T

Accepted forms of valid reasoning (Rules of Inference):

The following forms of logical statements have been recognized as accepted forms of valid reasoning.

1. Modus Ponens (The Law of Detachment):

If $p \to q$ is true and p is true, therefore q is true.

In logical notation:

$$\begin{array}{l} p \to q \\ \underline{p \qquad} \\ \therefore q \end{array}$$

Example:

If I add 4 and 3, then my answer is 7.

I add 4 and 3.

“ Therefore, my answer is 7. ” is a valid conclusion.

2. Modus Tollens:

If $p \rightarrow q$ is true and $\sim q$ is true, therefore $\sim p$ is true.

In logic notation:

$$\begin{array}{l} p \rightarrow q \\ \underline{\sim q \quad\quad} \\ \therefore \sim p \end{array}$$

Example:

If I add 4 and 3, then my answer is 7.
My answer is not 7.
" Therefore, I did not add 4 and 3 " is a valid conclusion.

If a logic statement does not fit the forms of valid reasoning "modus ponens" or "modus tollens", we simple say " no valid conclusion (statement)".

3. The Law of Syllogism (Chain Rule):

If $p \rightarrow q$ is true and $q \rightarrow r$ is true, therefore $p \rightarrow r$ is true.

In logic notation:

$$\begin{array}{l} p \rightarrow q \\ \underline{q \rightarrow r} \\ \therefore p \rightarrow r \end{array}$$

Example:

If you study hard, then you will have good grade.
If you have good grade, then you will be selected by UCLA.
" Therefore, if you study hard, then you will be selected by UCLA." is a valid conclusion.

Examples

1. Find the valid conclusion of the following statements.

If $x = 5$ for this equation, then $y = 25$.
$x = 5$ for this equation.

Solution:

$$\begin{array}{l} p \rightarrow q \\ \underline{p \quad\quad} \\ \therefore q \end{array}$$

It fits the law of detachment. We have the valid conclusion:
" Therefore, $y = 25$."

2. Find the valid conclusion of the following statements.

If $x = 5$ for an equation, then $y = 25$.

$y = 25$.

Solution:

$p \to q$

q

There is no valid conclusion.

(Hint: $y = x^2$, x is either 5 or -5.)

3. Determine whether the conclusion of the following statements is valid.

If n is an even number, then it is divisible by 2.

n is not divisible by 2.

Therefore, n is not an even number.

Solution:

$$\begin{array}{l} p \to q \\ \underline{\sim q} \\ \therefore \sim p \end{array}$$

It fits modus tollens. The conclusion is valid.

4. Determine whether the conclusion of the following statement is valid.

If n is divisible by 4, then it is divisible by 2.

If it is divisible by 2, then it is an even number.

Therefore, if n is divisible by 4, then it is an even number.

Solution:

$$\begin{array}{l} p \to q \\ \underline{q \to r} \\ \therefore p \to r \end{array}$$

It fits the law of syllogism. The conclusion is valid.

1-31 Eigenvalues and Eigenvectors

$$A=\begin{bmatrix} a_{11} & a_{12} & \cdots & a_{1n} \\ a_{21} & a_{22} & \cdots & a_{2n} \\ \cdots & \cdots & \cdots & \cdots \\ a_{n1} & a_{n2} & \cdots & a_{nn} \end{bmatrix} \text{ is a square matrix .}$$

The equation of the following form is called an **Eigen Equation** (or **characteristic equation**) of the matrix A:

$$f(\lambda)=\begin{vmatrix} a_{11}-\lambda & a_{12} & \cdots & a_{1n} \\ a_{21} & a_{22}-\lambda & \cdots & a_{2n} \\ \cdots & \cdots & \cdots & \cdots \\ a_{n1} & a_{n2} & \cdots & a_{nn}-\lambda \end{vmatrix}=0 \qquad \textbf{or: } \det(A-\lambda I)=0$$

where I is an identity matrix and λ is a scalar.

The root λ (a scalar) of the equation are called the **eigenvalue** (or **characteristic value**) of the matrix A . The eigenvalues of A are the values of λ that satisfy the the eigen equation. A nonzero column vector $\boldsymbol{x}$ is a corresponding **eigenvector** (or **characteristic vector**) to an eigenvalue.

If k is a real number (a scalar) and A is a matrix, The matrix kA is formed by multiplying each entry in A by k. The matrix kA is called a scalar multiple of A .

If we multiply a square matrix A by a nonzero column vector $\boldsymbol{x}$, the matrix $A\boldsymbol{x}$ is generally not a scalar multiple of $\boldsymbol{x}$.

An **eigenvector** $\boldsymbol{x}$ corresponding to each eigenvalue of the matrix A is a nonzero column vector $\boldsymbol{x}$ that satisfy $(A-\lambda I)\boldsymbol{x}=0$ and makes $A\boldsymbol{x}$ equal a scalar multiple of $\boldsymbol{x}$. $(A-\lambda I)\boldsymbol{x}=0$ is equivalent to $A\boldsymbol{x}=\lambda\boldsymbol{x}$.

The eigen equation of $A=\begin{bmatrix} a & b \\ c & d \end{bmatrix}$ is $\begin{vmatrix} a-\lambda & b \\ c & d-\lambda \end{vmatrix}=0$

Or: $\lambda^2-(a+d)\lambda+(ad-bc)=0$

Where $\lambda_1+\lambda_2=a+d$, $\lambda_1\lambda_2=\begin{vmatrix} a & b \\ c & d \end{vmatrix}=ad-bc$

The eigen equation of $A=\begin{bmatrix} a_{11} & a_{12} & a_{13} \\ a_{21} & a_{22} & a_{23} \\ a_{31} & a_{32} & a_{33} \end{bmatrix}$ is $\begin{vmatrix} a_{11}-\lambda & a_{12} & a_{13} \\ a_{21} & a_{22}-\lambda & a_{23} \\ a_{31} & a_{32} & a_{33}-\lambda \end{vmatrix}=0$

$$\text{Or: } \lambda^3-(a_{11}+a_{22}+a_{33})\lambda^2+\left(\begin{vmatrix} a_{11} & a_{12} \\ a_{21} & a_{22} \end{vmatrix}+\begin{vmatrix} a_{11} & a_{13} \\ a_{31} & a_{33} \end{vmatrix}+\begin{vmatrix} a_{22} & a_{23} \\ a_{32} & a_{33} \end{vmatrix}\right)\lambda-\begin{vmatrix} a_{11} & a_{12} & a_{13} \\ a_{21} & a_{22} & a_{23} \\ a_{31} & a_{32} & a_{33} \end{vmatrix}=0$$

Where $(\lambda_1+\lambda_2+\lambda_2)$ equals the coefficient of λ^2.

$(\lambda_1\lambda_2+\lambda_2\lambda_3+\lambda_1\lambda_3)$ equals the coefficient of λ.

$\lambda_1\lambda_2\lambda_3$ equals the determinant of the matrix A .

Examples:

1. Find the eigenvalues and eigenvectors of the matrix: $A=\begin{bmatrix}1 & -4\\ -2 & 3\end{bmatrix}$

Solution:

$$A-\lambda I=\begin{vmatrix}1-\lambda & -4\\ -2 & 3-\lambda\end{vmatrix}=0$$

$$(1-\lambda)(3-\lambda)-8=0$$

$$\lambda^2-4\lambda-5=0$$

$$(\lambda+1)(\lambda-5)=0,\ \lambda=-1 \text{ and } 5.$$

The eigenvalues of A are -1 and 5.

Find the eigenvector corresponding to $\lambda=-1$, we solve the system of equations:

Solve $(A+I)\boldsymbol{x}=\begin{bmatrix}2 & -4\\ -2 & 4\end{bmatrix}\cdot\begin{bmatrix}x_1\\ x_2\end{bmatrix}=0$, we have $x_1=2x_2$. $\therefore \boldsymbol{x}=k\begin{bmatrix}2\\ 1\end{bmatrix}$

The eigenvector corresponding to $\lambda=-1$ are all the nonzero multiples of $(2, 1)^T$.

Find the eigenvector corresponding to $\lambda=5$, we solve the system of equations:

Solve $(A-5I)\boldsymbol{x}=\begin{bmatrix}-4 & -4\\ -2 & -2\end{bmatrix}\cdot\begin{bmatrix}x_1\\ x_2\end{bmatrix}=0$, we have $x_1=x_2$. $\therefore \boldsymbol{x}=k\begin{bmatrix}1\\ 1\end{bmatrix}$

The eigenvector corresponding to $\lambda=5$ are all the nonzero multiples of $(1, 1)^T$

2. Find the eigen equation and eigenvalues of the matrix: $A=\begin{bmatrix}1 & 0 & 1\\ 0 & 1 & 0\\ 1 & 1 & 1\end{bmatrix}$

Solution:

$$A-\lambda I=\begin{vmatrix}1-\lambda & 0 & 1\\ 0 & 1-\lambda & 0\\ 1 & 1 & 1-\lambda\end{vmatrix}=0$$

$$(1-\lambda)^3-(1-\lambda)=0$$

We have the eigen equation: $\lambda^3-3\lambda^2+2\lambda=0$

Solve the eigen equation:

$$\lambda(\lambda^2-3\lambda+2)=0$$

$$\lambda(\lambda-2)(\lambda-1)=0$$

$$\therefore \lambda=0,\ 1, \text{ and } 2$$

The eigenvalues of A are 0, 1, and 2.

Chapter Two : Algebra

2-1 Algebraic Expressions

Variable: It is a letter that can be used to represent one or more numbers.

If $x=1$, then $2x=2(1)=2$. If $x=2$, then $2x=2(2)=4$. x is a variable.

We use variables to form algebraic expressions and equations.

Numerical Expression: It is an expression that includes only numbers.

Algebraic Expression: It is an expression that combines numbers and variables by four operations (add, subtract, multiply or divide).

$xy+x+2y-3$ is an algebraic expression.

Algebraic Equation: It is a statement by placing an equal sign " = " between two expressions. $2x+5=7$ is an equation. $2x+5=9$ is an equation.

Term: A single number or a product of numbers and variables in an expression.

A monomial: An expression with only one term: $5x$ is a monomial.

A polynomial: An expression formed by two or more terms: $xy+x+2y-3$

A polynomial or binomial: An expression formed by only two terms: $3x^2+6$

A polynomial or trinomial: An expression formed by three terms: $3x^2-5x-2$

Constant: It is a number only. In the expression $x+3$, 3 is a constant.

Coefficient: It is a number that is multiplied by a variable.

In the term $2y$, 2 is the coefficient. The coefficient of the term x is 1.

Degree of an expression: It is the greatest of the degrees of its terms after it has been simplified and its like terms are combined.

The degree of $2x+1$ is 1, a linear form.

The degree of $3x^2+x-1$ is 2, a quadratic form.

The degree of $4x^3+5x$ is 3, a cubic form. The degree of x^4-1 is 4, a quartic form.

The degree of $4a^2b+2a^3b^4-2$ is 7, a polynomial form.

Evaluating an Expression: The process of replacing variables with numbers in an algebraic expression and finding its value.

Examples

1. Evaluate $2x+15$ when $x=4$.

Solution:

$2x+15=2(4)+15=23$.

2. Evaluate a^2+b-2 when $a=3$, $b=1$.

Solution:

$a^2+b-2=3^2+1-2=9+1-2=8$.

3. Evaluate $\frac{1}{2}x+3$ if $x=3$.

Solution: $\frac{1}{2}x+3=\frac{1}{2}(3)+3=4\frac{1}{2}$.

4. Evaluate $\frac{1}{2}x-3$ if $x=3$.

Solution: $\frac{1}{2}x-3=\frac{1}{2}(3)-3=-1\frac{1}{2}$.

Adding and Subtracting Expressions

Like terms (Similar terms): Terms that contain the same form of variables in an algebraic expression.

$5x$ and $9x$ are like terms. $4a$ and $10a$ are like terms.

$3x^2y$ and $7x^2y$ are like terms.

Simplifying an expression: To simplify an expression, we combine the like terms.

To simplify an expression, we also follow the **order of operations**. (See Section 1 ~ 9)
To simplify an expression, we must use the distributive property to remove the parentheses.
If there are like terms inside the parentheses, we combine them first.

Distributive Property: $a(b+c)=ab+ac$

$a(b-c)=ab-ac$

If there is only a negative sign in front of the parenthesis, we remove the parenthesis by writing the opposite of each term inside the parenthesis. Or, consider the negative sign as "–1".

Examples: $-(3x+4-2x)=-(x+4)=-1(x)+(-1)(4)=-x+(-4)=-x-4$.

$-(x-4)=-1(x)+(-1)(-4)=-x+4$.

$-(-x-4)=-1(-x)+(-1)(-4)=x+4$.

$(-x-4)(-1)=-x(-1)+(-4)(-1)=x+4$.

To add or subtract expressions, we remove the parentheses using the Distributive Property and then regroup like terms.

When we write an expression (polynomial), we always write it in **standard form**. The terms are ordered from the greatest exponent to the least.

Examples

1. $2x+3x=5x$. **2.** $2x-3x=-x$. **3.** $-(2x-3x)=-(-x)=x$.

4. $-(2x+5)=-2x-5$. **5.** $-(2x-5)=-2x+5$. **6.** $-(-2x-5)=2x+5$.

7. $(3x-5)+(5x-3)=3x-5+5x-3=(3x+5x)+(-5-3)=8x+(-8)=8x-8$.

8. $(3x-5)-(5x-3)=3x-5-5x+3=(3x-5x)+(-5+3)=-2x+(-2)=-2x-2$.

9. $(4x+3y)+(6x-2y)=4x+3y+6x-2y=(4x+6x)+(3y-2y)=10x+y$.

10. $(4x+3y)-(6x-2y)=4x+3y-6x+2y=(4x-6x)+(3y+2y)=-2x+5y$.

11. $3a^2-2a+4-(2a^2+3a)=3a^2-2a+4-2a^2-3a=(3a^2-2a^2)+(-2a-3a)+4$.

$=a^2+(-5a)+4=a^2-5a+4$.

12. $x^3-2x^2+x-4+2x^3+5x^2+7=(x^3+2x^3)+(-2x^2+5x^2)+x+(-4+7)$

$=3x^3+3x^2+x+3$.

Multiplying and Dividing Expressions

To multiply any two expressions, we use the **distributive property** and **rules of exponents.** To divide an expression by a monomial (a divisor), each term in the expression is divided by the divisor.

1) $\dfrac{a+b}{c}=\dfrac{a}{c}+\dfrac{b}{c},\ c\neq 0$ **2)** $\dfrac{a-b}{c}=\dfrac{a}{c}-\dfrac{b}{c},\ c\neq 0$

Rules of Exponents (Powers): To multiply two powers having the same base, we add the exponents.
To divide two powers having the same base, we subtract the exponents.

1) $a^m\cdot a^n=a^{m+n}$ **2)** $\dfrac{a^m}{a^n}=a^{m-n}$

Examples:

1. $x^5\cdot x^2=x^{5+2}=x^7$.

2. $\dfrac{x^5}{x^2}=x^{5-2}=x^3$.

3. $4(3x-5)=4(3x)+4(-5)=12x-20$.

4. $5(-2a+1)=5(-2a)+5(1)=-10a+5$.

5. $3x(2x^2-4x-6)=3x(2x^2)+3x(-4x)+3x(-6)=6x^3-12x^2-18x$.

6. $\dfrac{4x+8}{2}=\dfrac{4x}{2}+\dfrac{8}{2}=2x+4$.

7. $\dfrac{12a-6}{3}=\dfrac{12a}{3}-\dfrac{6}{3}=4a-2$.

8. $\dfrac{4x^2-8x}{2x}=\dfrac{4x^2}{2x}-\dfrac{8x}{2x}=2x-4$.

9. $\dfrac{-28x^3+7x^2}{7x}=-\dfrac{28x^3}{7x}+\dfrac{7x^2}{7x}=-4x^2+x$.

10. $\dfrac{20n^3-16n^2-12n}{4n}=\dfrac{20n^3}{4n}-\dfrac{16n^2}{4n}-\dfrac{12n}{4n}=5n^2-4n-3$.

Notes

2-2 Polynomial Expressions

We have learned how to add, subtract, multiply, and divide expressions. Now we will learn how to multiply a polynomial by a monomial, and how to multiply two polynomials. It is helpful to arrange the terms of the polynomials in either increasing degree or decreasing degree of a particular variable.

A) To multiply a polynomial by a monomial, we use the **distributive property** and the **Rules of exponents**.

1. $4(x+3)=4(x)+4(3)=4x+12$

2. $-4x(x+3)=-4x(x)-4x(3)=-4x^2-12x$

3. $(x+3)(5x)=x(5x)+3(5x)=5x^2+15x$

4. $-2x(3x^2-4x+1)=-2x(3x^2)-2x(-4x)-2x(1)=-6x^3+8x^2-2x$

5. $4x^2y(2x^2-3xy+y^2)=4x^2y(2x^2)+4x^2y(-3xy)+4x^2y(y^2)$
$=8x^4y-12x^3y^2+4x^2y^3$

B) To multiply two polynomials, we use the **distributive property** and the **Rules of exponents**. Combine like terms.

6. $(x+4)(x+3)=(x+4)(x)+(x+4)(3)$
$=x^2+4x+3x+12$
$=x^2+7x+12$

7. $(2x+4)(5x-3)=(2x+4)(5x)+(2x+4)(-3)$
$=10x^2+20x-6x-12$
$=10x^2+14x-12$

8. $(2x-3)(3x^2-2x-5)=(2x-3)(3x^2)+(2x-3)(-2x)+(2x-3)(-5)$
$=6x^3-9x^2-4x^2+6x-10x+15$
$=6x^3-13x^2-4x+15$

Multiplying Binomials (FOIL & Patterns)

We have learned how to multiply two binomials by using the **distributive property**.

$$(2x-5)(3x-4) = (2x-5)(3x) + (2x-5)(-4)$$
$$= 6x^2 - 15x - 8x + 20$$
$$= 6x^2 - 23x + 20$$

Now we will learn a short way to multiply two binomials mentally. This short way is called the **FOIL Method** (First Outside, Inside Last).

FOIL Method:

1. Multiply the first terms: $6x^2$
2. Multiply the last terms: 20
3. Multiply the outside terms: $-8x$
   Multiply the inside terms: $-15x$
   Add these two products: $(-8x)+(-15x) = -23x$

$6x^2$ $+20$ ← **mentally**

$(2x-5)(3x-4) = 6x^2 - 23x + 20$

$-15x$

$-8x$ ← **add**

The following formulas are very useful to shorten the steps in multiplying two binomials which are in special multiplication patterns.

Formulas

1)	$(a+b)^2 = a^2 + 2ab + b^2$
2)	$(a-b)^2 = a^2 - 2ab + b^2$
3)	$(a+b)(a-b) = a^2 - b^2$

Examples

1. $(x+4)^2 = x^2 + 2\cdot x\cdot 4 + 4^2 = x^2 + 8x + 16$
2. $(x-4)^2 = x^2 - 2\cdot x\cdot 4 + 4^2 = x^2 - 8x + 16$
3. $(x+4)(x-4) = x^2 - 4^2 = x^2 - 16$

4. $(3x+2y)^2 = (3x)^2 + 2\cdot 3x\cdot 2y + (2y)^2 = 9x^2 + 12xy + 4y^2$
5. $(3x-2y)^2 = (3x)^2 - 2\cdot 3x\cdot 2y + (2y)^2 = 9x^2 - 12xy + 4y^2$
6. $(3x+2y)(3x-2y) = (3x)^2 - (2y)^2 = 9x^2 - 4y^2$

7. $(x^3+4)^2 = (x^3)^2 + 2\cdot x^3\cdot 4 + 4^2 = x^6 + 8x^3 + 16$
8. $(a^n - b^n)^2 = (a^n)^2 - 2\cdot a^n\cdot b^n + (b^n)^2 = a^{2n} - 2a^nb^n + b^{2n}$
9. $(a^2-b^2)(a^2+b^2) = (a^2)^2 - (b^2)^2 = a^4 - b^4$
10. $(xy+2z)(xy-2z) = (xy)^2 - (2z)^2 = x^2y^2 - 4z^2$

2-3 Rational Expressions

Rational number: A number that is a quotient (ratio) of two integers.
2 and 4.6 are rational numbers because $2=\frac{4}{2}$ and $4.6=\frac{46}{10}$.
$\sqrt{2}\approx 1.414$ is not a rational number. It is an irrational number.

Rational Expression (algebraic fraction):
An algebraic expression that is a quotient (ratio) of two polynomials.
It is also referred to as **algebraic fraction**.
$\frac{x+1}{2x}$, $\frac{2x^2+3x-5}{x^2+4}$, and $\frac{xy^2}{x^2+xy}$ are rational expressions.

All rules for operating with rational numbers that we have learned apply to rational expressions as well.
A rational expression is in simplest form if the numerator and the denominator have no common factors (except 1 and –1).
To simplify a simple rational expression, we reduce the fraction to lowest terms by using their greatest common factor (GCF) or the laws of exponents.
To simplify a rational expression, we factor the numerator and the denominator and then cancel any common factor.
To simplify a rational fraction, we also **restrict** the variables by **excluding** any values that make the denominator equal to zero. It is **undefined** if the denominator equals 0.
Important: If factors of the numerator and the denominator are opposites of one another, we take the negative of the numerator or the denominator. (See examples 4 and 5)

Examples (simplify each expression)

1. $\frac{16x^2}{24x}=\frac{16x^2/8x}{24x/8x}=\frac{2x}{3},\ x\neq 0$ **or:** $\frac{16x^2}{24x}=\frac{16}{24}\cdot x^{2-1}=\frac{2}{3}x,\ x\neq 0$

2. $\frac{18x^4y^2}{27xy^6}=\frac{18}{27}\cdot\frac{x^4}{x}\cdot\frac{y^2}{y^6}=\frac{2x^3}{3y^4}, x\neq 0, y\neq 0$ **3.** $\frac{3xy}{3x+3y}=\frac{3xy}{3(x+y)}=\frac{xy}{x+y}, x\neq -y$

4. $\frac{x-1}{1-x}=\frac{\cancel{x-1}}{-\cancel{(x-1)}}=-1,\ x\neq 1$ **5.** $\frac{3-x}{4x-12}=\frac{3-x}{4(x-3)}=\frac{-\cancel{(x-3)}}{4\cancel{(x-3)}}=-\frac{1}{4},\ x\neq 3$

6. $\frac{x^2-4}{x+2}=\frac{\cancel{(x+2)}(x-2)}{\cancel{x+2}}=x-2,\ x\neq -2$

7. $\frac{x^2-5x+6}{x^2-x-2}=\frac{\cancel{(x-2)}(x-3)}{(x+1)\cancel{(x-2)}}=\frac{x-3}{x+1},\ x\neq -1,\ x\neq 2$

Adding and Subtracting Rational Expressions

1) To add or subtract rational expressions (algebraic fractions) having the same denominators, we combine their numerators.

$$\frac{b}{a}+\frac{c}{a}=\frac{b+c}{a} \quad ; \quad \frac{b}{a}-\frac{c}{a}=\frac{b-c}{a}$$

2) To add or subtract rational expressions (algebraic fractions) having different denominators, we rewrite each fraction having the least common denominator (**LCD**), and then combine the resulting fractions. **LCD** is the least common multiple (**LCM**) of their denominators.

$$\frac{c}{a}+\frac{d}{b}=\frac{bc}{ab}+\frac{ad}{ab}=\frac{bc+ad}{ab} \quad ; \quad \frac{c}{a}-\frac{d}{b}=\frac{bc}{ab}-\frac{ad}{ab}=\frac{bc-ad}{ab}$$

Examples

1. $\frac{3x}{5}+\frac{7x}{5}=\frac{3x+7x}{5}=\frac{10x}{5}=2x$

2. $\frac{x}{5}-\frac{3x}{2}=\frac{2x}{10}-\frac{15x}{10}=\frac{2x-15x}{10}=-\frac{13x}{10}$

3. $\frac{5a+3}{4}-\frac{a-1}{4}=\frac{5a+3-(a-1)}{4}=\frac{5a+3-a+1}{4}=\frac{4a+4}{4}=\frac{\cancel{4}(a+1)}{\cancel{4}}=a+1$

4. $\frac{1}{x}+\frac{1}{2x}=\frac{2}{2x}+\frac{1}{2x}=\frac{2+1}{2x}=\frac{3}{2x},\ x\neq 0$

5. $\frac{1}{x^2}-\frac{2}{x^3}=\frac{x}{x^3}-\frac{2}{x^3}=\frac{x-2}{x^3},x\neq 0$

6. $\frac{1}{2x^2}-\frac{5}{6x}=\frac{3}{6x^2}-\frac{5x}{6x^2}=\frac{3-5x}{6x^2},x\neq 0$

7. $\frac{3}{x-1}+\frac{5}{x-1}=\frac{3+5}{x-1}=\frac{8}{x-1},x\neq 1$

8. $\frac{7}{x-2}+\frac{2}{2-x}=\frac{7}{x-2}-\frac{2}{x-2}=\frac{5}{x-2},x\neq 2$

9. $\frac{3}{x}-\frac{2}{x-2}=\frac{3(x-2)}{x(x-2)}-\frac{2x}{x(x-2)}=\frac{3(x-2)-2x}{x(x-2)}=\frac{3x-6-2x}{x(x-2)}=\frac{x-6}{x^2-2x},x\neq 0,2$

10. $\frac{2a}{a^2-4}-\frac{5}{a+2}=\frac{2a}{(a+2)(a-2)}-\frac{5(a-2)}{(a+2)(a-2)}=\frac{2a-5(a-2)}{(a+2)(a-2)}=\frac{2a-5a+10}{(a+2)(a-2)}$

$=\frac{-3a+10}{a^2-4},\ a\neq -2,2$

11. $\frac{-2}{x^2-5x+6}+\frac{2}{x-3}=\frac{-2}{(x-3)(x-2)}+\frac{2}{x-3}=\frac{-2}{(x-3)(x-2)}+\frac{2(x-2)}{(x-3)(x-2)}$

$=\frac{-2+2(x-2)}{(x-3)(x-2)}=\frac{-2+2x-4}{(x-3)(x-2)}=\frac{2x-6}{(x-3)((x-2)}=\frac{2\cancel{(x-3)}}{\cancel{(x-3)}(x-2)}=\frac{2}{x-2},x\neq 3,2$

Multiplying and Dividing Rational Expressions

1) To multiply two rational expressions (algebraic fractions), we multiply their numerators and multiply their denominators. We can multiply first and then simplify, or simplify first and then multiply.

$$\frac{x^2}{3y}\cdot\frac{6}{5x}=\frac{\cancel{6x^2}}{\cancel{15xy}}=\frac{2x}{5y}\ ; \qquad \text{Or: } \frac{\cancel{x^2}}{\cancel{3y}}\cdot\frac{\cancel{6}}{5\cancel{x}}=\frac{2x}{5y}$$

2) To divide two rational expressions (algebraic fractions), we multiply the reciprocal of the divisor.

$$\frac{4x}{y}\div\frac{8x^2}{3y}=\frac{\cancel{4x}}{\cancel{y}}\cdot\frac{3\cancel{y}}{\cancel{8x^2}}=\frac{3}{2x}$$

3) To divide long polynomials, we follow the same ways of dividing real numbers. We must arrange the terms in both polynomials in order of decreasing degree of one variable. Keep space (using 0) on missing terms in degree (see example **10**).

Examples (Assume that no variable has a value for which the denominator is zero.)

1. $\frac{x^2}{4}\cdot\frac{x}{2}=\frac{x^2\cdot x}{4\cdot 2}=\frac{x^3}{8}$

2. $\frac{x^2}{4}\div\frac{x}{2}=\frac{\cancel{x^2}}{\cancel{4}}\cdot\frac{\cancel{2}}{\cancel{x}}=\frac{x}{2}$

3. $\frac{\cancel{2x}}{3\cancel{y}}\cdot\frac{x\cancel{y}}{\cancel{8}}=\frac{x^2}{12}$

4. $\frac{2x}{3y}\div\frac{xy}{8}=\frac{2\cancel{x}}{3y}\cdot\frac{8}{\cancel{x}y}=\frac{16}{3y^2}$

5. $\frac{a-3}{a}\cdot\frac{a^3}{a^2-9}=\frac{\cancel{a-3}}{\cancel{a}}\cdot\frac{\cancel{a^3}}{(a+3)\cancel{(a-3)}}=\frac{a^2}{a+3}$

6. $\frac{a^2-5a+6}{a^2}\div\frac{a^2-9}{a}=\frac{a^2-5a+6}{a^2}\cdot\frac{a}{a^2-9}=\frac{(a-2)\cancel{(a-3)}}{\cancel{a^2}}\cdot\frac{\cancel{a}}{(a+3)\cancel{(a-3)}}=\frac{a-2}{a(a+3)}$

7. $\frac{x^2-4x-5}{x}\cdot\frac{x^2+x}{x-5}=\frac{(x+1)\cancel{(x-5)}}{\cancel{x}}\cdot\frac{\cancel{x}(x+1)}{\cancel{x-5}}=(x+1)^2$

8. $\frac{x^2-4x-5}{x}\div\frac{x^2+x}{x-5}=\frac{x^2-4x-5}{x}\cdot\frac{x-5}{x^2+x}=\frac{\cancel{(x+1)}(x-5)}{x}\cdot\frac{x-5}{x\cancel{(x+1)}}=\frac{(x-5)^2}{x^2}$

9. Divide: $\frac{x^3+4x^2+6x+2}{x+3}$

Solution:

$$\begin{array}{r l}
 & x^2+\ x+3 \\
x+3\,\big) & x^3+4x^2+6x+2 \\
-) & x^3+3x^2 \\ \hline
 & x^2+6x \\
-) & x^2+3x \\ \hline
 & 3x+2 \\
-) & 3x+9 \\ \hline
 & -7 \leftarrow \text{Remainder}
\end{array}$$

Ans: $\frac{x^3+4x^2+6x+2}{x+3}=x^2+x+3-\frac{7}{x+3}$

10. Divide: $\frac{1+x+12x^3}{1+2x}$

Solution:

$$\begin{array}{r l}
 & 6x^2-3x+2 \\
2x+1\,\big) & 12x^3+0\ +\ x+1 \\
-) & 12x^3+6x^2 \\ \hline
 & -6x^2+\ x \\
-) & -6x^2-3x \\ \hline
 & 4x+1 \\
-) & 4x+2 \\ \hline
 & -1 \leftarrow \text{Remainder}
\end{array}$$

Ans: $\frac{1+x+12x^3}{1+2x}=6x^2-3x+2-\frac{1}{2x+1}$.

Notes

2-4 Rules of Exponents and Radicals

Rules of Exponents (Rules of Powers)

If a and b are real numbers, m and n are positive integers, $a \neq 0$, $b \neq 0$, then:

1) $a^0 = 1$ **2)** $a^m \cdot a^n = a^{m+n}$ **3)** $\dfrac{a^m}{a^n} = a^{m-n}$ **4)** $\dfrac{a^m}{a^n} = \dfrac{1}{a^{n-m}}$

5) $(a^m)^n = a^{mn}$ **6)** $(ab)^m = a^m b^m$ **7)** $a^{-m} = \dfrac{1}{a^m}$ **8)** $\dfrac{1}{a^{-m}} = a^m$

9) $\left(\dfrac{b}{a}\right)^m = \dfrac{b^m}{a^m}$ **10)** $\left(\dfrac{b}{a}\right)^{-m} = \left(\dfrac{a}{b}\right)^m$ **11)** $a^{\frac{1}{m}} = \sqrt[m]{a}$ **12)** $a^{\frac{n}{m}} = (\sqrt[m]{a})^n$

Examples

1. $x^7 \cdot x^3 = x^{7+3} = x^{10}$

2. $\dfrac{x^7}{x^3} = x^{7-3} = x^4$, $\dfrac{x^7}{x^3} = \dfrac{1}{x^{3-7}} = \dfrac{1}{x^{-4}} = x^4$

3. $\dfrac{x^3}{x^7} = \dfrac{1}{x^{7-3}} = \dfrac{1}{x^4}$, $\dfrac{x^3}{x^7} = x^{3-7} = x^{-4} = \dfrac{1}{x^4}$

4. $\dfrac{x^3}{x^3} = 1$, $\dfrac{x^3}{x^3} = x^{3-3} = x^0 = 1$

5. $a^{\frac{1}{2}} = \sqrt{a}$, $a^{\frac{1}{3}} = \sqrt[3]{a}$; $a^{\frac{1}{5}} = \sqrt[5]{a}$

6. $1^0 = 1$, $2^0 = 1$, $10^0 = 1$, $x^0 = 1$ if $x \neq 0$

7. $(a^3)^{-2} = a^{-6} = \dfrac{1}{a^6}$

8. $\dfrac{ac^5}{a^3c^2} = a^{1-3}c^{5-2} = a^{-2}c^3 = \dfrac{c^3}{a^2}$

9. $\dfrac{x^2}{y^2}\left(\dfrac{y^3}{x^2}\right)^4 = \dfrac{x^2}{y^2} \cdot \dfrac{y^{12}}{x^8} = \dfrac{y^{10}}{x^6}$

10. $25^{\frac{1}{2}} = \sqrt{25} = 5$, $32^{\frac{3}{5}} = (\sqrt[5]{32})^3 = 2^3 = 8$

11. $25^{-\frac{1}{2}} = \dfrac{1}{25^{\frac{1}{2}}} = \dfrac{1}{\sqrt{25}} = \dfrac{1}{5}$

12. $32^{-\frac{3}{5}} = \dfrac{1}{32^{\frac{3}{5}}} = \dfrac{1}{(\sqrt[5]{32})^3} = \dfrac{1}{2^3} = \dfrac{1}{8}$

13. $9^{2-\pi} \cdot 3^{2+2\pi} = (3^2)^{2-\pi} \cdot 3^{2+2\pi} = 3^{4-2\pi} \cdot 3^{2+2\pi} = 3^{4-2\pi+2+2\pi} = 3^6 = 726$

Rules of Radicals

Rules of Radicals

For all nonnegative real numbers a and b:

1) $\sqrt{a^2}=a$ **2)** $\sqrt{ab}=\sqrt{a}\cdot\sqrt{b}$ **3)** $\sqrt{\frac{b}{a}}=\frac{\sqrt{b}}{\sqrt{a}}$, where $a\neq 0$

4) $\sqrt[n]{a}=a^{\frac{1}{n}}$ **5)** $\sqrt[n]{a^m}=a^{\frac{m}{n}}$

For all real numbers a, b, and x:

6) $\sqrt[n]{a^n}=|a|$ **if** n is even. **7)** $\sqrt{x^2}=|x|$

$\sqrt[n]{a^n}=a$ **if** n is odd. **8)** $\sqrt{x^3}=x\sqrt{x}$

The formula $\sqrt{x^2}=|x|$ shows that when we are finding **square root** of a variable expression, we must use absolute value sign to **ensure** the answer is positive.

$\sqrt{(-3)^2}=\sqrt{9}=3$, not -3. $\sqrt[3]{-27}=\sqrt[3]{(-3)^3}=-3$. (Cubed root could be negative.)

$\sqrt{4x^2}=\sqrt{4}\cdot\sqrt{x^2}=2|x|$. $\sqrt{4x^6}=\sqrt{4}\cdot\sqrt{(x^3)^2}=2|x^3|$. $\sqrt[3]{8x^3}=\sqrt[3]{8}\cdot\sqrt[3]{x^3}=2x$.

$\sqrt{4x^4}=\sqrt{4}\cdot\sqrt{(x^2)^2}=2|x^2|=2x^2$. ($x^2$ is always nonnegative. Omit the symbol $|\ |$.)

$\sqrt{4x^8}=\sqrt{4}\cdot\sqrt{(x^4)^2}=2|x^4|=2x^4$. ($x^4$ is always nonnegative. Omit the symbol $|\ |$.)

In the formula $\sqrt{x^3}=x\sqrt{x}$, we must assume that x is nonnegative. The radical $\sqrt{x^3}$ has no meaning (not a real number) if $x<0$. Therefore, we omit the symbol $|\ |$.

$\sqrt{x^3}=\sqrt{x^2\cdot x}=\sqrt{x^2}\cdot\sqrt{x}=|x|\cdot\sqrt{x}=x\sqrt{x}$ Similarly: $\sqrt[4]{x^5}=x\sqrt[4]{x}$

To simplify a square root in simplest radical form, we factor the radicand by using a perfect-square factor and leave any factor that is not a perfect-square in the radical form To simplify a radical, we must **rationalize the denominator** by eliminating the radical from the denominator. No radical is allowed in denominator (see example **14**).

Examples. (Simplify each radical in simplest radical form)

1. $\sqrt{75}=\sqrt{25\cdot 3}=5\sqrt{3}$ **2.** $\sqrt{32}=\sqrt{16\cdot 2}=4\sqrt{2}$ **3.** $\sqrt[3]{16}=\sqrt[3]{2^3\cdot 2}=2\sqrt[3]{2}$

4. $\sqrt{25x^2}=5|x|$ **5.** $\sqrt{81y^2}=9|y|$ **6.** $\sqrt{24a^2}=\sqrt{4\cdot 6\cdot a^2}=2\sqrt{6}|a|$

7. $\sqrt{16a^3}=4\sqrt{a^2\cdot a}=4a\sqrt{a}$ **8.** $\sqrt{a^2-6a+9}=\sqrt{(a-3)^2}=|a-3|$

9. $\sqrt[4]{x^4y^8}=\sqrt[4]{x^4}\cdot\sqrt[4]{(y^2)^4}=|x|y^2$ **10.** $\sqrt[3]{-8x^6}=\sqrt[3]{(-2x^2)^3}=-2x^2$

11. $\sqrt[5]{-x^{10}}=\sqrt[5]{(-x^2)^5}=-x^2$ **12.** $\sqrt{18a^4b^5}=\sqrt{9\cdot 2\cdot(a^2)^2\cdot(b^2)^2\cdot b}=3a^2b^2\sqrt{2b}$

13. $\sqrt{\frac{18x^2y^3}{z^2}}=\frac{\sqrt{9\cdot 2\cdot x^2\cdot y^2\cdot y}}{\sqrt{z^2}}=\frac{3|x|y\sqrt{2y}}{|z|}$, $z\neq 0$ (Hint: y is nonnegative.)

14. $\sqrt{\frac{2}{5}}=\frac{\sqrt{2}}{\sqrt{5}}\cdot\frac{\sqrt{5}}{\sqrt{5}}=\frac{\sqrt{10}}{\sqrt{25}}=\frac{\sqrt{10}}{5}$ **15.** $\sqrt[4]{9}=\sqrt[4]{3^2}=3^{\frac{2}{4}}=3^{\frac{1}{2}}=\sqrt{3}$

Simplifying Radical Expressions

To simplify radical expressions, we use the following steps:

1. Simplify each radical in simplest form.
2. Combine (add or subtract) with like radicands.
3. Use the **Rules of Radicals** to multiply or divide two radicals:

$$\sqrt[n]{a}\cdot\sqrt[n]{b}=\sqrt[n]{ab}, \qquad \frac{\sqrt[n]{b}}{\sqrt[n]{a}}=\sqrt[n]{\frac{b}{a}}$$

4. Multiply binomials by using the distributive property, FOIL, or formulas.
5. Rationalize the denominator. (No radicals are in the denominator.)

Examples (Assume that all variables are nonnegative real numbers.)

1. $4\sqrt{2}-\sqrt{2}=3\sqrt{2}$ **2.** $6\sqrt{5}-4\sqrt{5}=2\sqrt{5}$ **3.** $3\sqrt{7}-2\sqrt{11}+5\sqrt{7}=8\sqrt{7}-2\sqrt{11}$

4. $5\sqrt{3}+\sqrt{12}=5\sqrt{3}+2\sqrt{3}=7\sqrt{3}$ **5.** $\sqrt{24}-\sqrt{54}=2\sqrt{6}-3\sqrt{6}=-\sqrt{6}$

6. $3\sqrt{18}+2\sqrt{8}-6\sqrt{50}=3\cdot3\sqrt{2}+2\cdot2\sqrt{2}-6\cdot5\sqrt{2}=9\sqrt{2}+4\sqrt{2}-30\sqrt{2}=-17\sqrt{2}$

7. $5\sqrt{7}+6\sqrt{3}-\sqrt{7}+2\sqrt{3}=(5\sqrt{7}-\sqrt{7})+(6\sqrt{3}+2\sqrt{3})=4\sqrt{7}+8\sqrt{3}$

8. $\sqrt{2}\cdot\sqrt{16}=\sqrt{2\cdot16}=\sqrt{32}=4\sqrt{2}$ **9.** $\sqrt[3]{2}\cdot\sqrt[3]{16}=\sqrt[3]{2\cdot16}=\sqrt[3]{32}=2\sqrt[3]{4}$

10. $2\sqrt{5}\cdot6\sqrt{12}=2\cdot6\cdot\sqrt{5\cdot12}=12\sqrt{60}=12\cdot2\sqrt{15}=24\sqrt{15}$

11. $\sqrt{3}(\sqrt{4}+\sqrt{6})=\sqrt{3}\cdot\sqrt{4}+\sqrt{3}\cdot\sqrt{6}=\sqrt{12}+\sqrt{18}=2\sqrt{3}+3\sqrt{2}$

12. $(4+\sqrt{2})(5-\sqrt{2})=20-4\sqrt{2}+5\sqrt{2}-2=18+\sqrt{2}$

13. $(\sqrt{2}+\sqrt{3})^2=(\sqrt{2})^2+2\cdot\sqrt{2}\cdot\sqrt{3}+(\sqrt{3})^2=2+2\sqrt{6}+3=5+2\sqrt{6}$

14. $(\sqrt{2}-\sqrt{3})^2=(\sqrt{2})^2-2\cdot\sqrt{2}\cdot\sqrt{3}+(\sqrt{3})^2=2-2\sqrt{6}+3=5-2\sqrt{6}$

15. $(\sqrt{2}+\sqrt{3})(\sqrt{2}-\sqrt{3})=(\sqrt{2})^2-(\sqrt{3})^2=2-3=-1$

16. $\sqrt{\frac{5}{3}}\cdot\sqrt{\frac{2}{5}}=\sqrt{\frac{5}{3}\cdot\frac{2}{5}}=\frac{\sqrt{2}}{\sqrt{3}}\cdot\frac{\sqrt{3}}{\sqrt{3}}=\frac{\sqrt{6}}{3}$ **17.** $\frac{\sqrt{15}}{\sqrt{6}}=\sqrt{\frac{15}{6}}=\sqrt{\frac{5}{2}}=\frac{\sqrt{5}}{\sqrt{2}}\cdot\frac{\sqrt{2}}{\sqrt{2}}=\frac{\sqrt{10}}{2}$

18. $\sqrt[3]{\frac{5}{3}}\cdot\sqrt[3]{\frac{2}{5}}=\sqrt[3]{\frac{5}{3}\cdot\frac{2}{5}}=\frac{\sqrt[3]{2}}{\sqrt[3]{3}}\cdot\frac{\sqrt[3]{9}}{\sqrt[3]{9}}=\frac{\sqrt[3]{18}}{\sqrt[3]{27}}=\frac{\sqrt[3]{18}}{3}$

19. $4\sqrt{x}+5\sqrt{x}=9\sqrt{x}$ **20.** $4x\sqrt{3x}+7x\sqrt{3x}=(4x+7x)\sqrt{3x}=11x\sqrt{3x}$

21. $\sqrt{63x}-\sqrt{7x}=3\sqrt{7x}-\sqrt{7x}=2\sqrt{7x}$ **22.** $\sqrt{2x}\cdot\sqrt{9x}=\sqrt{18x^2}=3\sqrt{2}\,x$

23. $\sqrt{ab^2}\cdot\sqrt{4a}=\sqrt{4a^2b^2}=2ab$ **24.** $\sqrt{x}(2-\sqrt{x})=2\sqrt{x}-\sqrt{x^2}=2\sqrt{x}-x$

25. $\sqrt{\frac{b}{a}}\cdot\sqrt{\frac{8b}{a}}=\sqrt{\frac{b}{a}\cdot\frac{8b}{a}}=\sqrt{\frac{8b^2}{a^2}}=\frac{\sqrt{8b^2}}{\sqrt{a^2}}=\frac{2\sqrt{2}b}{a},\ a\neq0$

Notes

2-5 Factoring Polynomials

We have learned how to **multiply** a polynomial by a monomial, and how to multiply two polynomials by using the distributive property.

Multiply: $4x(x+5) = 4x \cdot x + 4x \cdot 5 = 4x^2 + 20x$

Working in the other direction, we can **factor** a polynomial by expressing it as a product of other **prime polynomials**. To check the factors of a polynomial, we can multiply the resulting factors to find the original expression.

Factor: $4x^2 + 20x = 4x(x+5)$

Hint: $4x$ is the greatest common factor of the two terms, $4x^2$ and $20x$.
Check: $4x(x+5) = 4x \cdot x + 4x \cdot 5 = 4x^2 + 20x$ ✓

To factor a polynomial, the first step is to find the **greatest common factor** of its terms, or **GCF.** Sometimes, we need to arrange factors that are opposite of each other (see example 13).

Examples:

1. $5x+10 = 5(x+2)$
2. $8a-6 = 2(4a-3)$
3. $5x^2+10x = 5x(x+2)$
4. $8a^2-6a = 2a(4a-3)$
5. $15x^3-20x^2 = 5x^2(3x-4)$
6. $12p^5-16p^3 = 4p^3(3p^2-4)$
7. $4xy+8x = 4x(y+2)$
8. $4c^3-2c^2+6c = 2c(2c^2-c+3)$
9. $2x^5+4x^3-12x^2 = 2x^2(x^3+2x-6)$
10. $8a^3b^2+20a^2b^3 = 4a^2b^2(2a+5b)$
11. $x(x-5)+4(x-5) = (x-5)(x+4)$
12. $a(a+4)-3(a+4) = (a+4)(a-3)$
13. $x(x-5)+4(5-x) = x(x-5)-4(x-5) = (x-5)(x-4)$

Factoring by Grouping: If a polynomial has four or more terms, we regroup the pairs that have a common factor. We may need to try several groups before we get the right groups.

Examples:

14. $x^3+x^2+4x+4 = (x^3+x^2)+(4x+4) = x^2(x+1)+4(x+1) = (x+1)(x^2+4)$
15. $m^2+2m+mn+2n = (m^2+2m)+(mn+2n) = m(m+2)+n(m+2)$
$= (m+2)(m+n)$
16. $xy+3y+x+3 = (xy+3y)+(x+3) = y(x+3)+(x+3) = (x+3)(y+1)$
17. $xy+3y-x-3 = (xy+3y)-(x+3) = y(x+3)-(x+3) = (x+3)(y-1)$
18. $a^2-bc-ab+ac = (a^2-ab)+(ac-bc) = a(a-b)+c(a-b) = (a-b)(a+c)$
19. $2x-ay+ax-2y = (2x+ax)-(ay+2y) = x(2+a)-y(a+2) = (a+2)(x-y)$
20. $ax+by+ay+bx = (ax+bx)+(ay+by) = x(a+b)+y(a+b) = (a+b)(x+y)$

Factoring Quadratic Trinomials

We have learned how to multiply two binomials by using distributive property or the FOIL method. The result is a quadratic trinomial.

Multiply: $(x+2)(x+3) = x^2 + 5x + 6$

In this section we will learn how to factor a quadratic trinomial as a product of two binomials.

Factor: $x^2 + 5x + 6 = (x+2)(x+3)$

An expression in the form $ax^2 + bx + c$, $a \neq 0$ is called a **quadratic polynomial.**

A quadratic trinomial can be factored as a product of the form $(px+r)(qx+s)$ by working backward with the FOIL method.

$$ax^2 + bx + c = (px+r)(qx+s)$$

We use the following steps to find the factors:

1) List all the pairs of factors of ax^2: $ax^2 = (px)(qx)$

2) List all the pairs of factors of c: $c = rs$

3) Test all possible factors of ax^2 and c to find out which factor produces the correct bx.

Examples

1. Factor $x^2 + 12x + 20$

Solution:

The pair of factors of x^2 : x and x.

The pairs of factors of 20 : 1 and 20, 2 and 10, 4 and 5.

Test all factors to see which produces the correct term $12x$.

$$x^2 + 12x + 20 = (x+2)(x+10).$$

2. Factor $14x^2 - 17x + 5$.

Solution:

The pairs of factors of $14x^2$: $14x$ and x, $2x$ and $7x$, $-14x$ and $-x$.

The pairs of factors of 5: 1 and 5, -1 and -5.

Test all factors to see which produces the correct term $-17x$.

$$14x^2 - 17x + 5 = (2x-1)(7x-5).$$

3. Factor $x^2 - x - 20$.

Solution:

$$x^2 - x - 20 = (x+4)(x-5).$$

4. Factor $x^2 - 4xy - 12y^2$

Solution:

$$x^2 - 4xy - 12y^2 = (x+2y)(x-6y).$$

Factoring Special Quadratic Patterns

The special quadratic multiplication patterns are useful in factoring.
Working backward with the formulas, we have the following patterns:

Pattern 1: Factoring a perfect-square trinomial.

$$a^2 + 2ab + b^2 = (a+b)^2$$
$$a^2 - 2ab + b^2 = (a-b)^2$$

Pattern 2: Factoring a difference of two perfect-squares.

$$a^2 - b^2 = (a+b)(a-b)$$

Rules: A trinomial is a perfect square if :

1) The first term is a perfect square a^2.
2) The last term is a perfect square b^2.
3) The middle term is twice of $a \times b$.

If a binomial is a difference of two perfect-squares, we apply the rule in pattern 2.
Sometimes, we need more than one method to factor a polynomial.
Sometimes, we need to group its terms in a polynomial before factoring it.

Examples

1. Decide whether it is a perfect-square trinomial.

a. $x^2 - 14x + 49$. **b.** $9x^2 + 30x + 25$

Solution:

a. $x^2 - 14x + 49$
x^2 is a perfect square: x^2
49 is a perfect square: 7^2
The middle term $= 2 \cdot x \cdot 7 = 14x$
It is a perfect square.
Hint: $x^2 - 14x + 49 = (x-7)^2$

b. $9x^2 + 30x + 25$
$9x^2$ is a perfect square: $(3x)^2$
25 is a perfect square: 5^2
The middle term $= 2 \cdot 3x \cdot 5 = 30x$
It is a perfect square.
Hint: $9x^2 + 30x + 25 = (3x+5)^2$

Factor each polynomial.

2. $x^2 + 6x + 9 = (x+3)^2$

3. $x^2 - 6x + 9 = (x-3)^2$

4. $a^2 + 18a + 81 = (x+9)^2$

5. $a^2 - 18x + 81 = (x-9)^2$

6. $4x^2 + 12x + 9 = (2x+3)^2$

7. $4x^2 - 12x + 9 = (2x-3)^2$

8. $x^2 - 9 = x^2 - 3^2 = (x+3)(x-3)$

9. $4x^2 - 9 = (2x)^2 - 3^2 = (2x+3)(2x-3)$

10. $x^2 - 8x + 16 - 4y^2$
$= (x^2 - 8x + 16) - 4y^2$
$= (x-4)^2 - (2y)^2$
$= (x-4+2y)(x-4-2y)$
$= (x+2y-4)(x-2y-4)$

11. $x^2 + y^2 - 2xy - 9$
$= (x^2 - 2xy + y^2) - 9$
$= (x-y)^2 - 3^2$
$= (x-y+3)(x-y-3)$

12. $a^4 - 81$
$= (a^2)^2 - 9^2$
$= (a^2+9)(a^2-9)$
$= (a^2+9)(a+3)(a-3)$

Notes

2-6 Equations and Functions

In this section, we will learn to solve equations that have two variables of degree 1.
Here is an example in different forms:

$$2x - y = 3, \quad \text{or } 2x = y + 3, \quad \text{or } y = 2x - 3$$

The solutions to equations in two variables are ordered pairs of numbers (x, y).

Example: Identify whether the ordered pair (0, –3) is a solution of $2x - y = 3$.

Solution: We substitute $x = 0$ and $y = -3$ in the equation:

$$2x - y = 2(0) - (-3) = 0 + 3 = 3 \ 4$$

Yes. It is a solution.

Therefore, $x = 0$ and $y = -3$ is a solution of $2x - y = 3$ and can be written as (0, –3).
The equation $2x - y = 3$ has infinitely many solutions. Here are some of its solutions:

(–2, –7), (–1, –5), (0, –3), (1, –1), (2, 1), ·······.

Each solution of the equation is a point (x, y) on the coordinate plane.
The graph of connecting all the solutions (points) of the equation $2x - y = 3$ forms a straight line. Therefore, we call the equation $2x - y = 3$ a **linear equation.**
We can write any linear equation into the form $ax + by = c$, where a, b, and c are real numbers with a and b not both zero.
The equation $ax + by = c$ is called the **standard form** of a linear equation. If a is negative, we change it to positive by multiplying each side of the equation by –1.

If we rewrite a linear equation by solving for y in terms of x, we have a **linear function.**

Example: Solve the equation $3x + 2y = 6$ for y in terms of x.

Solution: $3x + 2y = 6$, $2y = -3x + 6$

Divide each side by 2: $\therefore y = -\frac{3}{2}x + 3$.

In the equation $y = -\frac{3}{2}x + 3$, x is called the **independent variable**, and y is called the **dependent variable.** Each value of y depends on the value chosen for x.
To find the value of y in the equation, we could substitute any value (any number) chosen for x into the equation: $x = 2$, $y = -\frac{3}{2}(2) + 3 = 0$. The solution is (2, 0).
The equation $y = -\frac{3}{2}x + 3$ has infinitely many solutions. The $x-$values are called the **domain** of the function, and the $y-$values are called the **range** of the function.
In the equation $y = -\frac{3}{2}x + 3$, each value of x is assigned exactly one value of y. We say that "y **varies directly with** x", or "y **is a function of** x".
"y is a function of x" is written by **functional notation** $y = f(x)$.
Therefore, the equation $y = -\frac{3}{2}x + 3$ can be written by **functional notation**:

$$f(x) = -\frac{3}{2}x + 3 \quad ; \quad f(x) = -\frac{3}{2}x + 3 \text{ is equivalent to } y = -\frac{3}{2}x + 3.$$

Since two points determine a straight line, we need to find only two points to graph a linear equation.
The points where the line crosses the $x-axis$ (let $y=0$) and the $y-axis$ (let $x=0$) are the easiest two points to find. We call them the $x-$intercept and the $y-$intercept.

Example: Graph $2x-y=4$.

Solution:

Let $y=0$	Let $x=0$
$2x-0=4$	$2(0)-y=4$
$2x=4$	$-y=4$
$x=2$	$y=-4$
Point (2, 0)	Point (0, –4)
$x-$intercept $=2$	$y-$intercept $=-4$

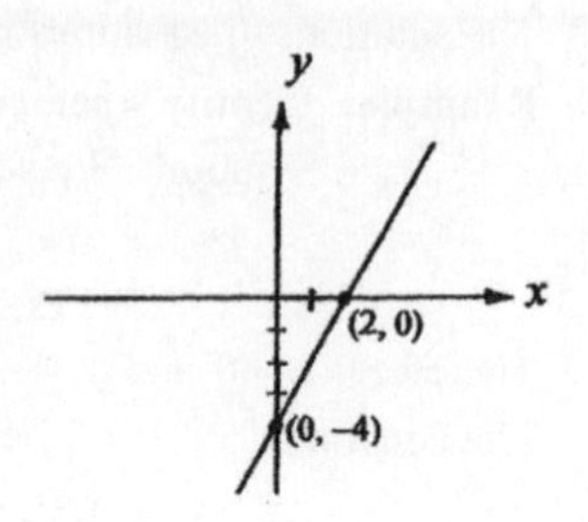

On the coordinate plane, an equation in one variable is a linear equation. Its graph is a vertical line or a horizontal line.
A vertical line has no restriction on y. A horizontal line has no restriction on x.

Examples:

a. Graph $x=4$ **b.** Graph $y=3$ **c.** Graph $x=-4$ **d.** Graph $y=-4$

Solution:

a. y x $x=4$

b.

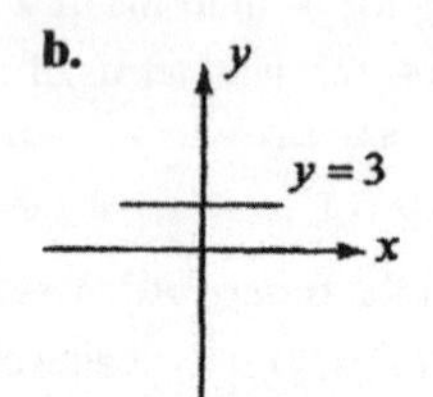

c.

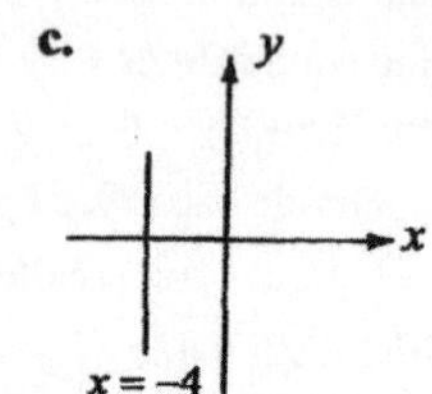

d.

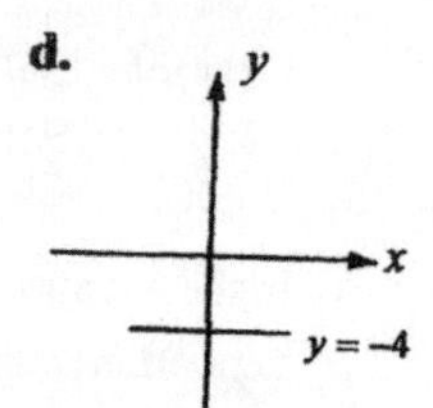

Using New Definitions (New Symbols, Defined Symbols)

A special symbol or sign (⊗, ⊕, ∗, Ω, #, !, ϒ, Δ, □, ……) is sometimes used to define a function for the particular case without using standard mathematical symbols.

Examples: 1. If $a^{\oplus}=a^2+3a-4$ is defined, then $3^{\oplus}=3^2+3(3)-4=14$.

2. If $a*b=(a+b)(a-b)$ is defined, then $5*3=(5+3)(5-3)=16$.

3. If $a\Delta b=a-2b$ and $a\square b=a+3b$ are defined, what is the value of x for $2\Delta(3x)=(4x)\square 5$?

Solution: $2-2(3x)=4x+3(5)$, $2-6x=4x+15$, $10x=-13$ $\therefore x=-1.3$

2-7 Slope of a Line

Slope is used to describe the **steepness** of a straight line. To find the slope of a line, we choose any two points on the line and form a right triangle which has the line as its hypotenuse. A line that rises more steeply has a greater slope. A line that runs more horizontally has a smaller slope. Therefore, the slope of a line is defined as the ratio of its vertical rise to horizontal run. The letter m is commonly used to represent the slope.

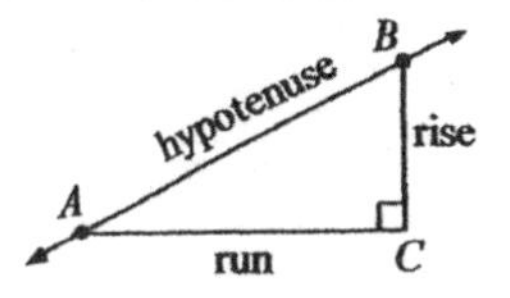

Definition of Slope:

$$\text{Slope of } \overleftrightarrow{AB} = \frac{rise}{run} = \frac{\overline{BC}}{\overline{AC}}$$

In the coordinate plane, the slope of a line between two points (x_1, y_1) and (x_2, y_2) is given in the following slope formula:

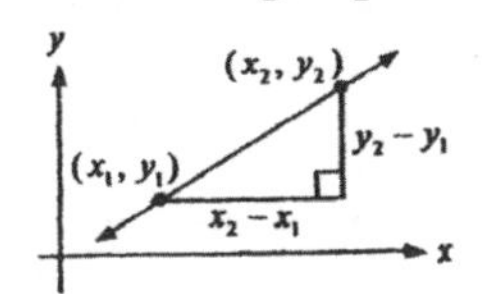

Slope Formula:

$$\text{Slope } (m) = \frac{y_2 - y_1}{x_2 - x_1}, \ (x_1 \neq x_2)$$

When we use the slope formula to find the slope of a line, we can choose any point as the first point and the other as the second point. The result will be the same.

Examples

1. Find the slope of the line passing through the points (2, 3) and (6, 5).
   Solution:

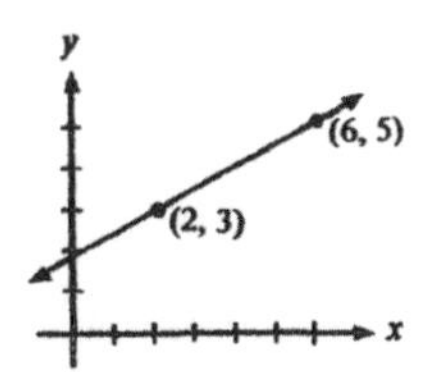

$$m = \frac{y_2 - y_1}{x_2 - x_1} = \frac{5-3}{6-2} = \frac{2}{4} = \frac{1}{2}.$$

Or: $$m = \frac{y_2 - y_1}{x_2 - x_1} = \frac{3-5}{2-6} = \frac{-2}{-4} = \frac{1}{2}.$$

2. Find the slope of the line containing the points (–2, 4) and (5, 2).
   Solution:

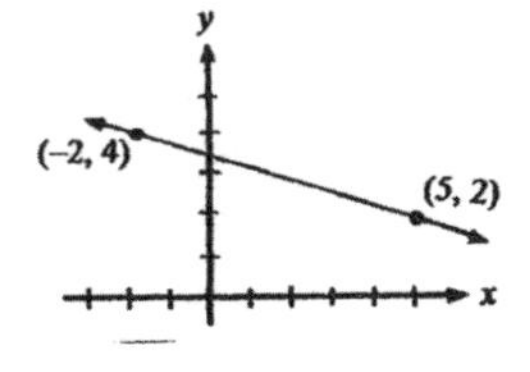

$$m = \frac{y_2 - y_1}{x_2 - x_1} = \frac{2-4}{5-(-2)} = \frac{-2}{7} = -\frac{2}{7}.$$

Or: $$m = \frac{y_2 - y_1}{x_2 - x_1} = \frac{4-2}{-2-5} = \frac{2}{-7} = -\frac{2}{7}.$$

Note that: 1. If a line slants up from left to right, the slope is positive.
2. If a line slants down from left to right, the slope is negative.

Examples

3. Find the slope of the line passing the points (–7, 4) and (3, 4).
Solution:

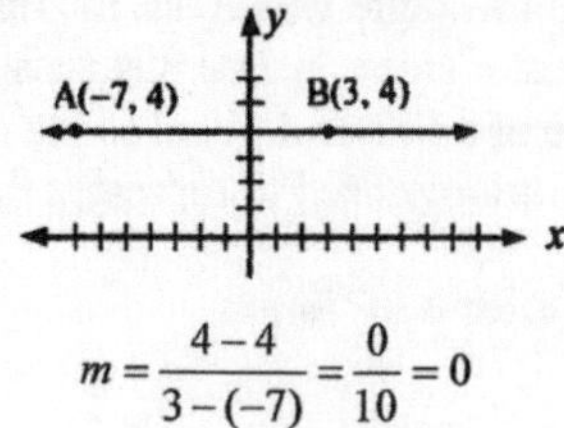

$$m=\frac{4-4}{3-(-7)}=\frac{0}{10}=0$$

Ans: The slope is 0.

4. Find the slope of the line passing through the points (7, 5) and (7, –3).
Solution:

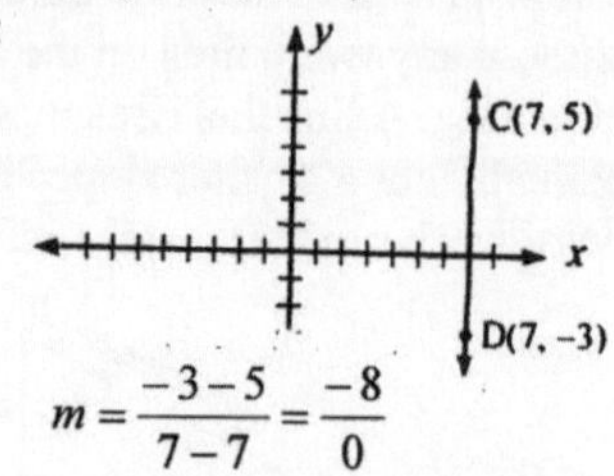

$$m=\frac{-3-5}{7-7}=\frac{-8}{0}$$

Ans: The slope is undefined.

Note that: 1) The slope of a horizontal line is 0.
2) The slope of a vertical line is undefined. It means that the slope of a vertical line does not exist.

Slope-Intercept Form of a Straight Line

We can choose any two points on a straight line and find its slope by the slope formula. However, an easier way to find the slope of a straight line is to rewrite the equation in the slope-intercept form:

$$y=mx+b$$

Then, the slope is m and $y-$ intercept is b .

5. Find the slope of the line $y+2x=3$.
Solution:

$$y+2x=3$$
$$y=-2x+3$$
$$\therefore m=-2.$$

6. Find the slope of the line $y-2x=3$
Solution:

$$y-2x=3$$
$$y=2x+3$$
$$\therefore m=2.$$

7. Find the slope and $y-$intercept of the line whose equation is $2x+3y-6=0$.
Solution:

$$2x+3y-6=0$$
$$3y=-2x+6$$
$$y=-\tfrac{2}{3}x+2$$

Ans: slope $=-\tfrac{2}{3}$.
$y-$intercept $=2$.

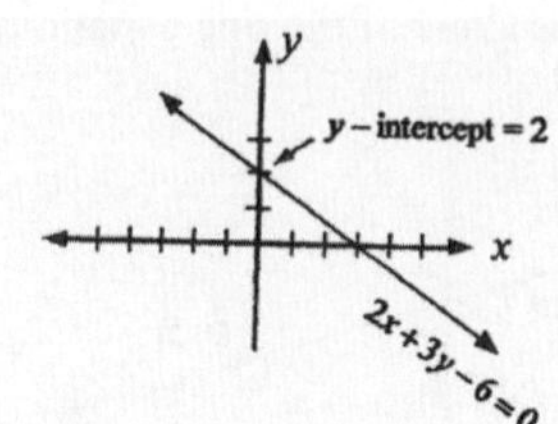

2-8 Finding the Equation of a Line

We have learned how to find the points, slope, and intercepts of a given line.
Now we will learn how to find the equation of a line under certain given information.
Such as, find the equation of a line:

1) having its slope and $y-$intercept.
2) having its slope and a point on the line.
3) having its two points on the line.

The following examples show how to find the equation of a line.
The equation of a line can be written in two different ways:

1) Slope-Intercept Form $y = mx + b$ **2) Standard Form** $ax + by = c$

Examples

1. Find the equation of the line having slope -5 and $y-$intercept 2.
Solution:

$m = -5$, $b = 2$
$y = mx + b$
$\therefore y = -5x + 2$. (slope-intercept form)
or: $5x + y = 2$. (standard form)

2. Find the equation of the line having slope $\frac{3}{4}$ and $y-$intercept $\frac{5}{2}$.
Solution: $m = \frac{3}{4}$, $b = \frac{5}{2}$

$y = mx + b$
$y = \frac{3}{4}x + \frac{5}{2}$. (slope-intercept form)
or: Multiply each side by 4:
$4y = 3x + 10$
$-3x + 4y = 10$ $\therefore 3x - 4y = -10$. (standard form)

3. Find the equation of the line having slope 4 and passing through the point (2, –3).
Solution:

$m = 4$
$y = mx + b$
$y = 4x + b$
To find b, we substitute the point (2, –3) into the above equation.
$-3 = 4(2) + b$
$-11 = b$
$\therefore y = 4x - 11$. (slope-intercept form)
or: $4x - y = 11$. (standard form)

Examples

4. Find the equation of the line passing through the points (–2, 3) and (4, 6).

Solution:

Find the slope: $m = \frac{y_2 - y_1}{x_2 - x_1} = \frac{6-3}{4-(-2)} = \frac{3}{6} = \frac{1}{2}$

$y = mx + b$

$y = \frac{1}{2}x + b$

To find b, substitute the point (–2, 3) into the above equation.

$3 = \frac{1}{2}(-2) + b$

$3 = -1 + b$

$4 = b$

$\therefore y = \frac{1}{2}x + 4$. (slope-intercept form)

or: $x - 2y = -8$. (standard form)

Point-Slope Form of a Line:

If m is the slope of a line and (x_1, y_1) is one of its points, we can write the equation of the line in point-slope form. To write the equation, we choose any other point (x, y) on the line.

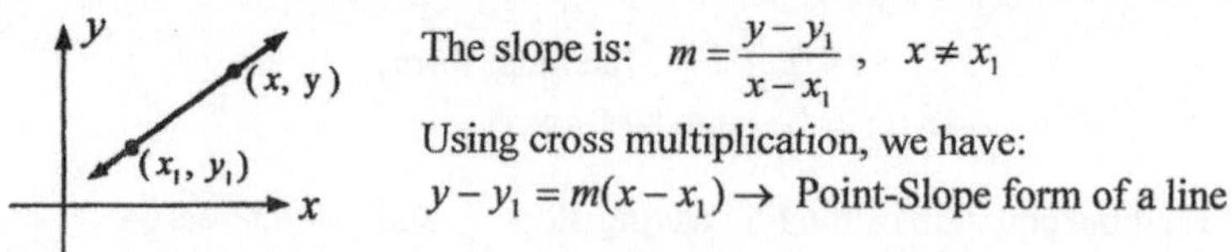

The slope is: $m = \frac{y - y_1}{x - x_1}$, $x \neq x_1$

Using cross multiplication, we have:

$y - y_1 = m(x - x_1) \rightarrow$ Point-Slope form of a line

Examples

5. Find the equation in point-slope form of a line having slope 4 and passing through the point (2, –3).

Solution:

$y - y_1 = m(x - x_1)$

$y - (-3) = 4(x - 2)$

$\therefore y + 3 = 4(x - 2)$. (point-slope form)

6. Find the equation in point-slope form of a line passing through the points (–2, 3) and (4, 6). (Hint: There are two answers.)

Solution: Find the slope: $m = \frac{y_2 - y_1}{x_2 - x_1} = \frac{6-3}{4-(-2)} = \frac{3}{6} = \frac{1}{2}$

Substitute one of the points, say (–2, 3), and the slope into the equation.

$y - y_1 = m(x - x_1)$

$y - 3 = \frac{1}{2}[x - (-2)]$

$\therefore y - 3 = \frac{1}{2}(x + 2)$. (point-slope form)

If we choose (4, 6), the other answer is: $y - 6 = \frac{1}{2}(x - 4)$.

Both answers have the same standard form: $x - 2y = -8$.

2-9 Parallel and Perpendicular Lines

Now we will learn how to identify parallel lines and perpendicular lines by comparing their slopes.
Two different lines that have the same slope are **parallel.** Two different lines that intersect to form a right angle (90° angle) are **perpendicular**.

Rules: 1. Two lines are parallel if and only if their slopes are equal: $m_1 = m_2$

2. Two lines are perpendicular if and only if the product of their slopes is -1**:**
(They are negative reciprocals of each other.)

$$m_1 \cdot m_2 = -1, \text{ or } m_1 = -\frac{1}{m_2}, \quad m_2 = -\frac{1}{m_1}$$

Examples

1. Find the slope of a line that is perpendicular to the line $8x - 4y = 3$.

Solution: $8x - 4y = 3$

$-4y = -8x + 3$

Divide each side by -4, we have the slope-intercept form:

$y = 2x - \frac{3}{4}$. slope $m = 2$

The slope of the perpendicular line is the negative reciprocal of 2.

$\therefore$ The slope of the perpendicular line is $-\frac{1}{2}$.

2. Tell whether or not the following two lines are parallel, perpendicular, or neither.

$3x - 4y = -5$ and $6x - 8y = 7$

Solution: To find their slopes, we rewrite each equation in slope-intercept form:

$3x - 4y = -5, \quad -4y = -3x - 5, \quad y = \frac{3}{4}x + \frac{5}{4} \quad \therefore m_1 = \frac{3}{4}$

$6x - 8y = 7, \quad -8y = -6x + 7, \quad y = \frac{3}{4}x - \frac{7}{8} \quad \therefore m_2 = \frac{3}{4}$

$m_1 = m_2$ $\therefore$ They are parallel.

3. Tell whether or not the following two lines are parallel, perpendicular, or neither.

$x - 2y = 4$ and $2x + y = 5$

Solution: $x - 2y = 4, \quad -2y = -x + 4, \quad y = \frac{1}{2}x - 2 \quad \therefore m_1 = \frac{1}{2}$

$2x + y = 5, \quad y = -2x + 5 \quad \therefore m_2 = -2$

$m_1 \cdot m_2 = \frac{1}{2} \cdot (-2) = -1$ $\therefore$ They are perpendicular.

4. Tell whether or not the following two lines are parallel, perpendicular, or neither.

$x + 3y = 4$ and $6y = 4x - 5$

Solution: $x + 3y = 4, \quad 3y = -x + 4, \quad y = -\frac{1}{3}x + \frac{4}{3} \quad \therefore m_1 = -\frac{1}{3}$

$6y = 4x - 5, \quad y = \frac{2}{3}x - \frac{5}{6} \quad \therefore m_2 = \frac{2}{3}$

$m_1 \neq m_2$ and $m_1 \cdot m_2 \neq -1$ $\therefore$ Neither

Notes

2-10 Direct and Inverse Variations

A linear equation (or function) with the form $y = kx$ is called an equation having a **direct variation** with x and y. k is a nonzero constant. k is called the **constant of variation** or **constant of proportionality**. In the equation $y = 3x$, 3 is a constant. We say that y varies directly as x. Therefore, $y = 3x$ is an example of a direct variation with x and y. The graph of $y = 3x$ is a straight line with slope 3 and passes through the origin (0, 0).
If (x_1, y_1) and (x_2, y_2) are two ordered pairs of an equation having a direct variation defined by $y = kx$ and that neither x_1 nor x_2 is zero, we have: $y_1 = kx_1$ and $y_2 = kx_2$,

$k = \frac{y_1}{x_1}$ and $k = \frac{y_2}{x_2}$. Therefore: $\frac{y_1}{x_1} = \frac{y_2}{x_2}$. **The ratios are equal.**

If $y = kx$, we say that y varies directly as x, or y is directly proportional to x.
In direct variation, the variables that vary may involve powers (exponents).

Note that $y = 3x - 2$ does not show x and y as a direct variation because the ratios of its ordered pairs $\frac{y_1}{x_1}$ and $\frac{y_2}{x_2}$ are not equal (not proportional).
However, the equation $y = 3x - 2$ shows x and $(y + 2)$ as a direct variation ($y + 2 = 3x$).
In science, there are many word problems which involve the concept of direct variations.

Joint Variation: An equation having one variable varies directly as the product of two or more other variables.
For $y = kxz$, we say that y varies jointly with x and y.

Examples

1. If y varies directly as x, and if $y = 6$ when $x = 2$, find the constant of variation.
Solution: Let $y = kx$, $6 = k \cdot 2$ $\therefore k = \frac{6}{2} = 3$.

2. If y varies directly as x, and if $y = 6$ when $x = 2$, find y when $x = 3$.
Solution: Let $y = kx$, $6 = k \cdot 2$ $\therefore k = \frac{6}{2} = 3$, $y = 3x$ is the equation.
When $x = 3$, $y = 3x = 3 \cdot 3 = 9$.

3. If (x_1, y_1) and (x_2, y_2) are ordered pairs of the same direct variation, find y_1.
$x_1 = 3$, $y_1 = ?$, $x_2 = 12$, $y_2 = 8$
Solution: The ratios are equal: $\frac{y_1}{x_1} = \frac{y_2}{x_2}$, $\frac{y_1}{3} = \frac{8}{12}$, $12y_1 = 24$ $\therefore y_1 = 2$.
Or: $y_2 = kx_2$, $8 = k \cdot 12$, $k = \frac{8}{12} = \frac{2}{3}$ $\therefore y_1 = kx_1 = \frac{2}{3}x_1 = \frac{2}{3} \cdot 3 = 2$.

An equation in the form $y = \frac{k}{x}$ (where $x \neq 0$), or $xy = k$ is called an equation having an an **inverse variation** with x and y. k is a nonzero constant. k is called the **constant of variation** or **constant of proportionality**. In the equation $y = \frac{3}{x}$, we say that y **varies inversely as** x, or y **is inversely proportional to** x. Therefore, $y = \frac{3}{x}$ is an example of an inverse variation with x and y. The graph of $y = \frac{3}{x}$, or $xy = 3$ is a **hyperbola.**
If (x_1, y_1) and (x_2, y_2) are two ordered pairs of an equation having an inverse variation defined by $y = \frac{3}{x}$, we have: $x_1 y_1 = k$ and $x_2 y_2 = k$. $\therefore x_1 y_1 = x_2 y_2$
In inverse variation, the variables that vary may involve powers (exponents).

Note that the equation $y = \frac{3}{x-1}$ does not show x and y as an inverse variation because the products of its ordered pairs $x_1 y_1$ and $x_2 y_2$ are not equal (not proportional).
However, the equation $y = \frac{3}{x-1}$ shows $(x-1)$ and y as an inverse variation $[y(x-1) = 3]$.

Combined Variation: The combination of direct and inverse variations.

In science, there are many word problems which involve the concept of inverse variations. The gravitational force (F) attracted each other between any two objects in the universe is given by $F = G\frac{m_1 m_2}{r^2}$ (**Newton's Law of Universal Gravitation**).

Examples

1. If y varies inversely as x, and if $y = 6$ when $x = 2$, find the constant of variation.
   Solution: Let $xy = k$ $\quad \therefore k = xy = 2 \cdot 6 = 12$.
2. If y varies inversely as x, and if $y = 6$ when $x = 2$, find y when $x = 3$.
   Solution: Let $xy = k$ $\quad \therefore k = xy = 2 \cdot 6 = 12$, $xy = 12$ is the equation.
   When $x = 3$, $y = \frac{12}{x} = \frac{12}{3} = 4$.
3. If (x_1, y_1) and (x_2, y_2) are ordered pairs of the same inverse variation, find y_1.
   $x_1 = 3$, $y_1 = ?$, $x_2 = 12$, $y_2 = 8$
   Solution: Let $x_1 y_1 = x_2 y_2$, $3y_1 = 12 \cdot 8$, $3y_1 = 96$ $\quad \therefore y_1 = 32$.

   Or: $k = x_2 y_2 = 12 \cdot 8 = 96$ $\quad \therefore x_1 y_1 = 96$, $y_1 = \frac{96}{x_1} = \frac{96}{3} = 32$.
4. The number of days needed to build a house varies inversely as the number of workers working on the job. It takes 140 days for 5 workers to finish the job. If the job has to be finished in 100 days, how many workers are needed ?
   Solution:
   Let $d_1 = 140$, $w_1 = 5$, $d_2 = 100$, $w_2 = ?$
   We have: $d_1 w_1 = d_2 w_2$, $140 \cdot 5 = 100 \cdot w_2$ $\quad \therefore w_2 = \frac{700}{100} = 7$ workers.

2-11 Solving One-Step Equations

An algebraic **equation** is a statement by placing an equal sign "=" between two expressions.

$x+5=9$ is an equation. $2y-3=7$ is an equation.

When a number which is substituted for (plugged into) the variable makes an equation a true statement, it is called a **solution** of the equation.

$x=4$ is a solution to the equation $x+5=9$.

$y=5$ is a solution to the equation $2y-3=7$.

To find the solution of an equation, we transfer the given equation into a simpler and equivalent equation that has the same solution. The last statement of the given equation after we transfer it should isolate the variable to one side of the equation in the form:

variable = a number, or **a number = variable**

To check a solution of an equation, we substitute the solution for the variable in the equation to see if we get a true statement.

The following four examples show the basic steps to solve simple **one-step equations**.

Examples

BY USING ADDITION

1. Solve $x-5=9$.

Solution:

$$x-5=9$$

Add 5 to each side:

$$x-\cancel{5}+\cancel{5}=9+5$$

$$\therefore\ x=14.$$

Check: $x-5=9$

$14-5=9$ ✓

BY USING SUBTRACTION

2. Solve $x+5=9$

Solution:

$$x+5=9$$

Subtract 5 from each side:

$$x+\cancel{5}-\cancel{5}=9-5$$

$$\therefore\ x=4.$$

Check: $x+5=9$

$4+5=9$ ✓

BY USING DIVISION

3. Solve $5x=40$.

Solution: $5x=40$

Divide each side by 5:

$$\frac{\cancel{5}x}{\cancel{5}}=\frac{40}{5}$$

$$x=8.$$

Check: $5x=40$

$5(8)=40$ ✓

BY USING MULTIPLICATION

4. Solve $\frac{x}{5}=40$.

Solution: $\frac{x}{5}=40$

Multiply each side by 5:

$$(\cancel{5})\frac{x}{\cancel{5}}=40(5)$$

$$x=200.$$

Check: $\frac{x}{5}=40$

$\frac{200}{5}=40$ ✓

Examples (Follow the same steps as shown in the examples 1 to 4.)

5. Solve $-12 = -3 + x$

Solution:

$$-12 = -3 + x$$
$$-12 + 3 = \not{-3} + x + \not{3}$$
$$-9 = x. \quad \text{Ans.}$$

6. Solve $15 = 3 - y$

Solution:

$$15 = 3 - y$$
$$15 - 3 = \not{3} - y - \not{3}$$
$$12 = -y \qquad \therefore -12 = y. \quad \text{Ans.}$$

7. Solve $24 = -8a$.

Solution:

$$24 = -8a$$
$$\frac{24}{-8} = \frac{\not{-8}a}{\not{-8}}$$
$$-3 = a. \quad \text{Ans.}$$

8. Solve $-24 = \frac{3}{5}n$.

Solution:

$$-24 = \frac{3}{5}n$$
$$\frac{5}{3} \cdot -24 = \frac{\not{3}}{\not{5}}n \cdot \frac{\not{5}}{\not{3}}$$
$$-40 = n. \quad \text{Ans.}$$

9. Solve $m + 2.5 = -3.8$.

Solution:

$$m + 2.5 = -3.8$$
$$m + \not{2.5} - \not{2.5} = -3.8 - 2.5$$
$$m = -6.3. \quad \text{Ans.}$$

10. Solve $y - 4.5 = -2.5$.

Solution:

$$y - 4.5 = -2.5$$
$$y - \not{4.5} + \not{4.5} = -2.5 + 4.5$$
$$y = 2. \quad \text{Ans.}$$

11. Solve $x - \frac{4}{5} = 2\frac{1}{5}$.

Solution:

$$x - \not{\frac{4}{5}} + \not{\frac{4}{5}} = 2\frac{1}{5} + \frac{4}{5}$$
$$x = \frac{11}{5} + \frac{4}{5} = \frac{15}{5} = 3$$
$$x = 3. \quad \text{Ans.}$$

12. Solve $x + 1\frac{3}{4} = \frac{2}{5}$.

Solution:

$$x + \not{1\frac{3}{4}} - \not{1\frac{3}{4}} = \frac{2}{5} - 1\frac{3}{4} = \frac{2}{5} - \frac{7}{4}$$
$$= \frac{8}{20} - \frac{35}{20} = -\frac{27}{20} = -1\frac{7}{20}$$
$$x = -1\frac{7}{20}. \quad \text{Ans.}$$

13. Solve $\frac{3}{7}x = \frac{5}{7}$.

Solution:

$$\frac{\not{7}}{\not{3}} \cdot \frac{\not{3}}{\not{7}}x = \frac{5}{\not{7}} \cdot \frac{\not{7}}{3}$$
$$x = \frac{5}{3} = 1\frac{2}{3}. \quad \text{Ans.}$$

14. Solve $-\frac{5y}{7} = \frac{1}{2}$

Solution:

$$-\frac{\not{7}}{\not{5}} \cdot -\frac{\not{5}y}{\not{7}} = \frac{1}{2} \cdot -\frac{7}{5}$$
$$y = -\frac{7}{10}. \quad \text{Ans.}$$

2-12 Solving Two-Step Equations

To solve some equations, it is necessary to use more than one step.
To solve an equation containing like terms, we simplify the equation by combining like terms on each side of the equation, and then apply the basic steps as we did before.
If an equation has a variable on both sides, we use the addition and the subtraction to get the variable alone on one side.
If an equation has numbers on both sides, we use the addition and subtraction to get the numbers alone to one side.
An easier way to simplify an equation is to transfer (isolate) all variables to one side of the equation and transfer (isolate) all numbers to the other side. Reverse (change) the signs in the process. (See Example 11 and Example 12.)

The following examples show the basic steps to solve simple **two-step equations**.

Examples

1. Solve $2x+3x=15$.
Solution:
$$2x+3x=15$$
$$5x=15$$
$$\frac{\cancel{5}x}{\cancel{5}}=\frac{15}{5}$$
$$\therefore x=3.$$

2. Solve $2x-3x=15$.
Solution:
$$2x-3x=15$$
$$-x=15$$
$$\frac{-x}{-1}=\frac{15}{-1}$$
$$\therefore x=-15.$$

3. Solve $4x+6x=-5+8$.
Solution:
$$4x+6x=-5+8$$
$$10x=3$$
$$\frac{\cancel{10}x}{\cancel{10}}=\frac{3}{10}$$
$$\therefore x=\frac{3}{10}.$$

4. Solve $-4x+6x=-5-8$.
Solution:
$$-4x+6x=-5-8$$
$$2x=-13$$
$$\frac{\cancel{2}x}{\cancel{2}}=\frac{-13}{2}$$
$$\therefore x=-6\frac{1}{2}.$$

5. Solve $6y=8-2y$.
Solution:
$$6y=8-2y$$
$$6y+2y=8\cancel{-2y}\cancel{+2y}$$
$$8y=8$$
$$\frac{\cancel{8}y}{\cancel{8}}=\frac{8}{8}$$
$$\therefore y=1.$$

6. Solve. $2.4n=1.8-1.26$
Solution:
$$2.4n=1.8n-1.26$$
$$2.4n-1.8n=\cancel{1.8n}-1.26\cancel{-1.8n}$$
$$0.6n=-1.26$$
$$\frac{\cancel{0.6}n}{\cancel{0.6}}=\frac{-1.26}{0.6}$$
$$\therefore n=-2.1.$$

Examples

7. Solve $8=7x-4x$.

Solution:

$$8=7x-4x$$
$$8=3x$$
$$\frac{8}{3}=\frac{\cancel{3}x}{\cancel{3}}$$
$$\frac{8}{3}=x \quad \therefore x=\frac{8}{3}=2\frac{2}{3}.$$

8. Solve $8=-7x-4x$.

Solution:

$$8=-7x-4x$$
$$8=-11x$$
$$\frac{8}{-11}=\frac{\cancel{-11}x}{\cancel{-11}}$$
$$\therefore -\frac{8}{11}=x.$$

9. Solve $\frac{1}{3}a-5=6$.

Solution:

$$\frac{1}{3}a-5=6$$
$$\frac{1}{3}a\cancel{-5}\cancel{+5}=6+5$$
$$\frac{1}{3}a=11$$
$$(\cancel{3})\frac{1}{\cancel{3}}a=11(3)$$
$$\therefore a=33.$$

10. Solve $\frac{y}{4}=-\frac{y}{2}-\frac{2}{5}$.

Solution:

$$\frac{y}{4}=-\frac{y}{2}-\frac{2}{5}$$
$$\frac{y}{4}+\frac{y}{2}=\cancel{-\frac{y}{2}}-\frac{2}{5}\cancel{+\frac{y}{2}}$$
$$\frac{y}{4}+\frac{2y}{4}=-\frac{2}{5} \rightarrow \frac{3y}{4}=-\frac{2}{5}$$
$$\frac{\cancel{4}}{\cancel{3}}\cdot\frac{\cancel{3}y}{\cancel{4}}=-\frac{2}{5}\cdot\frac{4}{3}$$
$$\therefore y=-\frac{8}{15}.$$

11. Solve $3x-12=18$.

Solution:

$$3x-12=18$$
$$3x=18+12$$
$$3x=30$$
$$\therefore x=10.$$

12. Solve $5y=7y-18$

Solution:

$$5y=7y-18$$
$$5y-7y=-18$$
$$-2y=-18$$
$$\therefore y=9.$$

2-13 Solving Multi-Step Equations

To solve an equation involving several steps, we simplify the equation by combining like terms on each side of the equation, and then apply the basic steps as we did before. To solve an equation containing parentheses, we combine likes terms inside the parentheses and use the Distributive Property to remove the parentheses.
An easier way to simplify an equation is to transfer (isolate) all variables to one side of the equation and transfer (isolate) all numbers to the other side. Reverse (change) the signs in the process. (See Example 5 and Example 6.)

Examples

1. Solve $2x+6=4x$.
Solution:

$$\begin{aligned} 2x+6&=4x \\ 2x+6-4x&=\cancel{4x}-\cancel{4x} \\ -2x+6&=0 \\ -2x\cancel{+6}\cancel{-6}&=-6 \\ -2x&=-6 \\ \frac{\cancel{-2}x}{\cancel{-2}}&=\frac{-6}{-2} \\ \therefore x&=3. \end{aligned}$$

2. Solve $2x-5=7-6x$.
Solution:

$$\begin{aligned} 2x-5&=7-6x \\ 2x-5+6x&=7\cancel{-6x}\cancel{+6x} \\ 8x-5&=7 \\ 8x\cancel{-5}\cancel{+5}&=7+5 \\ 8x&=12 \\ \frac{\cancel{8}x}{\cancel{8}}&=\frac{12}{8} \\ \therefore x&=1\tfrac{1}{2}. \end{aligned}$$

3. Solve $3(a-5+2a)=6$.
Solution:

$$\begin{aligned} 3(a-5+2a)&=6 \\ 3(3a-5)&=6 \\ 9a-15&=6 \\ 9a\cancel{-15}\cancel{+15}&=6+15 \\ 9a&=21 \\ \frac{\cancel{9}a}{\cancel{9}}&=\frac{21}{9} \\ \therefore a&=2\tfrac{1}{3}. \end{aligned}$$

4. Solve $3(2y+3)=4y-7$.
Solution:

$$\begin{aligned} 3(2y+3)&=4y-7 \\ 6y+9&=4y-7 \\ 6y+9-4y&=\cancel{4y}-7\cancel{-4y} \\ 2y+9&=-7 \\ 2y\cancel{+9}\cancel{-9}&=-7-9 \\ 2y=-16,\ \frac{\cancel{2}y}{\cancel{2}}&=\frac{-16}{2} \\ \therefore y&=-8. \end{aligned}$$

5. Solve $2x+6=4x$.
Solution:

$$\begin{aligned} 2x+6&=4x \\ 2x-4x&=-6 \\ -2x&=-6 \\ \therefore x&=\frac{-6}{-2}=3. \end{aligned}$$

6. Solve $2x-5=7-6x$.
Solution:

$$\begin{aligned} 2x-5&=7-6x \\ 2x+6x&=7+5 \\ 8x&=12 \\ \therefore x&=\frac{12}{8}=1\tfrac{1}{2}. \end{aligned}$$

Notes

2-14 Solving Exponential Equations

Equations that contains the terms a^x, $a>0$, $a\neq 1$ are called **exponential equations.**

To solve **exponential equations**, we apply the Rules of Exponents (See Section 2 ~ 4).

1. Rewrite the equation using the same base on both sides.
2. Make a substitution.

To solve an exponential equation we must check each solution in the original equation and discard solutions that are not permissible.

Examples

1. Solve $3^{2x}=81$.

Solution:

$$3^{2x}=3^4$$
$$2x=4$$
$$\therefore x=4.$$

2. Solve $3^{x+1}=9^{2x-1}$.

Solution:

$$3^{x+1}=(3^2)^{2x-1}$$
$$3^{x+1}=3^{4x-2}$$
$$x+1=4x-2$$
$$3=3x$$
$$\therefore x=1.$$

3. Solve $3^{4x+1}=\left(\frac{1}{9}\right)^{x-1}$

Solution:

$$3^{4x+1}=(3^{-2})^{x-1}$$
$$3^{4x+1}=3^{-2x+2}$$
$$4x+1=-2x+2$$
$$6x=1$$
$$\therefore\ x=\tfrac{1}{6}.$$

4. Solve $4^x-3\cdot 2^x-4=0$

Solution:

$$2^{2x}-3\cdot 2^x-4=0$$

Let $a=2^x$ (make a substitution):

$$a^2-3a-4=0$$
$$(a-4)(a+1)=0$$
$$a-4=0 \text{ or } a+1=0$$

$a=4$	$a=-1$
$2^x=4$	$2^x=-1$
$\therefore x=2$.	Not permissible ($2^x>0$)

5. Solve $x^{\frac{1}{2}}-6x^{-\frac{1}{2}}-5=0$

Solution:

Let $a=x^{\frac{1}{2}}$

Let $a=x^{\frac{1}{2}}$ (make a substitution)

$$a-\frac{6}{a}-5=0$$
$$a^2-5a-6=0$$
$$(a-6)(a+1)=0$$
$$a-6=0 \text{ or } a+1=0$$

$a=6$	$a=-1$
$x^{\frac{1}{2}}=6$	$x^{\frac{1}{2}}=-1$
$\therefore x=36$.	Not permissible

Notes

2-15 Solving Inequalities

An **inequality** is formed by placing an inequality sign or symbol (>, <, ≤, ≥) between numerical or variable expressions.

$5 > 3$ is read " 5 is greater than 3."

$3 < 5$ is read " 3 is less than 5."

$x > -4$ is read " x is greater than –4."

An inequality containing a variable is called **an open sentence.**

The inequality $x > -4$ indicates that all numbers greater than –4 are solutions of x.

$x \geq -3$ is read "x is greater than or equal to –3."

$x \leq 2$ is read "x is less than or equal to 2."

$-3 < x < 3$ is read "x is greater than –3 and less than 3."

$-3 \leq x < 2$ is read "x is greater than or equal to –3 and less than 2."

We can graph an inequality on a number line.
On a number line, the number on the right is greater than the number on the left.
To graph an inequality on a number line, we use a **close circle** " • " to show "included", and use an **open circle** " O " to show "not included".

Properties for inequalities:

1. Adding or subtracting the same number to or from each side of an inequality does not change the direction (order) of its inequality sign.
2. Multiplying or dividing each side of an inequality by the same positive number does not change the direction (order) of its inequality sign.
3. Multiplying or dividing each side of an inequality by the same negative number reverses (changes) the direction (order) of its inequality sign.

Examples

1. Add 6 to each side of $5 > 2$.
Then write a true inequality.
Solution:

$$5 > 2$$
$$5 + 6 > 2 + 6$$
$$11 > 8 \text{ (true)}$$

2. Multiply each side of $5 > 2$ by -6.
Then write a true inequality.
Solution:

$$5 > 2$$
$$5(-6) < 2(-6)$$
$$-30 < -12 \text{ (true)}$$

To solve an inequality, we use similar methods which are used to solve equations. Simply transfer all variables to one side of the inequality (usually to the left side), and transfer the others to the right side. Reverses (changes) the signs (+, –) in the process. Reverses the direction (order) of the inequality sign if we **multiply** (or **divide**) each side by **the same negative number**.

$$-\tfrac{1}{2}x > 4$$
$$(-2) \cdot -\tfrac{1}{2}x < 4 \cdot (-2) \ , \quad x < -8$$

Examples

1. Solve $4x-1<11$.

Solution:

$$4x-1<11$$
$$4x<11+1$$
$$4x<12$$
$$\therefore x<3.$$

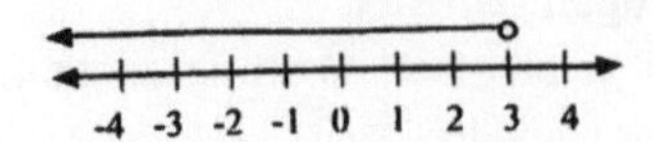

2. Solve $4x-1\geq 11$.

Solution:

$$4x-1\geq 11$$
$$4x\geq 11+1$$
$$4x\geq 12$$
$$\therefore x\geq 3.$$

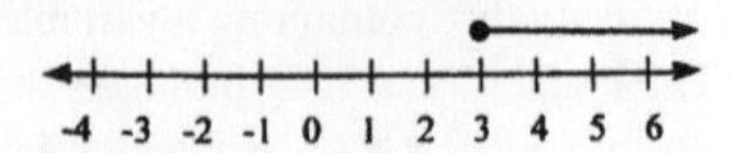

3. Solve $-4x-1<11$.

Solution:

$$-4x-1<11$$
$$-4x<12$$

Divide each side by –4:

$$\therefore x>-3.$$

4. Solve $-\frac{1}{4}x-1\geq 11$.

Solution:

$$-\tfrac{1}{4}x-1\geq 11$$
$$-\tfrac{1}{4}x\geq 12$$

Multiply each side by –4:

$$\therefore x\leq -48.$$

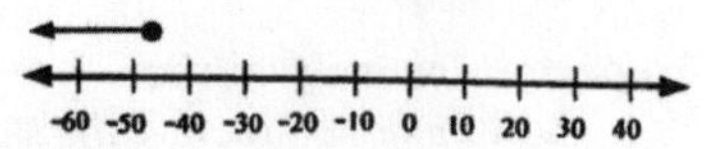

If the final statement is a "true statement", the inequality has solutions for all real numbers.
If the final statement is a "false statement", the inequality has no solution.

Examples

5. Solve $3x<3(x+4)$.

Solution:

$$3x<3(x+4)$$
$$3x<3x+12$$
$$3x-3x<12$$
$$0<12 \text{ (true)}$$

Ans: The inequality is true for all real numbers.

6. Solve $3x>3(x+4)$.

Solution:

$$3x>3(x+4)$$
$$3x>3x+12$$
$$3x-3x>12$$
$$0>12 \text{ (false)}$$

Ans: The inequality has no solution. The solution set is ϕ (empty).

2-16 Solving Combined Inequalities

A **combined inequality** (or **compound inequality**) is an inequality formed by joining two inequalities with "**and**" or "**or**".
There are two forms of combined inequalities, **conjunction** and **disjunction**.

1. **Conjunction of Inequality:** It is an inequality formed by two inequalities with the word "**and**". A conjunction is true when **both inequalities are true.**
   Example:
   $-5 < n$ and $n < 3$
   (or $-5 < n < 3$)

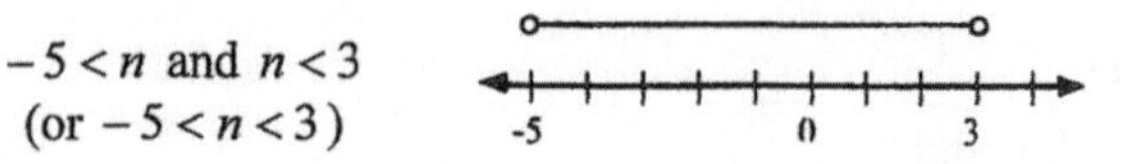

To solve a conjunction of inequality, we isolate the variable between the two inequality signs. Or, solve each part separately.

2. **Disjunction of Inequality:** It is an inequality forms by two inequalities with the word "**or**". A disjunction is true when **at least one of the inequalities is true.**
   Example:
   $n < -1$ or $n \geq 3$

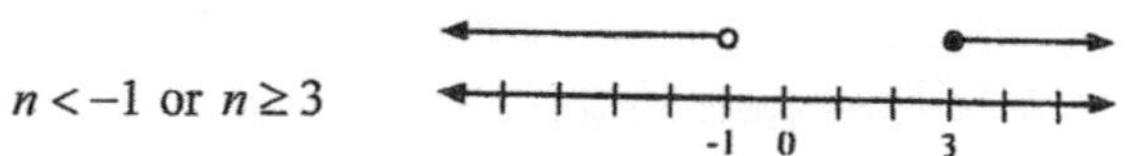

To solve a disjunction of inequality, we solve each part separately.

Examples

1. Solve $2 < \frac{1}{3}x + 4 \leq 6$.
   Solution:

Method 1:
(Solve each part separately.)
$2 < \frac{1}{3}x + 4$ and $\frac{1}{3}x + 4 \leq 6$

$-2 < \frac{1}{3}x$	$\frac{1}{3}x \leq 2$
$-6 < x$	$x \leq 6$

$\therefore -6 < x \leq 6$.

Method 2:
(Solve between the inequality signs.)
$2 < \frac{1}{3}x + 4 \leq 6$
Subtract 4 to each expression:
$-2 < \frac{1}{3}x \leq 2$
Multiply each expression by 3:
$\therefore -6 < x \leq 6$.

2. Solve $1 - 2x < -3$ or $3x + 14 \leq 2 - x$.
   Solution:

Solve each part separately:
$1 - 2x < -3$ or $3x + 14 \leq 2 - x$

$-2x < -4$	$4x \leq -12$
$x > 2$	$x \leq -3$

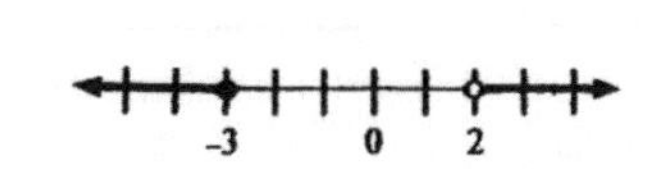

Ans: $x \leq -3$ or $x > 2$

Examples

3. Solve $n-2 \le 2n \le 3n-1$.

Solution:

$n-2 \le 2n$ and $2n \le 3n-1$

$-2 \le n$ $\quad$ $1 \le n$

It is a conjunction "and" inequality. Both must be true.

Ans: $n \ge 1$.

4. Solve $3x-1 \le 2x$ or $3x+10 \ge -2x$.

Solution:

$3x-1 \le 2x$ or $3x+10 \ge -2x$

$x \le 1$ $\quad$ $5x \ge -10$

$x \ge -2$

It is a disjunction "or" inequality. At least one is true.

Ans: All real numbers of x.

5. Solve $3a+5 < a+7$ and $3-a < 1$.

Solution:

$3a+5 < a+7$ and $3-a < 1$

$2a < 2$ $\quad$ $-a < -2$

$a < 1$ $\quad$ $a > 2$

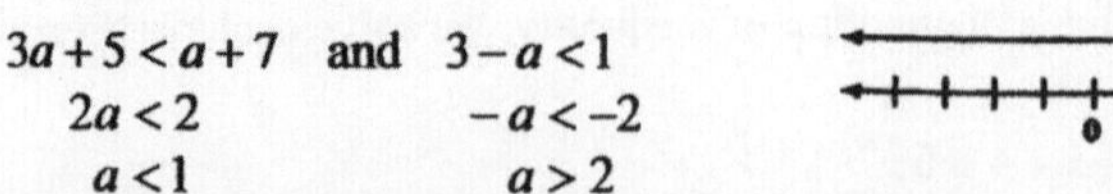

It is a conjunction "and" inequality. Both must be true.

Ans: No solution. ϕ (The solution set is empty.)

6. Solve $x-1 \le 0$ and $x+2 \ge 0$.

Solution:

$x-1 \le 0$ and $x+2 \ge 0$

$x \le 1$ $\quad$ $x \ge -2$

It is a conjunction "and" inequality.
Both must be true.

Ans: $-2 \le x \le 1$.

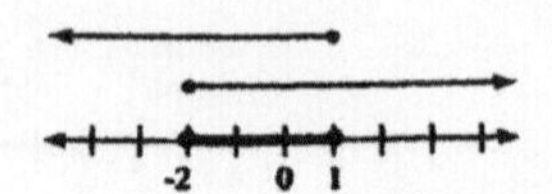

7. Solve $x-1 \le 0$ or $x+2 \ge 0$

Solution:

$x-1 \le 0$ or $x+2 \ge 0$

$x \le 1$ $\quad$ $x \ge -2$

It is a disjunction "or" inequality.
At least one is true.

Ans: All real numbers of x.

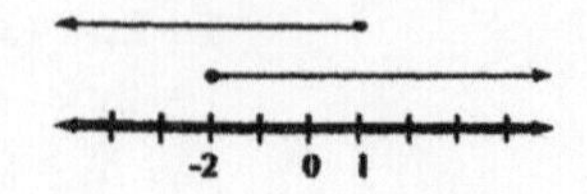

2-17 Absolute-Value Equations and Inequalities

To solve equations and inequalities involving absolute values, we start with the following most basic patterns. Following these patterns, we can solve all inequalities easily.

Rules: On the number line:

$|x| = 5$ means " the distance from x to 0 is 5 ".

$|x - c| = 5$ means " the distance from x to c is 5 ".

$|x + c| = 5$ means " the distance from x to $-c$ is 5 ".

$|2x - c| = 5$ means " the distance from $(2x)$ to c is 5 ". (See Example 4)

These rules are also valid for most absolute-value inequalities ($<, \leq, >, \geq$).

1. Solve $|x| = 5$.

Solution: $|x| = 5$

Ans: $x = 5$ or -5
(or $x = \pm 5$)

2. Solve $|x| > 5$.

Solution: $|x| > 5$

Ans: $x < -5$ or $x > 5$

3. Solve $|x| < 5$.

Solution: $|x| < 5$

Ans: $-5 < x < 5$

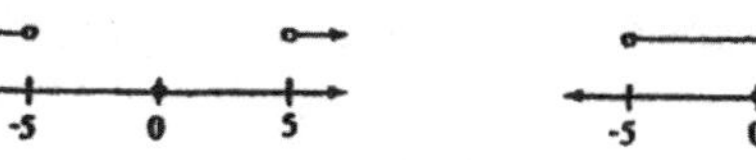

Examples

1. Solve $|x - 2| = 8$.

Solution: $x - 2 = \pm 8$

$x = \pm 8 + 2$

Ans: $x = 10$ or -6

2. Solve $|x - 2| > 8$.

Solution: $x - 2 < -8$ or $x - 2 > 8$

$x < -6 \quad | \quad x > 10$

Ans: $x < -6$ or $x > 10$

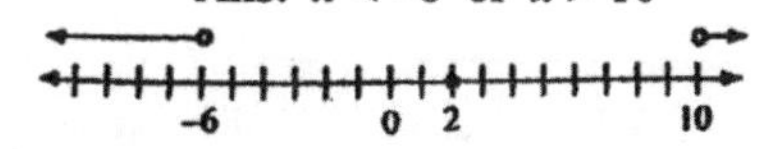

3. Solve $|x + 2| < 8$.

Solution: $-8 < x + 2 < 8$

$\therefore -10 < x < 6$.

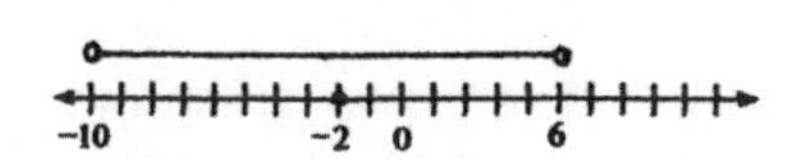

4. Solve $|2x - 7| = 5$.

Solution: $2x - 7 = \pm 5$

$2x = \pm 5 + 7$

$2x = 12$ or 2

Ans. $x = 6$ or 1

5. Solve $|3n + 5| - 3 \geq 7$.

Solution: $|3n + 5| \geq 10$

$3n + 5 \leq -10$ or $3n + 5 \geq 10$

$3n \leq -15 \quad | \quad 3n \geq 5$

$\therefore n \leq -5$ or $n \geq \frac{5}{3}$.

6. Solve $|20 - 5y| \leq 25$.

Solution: $-25 \leq 20 - 5y \leq 25$

$-45 \leq -5y \leq 5$

$9 \geq y \geq -1$

$\therefore -1 \leq y \leq 9$

Rational Inequalities and Absolute Values

Since we don't know whether its variable is positive or negative, a rational inequality in which a variable appears in a denominator can change signs of inequality if we multiply (or divide) each side by its common denominator. Therefore, we rewrite the inequality with **0** on the right side and consider two cases. **<u>Testing the intervals</u>** of the critical points is a good method to identify the answers.

Examples

1. Solve $\frac{3}{x} < 5$.

Solution: **Method 1:** $\frac{3}{x} - 5 < 0$, $\therefore \frac{3-5x}{x} < 0$

$x > 0$ and $3 - 5x < 0$	$x < 0$ and $3 - 5x > 0$
$-5x < -3$	$-5x > -3$
$x > \frac{3}{5}$	$x < \frac{3}{5}$
$\therefore x > \frac{3}{5}$	$\therefore x < 0$

Ans. $x < 0$ or $x > \frac{3}{5}$

Method 2: $\frac{3-5x}{x} < 0$. It is negative.

Critical points: $x = 0,\ \frac{3}{5}$

Test intervals:

$x < 0$, $\frac{(+)}{(-)} = -$ yes

$0 < x < \frac{3}{5}$, $\frac{(+)}{(+)} = +$ no

$x > \frac{3}{5}$, $\frac{(-)}{(+)} = -$ yes

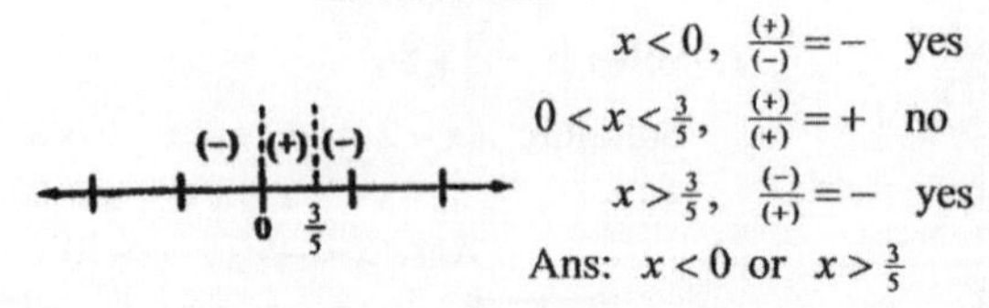

Ans: $x < 0$ or $x > \frac{3}{5}$

2. Solve $\frac{x+1}{x} > 3$.

Solution:

$\frac{x+1}{x} - 3 > 0$, $\frac{x+1-3x}{x} > 0$

$\therefore \frac{1-2x}{x} > 0$ It is positive.

Critical points: $x = 0,\ \frac{1}{2}$

Test intervals:

$x < 0$, $\frac{(+)}{(-)} = -$ no

$0 < x < \frac{1}{2}$, $\frac{(+)}{(+)} = +$ yes

$x > \frac{1}{2}$, $\frac{(-)}{(+)} = -$ no

Ans: $0 < x < \frac{1}{2}$

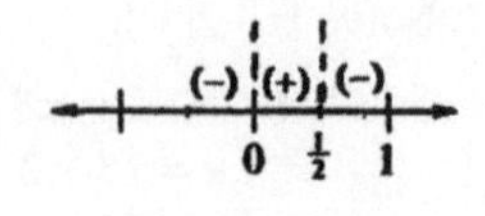

We have learned how to solve simple inequalities involving absolute values. To solve a rational inequality involving absolute value, we use this fact:

$$|x|^2 = x^2$$

3. Solve $\left|\frac{x-2}{4x+1}\right| \geq 1$.

Solution:

$|x-2| \geq |4x+1|$

$(x-2)^2 \geq (4x+1)^2$

$x^2 - 4x + 4 \geq 16x^2 + 8x + 1$

$-15x^2 - 12x + 3 \geq 0$

$5x^2 + 4x - 1 \leq 0$

$(5x-1)(x+1) \leq 0$

$5x - 1 \geq 0$ and $x + 1 \leq 0$	$5x - 1 \leq 0$ and $x + 1 \geq 0$
$x \geq \frac{1}{5}$, $x \leq -1$	$x \leq \frac{1}{5}$, $x \geq -1$
Not permissible	$\therefore -1 \leq x \leq \frac{1}{5}$, $x \neq -\frac{1}{4}$.

(We can also test the intervals to find the answer.)

Ans: $-1 \leq x \leq \frac{1}{5}$, $x \neq -4$

2-18 Solving Quadratic Equations

Steps for solving a quadratic equation:

1. Solve the equation in the form of perfect-square.
   If $x^2 = k$, then $x = \pm\sqrt{k}$.
2. Solve by factoring if it can be factored.
   If $ax^2 + bx + c = 0$ can be factored, then $(px + r)(qx + s) = 0$.
3. Solve by using the quadratic formula if factoring is not possible.
   If $ax^2 + bx + c = 0$, then $x = \dfrac{-b \pm \sqrt{b^2 - 4ac}}{2a}$ **(Quadratic Formula)**
4. Some equations are not quadratic equations, but they can be written in quadratic form by factoring, or by making a substitution.

Examples

1. Solve $2x^2 - 5x - 3 = 0$ by factoring.
Solution:

$$(2x + 1)(x - 3) = 0$$

$2x + 1 = 0$, $2x = -1$, $x = -\frac{1}{2}$ | $x - 3 = 0$, $x = 3$

Ans: $x = -\frac{1}{2}$ or 3

2. Solve $2x^2 - 5x - 3 = 0$ by formula.
Solution: $a = 2$, $b = -5$, $c = -3$

$$x = \frac{-b \pm \sqrt{b^2 - 4ac}}{2a} = \frac{-(-5) \pm \sqrt{(-5)^2 - 4 \cdot 2 \cdot (-3)}}{2(2)} = \frac{5 \pm \sqrt{25 + 24}}{4} = \frac{5 \pm 7}{4}$$

$= \frac{5+7}{4}$ or $\frac{5-7}{4} = 3$ or $-\frac{1}{2}$. Ans.

3. Solve $x^4 - 2x^2 - 8 = 0$.
Solution: (By factoring)

$$(x^2 - 4)(x^2 + 2) = 0$$

$x^2 - 4 = 0$, $x^2 = 4$, $x = \pm 2$ | $x^2 + 2 = 0$, $x^2 = -2$, $x = \pm\sqrt{-2} = \pm\sqrt{2}\, i$

Ans: $x = \pm 2$ or $\pm\sqrt{2}\, i$

4. Solve $x^2 + 4x + 8 = 0$
Solution: $a = 1$, $b = 4$, $c = 8$

$$x = \frac{-b \pm \sqrt{b^2 - 4ac}}{2a} = \frac{-4 \pm \sqrt{4^2 - 4 \cdot 1 \cdot 8}}{2(1)} = \frac{-4 \pm \sqrt{16 - 32}}{2} = \frac{-4 \pm \sqrt{-16}}{2} = \frac{-4 \pm 4i}{2}$$

$= -2 \pm 2i$. Ans.

5. Solve $2x - 2\sqrt{2x} - 8 = 0$.
Solution: (By making a substitution)

Let $u = \sqrt{2x}$, $u^2 = 2x$

$$u^2 - 2u - 8 = 0$$

$$(u - 4)(u + 2) = 0$$

$u - 4 = 0$, $u = 4$, $\sqrt{2x} = 4$, $x = 8$ | $u + 2 = 0$, $u = -2$, $\sqrt{2x} = -2$, no solution

Ans: $x = 8$

Notes

2-19 Graphing Quadratic Functions

Quadratic Equation: An equation of the form $ax^2+bx+c=0$ $(a\neq 0)$.
Quadratic Function: A quadratic equation of the form $y=ax^2+bx+c$ $(a\neq 0)$.
The graph of a quadratic function $y=ax^2+bx+c$ $(a\neq 0)$ on the coordinate plane is a **parabola.** The central line of a parabola is called **the axis of symmetry**, or simply the **axis.** The point where the parabola crosses its axis is the **vertex.**

General Form of a parabola: $y-k=a(x-h)^2$. The vertex is $(h,\ k)$.
Intercept Form of a parabola: $y=a(x-p)(x-q)$. The $x-$intercepts are p and q.
The $x-$coordinate of the vertex is the midpoint between the two $x-$intercepts.

If $a>0$ (positive), the parabola opens upward. Vertex is a minimum point.
If $a<0$ (negative), the parabola opens downward. Vertex is a maximum point.
$y=x^2$ is the simplest quadratic function. Its vertex is located on the origin (0, 0).

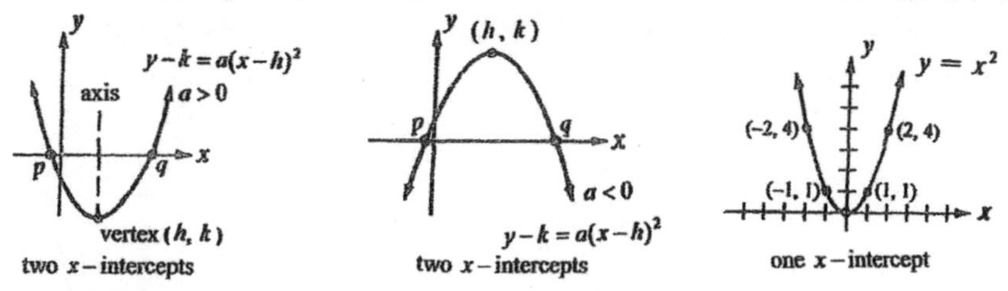

A parabola can have two, one, or no $x-$intercepts.
To find the $x-$intercepts of a parabola, we find the **zeros** (roots) of the related quadratic equation (let $y=0$):

Solve $ax^2+bx+c=0$ by factoring, or by the formula: $x=\dfrac{-b\pm\sqrt{b^2-4ac}}{2a}$.

If $b^2-4ac>0$, the equation has two roots. The parabola has two $x-$intercepts.
If $b^2-4ac=0$, the equation has one root. The parabola has one $x-$intercept.
If $b^2-4ac<0$, the equation has no real-number root. The parabola has no $x-$intercept.
The value of b^2-4ac indicates the differences among the three cases. Therefore, we call the value of b^2-4ac **"the discriminant"** of the quadratic equation.
Steps for graphing a quadratic function: $y=ax^2+bx+c$ $(a\neq 0)$

1. Let $y=0$. Find the zeros (roots). The roots are the $x-$intercepts.
2. Find the vertex $(h,\ k)$ by matching the general form $y-k=a(x-h)^2$.
3. If $a>0$, the graph opens upward. If $a<0$, the graph opens downward.

The graph of a parabola is symmetrical. One half of the graph on one side of the axis is the **mirror image** of the other half. The $x-$coordinate of the vertex is $-\frac{b}{2a}$.

Examples

1. Graph $y = 2x^2$.

Solution:

One x – intercept = 0

Vertex (0, 0)

$a = 2 > 0$

Open upward

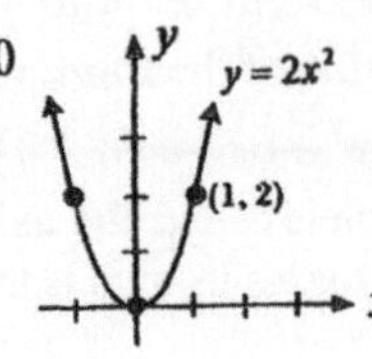

2. Graph $y = -2x^2$.

Solution:

One x – intercept = 0

Vertex (0, 0)

$a = -2 < 0$

Open downward

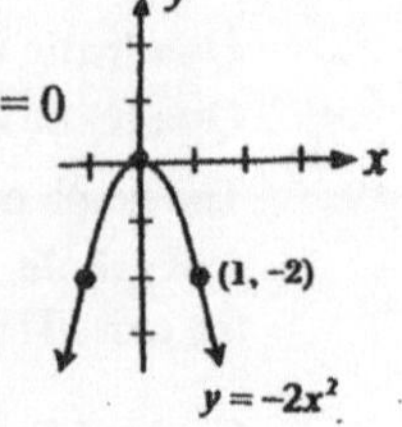

3. Graph $f(x) = x^2 - 2$.

Solution:

$x^2 - 2 = 0$, $x^2 = 2$

$x = \pm\sqrt{2} \approx \pm 1.4$

$\therefore$ Two x – intercepts = 1.4 and –1.4

$y + 2 = x^2$, $y - (-2) = (x - 0)^2$

$\therefore$ Vertex (0, –2)

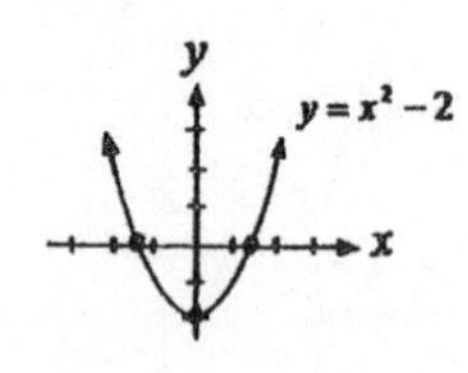

4. Graph $f(x) = x^2 - 4x$.

Solution:

$x^2 - 4x = 0$, $x(x - 4) = 0$

$x = 0$ and 4

$\therefore$ Two x – intercepts = 0 and 4

$y = (x^2 - 4x + 4) - 4$

$y + 4 = x^2 - 4x + 4$

$y - (-4) = (x - 2)^2$ $\therefore$ Vertex (2, –4)

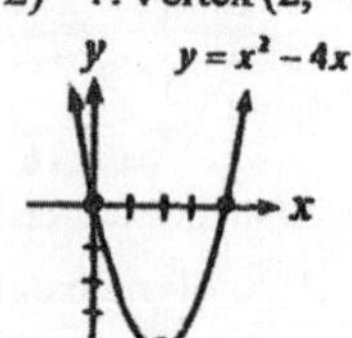

5. Graph $f(x) = x^2 - 2x + 1$.

Solution:

$x^2 - 2x + 1 = 0$, $(x - 1)^2 = 0$

$x = 1$

$\therefore$ One x – intercept = 1

$y = x^2 - 2x + 1$

$y - 0 = (x - 1)^2$

$\therefore$ Vertex (1, 0)

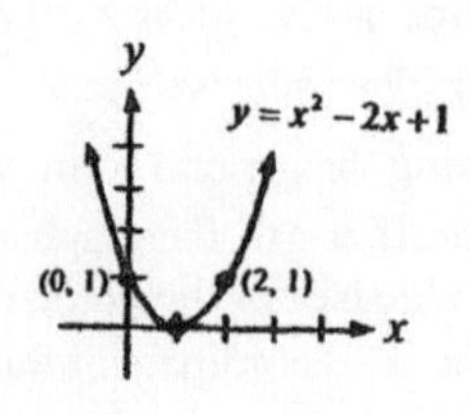

6. Graph $f(x) = (x - 3)^2 + 1$

Solution:

$(x - 3)^2 + 1 = 0$, $(x - 3)^2 = -1$

No real-number root

$\therefore$ No x – intercept

$y = (x - 3)^2 + 1$

$y - 1 = (x - 3)^2$

$\therefore$ Vertex (3, 1)

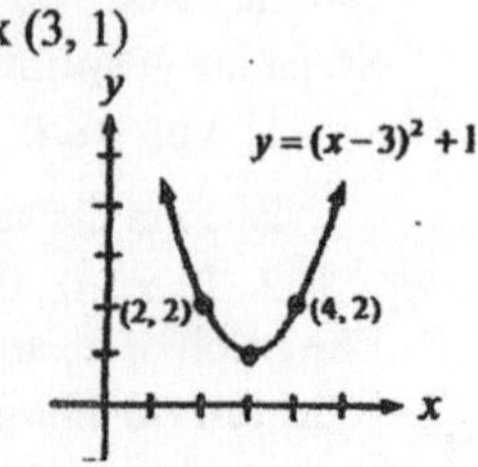

Graphing any Quadratic Function – Completing The Square

To graph any quadratic function $f(x) = ax^2 + bx + c\ (a \neq 0)$, we need to rewrite it to match the general form of parabola by the method of **completing the square.**

General Form of Parabola: $f(x) - k = a(x-h)^2$ or $y - k = a(x-h)^2$

1) Vertex is (h, k). **2)** If $a > 0$, open upward. **3)** If $a < 0$, open downward.

In science, we use the quadratic function to find the **extreme values** (maximum or minimum) of its application. There is a maximum value if it is open downward, a minimum value if it is open upward.

Rules of Completing the square for $y = ax^2 + bx + c$:

1. If $a = 1$, we add $\left(\frac{b}{2}\right)^2$ on right side to complete the square.

2. If $a \neq 1$, we rewrite the right side by $a(x^2 + \frac{b}{a}x + \frac{c}{a})$ so that the coefficient of x^2 is 1. Then follow Rule 1 to complete the square on the right side.

Examples

1. Graph $y = -x^2 - 2x + 1$.

Solution:

$$y = -x^2 - 2x + 1$$
$$y - 1 = -(x^2 + 2x)$$

Completing the square

$$y - 1 = -(x^2 + 2x + 1 - 1)$$
$$y - 1 = -(x^2 + 2x + 1) + 1$$
$$y - 1 = -(x+1)^2 + 1$$

$\therefore\ y - 2 = -(x+1)^2$ is the general form.

Vertex $(-1, 2)$, $a = -1 < 0$ open downward.

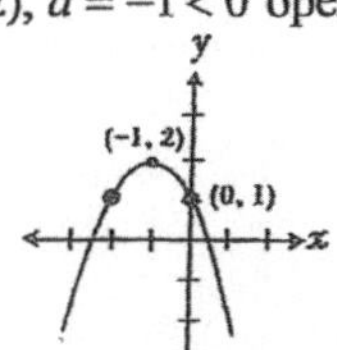

2. Graph $f(x) = 4x^2 + 12x + 10$.

Solution:

$$f(x) = 4x^2 + 12x + 10$$
$$f(x) - 10 = 4x^2 + 12x$$
$$f(x) - 10 = 4(x^2 + 3x)$$
$$f(x) - 10 = 4(x^2 + 3x + \tfrac{9}{4} - \tfrac{9}{4})$$
$$f(x) - 10 = 4(x^2 + 3x + \tfrac{9}{4}) - 9$$
$$f(x) - 1 = 4(x^2 + 3x + \tfrac{9}{4})$$

$\therefore\ f(x) - 1 = 4(x + \frac{3}{2})^2$ is the general form.

Vertex $(-\frac{3}{2}, 1)$, $a = 4 > 0$ open upward.

y
(-2, 2)
$(-\frac{3}{2}, 1)$
x

3. Find the extreme value of $f(x) = -2x^2 + 16x - 2$ and determine whether it is a maximum or a minimum.

Solution: $f(x) = -2x^2 + 16x - 2$

$$f(x) + 2 = -2(x^2 - 8x)$$
$$f(x) + 2 = -2(x^2 - 8x + 16 - 16)$$
$$f(x) + 2 = -2(x^2 - 8x + 16) + 32$$
$$f(x) - 30 = -2(x^2 - 8x + 16)$$

$f(x) - 30 = -2(x-4)^2$ is the general form.

Vertex $(4, 30)$, $a = -2 < 0$ open downward

There is a maximum value of $f(x)$ at $x = 4$

$\therefore\ f(x)_{max.} = 30$ at $x = 4$. .

Examples

1. **Find the dimensions of the rectangular flower bed that can be built by the 12-feet fence to have as much area as possible ?**

Solution:

Let y = the area

x = the length of one side

We have the equation:

$$y = x(6-x)$$

$$y = -x^2 + 6x$$

$$y = -(x^2 - 6x)$$

Completing the square:

$$y = -(x-3)^2 + 9$$

$$y - 9 = -(x-3)^2$$

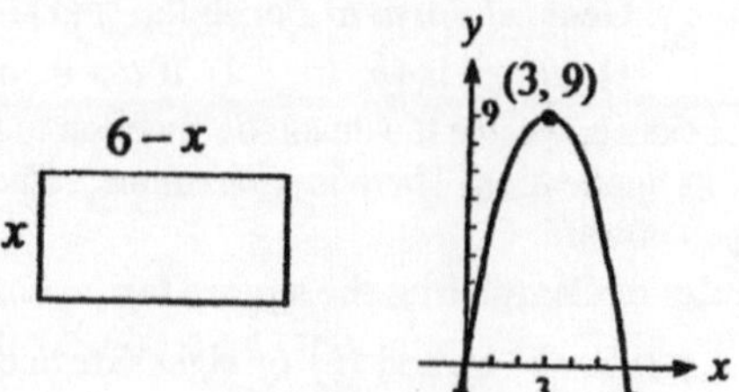

Ans: Dimensions = 3 × 3 (It is a square.)

Maximun area = 9 ft^2

2. **Prove that it must be a square to have the greatest area that can be made by a perimeter of length p.**

Proof:

Let y = the area

x = the length of one side

The equation is:

$$y = x(\tfrac{p}{2} - x)$$

$$y = -x^2 + \tfrac{p}{2}x$$

$$y = -(x^2 - \tfrac{p}{2}x)$$

$\frac{p}{2} - x$

x

Completing the square:

$$y = -(x - \tfrac{p}{4})^2 + \tfrac{p^2}{16}$$

$$y - \tfrac{p^2}{16} = -(x - \tfrac{p}{4})^2$$

It is a parabola (open downward) with vertex $(\frac{p}{4}, \frac{p^2}{16})$.

Max. area $y = \frac{p^2}{16}$ when $x = \frac{p}{4}$. It is a square.

2-20 Synthetic Division

Dividing a polynomial in x by a binomial of the form $x-c$ is very much like dividing real numbers.

When we divide a polynomial by $x-c$, we must arrange the terms of the polynomial in order of decreasing degree of the variable. Keep space (using 0) on the missing terms in degrees of x.

The common process of dividing a polynomial by $x-c$ is shown in Example 1. In this section we use the easier method called **Synthetic Division** to get the same answer.

Synthetic Division (Horner's Algorithm):

Dividing a polynomial by $x-c$, we proceed only the coefficients of the terms of the polynomial.

1) If the divisor is $x+c$, rewrite it as $x-(-c)$.

2) If the divisor is $ax-b$, rewrite it as $a(x-\frac{b}{a})$, use the divisor $x-\frac{b}{a}$ to proceed the synthetic division. Then, multiply the result by $\frac{1}{a}$.

Examples

1. Divide $\dfrac{2x^3-8x^2-4x-6}{x-3}$.

Solution: (by common division)

$$
\begin{array}{r}
2x^2-2x\ -10 \\
x-3\,\overline{)\,2x^3-8x^2-4x-6} \\
-)\ 2x^3-6x^2 \\
\hline
-2x^2-4x-6 \\
-)\ -2x^2+6x \\
\hline
-10x-6 \\
-)\ -10x+30 \\
\hline
-36 \\
\text{(Remainder)}
\end{array}
$$

Ans: $\dfrac{2x^3-8x^2-4x-6}{x-3}$

$=2x^2-2x-10-\dfrac{36}{x-3}$.

2. Divide $\dfrac{2x^3-8x^2-4x-6}{x-3}$.

Solution: (by synthetic division)

$$
\begin{array}{r|rrrr}
3 & 2 & -8 & -4 & -6 \\
+) & \downarrow & 6 & -6 & -30 \\
\hline
 & 2 & -2 & -10, & -36 \\
\end{array}
$$

-36 ← Remainder

Ans: $\dfrac{2x^3-8x^2-4x-6}{x-3}=2x^2-2x-10-\dfrac{36}{x-3}$.

3. Divide $\dfrac{1+x+12x^3}{1+2x}$.

Solution:

$$
\begin{array}{r}
6x^2-3x+2 \\
2x+1\,\overline{)\,12x^3+0\ +\ x+1} \\
-)\ 12x^3+6x^2 \\
\hline
-6x^2+\ x \\
-)\ -6x^2-3x \\
\hline
4x+1 \\
-)\ 4x+2 \\
\hline
-1
\end{array}
$$

-1 ← Remainder

Ans: $\dfrac{1+x+12}{1+2x}=6x^2-3x+2-\dfrac{1}{1+2x}$.

2-21 Remainder and Factor Theorems

We already know : $\frac{7}{2}=3+\frac{1}{2}$ where "1" is the remainder.

$\therefore\ 7=2\ (3+\frac{1}{2})=2\cdot 3+1$

The above example shows the following fact:

Dividend = divisor × quotient + remainder.

In general, when a polynomial $p(x)$ is divided by $x-c$, $P(x)$ is the dividend, $x-c$ is the divisor, $Q(x)$ is the quotient, R is the remainder, we have:

$$P(x)=Q(x)(x-c)+R$$

Let $x=c$

$$P(c)=Q(c)\cdot(c-c)+R$$

$\therefore\ P(c)=R$ --------------- Called **Remainder Theorem.**

Remainder Theorem: The remainder on dividing $P(x)$ by $x-c$ is $P(c)$.

Factor Theorem: If and only if $P(c)=0$ (i.e. no remainder), then $x-c$ is a factor of $p(x)$ and c is a root of the equation $P(x)=0$.

This is an easier way to find the value of polynomial $P(x)$ at $x=c$. We apply the **remainder theorem** and use synthetic division to find the remainder on dividing $P(x)$ by $x-c$. Then , the remainder is $P(c)$.
We apply the **factor theorem** to identify and find possible rational roots of a polynomial equation $P(x)=0$. If $P(c)=0$, then c is a root of $P(x)=0$.

Examples

1. Find the value of $P(x)=2x^3-8x^2-4x-6$ when $x=3$.
   Solution:

Method 1): Direct substitution.

$$P(x)=2x^3-8x^2-4x-6$$
$$P(3)=2(3)^3-8(3)^2-4(3)-6$$
$$=54-72-12-6$$
$$=-36.$$

Ans: $P(3)=-36$.

Method 2): Remainder theorem & Synthetic Division.

```
 3| 2 - 8 - 4   - 6
+)      6 - 6  - 30
-------------------
    2 - 2 - 10, - 36
              (Remainder)
```

Ans: $P(3)=-36$.

Examples

2. Divide $\frac{2x^3-8x^2-4x-6}{x-3}$, find remainder.

Solution:

Method 1: Common division.

$$\begin{array}{r} 2x^2-2x\ -10 \\ x-3\overline{)2x^3-8x^2-4x-6} \\ -)\ 2x^3-6x^2 \qquad\qquad \\ \hline -2x^2-4x-6 \\ -)\ -2x^2+6x \qquad \\ \hline -10x-6 \\ -)\ -10x+30 \\ \hline -36 \end{array}$$

(Remainder)

Method 2: Synthetic division:

$$\begin{array}{r} 3\ \ 2-8-4-6 \\ +)\quad 6-6-30 \\ \hline 2-2-10,-36 \end{array}$$

(Remainder)

Method 3: Remainder theorem.

$P(x)=2x^3-8x^2-4x-6$

$P(3)=2(3)^3-8(3)^2-4(3)-6$

$=54-72-12-6$

$=-36$ (Remainder)

3. Is $x-3$ a factor of

$P(x)=2x^3-8x^2-4x-6$?

Solution: $P(x)=2x^3-8x^2-4x-6$

$P(3)=2(3)^3-8(3)^2-4(3)-6$

$=54-72-12-6=-30\neq 0.$

Ans: $x-3$ is not a factor of $P(x)$.

(3 is not a root of $P(x)=0$.)

4. Is $x+1$ a factor of

$P(x)=x^3+6x^2+11x+6$?

Solution:

$P(x)=x^3+6x^2+11x+6$

$P(-1)=(-1)^3+6(-1)^2+11(-1)+6$

$=-1+6-11+6$

$=0$

Ans: $x+1$ is a factor of $P(x)$.

(-1 is a root of $P(x)=0$.)

5. Is $x-\sqrt{3}$ a factor of

$P(x)=x^3-3x^2-3x+9$?

Solution:

$P(x)=x^3-3x^2-3x+9$

$P(\sqrt{3})=(\sqrt{3})^3-3(\sqrt{3})^2-3(\sqrt{3})+9$

$=3\sqrt{3}-9-3\sqrt{3}+9$

$=0$

Ans: $x-\sqrt{3}$ is a factor of $P(x)$.

($\sqrt{3}$ is a root of $P(x)=0$.)

6. Solve $x^3-4x^2+8x-5=0$ if one root is 1.

Solution: By synthetic division:

$\frac{x^3-4x^2+8x-5}{x-1}=x^2-3x+5$

$x^3-4x^2+8x-5=0$

$(x-1)(x^2-3x+5)=0$

$x-1=0$, $x^2-3x+5=0$

$x=1$, $x=\frac{3\pm\sqrt{9-20}}{2}=\frac{3\pm\sqrt{11}i}{2}$

Ans: $x=1$ and $\frac{3\pm\sqrt{11}i}{2}$

Notes

2-22 Solving Polynomial Equations

Many polynomial equations can be solved by **factoring** and the **zero-product theory**.

Zero-Product Theory: For all real numbers a and b:

$a \cdot b = 0$ **if and only if** $a = 0$ or $b = 0$.

The zero-product theory tells us when an equation is written as a product of factors, we can find its solutions by setting each of the factors equal to 0 and solve for its roots.

Double Roots: If an equation has two identical factors, the equation is said to have a **double root** (see example 4).

The following formulas of special patterns are useful in the process to factor polynomials:

Formulas:

1. $a^2 + 2ab + b^2 = (a+b)^2$	**4.** $a^3 + b^3 = (a+b)(a^2 - ab + b^2)$
2. $a^2 - 2ab + b^2 = (a-b)^2$	**5.** $a^3 - b^3 = (a-b)(a^2 + ab + b^2)$
3. $a^2 - b^2 = (a+b)(a-b)$	

An $x-$intercept of a function $y = f(x)$ is a point $(x, 0)$ at which the graph crosses the $x-axis$ ($y = 0$). To find the $x-$intercepts of a function, we find the zeros of the function $y = f(x)$ by letting $f(x) = 0$ (see example 1).

Examples

1. Find the zeros of $f(x) = x^2 + x - 2$.

Solution: Let $f(x) = 0$, $x^2 + x - 2 = 0$

$(x+2)(x-1) = 0$

$x + 2 = 0$ | $x - 1 = 0$

$x = -2$ | $x = 1$

Ans: $x = -2$ and 1.

2. Solve $(2x-4)(x+5) = 0$.

Solution: $(2x-4)(x+5) = 0$

$2x - 4 = 0$ | $x + 5 = 0$

$2x = 4$ | $x = -5$

$x = 2$

Ans: $x = 2$ and -5.

3. Solve $2x^2 + 3x = 9$.

Solution: $2x^2 + 3x - 9 = 0$

$(2x-3)(x+3) = 0$

$2x - 3 = 0$ | $x + 3 = 0$

$2x = 3$ | $x = -3$

$x = 1\frac{1}{2}$ | Ans: $x = 1\frac{1}{2}$ and -3.

4. Solve $x^2 - 14x + 49 = 0$.

Solution: $x^2 - 14x + 49 = 0$

$(x-7)(x-7) = 0$

$x - 7 = 0$ | $x - 7 = 0$

$x = 7$ | $x = 7$

Ans: $x = 7$ (**a double root**).

5. Solve $2a^3 - 6a^2 - 20a = 0$.

Solution: $2a^3 - 6a^2 - 20a = 0$

$2a(a^2 - 3a - 10) = 0$

$2a(a-5)(a+2) = 0$

$2a = 0$ | $a - 5 = 0$ | $a + 2 = 0$

$a = 0$ | $a = 5$ | $a = -2$

Ans: $a = 0$, 5, and -2.

6. Solve $x^4 - 8x = 0$.

Solution: $x^4 - 8x = 0$, $x(x^3 - 8) = 0$

$x(x-2)(x^2 + 2x + 4) = 0$

$x = 0$, $x = 2$, $x^2 + 2x + 4 = 0$

$x = \frac{-b \pm \sqrt{b^2 - 4ac}}{2a} = \frac{-2 \pm \sqrt{2^2 - 4 \cdot 1 \cdot 4}}{2(1)} = -1 \pm \sqrt{3}\, i$

Ans: $x = 0, 2$, and $-1 \pm \sqrt{3}\, i$.

Solving Polynomial Equations with Degree *n*

Polynomial Function: It is an algebraic function in the following form: $y = p(x)$

$$p(x) = a_n x^n + a_{n-1} x^{n-1} + \cdots\cdots + a_1 x + a_0 .$$

(Where a_n, a_{n-1}, ·····, a_1, a_0 are real numbers and n is a nonnegative integer.)

Finding the zeros of a polynomial function is to find any value of x that makes $p(x) = 0$. Therefore, finding the zeros of a polynomial function is to solve the following polynomial equation. $a_n x^n + a_{n-1} x^{n-1} + \cdots\cdots + a_1 x + a_0 = 0$

The **real zeros** of a polynomial function $y = p(x)$ (the real roots of a polynomial equation $p(x) = 0$) are also the x – intercepts of the graph of $y = p(x)$ on the coordinate plane.

Rules for finding the zeros of a polynomial function with degree $n\,(n \ge 3)$:

1) **Fundamental Theorem of Polynomial**
   Every polynomial function of positive degree n has exactly n zeros.
2) **Conjugate Pairs Theorem:** a and b are real numbers, $b \ne 0$.
   If $a + bi$ is a zero of polynomial function, then $a - bi$ is also a zero.
3) **Newton's Rule (Rational Zeros Theorem):** It is used to find the rational zeros of a polynomial function $p(x) = a_n x^n + a_{n-1} x^{n-1} + \cdots\cdots + a_1 x + a_0$.
   a) If $a_n = 1$ and r is an integer zero, then r must be a factor of a_0.
   b) If $a_n \ne 1$ and $\frac{b}{a}$ is a zero (a, b are integers, $x = \frac{b}{a}$ and $ax - b = 0$), then b must be a factor of a_0, a must be a factor of a_n.

To solve a polynomial equation in degree n, we find certain roots by Newton's Rule and factor it as a product of its factors in degree 1 and quadratic forms.

If a polynomial $p(x)$ has the factor $(x - c)^n$, we say that c is a zero of **multiplicity** n of $p(x)$, or a root of multiplicity n of $p(x) = 0$. It has an x – intercept at $x = c$.

Examples

1. Solve $x(x+3)^2(x-1) = 0$.

Solution:

$x = 0$ | $(x+3)^2 = 0$, $x = -3$ | $x - 1 = 0$, $x = 1$

Ans: $x = 0, 1$, and –3 (multiplicity 2, or a double root)

2. If two roots of a quadratic equations are –3 and 1, find such an equation.

Solution: the equation is :

$$(x+3)(x-1) = 0$$

$\therefore x^2 + 2x - 3 = 0$ is the equation.

3. Solve $x^3 + 6x^2 + 11x + 6 = 0$.

Solution: By Newton's Rule: (factors of 6)

Possible roots: $\pm 1, \pm 2, \pm 3, \pm 6$

By factor theorem:

$P(-1) = (-1)^3 + 6(-1)^2 + 11(-1) + 6 = 0$

$\therefore x + 1$ is a factor.

By synthetic division:

$$\frac{x^3 + 6x^2 + 11x + 6}{x+1} = x^2 + 5x + 6$$

$$x^3 + 6x^2 + 11x + 6 = 0$$
$$(x+1)(x^2 + 5x + 6) = 0$$
$$(x+1)(x+2)(x+3) = 0$$

Ans: $x = -1, -2$, and -3

Finding Complex Zeros of Polynomial Functions with Degree n

We have learned how to find the real zeros of a polynomial function. Here, we will learn how to find the **complex zeros** in the form $a+b\,i$ of a polynomial function with degree n.
In the complex number $a+b\,i$, it is a real number if $b=0$.
Therefore, Finding the complex zeros of a polynomial function requires finding all zeros of the form $a+b\,i$ including the real zeros ($b=0$).
We have also learned to solve a polynomial equation in degree n by factoring it as a product of its factors in degree 1 and quadratic forms.
In the real number system, a quadratic equation is said to be **no solution** (no zeros) if the equation has no real number solution.
In the complex number system, every quadratic equation has a solution (zeros), either real or complex.
In the complex number system, every polynomial function $p(x)$ of degree $n \geq 1$ can be factored into n linear factors of the form:

$p(x)=a_n(x-r_1)(x-r_2)\cdots\cdots(x-r_n)$: $r_1, r_2, \cdots\cdots, r_n$ are the zeros, either real or complex.

Fundamental Theorem of Algebra: In the complex number system, a polynomial function $p(x)$ of degree $n \geq 1$ has at least one zero.

In the complex number system, every polynomial function of degree $n \geq 1$ has exactly n zeros, either real or complex.

Conjugate Pairs Theorem: In the complex number system, if $a+b\,i\,(b \neq 0)$ is a complex zero of a polynomial function $p(x)$ whose coefficients are real numbers, then the conjugate $a-b\,i$ is also a complex zero of $p(x)$.

Examples

4. Finding all zeros of the function $f(x)=x^4-5x^3+3x^2+19x-30$.

Solution:

Using the Newton's Rule (Rational Zeros Theorem), we can find two rational zeros of $f(x)$, 3 and –2.

Using synthetic division, we have:

$$f(x)=x^4-5x^3+3x^2+19x-30=(x-3)(x+2)(x^2-4x+5)$$

To find zeros(roots), let $f(x)=0$.

$$(x-3)(x+2)(x^2-4x+5)=0$$

We have the following four zeros(roots):

$$x=3,\ -2$$

$$x=\frac{-(-4)\pm\sqrt{(-4)^2-4\cdot1\cdot5}}{2(1)}=\frac{4\pm\sqrt{-4}}{2}=\frac{4\pm2i}{2}=2\pm i$$

Ans: $x=3$, –2, and $2\pm i$.

The Relationships between Roots and Coefficients

The standard form of the quadratic equation $ax^2+bx+c=0$ ($a\neq 0$) can be written as:

$$x^2+\frac{b}{a}x+\frac{c}{a}=0 \ (a\neq 0)$$

If r_1 and r_2 are the roots (solutions) of the quadratic equation, we can write the equation in the form $(x-r_1)(x-r_2)=0$. We have the equation $x^2-(r_1+r_2)x+r_1r_2=0$.
Therefore, the relationships between the roots and coefficients of a quadratic equation are:

1) $r_1+r_2=-\frac{b}{a}$; **2)** $r_1\cdot r_2=\frac{c}{a}$

We can use the above two relationships to find a quadratic equation having the given roots.
If $r_1, r_2, r_3, \ldots, r_n$ are roots of a polynomial equation $p(x)=a_x x^n+a_{n-1}x^{n-1}+\cdots+a_1x+a_0=0$, then **the relations between roots and coefficients of a polynomial equation are:**

1) $r_1+r_2+r_3+\cdots+r_n=-\frac{a_{n-1}}{a_n}$ **2)** $r_1\cdot r_2\cdot r_3\cdots r_n=(-1)^k\frac{a_0}{a_n}$, *k* is the number of roots.

3) $r_1r_2r_3\cdots r_k+r_1r_2r_4\cdots r_k+\cdots\cdots=(-1)^k\frac{a_{n-k}}{a_n}$, *k* is the number of roots.

Examples

1. Find the quadratic equation whose roots are 1 and –3.

Solution: $r_1+r_2=1+(-3)=-2=-\frac{b}{a}$; $\frac{b}{a}=2$

$r_1\cdot r_2=1\cdot(-3)=-3=\frac{c}{a}$; $\frac{c}{a}=-3$ $\therefore x^2+2x-3=0$ is the equation. Ans.

2. Find the quadratic equation whose roots are $\frac{-1+\sqrt{13}}{3}$ and $\frac{-1-\sqrt{13}}{3}$.

Solution: $r_1+r_2=\frac{-1+\sqrt{13}}{3}+\frac{-1-\sqrt{13}}{3}=-\frac{2}{3}=-\frac{b}{a}$; $\frac{b}{a}=\frac{2}{3}$

$r_1\cdot r_2=\frac{-1+\sqrt{13}}{3}\cdot\frac{-1-\sqrt{13}}{3}=\frac{(-1)^2-(\sqrt{13})^2}{9}=\frac{-12}{9}=-\frac{4}{3}=\frac{c}{a}$; $\frac{c}{a}=-\frac{4}{3}$

$x^2+\frac{2}{3}x-\frac{4}{3}=0$ $\therefore 3x^2+2x-4=0$ is the equation. Ans.

3. Given the equation $2x^4-5x^3+5x-2=0$, find: a) the sum and the product of the roots.
b) the sum of the product of the roots, take 3 at a time. (Hint: $x=1,-1,2,\frac{1}{2}$)

Solution:

a) $r_1+r_2+r_3+r_4=-\frac{a_{n-1}}{a_n}=-\frac{-5}{2}=\frac{5}{2}$; $r_1r_2r_3r_4=(-1)^4\frac{a_0}{a_n}=(-1)^4\frac{-2}{2}=-1$

b) $r_1r_2r_3+r_1r_2r_4+r_1r_3r_4+r_2r_3r_4=(-1)^3\frac{a_{n-3}}{a_n}=-\frac{5}{2}$. (Hint: $a_{n-2}=0$)

Notes

Notes

2-23 Graphing Polynomial Functions

The graph of a polynomial function with degree n has the following characteristics:

1. The graph of every polynomial function $f(x) = a_n x^n + a_{n-1}x^{n-1} + \cdots\cdots + a_1 x + a_0$ is a smooth and continuous curve. Its turning points are rounded. It has no holes, gaps, or sharp turns.
2. The points at which a graph changes direction are called **turning points**. The turning points are the **local maxima or minima (relative maxima or minima)** of the function. The maximum number of turning points of the graph is $n-1$.
3. The maximum number of real zeros ($x-$intercepts) is n.
4. If r is a repeated zero of **even** multiplicity (the sign of the function does not change from one side to the other side of r), the graph **touches** the $x-$intercept at r.
5. If r is a repeated zero of **odd** multiplicity (the sign of the function changes from one side to the other side of r), the graph **crosses** the $x-$intercept at r.
6. **End Behavior**: If n is even and the **leading coefficient** is positive ($a_n > 0$), the graph rises to the left and right. If n is even and the leading coefficient is negative ($a_n < 0$), the graph falls to the left and right.
7. **End Behavior**: If n is odd and the **leading coefficient** is positive ($a_n > 0$), the graph falls to the left and rises to the right. If n is odd and the leading coefficient is negative ($a_n < 0$), the graph rises to the left and falls to the right.

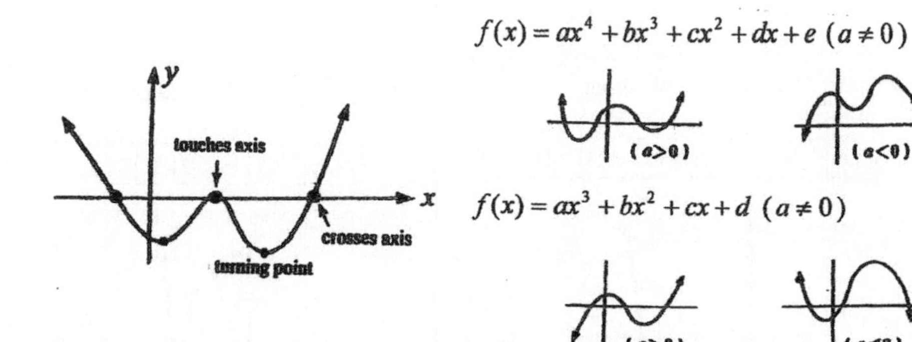

Steps for graphing a polynomial function $y = f(x)$

1. Solve the equation $f(x) = 0$ for real zeros (x – intercepts).
2. Decide whether the graph crosses or touches the x – axis at each x – intercept.
3. Decide the maximum number of turning points.
4. Apply the Leading Coefficient Test to decide whether the graph rises or falls. (Optional)
5. Test each interval between two x – intercepts to find each interval on which the graph is either above the x – axis or below the x – axis.
6. Plot a few additional points and connect the points with a smooth and continuous curve.

Examples

1. Graph: $f(x) = x^3 - 8x^2 + 20x - 16$.

Solution:

Factor the function and solve for real zeros:

$$f(x) = (x-2)^2(x-4) = 0$$

The x – intercepts are 2 (multiplicity 2) and 4
The graph touches the x – axis at 2 and crosses the x – axis at 4.
It has at most two turning points.
Test each interval: The two x – intercepts divide the graph into 3 intervals.

$x < 2$, $f(x)$: $+ \cdot - = -$ below x – axis
$2 < x < 4$, $f(x)$: $+ \cdot - = -$ below x – axis
$x > 4$, $f(x)$: $+ \cdot + = +$ above x – axis

Compute a few additional points:

x	1	1.5	2.5	3	3.5	4.5
$f(x)$	-3	-0.63	-0.38	-1	-1.13	3.13

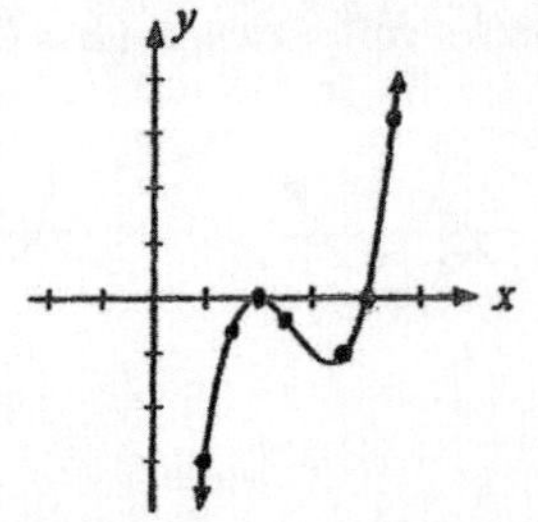

Hint: In calculus, we can find the location of the turning points, a local maximum at $x = 2$ and a local minimum at $x = 3.3$.

2. Graph: $f(x) = x^3(x-4)^2$.

Solution:

$$f(x) = x^3(x-4)^2 = 0$$

The x – intercepts are 0 (multiplicity 3) and 4 (multiplicity 2).
The graph crosses the origin and touches the x – axis at 4
It has at most four turning points.
Test each interval: (3 intervals)

$x < 0$, $f(x)$: $- \cdot + = -$ below x – ax
$0 < x < 4$, $f(x)$: $+ \cdot + = +$ above x – ax
$x > 4$, $f(x)$: $+ \cdot + = +$ above x – axi

Compute a few additional points:

x	-1	1	2	3	5
$f(x)$	-25	9	32	27	125

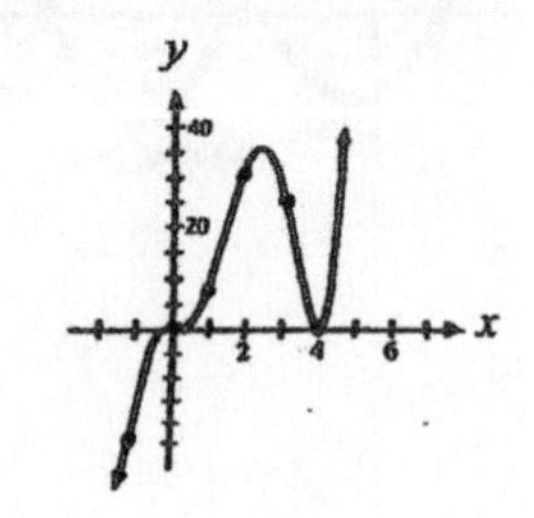

2-24 Graphing Techniques – Transformations

We have already learned how to graph the most commonly used polynomial functions in algebra. Understanding the basic shapes of the following graphs will help us to analyze and graph more complicated polynomial functions.

1. $f(x)=0$. It is a zero function. The graph is the x-axis.
2. $f(x)=c$. It is a constant function.
   The graph is a horizontal line with $y-$intercept c.
3. $f(x)=x$. It is an identity function.
   The graph is a straight line passing through the origin.
4. $f(x)=mx+b$. It is a linear function.
   The graph is a straight line with slope m and $y-$intercept b.
5. $f(x)=x^2$. It is a quadratic function. It is a parabola.
   The graph is symmetric with respect to the $y-$axis.
6. $f(x)=x^3$. It is a cubic function.
   The graph is symmetric with respect to the origin.

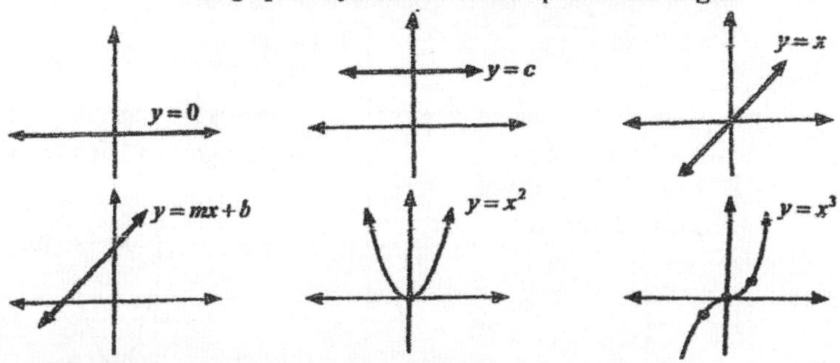

Graphing Power Functions of the form $f(x)=x^n$

As n increases, the graph tends to be flat near the origin and closer to the $x-$axis in the interval $[-1, 1]$. The graph increases rapidly when $x<-1$ or $x>1$.

A. $f(x)=x^n$ if n is even. **B.** $f(x)=x^n$ if n is odd.

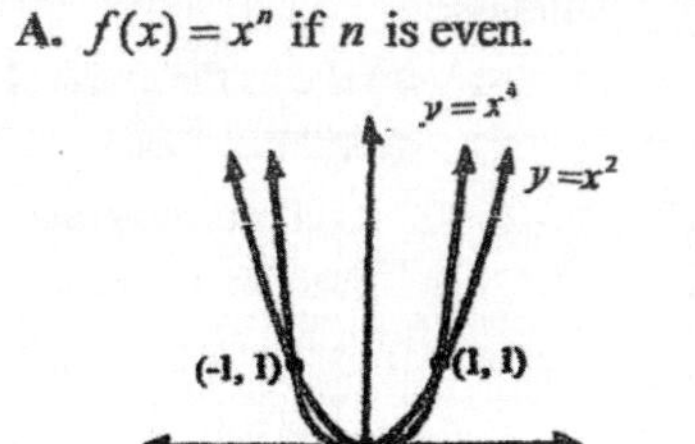

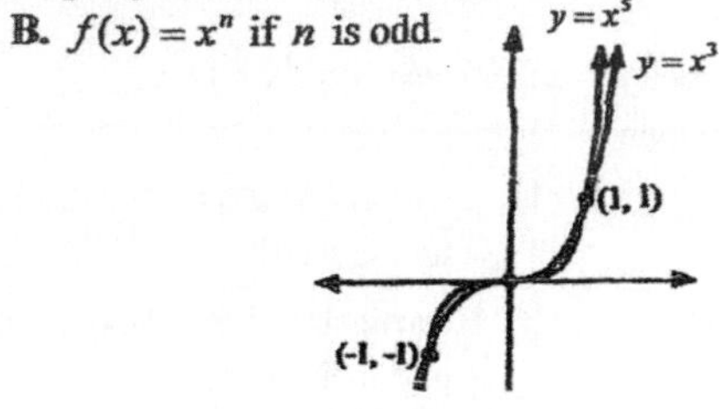

Transformations (Shifts, Reflects, Stretches, and Compressions)

Knowing the graphs of common functions $y = f(x)$ can help us to graph a wide variety of polynomial functions by transformations (shifts, reflects, stretches, compressions). Shift and reflection translate only the position of the graph. Stretch and compression distort the basic shape of the original graph.

The transformations of the original graph of $y = f(x)$ have the following types:

1. $y = f(x)+c$. It moves (vertical shift) the original graph c units upward.
2. $y = f(x)-c$. It moves (vertical shift) the original graph c units downward.
3. $y = f(x+c)$. It moves (horizontal shift) the original graph c units to the left.
4. $y = f(x-c)$. It moves (horizontal shift) the original graph c units to the right.
5. $y = -f(x)$. It reflects the original graph over the $x-axis$.
6. $y = f(-x)$. It reflects the original graph over the $y-axis$.
7. $y = c\ f(x)$. It stretches vertically if $c > 1$ (narrower).
   It compresses vertically if $0 < c < 1$ (wider).
8. $y = f(cx)$. It compresses horizontally if $c > 1$ (narrower).
   It stretches horizontally if $0 < c < 1$ (wider).

Examples (Show how the graphs are transformed with the above types.)

1. Graph: $y = x+3$ and $y = x-3$.

   Solution:

   $y = x+3$ It moves the graph of $y = x$, 3 units upward.

   $y = x-3$ It moves the graph of $y = x$, 3 units downward.

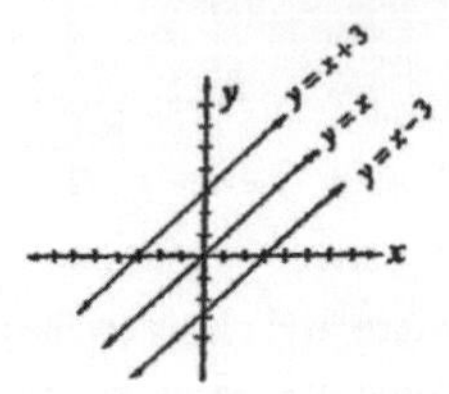

2. Graph: $y = x^2+1$ and $y = (x+1)^2$.

   Solution:

   $y = x^2+1$ It moves the graph of $y = x^2$, 1 unit upward.

   $y = (x+1)^2$ It moves the graph of $y = x^2$, 1 unit to the left.

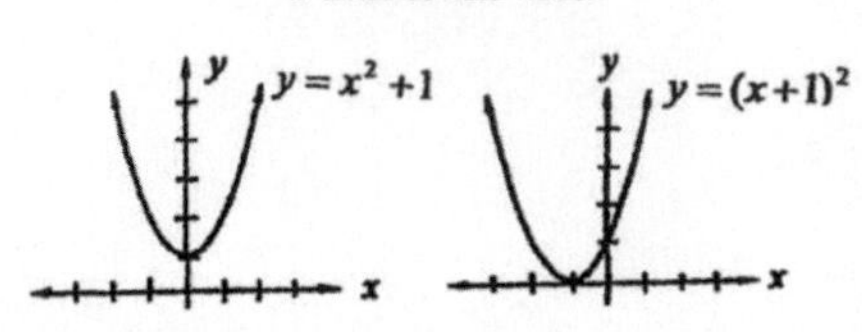

3. Graph: $y = (x-2)^2-3$.

   Solution:

   It moves the graph of $y = x^2$, 2 units to the right and 3 units downward.

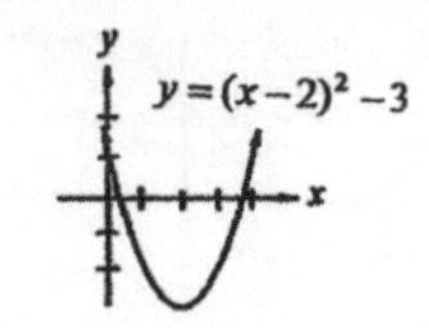

4. Graph: $y = -x^3$ and $y = (-x)^3$

   Solution:

   $y = -x^3$ It reflects the graph of $y = x^3$ over the $x-$axis.

   $y = (-x)^3$ It reflects the graph of $y = x^3$ over the $y-$axis.

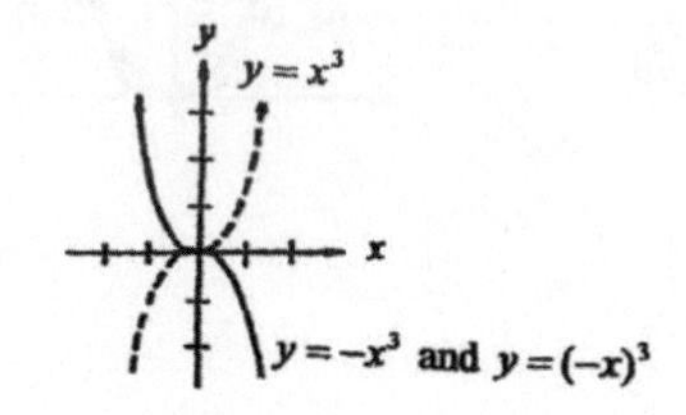

Examples

5. Graph: $y = 2x^2$ and $y = \frac{1}{2}x^2$.

Solution:

$y = 2x^2$. It stretches the graph of $y = x^2$ vertically from (2, 4) to(2, 8) and give a narrower parabola.

$y = \frac{1}{2}x^2$. It compresses the graph of $y = x^2$ vertically from (2, 4) to (2, 2) and gives a wider parabola.

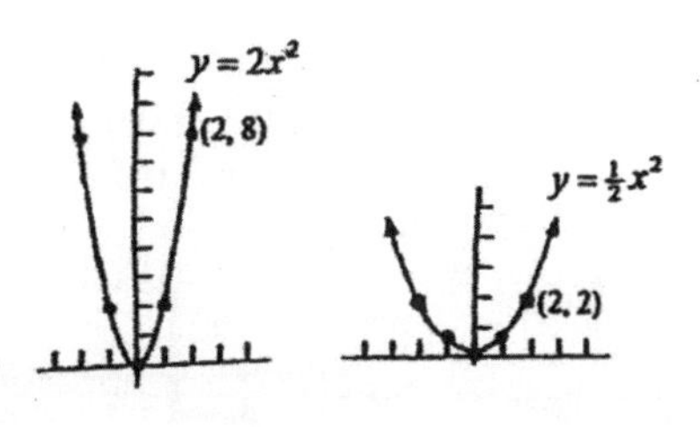

6. Graph: $y = (2x)^2$ and $y = (\frac{1}{2}x)^2$.

Solution:

$y = (2x)^2$. It compresses the graph of $y = x^2$ horizontally from (2, 4) to (1, 4) and give a narrower parabola.

$y = (\frac{1}{2}x)^2$. It stretches the graph of $y = x^2$ horizontally from (2, 4) to (4, 4) and give a wider parabola.

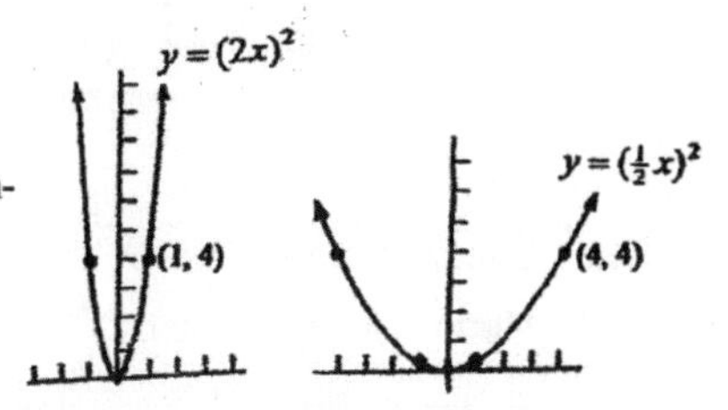

7. The graph of a function $y = f(x)$ is given. Graph each of the following function using the techniques of transformations.

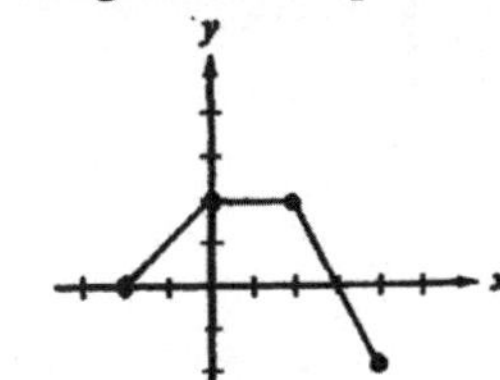

1. $y = -f(x)$ **2.** $y = f(-x)$ **3.** $y = f(x) + 2$

4. $y = f(x+2)$ **5.** $y = f(x+2) - 3$ **6.** $y = 2f(x)$

7. $y = f(2x)$ **8.** $y = f(\frac{1}{2}x)$ **9.** $y = f(|x|)$

1. $y = -f(x)$
Reflect about the $x-axis$.

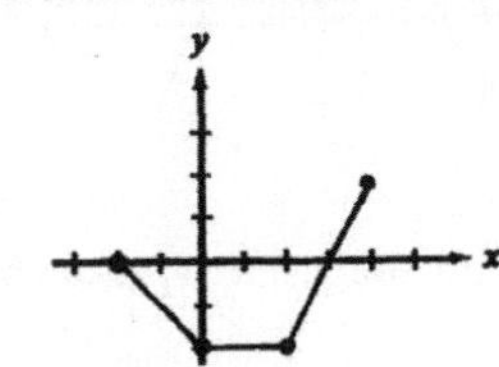

2. $y = f(-x)$
Reflect about the $y-axis$.

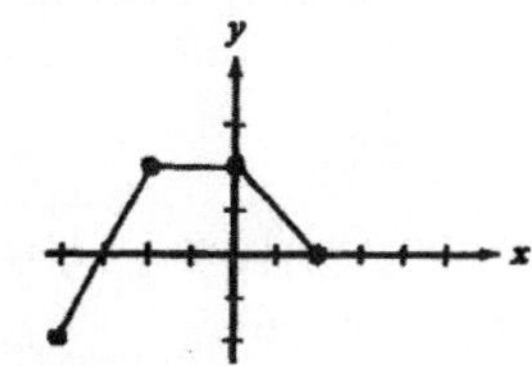

3. $y = f(x) + 2$
Vertical shift up 2 units.

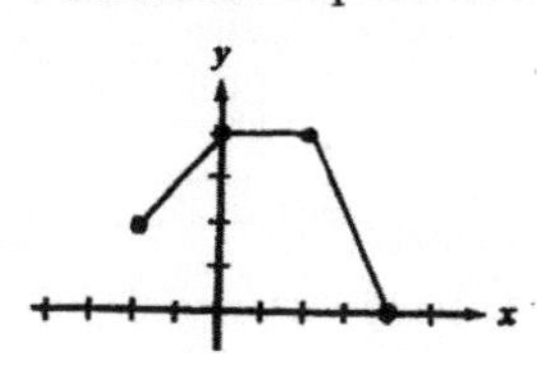

4. $y = f(x+2)$
Horizontal shift left 2 units.

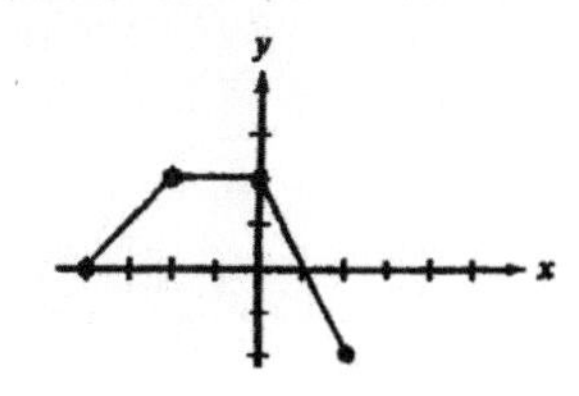

5. $y = f(x+2) - 3$

Horizontal shift left 2 units.

Vertical shift down 3 units.

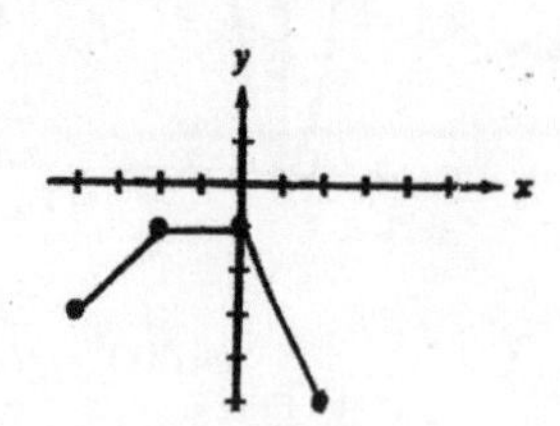

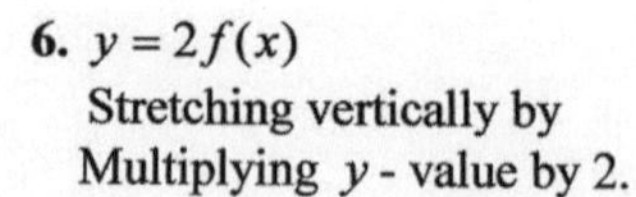

6. $y = 2f(x)$

Stretching vertically by

Multiplying y - value by 2.

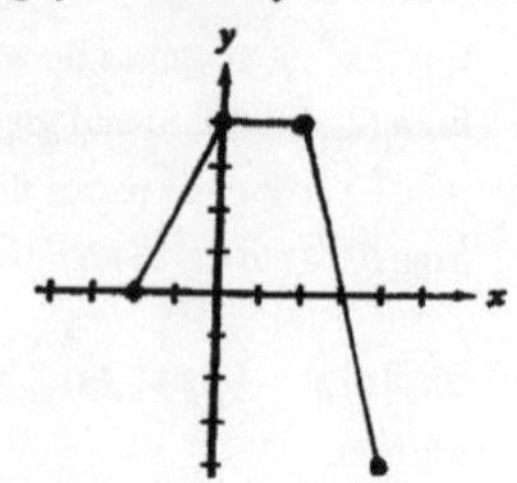

7. $y = f(2x)$

Compressing vertically.

Mutiplying x - value by $\frac{1}{2}$.

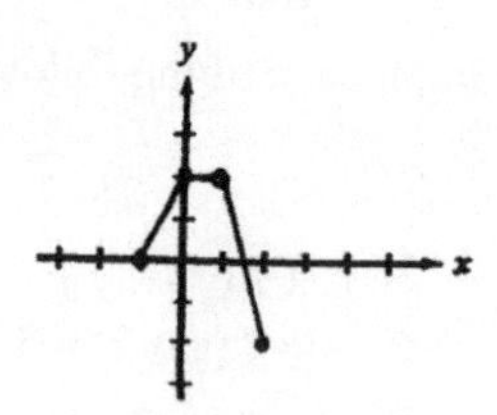

8. $y = f(\frac{1}{2}x)$

Stretching horizontally.

Multiplying x - value by 2.

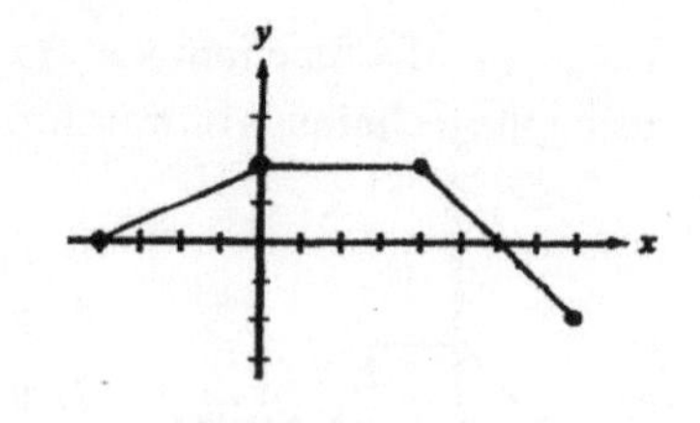

9. $y = f(|x|)$

Replace the graph on the left of the $y-axis$ by reflecting the part that is on the right of the $x-axis$ about the $y-axis$.

Hint: Graphing $y = |f(x)|$, reflect the part that is below the $x-axis$ through the $x-axis$.

2-25 Solving Rational Equations

If a rational equation consists of one fraction equal to another (it is a proportion), then the **cross products** are equal.
To solve a rational equation consisting of two or more fractions on one side, we multiply both sides of the equation by their **least common denominator (LCD)**.

Example 1:

Solve $\frac{2}{x}=\frac{4}{5}$.

Solution:

$$2\cdot 5=4\cdot x$$
$$10=4x$$
$$\therefore x=\tfrac{10}{4}=2\tfrac{1}{2}.$$

Example 2:

Solve $\frac{x}{3}+\frac{x}{4}=7$.

Solution: LCD = 12

$$12\left(\frac{x}{3}+\frac{x}{4}\right)=12(7)$$
$$4x+3x=84$$
$$7x=84$$
$$\therefore x=12.$$

Example 3:

Solve $\frac{3}{x}-\frac{1}{2x}=\frac{1}{3}$.

Solution: LCD = $6x$

$$6x\left(\frac{3}{x}-\frac{1}{2x}\right)=6x\left(\frac{1}{3}\right)$$
$$18-3=2x$$
$$15=2x$$
$$\therefore x=\tfrac{15}{2}=7\tfrac{1}{2}.$$

To solve a rational equation, we always test each root in the original equation. A root is **not permissible** if it makes the denominator of the original equation equal to 0.

Example 4: Solve $\frac{18}{x^2-9}=\frac{x}{x-3}$.

Solution:
$$18(x-3)=x(x^2-9)$$
$$18\cancel{(x-3)}=x(x+3)\cancel{(x-3)}$$
$$18=x(x+3)$$
$$18=x^2+3x$$
$$0=x^2+3x-18$$
$$x^2+3x-18=0$$
$$(x+6)(x-3)=0$$
$x=-6$ | $x=3$ **(not permissible)** $\quad\therefore x=-6$.

If the final statement is a "**true**" statement, the equation has roots for all real numbers.
If the final statement is a "**false**" statement, the equation has no root.

Example 5: Solve $\frac{4}{2x-2}=\frac{2}{x-1}$.

Solution:
$$4(x-1)=2(2x-2)$$
$$\cancel{4x}-\cancel{4}=\cancel{4x}-\cancel{4}$$
$0=0$ **(true)**

Ans: The equation has roots for all real numbers except 1.

6: Solve $\frac{4}{2x+1}=\frac{2}{x}$.

Solution:
$$4x=2(2x+1)$$
$$\cancel{4x}=\cancel{4x}+2$$
$0=2$ **(false)**

Ans: No solution.

Notes

2-26 Graphing Rational Functions

A **rational expression (algebraic fraction)** is a quotient (ratio) of two polynomials.
A **rational (fractional) function** is a function with the quotient (ratio) of two polynomials.
$y = \frac{1}{x}$ is the most popular and simplest rational function. Its graph is a **hyperbola**.

The graph of a rational function has a **vertical asymptote** at $x = a$ if the values for y increase (or decrease) without bound as x approaches a, either from the right or from the left.
The graph of a rational function has a **horizontal asymptote** at $y = b$ if the values for y approach b as x increases (or decreases) without bound.
The graph of a rational function may have several vertical asymptotes, but at most one horizontal asymptote.

Rules for graphing a rational function:

$$y = f(x) = \frac{a_n x^n + a_{n-1} x^{n-1} + \cdots\cdots + a_0}{b_m x^m + b_{m-1} x^{m-1} + \cdots\cdots + b_0}$$

1. Find $x-$intercept (let $y = 0$) and $y-$intercept (let $x = 0$).
2. Find the vertical asymptotes by setting the denominator equal to zero.
3. If $n < m$, then $y = 0$ (it is the $x-axis$) is a horizontal asymptote.
4. If $n > m$, then there is no horizontal asymptote.
5. If $n = m$, then $y = \frac{a_n}{b_m}$ is a horizontal asymptote.
6. If n is more than m only by 1 degree, then there is a slant (oblique) asymptote which can be determined by dividing the denominator into the numerator.
7. Complete the missing portions of the graph by plotting a few additional points.
8. The graphs of all rational functions of the form $y = \frac{ax + a_0}{bx + b_0}$ are hyperbolas.

Examples

1. Graph $y = \dfrac{1}{x}$.

Solution:

1. $f(-x) = -f(x)$. It is symmetric about the origin.
2. There is no $x-$intercept and $y-$intercept.
3. $x = 0$ (It is the $y-axis$) is a vertical asymptote.
4. The degree of the numerator is less than the degree of denominator. Therefore $y = 0$ (It is the $x-axis$) is a horizontal asymptote.
5. The domain and range are all nonzero real numbers.

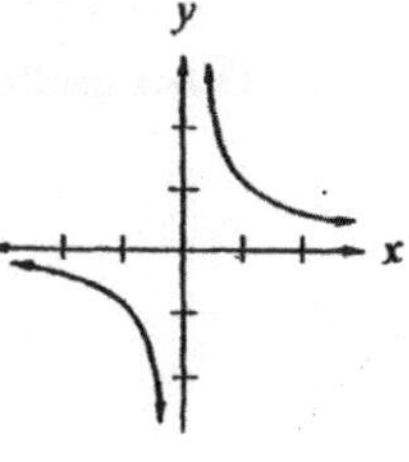

Examples

2. Graph $y=\dfrac{3}{x-2}$.

Solution:

1. Let $x=0$, $y=-\frac{3}{2}$, $(0,\ -\frac{3}{2})$ is the $y-$intercept.
2. $x=2$ is a vertical asymptote.
3. The degree of the numerator is less than the degree of the denominator. Therefore, $y=0$ (it is the $x-axis$) is a horizontal asymptote.
4. The domain is all real numbers except 2. The range is all nonzero real numbers.

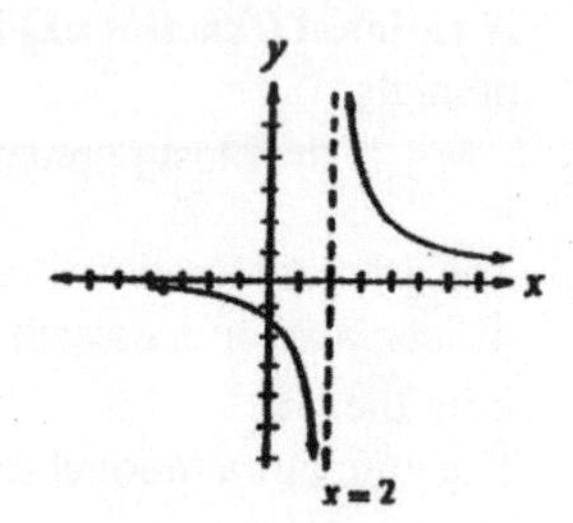

3. Graph $y=\dfrac{1}{x^2}$.

Solution:

1. $f(-x)=f(x)$. It is symmetric about the $y-axis$.
2. There is no $x-$intercept and $y-$intercept.
3. $x=0$ (it is the $y-axis$) is a vertical asymptote.
4. $y=0$ (it is the $x-axis$) is a horizontal asymptote.
5. The domain is all nonzero real numbers. The range is all real numbers of $y>0$.

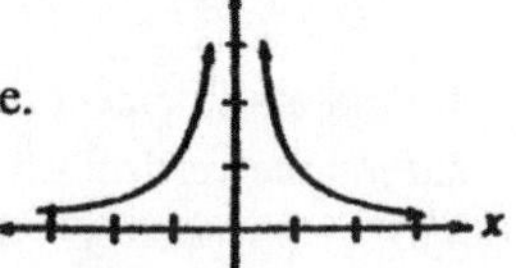

4. Graph $y=\dfrac{1}{(x-1)^2}-4$.

Solution:

1. Let $x=0$, then $y=-3$. $(0, -3)$ is the $y-$intercept.
   Let $y=0$, then $x=\frac{1}{2}$ and $1\frac{1}{2}$. $(\frac{1}{2}, 0)$ and $(1\frac{1}{2}, 0)$ are the $x-$intercepts.
2. The line $x=1$ is a vertical asymptote.
3. The line $y=-4$ is a horizontal asymptote because y approaches -4 as x increases (or decreases) without bound.
4. The domain is all real numbers except 1. The range is all real numbers of $y>-4$.

Other method: It shifts the graph of $y=\frac{1}{x^2}$ by 1 unit to the right and 4 units downward.

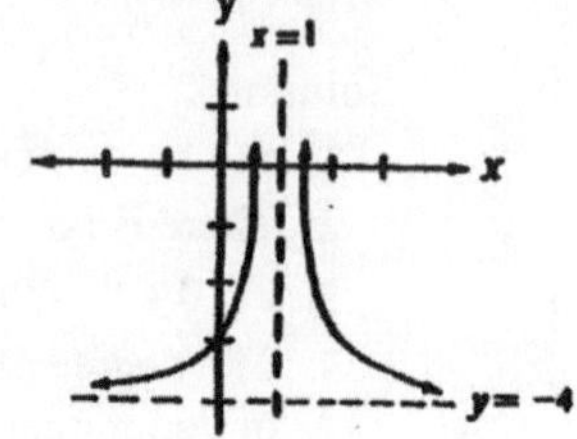

Examples

5. Graph $y = \dfrac{2}{x^2 - 2x + 2}$.

Solution:

1. Let $x = 0$, then $y = 1$. (0, 1) is the $y-$intercept.
2. $x^2 - 2x + 2 = x^2 - 2x + 1 + 1 = (x-1)^2 + 1 \neq 0$.
   There is no vertical asymptote.
3. The degree of the numerator is less than the degree of the denominator. Therefore, $y = 0$ is the horizontal asymptote.

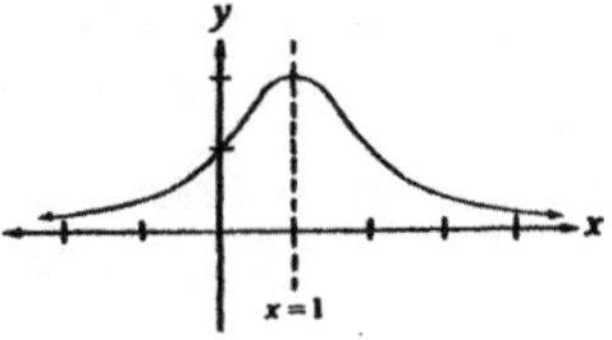

4. $y = \dfrac{2}{(x-1)^2 + 1} > 0$. The graph is above the $x-axis$.
5. $f(0) = f(2)$. The graph is symmetric about the line $x = 1$.

6. Graph $y = \dfrac{x^2 - 7x + 12}{x - 3}$.

Solution:

1. $y = \dfrac{x^2 - 7x + 12}{x - 3} = \dfrac{(x-3)(x-4)}{x-3} = x - 4$, $x \neq 3$
2. The value of y at $x = 3$ is undefined. The graph of the function is a straight line that has an open circle (hole) at $(3, -1)$.

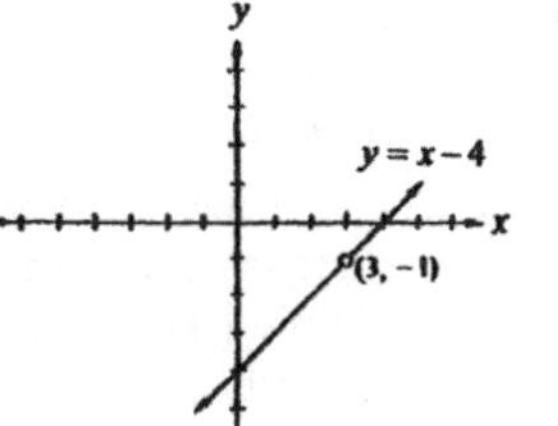

7. Graph $y = \dfrac{x-3}{x^2 + x - 2}$.

Solution:

$$y = \frac{x-3}{x^2 + x - 2} = \frac{x-3}{(x+2)(x-1)} .$$

1. Let $x = 0$, then $y = \frac{-3}{-2} = \frac{3}{2}$. $(0, \frac{3}{2})$ is the $y-$intercept.
   Let $y = 0$, then $x = 3$. (3, 0) is the $x-$intercept.
2. Two vertical asymptotes: $x = -2$ and $x = 1$.
3. One horizontal asymptote: $y = 0$
4. Critical points: $x = -2, 1, 3$

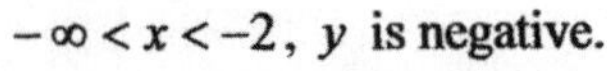

$-\infty < x < -2$, y is negative.
$-2 < x < 1$, y is positive.
$1 < x < 3$, y is negative.
$3 < x < \infty$, y is positive.

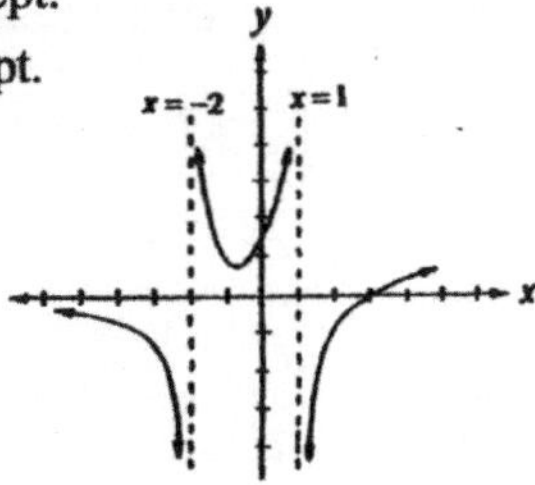

Examples

8. Graph $y=\frac{1}{|x|}$.

Solution:

If $x>0$, $y=\frac{1}{x}$.

If $x<0$, $y=-\frac{1}{x}$.

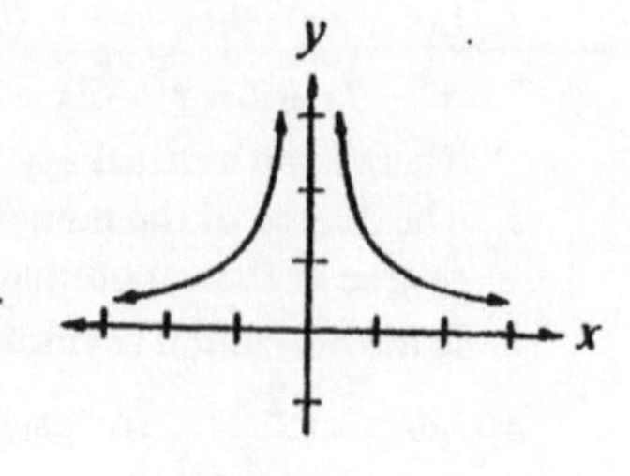

1. $f(-x)=f(x)$. It is symmetric about the $y-axis$.
2. There is no $x-$intercept and $y-$intercept.
3. $x=0$ is a vertical asymptote.
4. $y=0$ is a horizontal asymptote.
5. $y>0$.

9. Graph $y=\frac{x^2}{x+1}$.

Solution:

1. Let $x=0$, then $y=0$. It passes the origin. There is no $y-$intercept.
   Let $y=0$, then $x=0$. It passes the origin. There is no $x-$intercept.
2. $x=-1$ is the vertical asymptote.
   There is no horizontal asymptote.
3. Dividing the denominator into the numerator, we have:

   $$y=\frac{x^2}{x+1}=x-1+\frac{1}{x+1}.$$

   $y=x-1$ is the slant asymptote.
   (oblique asymptote)

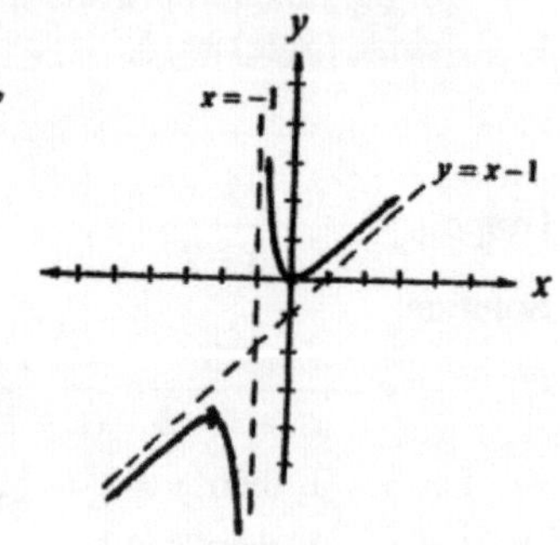

Notes

Notes

2-27 Systems of Equations and Inequalities

The graph of a linear equation $ax + by = c$ is a **straight line.**

The graphs of two linear equations (two straight lines) in the same coordinate plane must show one of the following:

1) Intersect in one point (they are consistent and independent).
2) Intersect in no point (parallel, they are inconsistent).
3) Intersect in unlimited number of points (coincide, they are consistent and dependent).

System of equations: Two or more equations with the same variables.
System of inequalities: Two or more inequalities with the same variables.
A solution to a system of equations in two variables is the intersection point (x, y) that satisfies both equations.
A system of equations is also called a **system of simultaneous equations.**

There are two algebraic methods for solving systems of equations.

1) **The substitution method**
2) **The elimination method** (or **The combination method**)

We may choose any method of the above two methods to find the solution of a system.
In general, it is good to choose the **substitution method** if the coefficient of a variable is 1 or -1.
The main idea of the **elimination method** is to eliminate one variable by **adding or subtracting** the two equations to obtain a new equation with only one variable.

Examples

1. Solve $\begin{cases} 2x - y = 4 \text{-----①} \\ 2x + 5y = 16 \text{---②} \end{cases}$

Solution:

Method 1: (Substitution Method)

From equation ①, we have $y = 2x - 4$
Substitute $y = 2x - 4$ in equation ②, we have:

$$2x + 5y = 16$$
$$2x + 5(2x - 4) = 16$$
$$2x + 10x - 20 = 16$$
$$12x = 36$$
$$\therefore x = 3$$

Substitute $x = 3$ in equation ①:

$$2x - y = 4$$
$$2(3) - y = 4 \ , \ 6 - y = 4 \quad \therefore y = 2$$

Ans: $x = 3$, $y = 2$ or $(3, 2)$

Method 2: (Combination Method)

equation ② − equation ①
We have: (x will be eliminated.)

$$\begin{array}{r} \cancel{2x} + 5y = 16 \\ -)\ \cancel{2x} - y = 4 \\ \hline 6y = 12 \end{array}$$
$$\therefore y = 2$$

Substitute $y = 2$ in equation ①

$$2x - y = 4$$
$$2x - 2 = 4$$
$$2x = 6$$
$$\therefore x = 3$$

Ans: $x = 3$, $y = 2$ or $(3, 2)$

Sometimes it is necessary to multiply one or both of the equations by a number to obtain the same coefficients(or differ in sign) for one of the variables. Then, we can eliminate this variable. If the final statement is a "true statement", the system of equations has unlimited number of solutions. The system is consistent and dependent.
If the final statement is a "false statement", the system of equations has no solution. The system is inconsistent.

Examples

2. Solve $\begin{cases} 3x-2y=10 ----① \\ 5x+4y=13 ----② \end{cases}$

Solution:

①×2: $6x-4y=20$

②: $+)\ 5x+4y=13$

$11x = 33$

$\therefore x=3$

Substitute $x=3$ in equation ①:

$3x-2y=10$

$3(3)-2y=10$

$9-2y=10$

$-2y=1$

$\therefore y=-\frac{1}{2}$

Ans: $x=3$, $y=-\frac{1}{2}$ or $(3, -\frac{1}{2})$.

3. Solve $\begin{cases} 3x-5y=12 ----① \\ 4x-2y=\frac{20}{3} ----② \end{cases}$

Solution:

①×4: $12x-20y=48$

②×3: $-)\ 12x-6y=20$

$-14y=28$

$\therefore y=-2$

Substitute $y=-2$ in ①:

$3x-5y=12$

$3x-5(-2)=12$

$3x+10=12$

$3x=2$

$\therefore x=\frac{2}{3}$

Ans: $x=\frac{2}{3}$, $y=-2$ or $(\frac{2}{3}, -2)$.

4. Solve $\begin{cases} 4x-2y=7 ----① \\ 10x-5y=6 ----② \end{cases}$

Solution:

①×5: $20x-10y=35$

②×2: $-)\ 20x-10y=12$

$0=23$ (false)

Ans: No solution. (They are parallel)

5. Solve $\begin{cases} 4x+6y=9 ----① \\ 12x+18y=27 ---② \end{cases}$

Solution:

①×3: $12x+18y=27$

②: $-)\ 12x+18y=27$

$0=0$ (true)

Ans: Unlimited number of solutions. (they coincide).

Graphing Linear inequalities

In a coordinate plane, the graph of a linear inequality in two variables is an open or close half-plane.

For example, the graph of the linear equation $x+2y=4$ (a straight line) separates the coordinate plane into two open half-planes.

The upper open half-plane is the region of the inequality $x+2y>4$.

The lower open half-plane is the region of the inequality $x+2y<4$.

$x+2y\geq 4$ and $x+2y\leq 4$ are called the close half-planes.

$x+2y=4$ is the associated equation of each linear inequality.

To find out which half-plane region (upper or lower) is the region of each linear inequality, we use the origin (0, 0) to test it. Simply substitute (0, 0) in the inequality and see if (0, 0) is included in the region of the inequality.

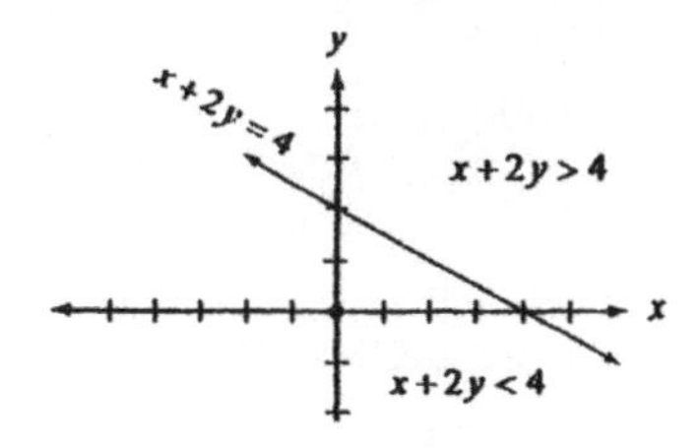

Substitute (0, 0) in $x+2y<4$

$0<4$ (True)

Therefore, (0, 0) is included in the region of $x+2y<4$.

Examples (Hint: Graph the boundary line first.)

6. Graph $4x+5y<20$.

Solution:

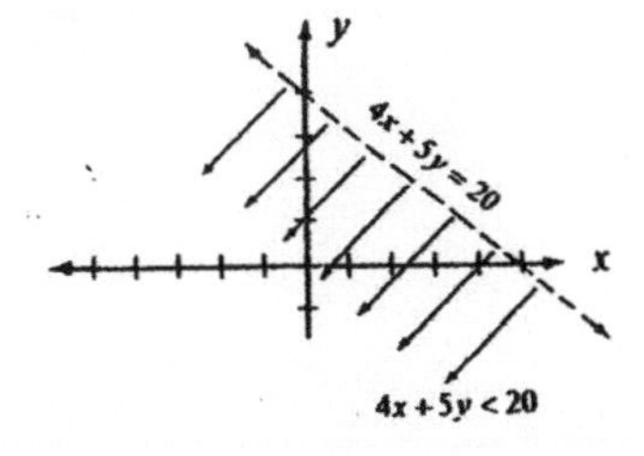

Its graph covers the lower region of the half-plane below the line. The boundary line $4x+5y=20$ is not included. The line is dashed.

7. Graph $4x+5y\geq 20$.

Solution:

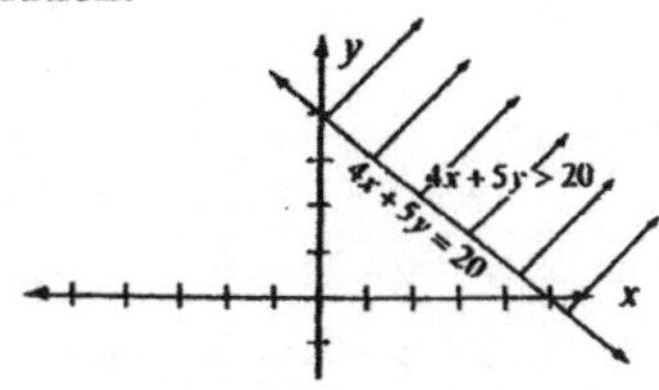

Its graph covers the upper region of the half-plane and the region on the line. The boundary line $4x+5y=20$ is included. The line is solid.

Solving a system of two linear inequalities

To solve a system of two linear inequalities, we graph the two boundary lines. Use a solid boundary line to represent ≤ or ≥ and a dashed boundary line to represent < or >. The solution of the system is the intersection of two overlap (combined) regions of the combined graph. Every point in the intersection region satisfies both inequalities.

Example

8. Solve $\begin{cases} 4x+5y \ge 20 \\ 4x-5y < 10 \end{cases}$

Solution:

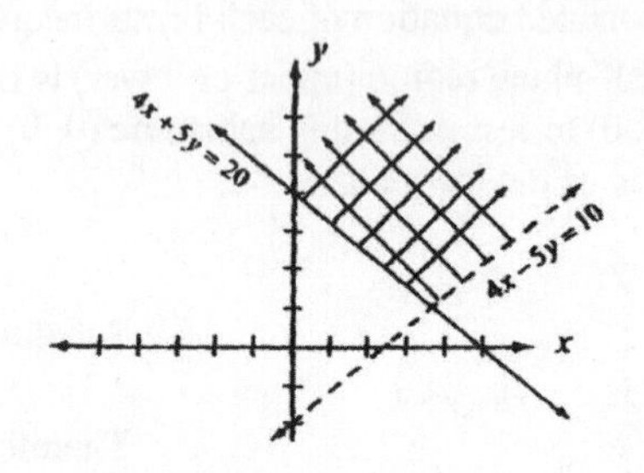

Its graph covers the overlap region above and on the line $4x+5y=20$ and above the line $4x-5y=10$.

Solving a system of three linear equations

Solving a system of three linear equations containing three variables, we eliminate one variable at a time by two of the three equations to obtain a system of two equations in two variables.

Example

9. Solve $\begin{cases} x+y+z=0 \text{-----} ① \\ 2x-y-2z=6 \text{-----} ② \\ x-y-3z=8 \text{-----} ③ \end{cases}$

Solution:

① + ② $3x - z = 6$ ····· ④

② − ③ $x + z = -2$ ·····⑤

④ + ⑤ $4x = 4$

$\therefore x = 1$

Substitute $x=1$ in ④

$3(1) - z = 6$

$\therefore z = -3$

Substitute $x=1$ and $z=-3$ in ①

$1 + y - 3 = 0$

$\therefore y = 2$

Ans. $x=1$, $y=2$, $z=-3$ or $(1, 2, -3)$.

2-28 Solving Polynomial Inequalities

We have learned how to solve inequalities in one variable of degree 1, and how to solve absolute-value equations and inequalities.
To solve a polynomial inequality which can be factored into factors of degree less than two, we find the critical points and test each interval between two sets of critical points. If the original inequality is satisfied by testing the interval, the interval is the answer. A sign graph is useful in the process.

Examples

1. Solve $x^2+2x-15>0$.

Solution:

$$x^2+2x-15>0$$
$$(x+5)(x-3)>0$$
$$\text{Let } y=(x+5)(x-3)=0$$

We have the **critical points**: $x=-5$ and 3

Test each interval between two sets of critical points if the inequality is satisfied. (Choose a convenient $x-$value in each interval.)

$y=(x+5)(x-3)>0$ (It is positive.)

$x<-5$, $-\cdot-=+$ yes

$-5<x<3$, $+\cdot-=-$ no

$x>3$, $+\cdot+=+$ yes

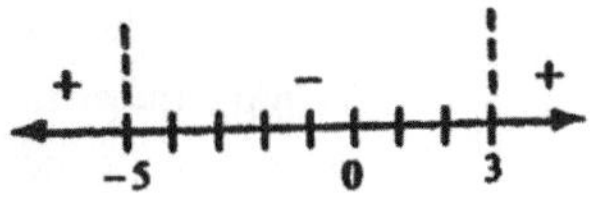

Ans: $x<-5$ or $x>3$. (disjunction inequality)

2. Solve $x^3+x^2-2x\leq 0$.

Solution:

$$x^3+x^2-2x\leq 0$$
$$x(x^2+x-2)\leq 0$$
$$x(x+2)(x-1)\leq 0$$
$$\text{Let } y=x(x+2)(x-1)=0$$

We have the **critical points**: $x=-2,\ 0,\ 1$

Test each interval between two sets of critical points if the original inequality is satisfied: (Choose a convenient $x-$value in the interval.)

$y=x(x+2)(x-1)\leq 0$. (It is negative.)

$x<-2$, $-\cdot-\cdot-=-$ yes

$-2<x<0$, $-\cdot+\cdot-=+$ no

$0<x<1$, $+\cdot+\cdot-=-$ yes

$x>1$, $+\cdot+\cdot+=+$ no

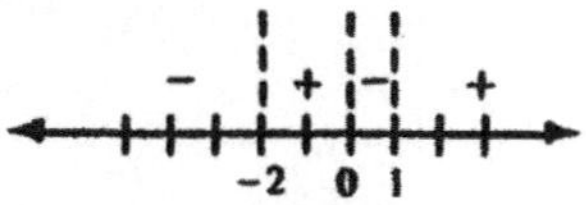

The critical points ($x=-2,\ 0,\ 1$) also satisfy the original inequality.

Ans: $x\leq -2$ or $0\leq x\leq 1$. (disjunction inequality)

Solving polynomial inequalities with Absolute Values

We have learned how to solve simple inequalities involving absolute values.
To solve a polynomial inequality with absolute value, we may use the following facts:

1. $|x|^2 = x^2$. **2.** If $x^2 > a$, then $|x| > \sqrt{a}$. **3.** If $x^2 < a$, then $|x| < \sqrt{a}$.

Examples

3. Solve $|x+2| < 5$.

Solution:

Method 1: $|x+2| < 5$

$$-5 < x+2 < 5$$

$$\therefore -7 < x < 3. \text{ Ans.}$$

Method 2: $|x+2| < 5$

$$(x+2)^2 < 5^2$$

$$x^2 + 4x + 4 < 25$$

$$x^2 + 4x - 21 < 0$$

$$(x+7)(x-3) < 0$$

We have the critical points:

$x = -7$ and 3

Test: $y = (x+7)(x-3) < 0$

$x < -7$, $- \cdot - = +$ no

$-7 < x < 3$, $+ \cdot - = -$ yes

$x > 3$, $+ \cdot + = +$ no

Ans: $-7 < x < 3$

4. Solve $|x^2 - 2x| < x$.

Solution:

$$(x^2 - 2x)^2 < x^2$$

$$x^4 - 4x^3 + 4x^2 - x^2 < 0$$

$$x^4 - 4x^3 + 3x^2 < 0$$

$$x^2(x^2 - 4x + 3) < 0$$

$$x^2 - 4x + 3 < 0$$

$$(x-3)(x-1) < 0$$

We have the critical points:

$x = 3$ and 1

Test: $y = (x-3)(x-1) < 0$

$x < 1$, $- \cdot - = +$ no

$1 < x < 3$, $- \cdot + = -$ yes

$x > 3$, $+ \cdot + = +$ no

Ans: $1 < x < 3$

5. Solve $|x^2 + 4x + 4| > 5$.

Solution:

$$|(x+2)^2| > 5$$

$$(x+2)^2 > 5$$

$$|x+2| > \sqrt{5}$$

$$x+2 < -\sqrt{5} \text{ or } x+2 > \sqrt{5}$$

Ans: $x < -1-\sqrt{5}$ or $x > -2+\sqrt{5}$

6. Solve $|x^2 + 6x + 9| < 25$

Solution:

$$|(x+3)^2| < 25$$

$$(x+3)^2 < 25$$

$$|x+3| < 5$$

$$-5 < x+3 < 5$$

Ans: $-8 < x < 2$

Solving inequalities if factoring is not possible, we apply the method of completing the square.

Examples

7. Solve $x^2+6x+9<25$.

Solution:

$$(x+3)^2<25$$
$$|x+3|<5$$
$$-5<x+3<5$$

Ans: $-8<x<2$.

(Hint: Same answer as example 6)

8. Solve $x^2+2x+1\ge\frac{9}{4}$.

Solution:

$$(x+1)^2\ge\tfrac{9}{4}$$
$$|x+1|\ge\tfrac{3}{2}$$
$$x+1\le-\tfrac{3}{2}\quad\text{or}\quad x+1\ge\tfrac{3}{2}$$
$$x\le-\tfrac{3}{2}-1\qquad x\ge\tfrac{3}{2}-1$$

Ans: $x\le-\frac{5}{2}$ or $x\ge\frac{1}{2}$

9. Solve $x^2-3x+1\le0$.

Solution:

Completing the square:

$$(x-\tfrac{3}{2})^2-\tfrac{9}{4}+1\le0$$
$$(x-\tfrac{3}{2})^2-\tfrac{5}{4}\le0$$
$$(x-\tfrac{3}{2})^2\le\tfrac{5}{4}$$
$$|x-\tfrac{3}{2}|\le\tfrac{\sqrt{5}}{2}$$
$$-\tfrac{\sqrt{5}}{2}\le x-\tfrac{3}{2}\le\tfrac{\sqrt{5}}{2}$$

Ans: $\frac{3-\sqrt{5}}{2}\le x\le\frac{3+\sqrt{5}}{2}$

10. Solve $2x^2-x-5\ge0$.

Solution:

Completing the square:

$$2(x^2-\tfrac{1}{2}x)-5\ge0$$
$$2[(x-\tfrac{1}{4})^2-\tfrac{1}{16}]-5\ge0$$
$$2(x-\tfrac{1}{4})^2-\tfrac{1}{8}-5\ge0$$
$$2(x-\tfrac{1}{4})^2\ge\tfrac{41}{8}\ ,\ (x-\tfrac{1}{4})^2\ge\tfrac{41}{16}$$
$$|x-\tfrac{1}{4}|\ge\tfrac{\sqrt{41}}{4}$$
$$x-\tfrac{1}{4}\le-\tfrac{\sqrt{41}}{4}\quad\text{or}\quad x-\tfrac{1}{4}\ge\tfrac{\sqrt{41}}{4}$$

Ans: $x\le\frac{1-\sqrt{41}}{4}$ or $x\ge\frac{1+\sqrt{41}}{4}$.

11. Solve $3x^2-6x\le4$.

Solution:

Completing the square:

$$3(x^2-2x)\le4$$
$$3[(x-1)^2-1]\le4$$
$$3(x-1)^2-3\le4$$
$$3(x-1)^2\le7$$
$$(x-1)^2\le\tfrac{7}{3}$$
$$|x-1|\le\sqrt{\tfrac{7}{3}}$$
$$-\sqrt{\tfrac{7}{3}}\le x-1\le\sqrt{\tfrac{7}{3}}$$

Ans: $1-\sqrt{\frac{7}{3}}\le x\le1+\sqrt{\frac{7}{3}}$

12. Solve $|x^2-2x-5|>2$.

Solution:

$$x^2-2x-5<-2\quad\text{or}\quad x^2-2x-5>2$$

Completing the squares:

$(x-1)^2-1-5<-2$	$(x-1)^2-1-5>2$
$(x-1)^2<4$	$(x-1)^2>8$
$\|x-1\|<2$	$\|x-1\|>\sqrt{8}$

$$-2<x-1<2,\ x-1<-\sqrt{8}\ \text{or}\ x-1>\sqrt{8}$$
$$-1<x<3,\ x<1-\sqrt{8}\ \text{or}\ x>1+\sqrt{8}$$

Ans: $x<1-\sqrt{8}$, or $-1<x<3$,

or $x>1+\sqrt{8}$

Notes

2-29 Translating Words into Algebraic Models

Mathematical Modeling: To write mathematical expressions, equations, or inequalities that represent real-life situations.

Before we learn how to solve word problems in algebra, we must learn how to translate or represent words into symbols containing variables.

A. Translate word phrases into variable expressions.

5 more than x	$\to x+5$	The sum of 5 and x	$\to 5+x$
5 less than x	$\to x-5$	The difference between 5 and x	$\to 5-x$
x less than 5	$\to 5-x$	The product of 5 and x	$\to 5x$
x minus 5	$\to x-5$	The quotient of 5 and x	$\to 5/x$
x is less than 5	$\to x<5$	x increased by 5	$\to x+5$
x is greater than 5	$\to x>5$	x decreased by 5	$\to x-5$
x is less than or equal to 5	$\to x\le 5$	x times 5	$\to 5x$
x is greater than or equal to 5	$\to x\ge 5$	x divided by 5	$\to x/5$

5 times the quantity x decreased by 3 $\to 5(x-3)$
5 times the difference of x and 3 $\to 5(x-3)$
5 times the sum of x and 3 $\to 5(x+3)$
3 more than 5 times x $\to 5x+3$
3 increased by 5 times x $\to 3+5x$
The difference between 5 times x and 3 $\to 5x-3$
3 decreased by 5 times x $\to 3-5x$
3 less than 5 times x $\to 5x-3$
3 less than half of x $\to \frac{1}{2}x-3$
Half the difference between x and 3 $\to \frac{1}{2}(x-3)$
Twice the sum of x and 3 $\to 2(x+3)$
Twice x, increased by 3 $\to 2x+3$

Three consecutive integers (n is an integer) $\to n,\ n+1,\ n+2$
Three consecutive even integers (n is an even integer) $\to n,\ n+2,\ n+4$
Three consecutive odd integers (n is an odd integer) $\to n,\ n+2,\ n+4$
Three consecutive positive multiples of 5 (n is a multiple of 5) $\to n,\ n+5,\ n+10$

Examples: Translate each statement into a variable expression.

1. There are 15 fewer girls than boys in a class. How many girls are there if the number of boys is b ?
   Solution: $b-15$
2. You paid \$50 cash and made 10 equal monthly payments. What is the total cost of the product if the amount in dollars of each of the monthly payments is x ?
   Solution: $50+10x$

B. Write an Algebraic Model

1. Eight less than a number n is 15.
Translation:
$$n-8=15$$

2. Four more than a number n is 12.
Translation:
$$n+4=12$$

3. Five more than a number n is equal to 9.
Translation:
$$n+5=9$$

4. 5 times the quantity x decreased by 3 is 25.
Translation: $5(x-3)=25$

5. Twice the sum of a number x and 3 is 22.
Translation:
$$2(x+3)=22$$

6. A number n decreased by 12 is 30.
Translation:
$$n-12=30$$

7. The result of a number x multiplied by 5, decreased by 3 is 57.
Translation:
$$5x-3=57$$

8. The sum of three consecutive numbers is 48.
Translation:
$$n+(n+1)+(n+2)=48$$

9. One-third of a pizza sold for \$3.50. Write an equation that represents the cost c of the whole pizza.
Solution: $\frac{1}{3}c=3.50$

10. You paid \$50 cash and made 10 equal monthly payments. The total cost of the product was \$400. If the amount in dollars of each of the monthly payments is x, write an equation that represents the amount of each of the monthly payments.
Solution:
$$400=50+10x$$

11. John picks up two-thirds of the books in the shelf. There are 12 books left in the shelf. Write an equation that represents the number n of the books in the shelf to begin with.
Solution: $n-\frac{2}{3}n=12$

2-30 Solving Word Problems in One Variable

To solve a word problem using an equation in algebra, we use the following steps.

Steps for solving a word problem:

1. Read the problem carefully.
2. Choose a variable and assign it to represent the unknown (answer).
3. Write an equation based on the given fact in the problem.
4. Solve the equation and find the answer.
5. Check your answer with the problem (usually on a scratch paper).

Examples

1. Eight less than a number is 18. Find the number.

Solution:

Let $n =$ the number

The equation is

$$n - 8 = 18$$
$$n - \not{8} + \not{8} = 18 + 8$$
$$\therefore n = 26.$$

Or:
$$n - 8 = 18$$
$$n = 18 + 8$$
$$\therefore n = 26.$$

Check:
$$n - 8 = 18$$
$$26 - 8 = 18 \checkmark$$

2. A number increased by 15 is 32. Find the number.

Solution:

Let $n =$ the number

The equation is

$$n + 15 = 32$$
$$n + \not{15} - \not{15} = 32 - 15$$
$$\therefore n = 17.$$

Or:
$$n + 15 = 32$$
$$n = 32 - 15$$
$$\therefore n = 17.$$

Check:
$$n + 15 = 32$$
$$17 + 15 = 32 \checkmark$$

3. Find two consecutive integers whose sum is 37.

Solution:

Let $n = 1^{st}$ integer

$n + 1 = 2^{nd}$ integer

The equation is

$$n + (n+1) = 37, \quad 2n + 1 = 37$$
$$2n = 37 - 1$$
$$2n = 36 \qquad \therefore n = 18 \quad 1^{st} \text{ integer}$$
$$n + 1 = 19 \quad 2^{nd} \text{ integer.}$$

Check: $18 + 19 = 37 \checkmark$

Examples

4. Twice the sum of a number and 3 is 22. Find the number.
Solution:

Let $n =$ the number
The equation is
$2(n+3) = 22$, $2n+6 = 22$
$2n = 22-6$
$2n = 16$
$\therefore n = 8$.

Check: $2(n+3) = 22$
$2(8+3) = 22$ √

5. Find three consecutive even integers whose sum is 120.
Solution:

Let $n = 1^{st}$ even integer
$n+2 = 2^{nd}$ even integer
$n+4 = 3^{rd}$ even integer
The equation is
$n+(n+2)+(n+4) = 120$, $3n+6 = 120$
$3n = 114$
$\therefore n = 38$ 1st integer
$n+2 = 40$ 2nd integer
$n+4 = 42$ 3rd integer

Ans: 38, 40, 42
Check: $38+40+42 = 120$ √

6. John has twice as much money as Carol. Together they have \$36. How much money does each have ?
Solution: Let $c =$ Carol's money
$2c =$ John's money
The equation is
$c+2c = 36$, $3c = 36$
$\therefore c = 12 \rightarrow$ Carol's money.
$2c = 24 \rightarrow$ John's money.

Check: $12+24 = 36$ √

7. A number divided by 2 is equal to the number increased by 2. Find the number.
Solution: Let $n =$ the number

The equation is $\frac{n}{2} = n+2$

Multiply each side by 2: $n = 2n+4$, $n-2n = 4$, $-n = 4$, $\therefore n = -4$.

Check: $\frac{-4}{2} = -4+2$ √

Examples

8. John picked up two-thirds of the books in the shelf. There are 12 books left in the shelf. How many books were in the shelf to begin with ?
Solution:

Let $n =$ the number of books to begin with
There are $(1-\frac{2}{3})\ n$ books left in the shelf after John picked up $\frac{2}{3}$ of the books.
The equation is

$$(1-\tfrac{2}{3})n = 12$$
$$\tfrac{1}{3}n = 12$$
$$\therefore n = 36 \text{ books.}$$

Check: $\frac{2}{3}(36) = 24$
$36 - 24 = 12\ \checkmark$

9. Roger picked up two-fifths of the books in the shelf. Maria picked up one-half of the remaining books. Jack picked up 6 books that were left. How many books were in the shelf to begin with ?
Solution:

Let $n =$ the number of books to begin with
There are $(1-\frac{2}{5})\ n$ remaining books in the shelf after Roger picket up $\frac{2}{5}$ of the books.
Then, Maria picked up $\frac{1}{2}(1-\frac{2}{5})n$ books after Roger picked up.
The equation is

$$(1-\tfrac{2}{5})n - \tfrac{1}{2}(1-\tfrac{2}{5})n = 6, \quad \tfrac{3}{5}n - \tfrac{1}{2}\cdot\tfrac{3}{5}n = 6$$
$$\tfrac{3}{5}n - \tfrac{3}{10}n = 6$$
$$\tfrac{6n-3n}{10} = 6$$
$$\tfrac{3n}{10} = 6$$
$$3n = 60$$
$$\therefore\ n = 20 \text{ books.}$$

Check: $\frac{2}{5}(20) = 8$
$20 - 8 = 12$
$\frac{1}{2}(12) = 6$
$12 - 6 = 6\ \checkmark$

10. There are one-dollar bills and five-dollar bills in the piggy bank. It has 60 bills with a total value of \$228. How many ones and how many fives in the piggy bank ?
Solution:

Let $x =$ number of one-dollar bills
$60 - x =$ number of five-dollar bills
The equation is

$$1x + 5(60-x) = 228, \quad x + 300 - 5x = 228$$
$$-4x = -72$$
$$\therefore x = 18 \rightarrow \text{one-dollar bills}$$
$$60 - x = 42 \rightarrow \text{five-dollar bills.}$$

Check: $\$18 + \$5(42) = \$228\ \checkmark$

Examples

11. The length of a rectangle is $14\,cm$ longer than the width. The perimeter is $76\,cm$. Find its length and width.

Solution:

$x+14$

x

Let x = the width

$x+14$ = the length

The equation is

$$2x+2(x+14)=76$$
$$2x+2x+28=76$$
$$4x=76-28$$
$$4x=48$$
$$\therefore x=12\ cm \rightarrow \text{the width.}$$
$$x+14=26\ cm \rightarrow \text{the length.}$$

Check: $(12+26)\times 2=76$ ✓

12. A suit is on sale for 20% discount. Roger paid \$8.82, including a 5% sales tax. Find the original price of the suit.

Solution:

Let x = the original price

The sales price = $80\% \cdot x = 0.80x$

The sales tax = $0.80x(5\%) = 0.80x(0.05)$

The equation is

$$0.80x+0.80x(0.05)=8.82$$
$$0.80x+0.04x=8.82$$
$$0.84x=8.82$$
$$\therefore x=\tfrac{8.82}{0.84}=\$10.50.$$

Check: $\$10.50\times 0.80=\$8.40$

$\$\ 8.40\times 0.05=\$0.42$

$\$8.40+\$0.42=\$8.82$ ✓

Notes

Notes

2-31 Solving Word Problems by Factoring

Many word problems can be written in polynomial equations and solved by factoring.

Steps for solving a word problem by factoring:

1. Read the Problem.
2. Assign a variable to represent the unknown.
3. Write an equation based on the given facts.
4. Solve the equation by factoring to find the answers.
5. Check your answers (usually on a scratch paper)

Examples

1. If a number adds its square, the result is 42. Find the number.

Solution:

Let $n =$ the number.

The equation is:

$$n + n^2 = 42$$

$$n^2 + n - 42 = 0$$

$$(n+7)(n-6) = 0$$

$n + 7 = 0$ | $n - 6 = 0$

$n = -7$ | $n = 6$

Ans: The number is -7 or 6.

2. Find two consecutive positive integers whose product is 110.

Solution:

Let $n =$ the 1st positive integer.

$n+1 =$ the 2nd positive integer.

The equation is:

$$n(n+1) = 110$$

$$n^2 + n - 110 = 0$$

$$(n+11)(n-10) = 0$$

$n = -11$, not permissible | $n = 10$ (1st integer), $n+1 = 11$ (2nd integer)

Ans: The integers are 10 and 11.

3. A box has a square bottom, and the top is 5 $in.$ high. Its total surface area is 192 $in.^2$. Find its volume.

Solution:

Let $x =$ one side of the equation.

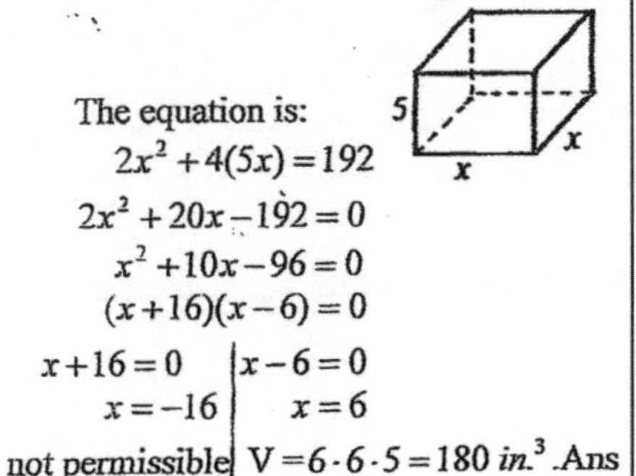

The equation is:

$$2x^2 + 4(5x) = 192$$

$$2x^2 + 20x - 192 = 0$$

$$x^2 + 10x - 96 = 0$$

$$(x+16)(x-6) = 0$$

$x + 16 = 0$ | $x - 6 = 0$

$x = -16$ | $x = 6$

not permissible | $V = 6 \cdot 6 \cdot 5 = 180$ $in.^3$ .Ans

4. The height in feet of an object reaches in t seconds after it is thrown upward with an initial speed of v feet per second is given by the formula $h = vt - 16t^2$.

A rocket is fired upward with an initial speed of 208 feet per second. When will the rocket be at a height of 576 feet ?

Solution:

$$h = vt - 16t^2$$

$$576 = 208t - 16t^2$$

$$16t^2 - 208t + 576 = 0$$

Divide each side by 16:

$$t^2 - 13t + 36 = 0$$

$$(t-4)(t-9) = 0$$

$\therefore$ $t = 4$ and 9 seconds. Ans.

Example

5. A rocket is fired upward with an initial speed of 160 feet per second. The height (h) of the rocket is given by the formula $h = vt - 16t^2$, where v is the initial speed and t is the time in seconds.

a) Find the height of the rocket reaches in 3 seconds after being fired upward.
b) When will the rocket be at a height of 384 feet ?
c) How many seconds will it take to reach the ground again ?
d) What will be the maximum height the rocket reaches ?

Solution:

a) $h = vt - 16t^2 = 160(3) - 16(3)^2 = 480 - 144 = 336$ feet. Ans.

b)
$h = vt - 16t^2$
$384 = 160t - 16t^2$
$16t^2 - 160t + 384 = 0$
Divide each side by 16:
$t^2 - 10t + 24 = 0$
$(t-4)(t-6) = 0$
$\therefore t = 4$ and 6 seconds.

c)
$h = vt - 16t^2$
$h = 160t - 16t^2$
Let $h = 0$
$0 = 160t - 16t^2$
$16t^2 - 160t = 0$
Factor out the GCF:
$16t(t-10) = 0 \ \therefore t = 10$ seconds.

d) It is the time halfway between the starting time and the time to reach the ground again. Therefore, it reaches the maximum height when $t = 5$.
$\therefore h = vt - 16t^2 = 160(5) - 16(5)^2 = 800 - 400 = 400$ feet.

2-32 Solving Word Problems involving Inequalities

Steps for Solving a word problem involving inequality:

1. Read the problem.
2. Assign a variable to represent the unknown.
3. Determine which inequality sign ($<, >, \leq, \geq$) needed.
4. Write an inequality based on the given facts.
5. Solve the inequality to find the answer.
6. Check your answer (usually on a scratch paper).

Examples

1. You want to rent a car for a rental cost of $50 plus $30 per day, or a van for $80 plus $25 per day. For how many days at most is it cheaper to rent a car ?
   Soluion:
   Let $d =$ days needed
   The inequality is:
   $50 + 30d \leq 80 + 25d$
   $5d \leq 30$
   $\therefore d \leq 6$
   Ans: At most 6 days (or less).
   Check: $50 + 30d = 50 + 30(6) = \$230\,4$
   $80 + 25d = 80 + 25(6) = \$230\,4$
   The rental costs are the same for 6 days. It is cheaper to rent the car for less than 6 days. It is cheaper to rent the van for more than 6 days.
2. Your scores on your 4 math tests were 71, 74, 75 and 86. What is the lowest score you need in your next test in order to have an average score of at least 80 ?
   Solution:
   Let $x =$ score for next test
   The inequality is: $\frac{71 + 74 + 75 + 86 + x}{5} \geq 80$, $\frac{306 + x}{5} \geq 80$,
   $306 + x \geq 400 \quad \therefore x \geq 94$ Ans: 94 or more.
3. John wants to rent a car and pay at most $250. The car rental costs $150 plus $0.25 a mile. How far can he travel ?
   Solution:
   Let $x =$ mileage he can travel
   The inequality is:
   $150 + 0.25x \leq 250$
   $0.25x \leq 100 \quad \therefore x \leq 400$ Ans: 400 miles or less.
4. The sum of two consecutive integers is less than 79. Find the integers with the greatest sum.
   Solution: Let $n = 1^{st}$ integer. $n + 1 = 2^{nd}$ integer.
   The inequality is:
   $n + (n + 1) < 79, \quad 2n < 78, \quad \therefore n < 39$
   The greatest integer in n is 38. Ans: 38 and 39.

Notes

2-33 Solving Word Problems with Rational Equations

Rational equations (fractional equations) are often used in solving word problems.

Examples:

1. One fifth of a number is 7 less than two thirds of the number. Find the number.

Solution:

Let $n =$ the number

The equation is:

$$\frac{n}{5}+7=\frac{2n}{3}$$

LCD = 15

$$3n+105=10n$$

$$105=7n \quad \therefore n=15$$

2. The sum of two numbers is 48 and their quotient is $\frac{3}{5}$. Find the numbers.

Solution:

Let $n = 1^{st}$ number

$48-n=2^{nd}$ number

The equation is: $\frac{n}{48-n}=\frac{3}{5}$

$$5n=3(48-n)$$

$$5n=144-3n$$

$$8n=144 \quad \therefore n=18$$

3. If two pounds of beef cost \$5.80, how much do 8 pounds cost ?

Solution:

Let $x =$ cost for 8 pounds

The equation is:

$$\frac{5.80}{2}=\frac{x}{8}$$

$$5.80(8)=2x$$

$$46.40=2x \quad \therefore x=\$23.20.$$

4. A bus uses 4.5 gallons of gasoline to travel 75 miles. How many gallons would the bus use to travel 95 miles ?

Solution:

Let $x =$ the number of gallons

The equation is: $\frac{75}{4.5}=\frac{95}{x}$

$$75x=4.5(95)$$

$$75x=427.5 \quad \therefore x=5.7 \text{ gallons.}$$

5. John can finish a job in 5 hours. Steve can finish the same job in 4 hours. How many hours do they need to finish the job if they work together ?

Solution:

Let $x =$ number of hours needed to do the job together

John can do $\frac{1}{5}$ of the job per hour.

Steve can do $\frac{1}{4}$ of the job per hour.

The equation is: $\frac{x}{5}+\frac{x}{4}=1$

$$4x+5x=20, \quad 9x=20$$

$$\therefore x=2\tfrac{2}{9} \text{ hours.}$$

6. A 16-gallon salt-water solution contains 25% pure salt. How much water should be added to produce the solution to 20% salt ?

Solution:

Let $x =$ gallons of water added

Pure salt in the original solution:

$16\times 25\%=16\times 0.25=4$ gal.

The equation is:

$$\frac{4}{16+x}=20\%, \quad \frac{4}{16+x}=\frac{1}{5}$$

$$20=16+x \quad \therefore x=4$$

$\therefore x=4$ gallons of water.

Notes

2-34 Solving Word Problems Using Systems

We have learned to solve a word problem using one equation in one variable.
Now, we can use a system of two equations to solve a word problem in two variables.
The steps for solving a word problem in two variables are similar to the steps for solving a word problem in one variable.
We assign two variables to represent the unknowns.

Examples

1. John has twice as much money as Carol. Together they have \$36. How much money does each have ?

Solution:

(Assign two variables)

Let $x =$ John's money

$y =$ Carol's money

The system of equations is:

$$\begin{cases} x = 2y \cdots\cdots ① \\ x + y = 36 \cdots\cdots ② \end{cases}$$

Substitute $x = 2y$ in ②:

$$2y + y = 36$$
$$3y = 36$$
$$\therefore y = 12$$

Substitute $y = 12$ in ①:

$$x = 2(12)$$
$$\therefore x = 24$$

Ans: John's money \$24.
Carol's money \$12.

2. There are one-dollar bills and five-dollar bills in the piggy bank. It has 60 bills with a total value of \$228. How many ones and how many fives in the piggy bank ?

Solution:

(Assign two variables)

Let $x =$ number of one-dollar bills

$y =$ number of five-dollar bills

The system of equations is:

$$\begin{cases} x + y = 60 \cdots\cdots ① \\ x + 5y = 228 \cdots\cdots ② \end{cases}$$

②–①: $4y = 168$

$$\therefore y = 42$$

Substitute $y = 42$ in ①:

$$x + 42 = 60$$
$$\therefore x = 18$$

Ans: 18 one-dollar bills.
42 five-dollar bills.

3. Flying with the tail wind, an airplane flies 600 km per hour. Flying with the same wind, it flies 500 km per hour to make the return flight against the wind. Find the speed of the wind and the speed of the airplane.

Solution:

Let $x =$ Airplane speed

$y =$ Wind speed

We have:

$$\begin{cases} x + y = 600 \cdots\cdots ① \\ x - y = 500 \cdots\cdots ② \end{cases}$$

① + ②: $2x = 1100$ $\therefore x = 550$

Substitute $x = 550$ in ①:

$550 + y = 600$ $\therefore y = 50$

Ans: Airplane speed = 550 km/hr.
wind speed = 50 km/hr.

Examples

4. **Two hamburgers and one bagel cost \$4.82. At the same price, one hamburger and two bagels cost \$3.70. How much does each hamburger and each bagel cost ?**

Solution:

Let $x=$ cost of each hamburger

$y=$ cost of each bagel

We have:

$$\begin{cases} 2x+y=4.82 \quad \cdots\cdots① \\ x+2y=3.70 \quad \cdots\cdots② \end{cases}$$

From ①: $y=4.82-2x$

Substitute $y=4.82-2x$ in ②:

$$x+2(4.82-2x)=3.70$$
$$x+9.64-4x=3.70$$
$$-3x=-5.94$$
$$\therefore x=1.98$$

Substitute $x=1.98$ in ①:

$2(1.98)+y=4.82 \quad \therefore y=0.86$

Ans: Each hamburger \$1.98.
Each bagel \$0.86.

5. **A teacher distributes a box of pencils to the students. If each student receives 5 pencils, there are 15 pencil left. If each students receives 6 pencils, there are 22 pencils short. How many students are there ? How many pencils are there ?**

Solution:

Let $x=$ number of students

$y=$ number of pencils

We have:

$$\begin{cases} y-5x=15 \quad \cdots\cdots① \\ 6x-y=22 \quad \cdots\cdots② \end{cases}$$

① + ②: $x=37$

Substitute $x=37$ in ①

$$y-5(37)=15$$
$$y-185=15$$
$$\therefore y=200$$

Ans: 37 students, 200 pencils.

6. **A student wants to mix a salt-water solution containing 10% pure salt with another salt-water solution containing 15% pure salt. The total mixture must at least 3 gallons and must contain at most 0.6 gallons of pure salt. Graph the possible solution region to find the amount of each solution needed.**

Solution:

Let $x=$ the amount of 10% solution

$y=$ the amount of 15% solution

We have:

$$x+y\geq 3$$
$$0.10x+0.15y\leq 0.6$$

Rewrite the above system:

$$\begin{cases} x+y\geq 3 \\ 2x+3y\leq 12 \end{cases}$$

Solve the system by graphing:

$2x+3y=12$, $x+y=3$

Since x and y are nonnegative, the solution region is in the first quadrant. Any point in the region will be the amounts of each solution to satisfy the required mixtur

2-35 Inverse Functions

One to One function: A function f is a one-to-one function if and only if for each range value there corresponds exactly one domain value. An increasing (or decreasing) function is a one-to-one function. A horizontal line intersects the graph of a one-to-one function in at most one point. Note that, both $y = x^3$ and $y = x^2$ are functions. However, $y = x^3$ is a one-to-one function, $y = x^2$ is not a one-to-one function.

Inverse of a function: The inverse of a function f is obtained by interchanging the variables x and y in f. The inverse of f has the set of ordered pairs (y, x).

Inverse Functions: If $y = f(x)$ is a one-to-one function, a function $y = g(x)$ is the inverse function of f **if and only if** $f(g(x)) = g(f(x)) = x$.

$y = x^2$ has an inverse $x = y^2$ (or $y = \pm\sqrt{x}$) which is not a function.

For the function $y = f(x)$ to have an inverse function, f must be one-to-one.

The inverse function of the function f is denoted by f^{-1}, which is read "f inverse".

If the functions $f(x)$ and $g(x)$ are inverses (functions) of each other, then:

1. the ordered pairs (x, y) of one equation are obtained by interchanging the variables x and y in the other equation.
2. their graphs are **mirror image** of each other with respect to the line $y = x$.
3. If $f(x)$ and $g(x)$ are inverses, then they have the property : $f(g(x)) = g(f(x)) = x$

   Example: $f(x) = x + 2$ and $g(x) = x - 2$ are inverse functions of each other.

   $f(g(x)) = f(x-2) = x - 2 + 2 = x$; $g(f(x)) = g(x+2) = x + 2 - 2 = x$

A function f must have an inverse (a relation) which can be a function or not a function. In certain textbooks, they conventional state that "this function has no inverse". It means "this function has no inverse function".

Examples

1. Let $f(x) = x + 2$, find $f^{-1}(x)$.

Solution: $y = x + 2$

Interchange x and y: $x = y + 2$

We have: $y = x - 2$

$\therefore f^{-1}(x) = x - 2$. Ans.

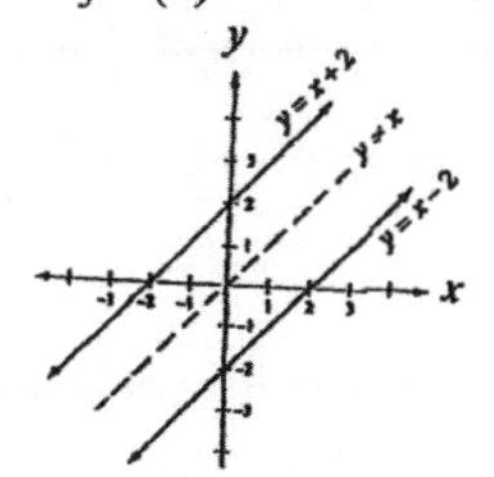

2. Let $f(x) = \frac{2}{3}x + 1$, find $f^{-1}(x)$.

Solution: $y = \frac{2}{3}x + 1$

Interchange x and y: $x = \frac{2}{3}y + 1$

We have: $y = \frac{3}{2}x - \frac{3}{2}$

$\therefore f^{-1}(x) = \frac{3}{2}x - \frac{3}{2}$. Ans.

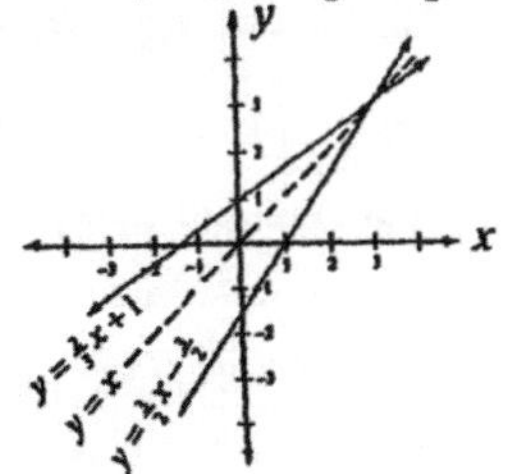

Examples

3. Let $f(x)=x^2$, find the inverse. If $f(x)$ has no inverse function, so state.

Solution: $y=x^2$

Interchange x and y :

We have: $x=y^2$ (It is a relation, not a function.)

$y=\pm\sqrt{x}$

Ans: $f(x)$ has an inverse $y=\pm\sqrt{x}$.

The inverse is not a function.

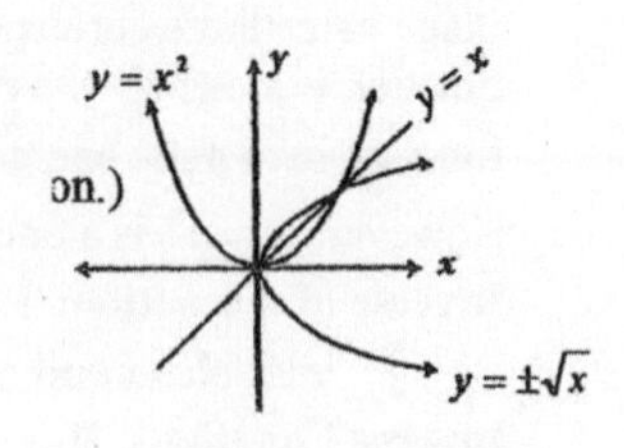

4. Let $f(x)=x^2$ and $x\le 0$, find $f^{-1}(x)$.

Solution: $y=x^2$ and $x\le 0$

Interchange x and y :

We have: $x=y^2$ and $y\le 0$ (It is a function.)

$y=-\sqrt{x}$ where $x\ge 0$

Ans: $f(x)$ has an inverse $f^{-1}(x)=-\sqrt{x}$ where $x\ge 0$.

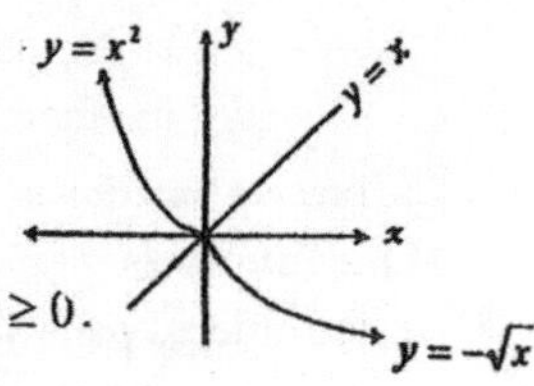

5. If $f(x)=\sqrt{x+1}$, find the inverse function by limiting the domain of $f^{-1}(x)$.

Solution: $y=\sqrt{x+1}$, $y^2=x+1$ and $x\ge -1$

Interchange x and y :

$x^2=y+1$ and $y\ge -1$ (it is a function)

$y=x^2-1$ and $x\ge 0$

Therefore $f^{-1}(x)=x^2-1$ where $x\ge 0$.

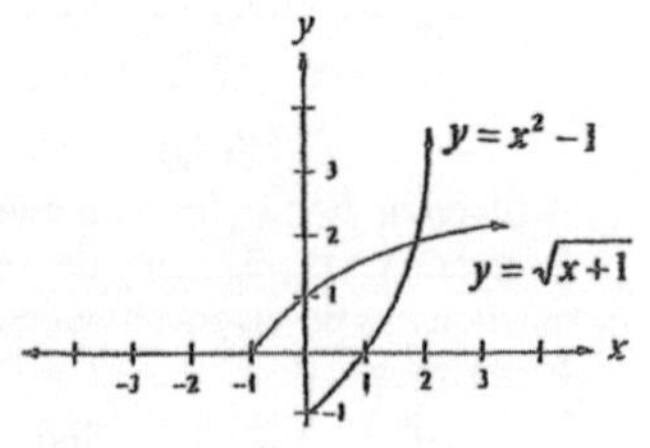

6. If $f=\{(3,1),(4,2),(5,2)\}$, find the inverse function by limiting the domain of f.

Solution:

The inverse {(1, 3), (2, 4), (2, 5)} is not a function.

The domain of the original function must be limited to "3 and 4" or "3 and 5"

Therefore $f^{-1}=\{(1,3),(2,4)\}$ by limiting the domain of f to 3 and 4.

Or $f^{-1}=\{(1,3),(2,5)\}$ by limiting the domain of f to 3 and 5.

7. Are the function $f(x)=\frac{2}{3}x+1$ and $g(x)=\frac{3}{2}x-\frac{3}{2}$ inverses of each other ?

Solution: $f(g(x))=\frac{2}{3}(\frac{3}{2}x-\frac{3}{2})+1=x-1+1=x$

$g(f(x))=\frac{3}{2}(\frac{2}{3}x+1)-\frac{3}{2}=x+\frac{3}{2}-\frac{3}{2}=x$

$f(g(x))=g(f(x))=x$ $\therefore$ $f(x)$ and $g(x)$ are inverses of each other.

2-36 Composition of Functions

When we apply one function after another function's values, such as $f(g(x))$, is called " **the composition of** the function f with the function g ". It is denoted as $f \circ g$.

$f(g(x))$ indicates that the domain of $f \circ g$ is all x-values **such that** x **is in the range of** g and g **is in the domain of** f.

The composition of f with g is: $(f \circ g)(x) = f(g(x))$
The composition of g with f is: $(g \circ f)(x) = g(f(x))$
The composition of a function $f(x)$ with its inverse $f^{-1}(x)$ is equal to x.

$$(f \circ f^{-1})(x) = f(f^{-1}(x)) = x, \quad (f^{-1} \circ f)(x) = f^{-1}(f(x)) = x$$

Examples

1. If $f(x) = 3x^2 - 6$ and $g(x) = x^2 - 4$, find $(f \circ g)(x)$ and $(g \circ f)(x)$.
Solution:

$$(f \circ g)(x) = f(g(x)) = 3(g(x))^2 - 6 = 3(x^2 - 4)^2 - 6 = 3(x^4 - 8x^2 + 16) - 6$$
$$= 3x^4 - 24x^2 + 42.$$
$$(g \circ f)(x) = g(f(x)) = (f(x))^2 - 4 = (3x^2 - 6)^2 - 4 = 9x^4 - 36x^2 + 36 - 4$$
$$= 9x^4 - 36x^2 + 32.$$

2. Given $f(x) = \frac{2}{3}x + 1$ and $f^{-1}(x) = \frac{3}{2}x - \frac{3}{2}$, find $f \circ f^{-1}$ and $f^{-1} \circ f$.
Solution:

$$f \circ f^{-1} = f(f^{-1}(x)) = \tfrac{2}{3}(\tfrac{3}{2}x - \tfrac{3}{2}) + 1 = x - 1 + 1 = x.$$
$$f^{-1} \circ f = f^{-1}(f(x)) = \tfrac{3}{2}(\tfrac{2}{3}x + 1) - \tfrac{3}{2} = x + \tfrac{3}{2} - \tfrac{3}{2} = x.$$

3. If $f(g(x)) = 2x - 8$ and $f(x) = 2x$, find $g(x)$ and $g(2)$.
Solution:

$$f(g(x)) = 2(g(x)) = 2x - 8 \quad \therefore g(x) = x - 4 \text{ and } g(2) = 2 - 4 = -2.$$

4. If $f(g(x)) = 2x - 8$ and $g(x) = x - 4$, find $f(x)$.
Solution:

$$f((g(x)) = 2x - 8 = 2(x - 4) = 2g(x) \quad \therefore f(x) = 2x.$$

5. If $f(g(x)) = 2x - 1$ and $f(x) = x - 4$, find $g(x)$.
Solution: $f(g(x)) = g(x) - 4 = 2x - 1 \quad \therefore g(x) = 2x + 3$.

6. If $f(g(x)) = 2x - 1$ and $g(x) = 2x + 3$, find $f(x)$.
Solution: $f(g(x)) = 2x - 1 = (2x + 3) - 4 = g(x) - 4 \quad \therefore f(x) = x - 4$.

Notes

2-37 Greatest Integer Functions

The greatest-integer function is the most common type of the **step functions.**
The greatest-integer function is in the form:

$f(x) = \text{int}(x)$ or $y = [x] = i$, i is an integer and $i \le x < i+1$

(i is the greatest integer that is less than or equal to the real number x.)

We have:

$\text{int}(1)=1$ or $[1]=1$, $\text{int}(2.7)=2$ or $[2.7]=2$, $[3]=3$, $[1.2]=1$, $[2.8]=2$,
$[0.4]=0$, $[0.9]=0$, $[\frac{7}{3}]=2$, $[-1]=-1$, $[-2]=-2$, $[-3]=-3$, $[-0.5]=-1$,
$[-1.5]=-2$, $[-2.1]=-3$, $[-3.2]=-4$, $[-3.5]=-4$, $[-3.9]=-4$.

To graph $f(x) = \text{int}(x)$, we plot several points. For values of $-3 \le x < -2$, the value of $f(x)$ is -3. The x-intercepts are in the interval $[0, 1)$. A solid dot is used to indicate the value of $f(x) = -3$ at $x = -3$. The graph has an infinite number of line segments (or steps) and suddenly steps from one value to another without continuity.
The domain of $f(x) = \text{int}(x)$ is all real numbers and its range is all integers.

Examples:

1. Graph $y = [x]$.

Solution:

$-3 \le x < -2$, $y = -3$
$-2 \le x < -1$, $y = -2$
$-1 \le x < 0$, $y = -1$
$0 \le x < 1$, $y = 0$
$1 \le x < 2$, $y = 1$
$2 \le x < 3$, $y = 2$
$3 \le x < 4$, $y = 3$

2. Graph $y = [2x]$.

Solution:

$-1.5 \le x < -1$, $y = -3$
$-1 \le x < -0.5$, $y = -2$
$-0.5 \le x < 0$, $y = -1$
$0 \le x < 0.5$, $y = 0$
$0.5 \le x < 1$, $y = 1$
$1 \le x < 1.5$, $y = 2$
$1.5 \le x < 2$, $y = 3$

3. The cost of using a long-distance call is \$1.25 for the first minutes and \$0.35 for each additional minute or portion of a minute.

a) Use the greatest integer function to write the cost (c) for a call lasting t minutes.
b) Find the cost of a call lasting 16 minutes and 15 seconds.
c) Graph the function.

Solution:

a) $c(t) = 1.25 - 0.35\,[1 - t]$

b) $c(16\frac{15}{60}) = 1.25 - 0.35\,[1 - 16.25]$
$= 1.25 - 0.35\,[-15.25]$
$= 1.25 - 0.35(-16)$
$= 1.25 + 5.60 = \$6.85$

Hint: $c(t) = 1.25 + 0.35[t - 1]$ is incorrect.

c)

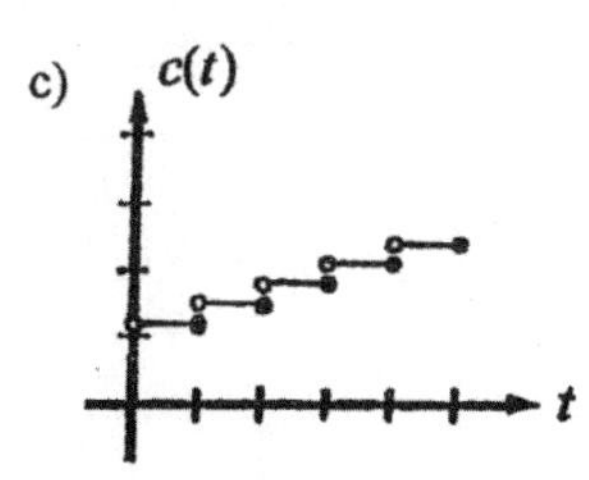

Notes

2-38 Piecewise-Defined Functions

The piecewise-defined function is a function described by two or more equations over a specified domain (the x-values).

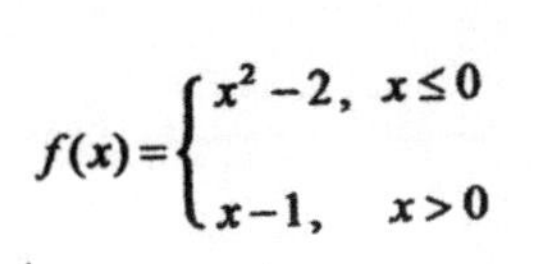

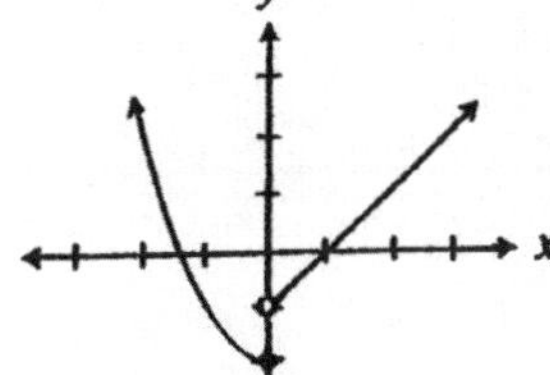

A close circle "•" shows "included". An open circle "o" shows "not included".

Examples

1. Evaluate the function when $x=-2, 0, 1$, and 2.

$$f(x)=\begin{cases} x^2-2, & x\le 0 \\ x-1, & x>0 \end{cases}$$

Solution: $f(x)=x^2-2$ when $x=-2$: $f(-2)=(-2)^2-2=2$.

$f(x)=x^2-2$ when $x=0$: $f(0)=0^2-2=-2$.

$f(x)=x-1$ when $x=1$: $f(1)=1-1=0$.

$f(x)=x-1$ when $x=2$: $f(2)=2-1=1$.

2. A publishing company pays income tax at a rate of 30% on the first income \$50,000, and all income over \$50,000 is taxed at a rate of 40%.

a) Write a piecewise-defined function $T(x)$ that shows the total tax on an income of x dollars.

b) What is the total tax on an income \$95,000 ?

Solution:

a) If $0<x\le 50000$, $T(x)=0.30x$

If $x>50000$, $T(x)=0.30(50{,}000)+0.40(x-50{,}000)$

$=15{,}000+0.40x-20{,}000$

$=0.40x-5{,}000$

We have the piecewise-defined function:

$$f(x)=\begin{cases} 0.30x & \text{if } 0<x\le 50{,}000 \\ 0.40x-5000 & \text{if } x>50{,}000 \end{cases}$$

b) $T(95{,}000)=0.40(95{,}000)-5{,}000=\$33{,}000$.

Notes

2-39 Square and Cubic Root Functions

We have learned how to graph the quadratic functions ($y = cx^2$) in Section 2~19 and the cubic functions ($y = cx^3$) in Section 2~24.

In this section we will learn to graph two types of radical functions, **the square root** and **cubic root functions.**

Understanding the basic shapes of the following two graphs will help us to analyze and graph more complicated square root and cubic root functions. (See page 150)

The graph of $y = \sqrt{x}$

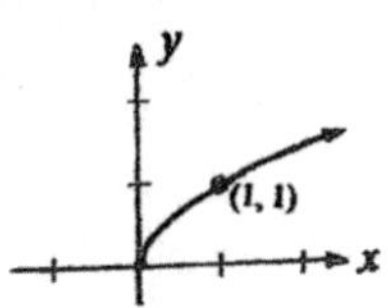

Starts at (0, 0) and passes through the point (1, 1).
Domain: $x \geq 0$. Range: $y \geq 0$
It is the upper half of the parabola $x = y^2$ having a horizontal axis.

The graph of $y = \sqrt[3]{x}$

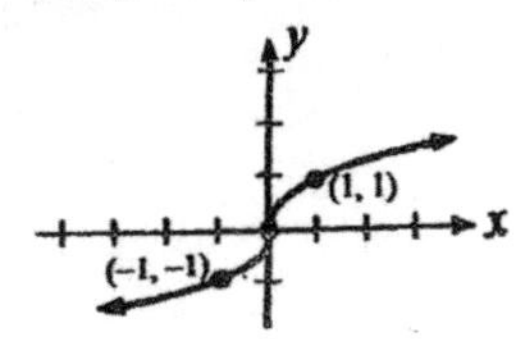

Passes through (0, 0) and the points (−1, −1) and (1, 1).
Domain: x are all real numbers.
Range: y are all real numbers.

1. The graph of $y = a\sqrt{x}$ starts at (0, 0) and passes through the point (1, a).
2. The graph of $y = a\sqrt[3]{x}$ passes through (0, 0) and the points $(-1, -a)$ and (1, a).

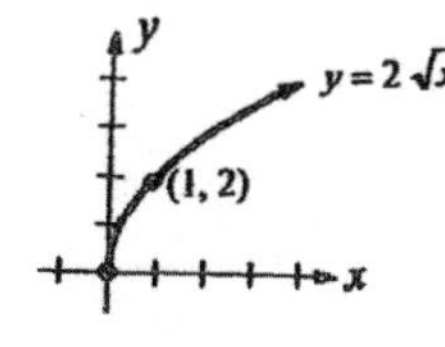

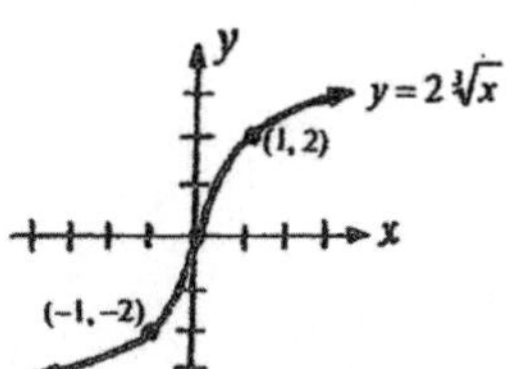

3. The graph of $y = a\sqrt{x-h} + k$ moves (shifts) the graph of $y = a\sqrt{x}$ by h units to the right and k units upward.
4. The graph of $y = a\sqrt[3]{x-h} + k$ moves (shifts) the graph of $y = a\sqrt[3]{x}$ by h units to the right and k units upward.

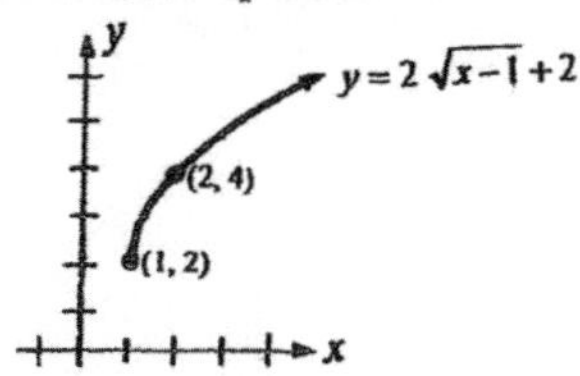

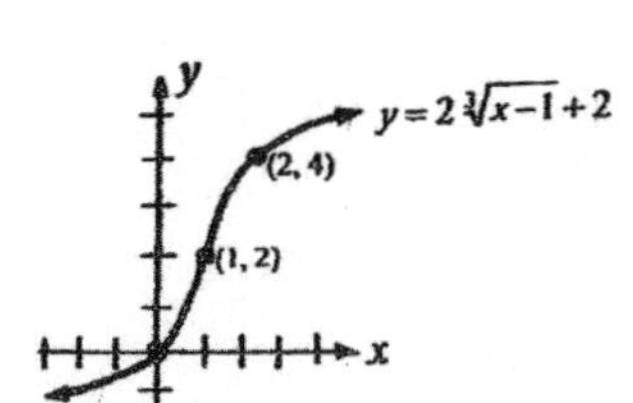

Notes

2-40 Common and Natural Logarithms

We have learned that the exponential function $y = a^x$, for $a > 0$ and $a \neq 1$, is a one-to-one function.

Since every one-to-one function has an inverse function, $y = a^x$ has the inverse function by the equation: $x = a^y$, $a > 0$ and $a \neq 1$

In order to solve the equation $x = a^y$ for the exponent y in terms of x, we create the word **logarithm** for the exponent y:

$y = \log_a x$ Read " y is the logarithm of x to the base a. ".

We simply say that $\log_2 8$ represents " 2 to what power gives 8. ". Therefore, $\log_2 8 = 3$.

$y = \log_a x$ is the equivalent logarithmic form of $x = a^y$.

$y = \log_a x$ and $y = a^x$ are inverse functions of each other.

The domains and ranges of $y = \log_a x$ and $y = a^x$ are interchanged.

Definition of a logarithmic function:

A logarithmic function $f(x)$ to a constant base a is a function of the form

$$f(x) = \log_a x \quad \text{or} \quad y = \log_a x$$

where $a > 0$, $a \neq 1$, and x is any real number. It is a one-to-one function.

Graphs of $y = \log_a x$:

1. The function is an increasing function if $a > 1$, y increases as x increases. The y – axis ($x = 0$) is a vertical asymptote to the graph as $x \to 0$.
2. The function is a decreasing function if $0 < a < 1$, y decreases as x increases. The y – axis ($x = 0$) is a vertical asymptote to the graph as $x \to 0$.
3. The graph lies to the right of y – axis and passes through the point (1, 0)

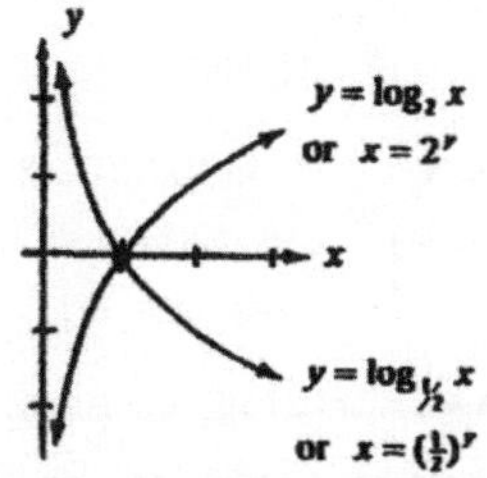

Domain: All positive real numbers ($x > 0$)
Range: All real numbers ($y \in R$)

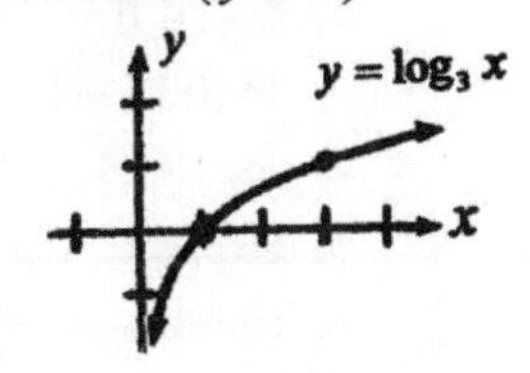

Graphing logarithmic functions using transformations

1. The graph of $y = \log_3(x-1)$ shifts the graph of $y = \log_3 x$, 1 unit to the right.
2. The graph of $y = \log_3 x - 1$ shifts the graph of $y = \log_3 x$, 1 unit downward.
3. The graph of $y = -\log_3 x$ reflects the graph of $y = \log_3 x$ over the x – axis.
4. The graph of $y = \log_3(-x)$ reflects the graph of $y = \log_3 x$ over the y – axis.

The Laws of Logarithms

We have learned the formulas about exponents. Now, we will learn the formulas about logarithms.

The Laws of Logarithms (Rules of Logarithms)

For all positive real numbers a, p, q with $a \neq 1$, and n is any real number:

1) $\log_a 1 = 0$ **2)** $\log_a a = 1$ **3)** $\log_a a^n = n$ **4)** $a^{\log_a n} = n$

5) $\log_a pq = \log_a p + \log_a q$ 6) $\log_a \frac{p}{q} = \log_a p - \log_a q$ 7) $\log_a p^n = n \log_a p$

8) Formula for Changing Base: $\log_a n = \dfrac{\log_b n}{\log_b a}$ where $b \neq 1$.

In practice, we use $b = 10$. Therefore, $\log_a n = \dfrac{\log_{10} n}{\log_{10} a}$

9) Formula for Antilogarithms: If $\log x = a$, then $x = anti\log a$.

Also: $anti\log_a p = a^p$ and $anti\log p = 10^p$

$anti\log_a (p+q) = a^{p+q} = a^p \cdot a^q = (anti\log_a p) \cdot (anti\log_a q)$

Examples: Given $\log_{10} 2 = 0.301$, $\log_{10} 3 = 0.477$, $\log_{10} 5 = 0.699$, we can find:

(Hint: We may omit the subscript 10.)

1. $\log_{10} 20 = \log_{10}(4 \times 5) = \log_{10} 4 + \log_{10} 5 = \log_{10} 2^2 + \log_{10} 5 = 2\log_{10} 2 + \log_{10} 5$
   $= 2(0.301) + 0.699 = 1.301$
2. $\log_{10} 0.2 = \log_{10} \frac{2}{10} = \log_{10} 2 - \log_{10} 10 = 0.301 - 1 = -0.699$
3. $\log_{10} 81 = \log_{10} 3^4 = 4\log_{10} 3 = 4(0.477) = 1.908$
4. $\log_{0.1} 3 = \dfrac{\log_{10} 3}{\log_{10} 0.1} = \dfrac{0.477}{-1} = -0.477$

Examples:

1. $\log 1 = 0$, $\log 100 = 2$, $\log 1000 = 3$, $\log 10000 = 4$
   $\log 0.1 = -1$, $\log 0.01 = -2$, $\log 0.001 = -3$, $\log 0.0001 = -4$
2. $\log_2 8 = \log_2 2^3 = 3$ **3.** $5^{\log_5 9} = 9$ **4.** $4^{\log_4 \pi} = \pi$
5. **Expand:** $\log_a(x^2\sqrt{x+1}) = \log_a x^2 + \log_a (x+1)^{1/2} = 2\log_a x + \frac{1}{2}\log_a(x+1)$
6. **Condense:** $4\log_a x + \frac{1}{2}\log_a(x-1) - 2\log_a(x+2) = \log_a x^4 + \log_a \sqrt{x-1} - \log_a(x+2)^2$
   $= \log_a \dfrac{x^4\sqrt{x-1}}{(x+2)^2}$
7. $anti\log 4 = 10^4 = 10{,}000$ **8.** $anti\log 0.9445 = 10^{0.9445} = 8.8$
9. $anti\log(-1.8763) = 10^{-1.8763} = 1/10^{1.8763} = \frac{1}{75.2142} = 0.0133$

The Number e and Natural Logarithms

When we graph the exponential function $y = a^x$, where $a > 0$ and $a \neq 1$, we notice the following two facts:

1. The slope (m) of the tangent to $y = 2^x$ through the point $p(0, 1)$ is less than 1.

2. The slope (m) of the tangent to $y = 3^x$ through the point $p(0, 1)$ is more than 1.

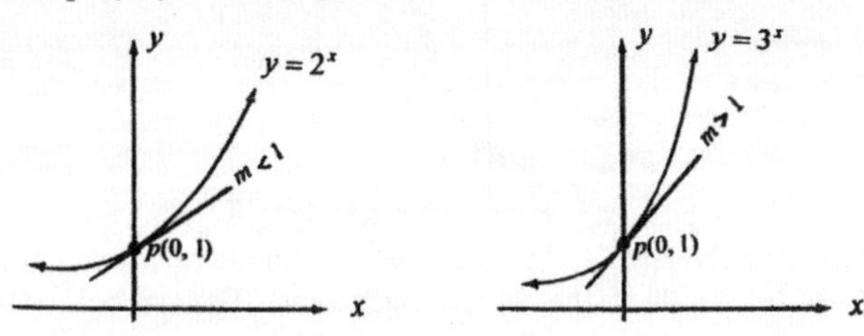

Therefore, we expect that there must be a number a, $2 < a < 3$, so that the slope of the tangent to $y = a^x$ through the point $p(0, 1)$ is exactly equal to 1. This number is designed by the letter e. The number e is called the **Euler number.**

The number e is one of the most famous numbers found in the history of mathematics. It is as famous as π and i. In advanced mathematics, we have found e from the following function in limit notation:

$e = \lim_{n \to \infty}(1 + \frac{1}{n})^n$. $e = 2.71828\cdots\cdots$. It is an irrational number.

$x = e^y$ is the inverse function of $y = e^x$. The **natural logarithmic function** of $x = e^y$ is written by $y = \log_e x$. (Note that : $y = \log_e x$ and $x = e^y$ are equivalent.)

Natural logarithms are abbreviated by ln, with the base understood to be e.

$y = \ln x \to$ read as " **y is the logarithm of x to the base e** ".

$y = \ln x$ and $y = e^x$ are inverses each other. The graph of $y = \ln x$ and $y = e^x$ are reflection of each other about the line $y = x$.

The domains and ranges of the two functions are interchanged.

$y = e^x \to x \in R,\ y > 0.$

$y = \ln x$ or $x = e^y \to y \in R,\ x > 0.$

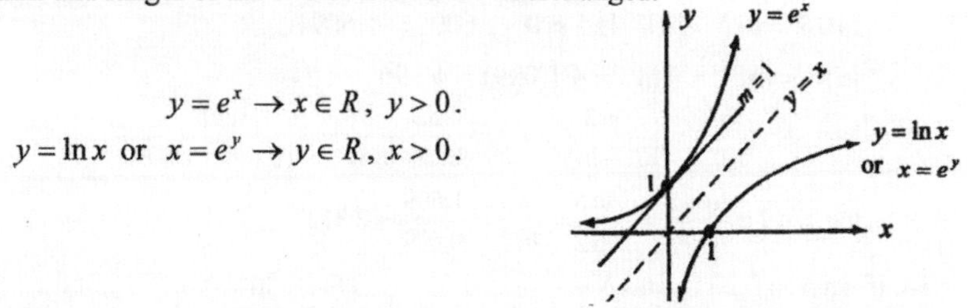

Natural logarithms are widely used in calculus and science. All laws of logarithms and the formulas for the antilogarithms that we have learned apply to the natural logarithms as well.

The function $y = e^x$ is called the **natural exponential function**. Its graph is shown below.

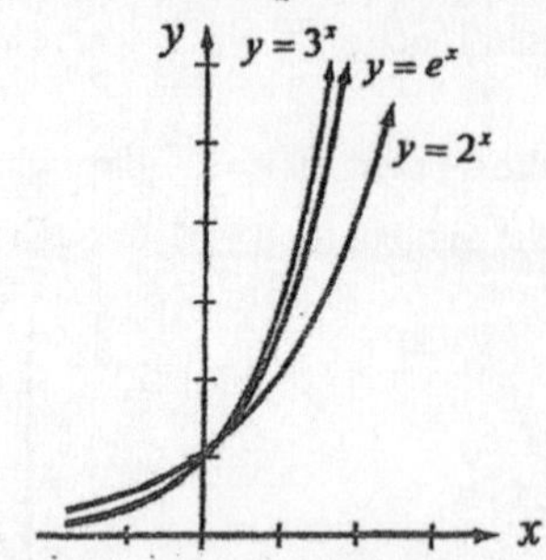

Most calculators have the key " e^x " to evaluate the exponential function for a given x-value and the key " LN " to evaluate the natural logarithms. The symbol for the antilogarithm of x to the base e is antiln x.

$$\text{If } \ln x = 0.6931, \text{ then } x = anti\ln 0.6931 = e^{0.6931} \approx 2.$$

Examples

1. $e^2 = (2.7182\cdots)^2 = 7.389$ **2.** $e^3 = (2.7182\cdots)^3 = 20.086$

3. $e^{1.2} = (2.7182\cdots)^{1.2} = 3.320$ **4.** $e^{5.8} = (2.7282\cdots)^{5.8} = 330.30$

5. $e^{-1.2} = \dfrac{1}{e^{1.2}} = \dfrac{1}{3.320} = 0.301$ **6.** $e^{-0.04} = \dfrac{1}{e^{0.04}} = \dfrac{1}{1.041} = 0.961$

7. $e^8 \cdot e^5 = e^{8+5} = e^{13}$ **8.** $3e^5 \cdot 4e^2 = 12e^{5+2} = 12e^7$

9. $\dfrac{12e^{10}}{2e^7} = 6e^{10-7} = 6e^3$ **10.** $\dfrac{12e^7}{2e^{10}} = 6e^{7-10} = 6e^{-3} = 6/e^3$

11. $(3e^5)^2 = 3^2e^{5(2)} = 9e^{10}$ **12.** $(2e^{-4x})^3 = 2^3e^{-4x\cdot 3} = 8e^{-12x} = 8/e^{12x}$

13. Given $\ln 2 = 0.693$, $\ln 3 = 1.099$, and $\ln 5 = 1.609$, Find:

$\ln 20 = \ln(4\times 5) = \ln 4 + \ln 5 = \ln 2^2 + \ln 5 = 2\ln 2 + \ln 5 = 2(0.693) + 1.609 = 2.995$.

$\ln 0.2 = \ln\frac{1}{5} = \ln 1 - \ln 5 = 0 - 1.609 = -1.609$.

$\ln 81 = \ln 3^4 = 4\ln 3 = 4(1.099) = 4.396$.

$\log_{0.1} 3 = \dfrac{\ln 3}{\ln 0.1} = \dfrac{\ln 3}{\ln 1 - \ln 10} = \dfrac{\ln 3}{0 - (\ln 2 + \ln 5)} = \dfrac{1.099}{-(0.693 + 1.609)} = -0.477$.

$\log_{\sqrt{2}} \sqrt{5} = \dfrac{\ln\sqrt{5}}{\ln\sqrt{2}} = \dfrac{\frac{1}{2}\ln 5}{\frac{1}{2}\ln 2} = \dfrac{\ln 5}{\ln 2} = \dfrac{1.609}{0.693} = 2.322$.

14. If $\ln N = 0.231$, find N.
Solution:
$N = anti\ln(0.231) = e^{0.231} = 1.26$.

15. If $\ln x = 0.492$, find x.
Solution:
$x = anti\ln(0.492) = e^{0.492} = 1.636$.

Solving Logarithmic Equations

Equations that contain the terms $\log_a x$, where a is a positive real number, with $a \neq 1$, are called **logarithmic equations.**
There are different ways to solve various types of logarithmic equations.

To solve **logarithmic equations,** we apply the laws of logarithms.

1. If write the exponential form using the same base on both side is not possible, take the logarithms on both sides.
2. Change the equation to exponential form if it has only a single logarithm.
3. Combine it as a single logarithm if it has multiple logarithms.

To solve a logarithmic equation, we must check each solution in the original equation and discard solutions that are not permissible.

Examples

1. Solve $3^x = 20$.

Solution:

$$\log 3^x = \log 20$$
$$x \log 3 = \log 20$$
$$\therefore x = {}^{\log 20}/_{\log 3} = {}^{1.301}/_{0.477} = 2.727.$$

2. Solve $\log_a 16 = 4$.

Solution:

$$a^4 = 16$$
$$a = 2 \text{ or } -2$$
$$(a = -2 \text{ is not permissible.})$$
$$\therefore a = 2.$$

3. Solve $\log_4 x + \log(x-6) = 2$.

Solution:

$$\log_4 x(x-6) = 2$$
$$x(x-6) = 4^2$$
$$x^2 - 6x - 16 = 0$$
$$(x-8)(x+2) = 0$$
$$x = 8 \text{ or } -2$$
$$(x = -2 \text{ is not permissible.})$$
$$\therefore x = 8.$$

4. Solve $\ln x + \ln(x-2) = 1$

Solution:

$$x(x-2) = e$$
$$x^2 - 2x - e = 0$$
$$x = \frac{2 \pm \sqrt{4+4e}}{2} = \frac{2 \pm 2\sqrt{1+e}}{2}$$
$$= 1 \pm \sqrt{1+e} \approx 2.3.$$

Notes

2-41 Domain and Range

The domain of a function $y = f(x)$ is the collection of all input values (x-values). The range of a function is the collection of all output values (y-values). For each x-value, there is exactly one y-value.

When a function is defined by an equation, its domain is restricted to **real numbers** for which the function can be evaluated. Otherwise, it must be excluded from the domain, such as:

1. In the Square-Root Function $y = \sqrt{x}$, the radicand x of a square root is always nonnegative. Both the domain and range are all nonnegative real numbers (greater than or equal to 0).

2. In the Absolute-Value Function $y = |x|$, the domain is all real numbers. The range is all nonnegative real numbers (greater than or equal to 0) because the absolute value of a number is never negative.

3. In the Rational Function $y = 1/x$, it is undefined if $x = 0$. A value that makes a division by zero must be excluded from the domain. Both the domain and range are all nonzero real numbers. $y = 0$ is the horizontal asymptote.

The graph of $y = \sqrt{x}$

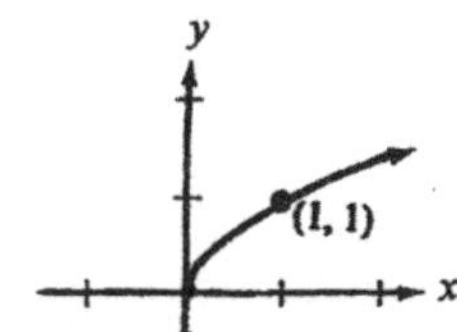

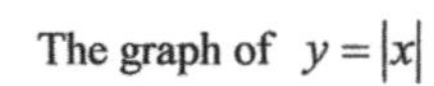

The graph of $y = |x|$

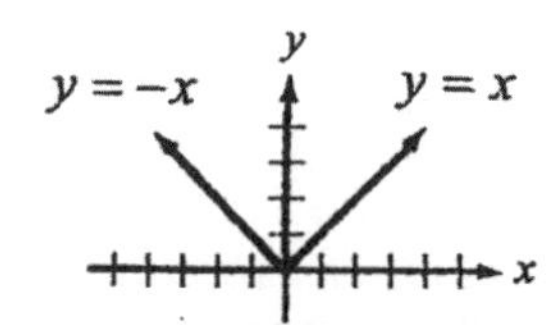

The graph of $y = \dfrac{1}{x}$

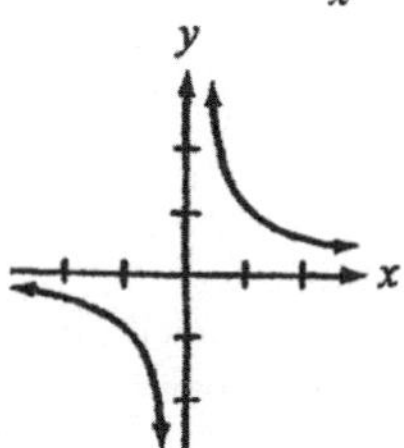

Write the domain and range of each function:

1. $y = \sqrt{x} + 2$
   D = All real numbers greater than or equal to 0.
   R = All real numbers greater than or equal to 2.

2. $y = \sqrt{x+3}$
   D = All real numbers greater than or equal to −3,
   R = All real numbers greater than or equal to 0

3. $y = 5 + \sqrt{x-2}$
   D = All real numbers greater than or equal to 2
   R = All real numbers greater than or equal to 5

4. $y = |x-2|$
   D = All real numbers.
   R = All real numbers greater than or equal to 0.

5. $y = |x| - 2$
   D = All real numbers.
   R = All real numbers greater than or equal to −2.

6. $y = -|x-2|$
   D = All real numbers.
   R = All real numbers less than or equal to 0.

Find the domain and range of each function:

7. $y=\dfrac{1}{x-2}$

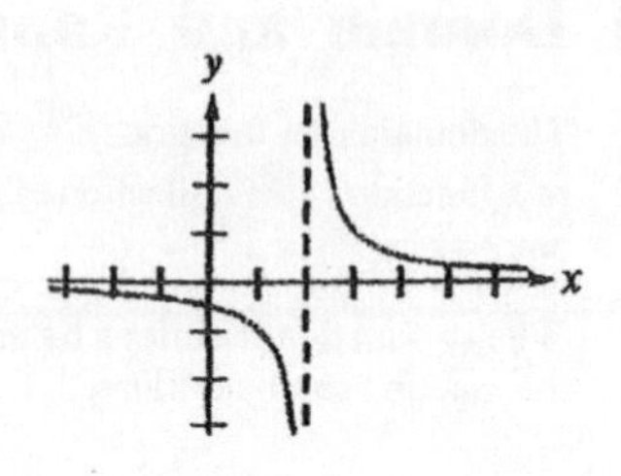

$y=0$ is the horizontal asymptote.

D = All real numbers except 2
R = All nonzero real numbers

8. $y=\dfrac{x-3}{x-2}$

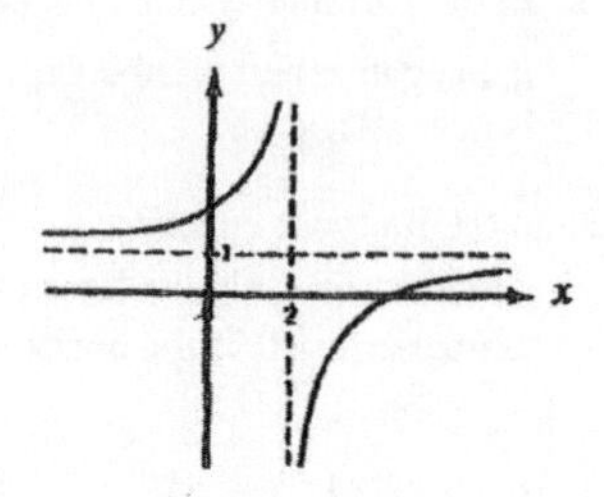

Dividing the denominator into the numerator, we have:

$$y=\frac{x-3}{x-2}=1-\frac{1}{x-2}$$

$y=1$ is the horizontal asymptote.

D = All real numbers except 2
R = All real numbers except 1

9. $y=\dfrac{x+1}{x+2}$

Dividing the denominator into the numerator, we have:

$$y=\frac{x+1}{x+2}=1-\frac{1}{x+2}$$

$y=1$ is the horizontal asymptote.

D = All real numbers except −2
R = All real numbers except 1

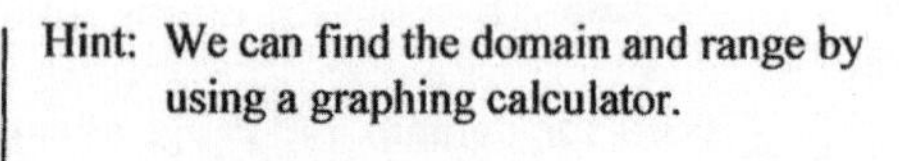
Hint: We can find the domain and range by using a graphing calculator.

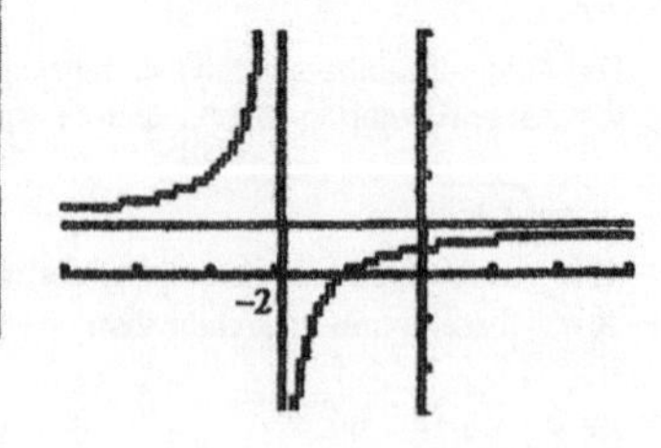

2-42 Imaginary Units and Complex Numbers

To find the solutions of the equation $x^2 - 1 = 0$, we have the solutions $x = \pm\sqrt{1} = \pm 1$.
However, there is no real number solution that satisfies the equation $x^2 + 1 = 0$.
The solutions of the equation $x^2 + 1 = 0$ is $x = \pm\sqrt{-1}$. The radical $\sqrt{-1}$ is not a real number. There is no real number x whose square is -1. Therefore, we introduce a new mathematical symbol using the **imaginary unit**, denoted by i, which is defined as:

Definition of i: $\quad i = \sqrt{-1}$ and $i^2 = -1$, where i is called "the square root of -1".

There is no real number square root of any negative number, such as:

$\sqrt{-1}$, $\sqrt{-2}$, $\sqrt{-4}$, $\sqrt{-9}$, $\sqrt{-10}$, ...→ $\sqrt[n]{b}$ if n is even and $b < 0$.

Examples: **1.** $\sqrt{-2} = \sqrt{2(-1)} = \sqrt{2} \cdot \sqrt{-1} = \sqrt{2}\, i$. **2.** $\sqrt{-4} = \sqrt{4(-1)} = \sqrt{4} \cdot \sqrt{-1} = 2i$.

3. $\sqrt{-9} = \sqrt{9(-1)} = \sqrt{9} \cdot \sqrt{-1} = 3i$. **4.** $\sqrt[3]{-8} = \sqrt[3]{-2 \cdot -2 \cdot -2} = -2$.

$i^2 = -1,\ i^3 = i^2 \cdot i = -i,\ i^4 = i^2 \cdot i^2 = (-1)(-1) = 1,\ i^7 = i^4 \cdot i^3 = 1 \cdot -i = -i.$

i and $-i$ are reciprocals: $\dfrac{1}{i} = \dfrac{1}{i} \cdot \dfrac{i}{i} = \dfrac{i}{i^2} = \dfrac{i}{-1} = -i$, $\dfrac{1}{-i} = \dfrac{1}{-i} \cdot \dfrac{i}{i} = \dfrac{i}{-i^2} = \dfrac{i}{-(-1)} = i$.

Before we simplify radical expressions, we must eliminate negative radicands by i.

$\sqrt{-2} \cdot \sqrt{-18} = \sqrt{2}\, i \cdot \sqrt{18}\, i = \sqrt{36}\, i^2 = 6(-1) = -6$ is correct.

$\sqrt{-2} \cdot \sqrt{-18} = \sqrt{(-2)(-18)} = \sqrt{36} = 6$ is incorrect.

$\sqrt{a} \cdot \sqrt{b} = \sqrt{ab}$ does not hold if a and b are negative.

Complex Numbers: Numbers in the form $a + b\, i$ or $a - b\, i$, such as:

$1 + 2i,\ 1 - 2i,\ 1 + \sqrt{2}\, i,\ 1 - \sqrt{2}\, i,\ 5 - 7\sqrt{3}\, i,\ 5\sqrt{2} + 7\sqrt{3}\, i$,......

Conjugates: Numbers in the form $a \pm b\, i$, such as: $1 \pm 2i$, $1 \pm \sqrt{2}\, i$, $5\sqrt{2} \pm 7\sqrt{3}\, i$ ....

The conjugate of $z = 2 - 5i$ is denoted by $\bar{z} = 2 + 5i$.

Rules: **1)** To add, subtract or multiply two complex numbers, we treat them the same ways as ordinary algebraic expressions.

2) To divide two complex numbers, we multiply both the numerator and the denominator by the conjugate of the denominator.

3) The product of two conjugate complex numbers is always a nonnegative real number (no imaginary part). $z\bar{z} = a^2 + b^2$

$(a + b\, i)(a - b\, i) = a^2 - (b\, i)^2 = a^2 - b^2 i^2 = a^2 - b^2(-1) = a^2 + b^2$.

Examples

1. $i^{30} = (i^2)^{15} = (-1)^{15} = -1$. **2.** $(2 + 3i) - (4 - 5i) = 2 + 3i - 4 + 5i = -2 + 8i$.

3. $(2 + 3i)(2 - 3i) = 2^2 - (3i)^2 = 4 - 9i^2 = 4 - 9(-1) = 4 + 9 = 13$.

4. $\dfrac{2+3i}{2-3i} = \dfrac{2+3i}{2-3i} \cdot \dfrac{2+3i}{2+3i} = \dfrac{4+12i+9i^2}{4-9i^2} = \dfrac{4+12i+9(-1)}{4-9(-1)} = \dfrac{-5+12i}{13} = -\dfrac{5}{13} + \dfrac{12}{13}i$.

Notes

2-43 Vectors

A quantity, such as a velocity or a force, that is given by both its magnitude and direction is represented by a **vector**.

A vector from point A to point B on a coordinate plane written by an arrow $\overrightarrow{AB}$.

We sometimes use a boldface letter to represent a vector. To write a vector, an arrow is placed over the letter.

A force of 10 units at a 40° angle.

$\mathbf{v} = \overrightarrow{AB}$

Point A is the **initial point** and Point B is the **terminal point**. The length of the arrow arrow indicates the **magnitude** (or **norm**) of the vector. The arrow-head of the arrow indicates the **direction** of the vector. The direction of the vector is the angle θ that the vector makes with the positive $x-axis$. The magnitude of a vector **v** is written as $\|\mathbf{v}\|$.

Zero Vector 0: A vector whose magnitude is zero. A zero vector has no direction.
Equivalent Vectors: They are vectors that have the same magnitude and direction.

$$\overrightarrow{AB} = \overrightarrow{CD}$$

The **Opposite** of a vector: $-\mathbf{v}$ is a vector with the same magnitude but the opposite direction as **v**.

Addition (Resultant) of two vectors: To add two vectors, we place the initial point of the second vector at the terminal point of the first vector, connect the initial point of the first vector to the terminal point of the second vector.

Subtraction (difference) of two vectors: To subtract two vectors, we add the opposite of the second vector. $\mathbf{u} - \mathbf{v} = \mathbf{u} + (-\mathbf{v})$, $\mathbf{v} - \mathbf{u} = \mathbf{v} + (-\mathbf{u})$

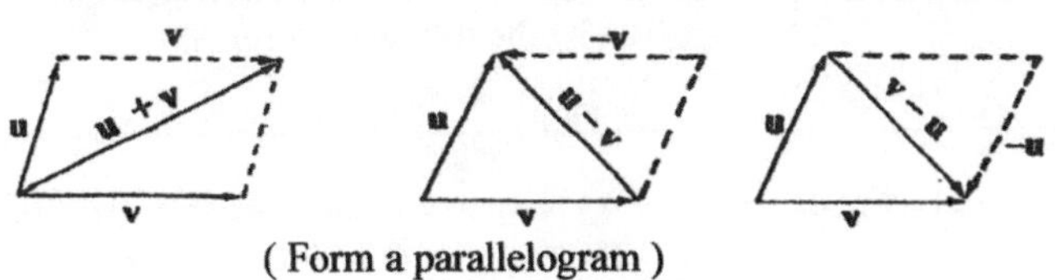

(Form a parallelogram)

The addition of vectors is **commutative** and **associative**:

$$\mathbf{v} + \mathbf{u} = \mathbf{u} + \mathbf{v}, \quad \mathbf{u} + (\mathbf{v} + \mathbf{w}) = (\mathbf{u} + \mathbf{v}) + \mathbf{w}$$

In a coordinate plane, a vector **v** with initial point at the origin and terminal point at $p(a, b)$ can be represented by a (the x – component) and b (the y – component) in the form of a **position vector** :

$$\mathbf{v} = \overrightarrow{op} = (a, b) \text{ or } \mathbf{v} = a\mathbf{i} + b\mathbf{j} \text{ and } \|v\| = \sqrt{a^2 + b^2}$$

where **i** is the **unit vector** in the same direction as the positive x – *axis* , **j** is the **unit vector in** the same direction as the positive y – *axis* , $\mathbf{i} = (1, 0)$ and $\mathbf{j} = (0, 1)$.

To add two vectors, we add the corresponding components. To subtract two vectors, we subtract the corresponding components.

If any vector **v** whose initial point $p_1(x_1, y_1)$ is not at the origin, and the terminal point is at $p_2(x_2, y_2)$, then we have:

$$\overrightarrow{p_1p_2} = \overrightarrow{op} \text{ (They have the same magnitude and direction.)}$$

$$\therefore \mathbf{v} = \overrightarrow{p_1p_2} = (x_2 - x_1)\mathbf{i} + (y_2 - y_1)\mathbf{j}$$

A vector **u** whose magnitude $\|u\| = 1$ is called a **unit vector**. To find the **unit vector** having the same direction as a given nonzero vector **v** , we use the following formula:

$$\textbf{unit vector} = \frac{v}{\|v\|}$$

Dot product (inner product) of two vectors

If $\mathbf{v} = a_1\mathbf{i} + b_1\mathbf{j}$ and $\mathbf{u} = a_2\mathbf{i} + b_2\mathbf{j}$ are two vectors, then the **dot product** $\mathbf{v} \cdot \mathbf{u} = a_1a_2 + b_1b_2$.

The dot product of two vectors are **commutative** and **distributive**:

$$\mathbf{v} \cdot \mathbf{u} = \mathbf{u} \cdot \mathbf{v} \quad ; \quad \mathbf{v} \cdot (\mathbf{u} + \mathbf{w}) = \mathbf{v} \cdot \mathbf{u} + \mathbf{v} \cdot \mathbf{w}$$

Two vectors **v** and **u** are **perpendicular** (or **orthogonal**) if and only if $\mathbf{v} \cdot \mathbf{u} = 0$.

Two vectors **v** and **u** are **parallel** if there is a nonzero scalar c and $\mathbf{v} = c\mathbf{u}$.

Angle between two vectors: If **v** and **u** are two nonzero vectors, the angle θ between **v** and **u** is given by the following formula:

$$\cos\theta = \frac{u \cdot v}{\|u\| \cdot \|v\|}$$

The vector $\mathbf{v} = 4\mathbf{i} - 2\mathbf{j}$ and $\mathbf{u} = 2\mathbf{i} + 4\mathbf{j}$ are perpendicular since $\mathbf{v} \cdot \mathbf{u} = (4)(2) + (-2)(4) = 0$. Or $\cos\theta = 0 \quad \therefore \theta = 90°$.

The vector $\mathbf{v} = 4\mathbf{i} - 2\mathbf{j}$ and $\mathbf{u} = 2\mathbf{i} - \mathbf{j}$ are parallel since $\mathbf{v} = 2\mathbf{u}$. Or

$$\cos\theta = \frac{(4)(2) + (-2)(-1)}{\sqrt{4^2 + (-2)^2} \cdot \sqrt{2^2 + (-1)^2}} = \frac{10}{\sqrt{20} \cdot \sqrt{5}} = \frac{10}{\sqrt{100}} = 1 \quad \therefore \theta = 0°.$$

Examples

1. If $\overrightarrow{AB} = (4,\ 1)$ and $\overrightarrow{BC} = (1,\ 2)$, find $\overrightarrow{AB} + \overrightarrow{BC}$.
   Solution:
   $$\overrightarrow{AB} + \overrightarrow{BC} = (4,\ 1) + (1, 2) = (4 + 1, 1 + 2) = (5, 3)$$

2. If $\mathbf{v} = (-4, 2)$ and $\mathbf{u} = (3, 4)$, find the **resultant** and the **norm** (magnitude) of $\mathbf{v} + \mathbf{u}$. Express $\mathbf{v}$ in terms of $\mathbf{i}$ and $\mathbf{j}$.
   Solution:
   Resultant $\mathbf{v} + \mathbf{u} = (-4 + 3, 2 + 4) = (-1, 6)$
   Norm $\|v + u\| = \sqrt{(-1)^2 + 6^2} = \sqrt{37}$
   $\mathbf{v} = -4\mathbf{i} + 2\mathbf{j}$

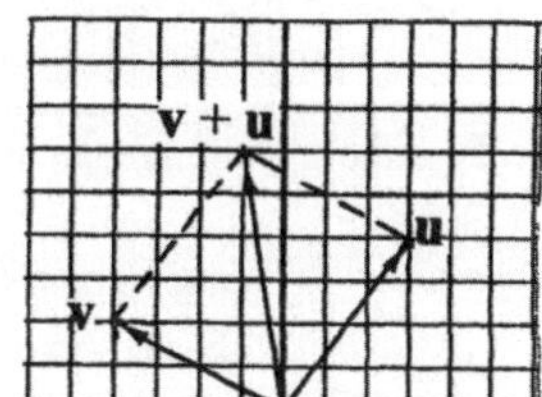

3. If $\mathbf{u} = (-4, 2)$ and $\mathbf{u} + \mathbf{v} = (-1, 6)$, find $\mathbf{v}$.
   Solution:
   $\mathbf{u} + \mathbf{v} = (-4 + v_1, 2 + v_2) = (-1, 6)$
   We have $v_1 = 3,\ v_2 = 4 \quad \therefore \mathbf{v} = (3, 4)$

4. Find the position vector of $\mathbf{v} = \overrightarrow{p_1 p_2}$ if its initial and terminal points are $p_1(5,\ 3)$ and $p_2(-1, -4)$.
   Solution: $\mathbf{v} = (-1 - 5)\mathbf{i} + (-4 - 3)\mathbf{j} = -6\mathbf{i} - 7\,\mathbf{j}$

5. If $\mathbf{v} = 3\mathbf{i} + 2\mathbf{j}$ and $\mathbf{u} = 5\mathbf{i} - 4\mathbf{j}$, find $\mathbf{v} + \mathbf{u}$, $\mathbf{v} - \mathbf{u}$, $\mathbf{u} - \mathbf{v}$, $4\mathbf{u}$, $3\mathbf{v} - 4\mathbf{u}$, and $\|v\|$.
   Solution:
   $$\mathbf{v} + \mathbf{u} = (3\mathbf{i} + 2\mathbf{j}) + (5\mathbf{i} - 4\mathbf{j}) = (3 + 5)\mathbf{i} + (2 - 4)\mathbf{j} = 8\mathbf{i} - 2\mathbf{j}$$
   $$\mathbf{v} - \mathbf{u} = \mathbf{v} + (-\mathbf{u}) = (3\mathbf{i} + 2\mathbf{j}) + (-5\mathbf{i} + 4\mathbf{j}) = (3 - 5)\mathbf{i} + (2 + 4)\mathbf{j} = -2\mathbf{i} + 6\mathbf{j}$$
   $$\mathbf{u} - \mathbf{v} = \mathbf{u} + (-\mathbf{v}) = (5\mathbf{i} - 4\mathbf{j}) + (-3\mathbf{i} - 2\mathbf{j}) = (5 - 3)\mathbf{i} + (-4 - 2)\mathbf{j} = 2\mathbf{i} - 6\mathbf{j}$$
   $$4\mathbf{u} = 4(5\mathbf{i} - 4\mathbf{j}) = 20\mathbf{i} - 16\mathbf{j}$$
   $$3\mathbf{v} - 4\mathbf{u} = (9\mathbf{i} + 6\mathbf{j}) + (-20\mathbf{i} + 16\mathbf{j}) = -11\mathbf{i} + 22\mathbf{j}$$
   $$\|v\| = \|3i + 2j\| = \sqrt{3^2 + 2^2} = \sqrt{13}$$

6. Find a **unit vector** having the same direction as $\mathbf{u} = 3\mathbf{i} + 4\mathbf{j}$.
   Solution:
   $$\|u\| = \|3i - 4j\| = \sqrt{3^2 + 4^2} = 5 \quad \therefore \textbf{unit vector} = \frac{u}{\|u\|} = \frac{3i + 4j}{5} = \frac{3}{5}i + \frac{4}{5}j$$
   Hint: Its magnitude (norm) $= \sqrt{(\frac{3}{5})^2 + (\frac{4}{5})^2} = \sqrt{\frac{9}{25} + \frac{16}{25}} = 1$

Examples

7. Express $\mathbf{v} = -4\mathbf{i} + 2\mathbf{j}$ in terms of $\mathbf{a} = (2, 3)$ and $\mathbf{b} = (4, 2)$.

Solution:

$$\mathbf{v} = (-4, 2)$$

$$\mathbf{v} = v_1\mathbf{a} + v_2\mathbf{b} = v_1(2, 3) + v_2(4, 2) = (2v_1, 3v_1) + (4v_2, 2v_2)$$
$$= (2v_1 + 4v_2, 3v_1 + 2v_2)$$

Solve $2v_1 + 4v_2 = -4$

$3v_1 + 2v_2 = 2$ we have $v_1 = 2$, $v_2 = -2$ $\therefore \mathbf{v} = 2\mathbf{a} - 2\mathbf{b}$

Check : $\mathbf{v} = 2\mathbf{a} - 2\mathbf{b} = 2(2, 3) - 2(4, 2) = (4, 6) - (8, 4) = (-4, 2)$

8. Find the x – component and y – component of a vector $\mathbf{v}$, 10 units long at a direction of 60°.

Solution:

$$x = 10\cos 60° = 10(0.5) = 5$$
$$y = 10\sin 60° = 10(0.866) = 8.66$$
$$\mathbf{v} = (5, 8.66)$$

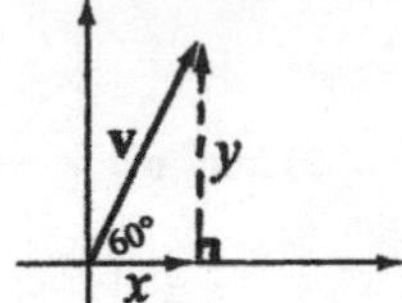

9. If $\mathbf{v} = 3\mathbf{i} + 4\mathbf{j}$ and $\mathbf{u} = 4\mathbf{i} + 3\mathbf{j}$, find the dot product $\mathbf{v} \cdot \mathbf{u}$ and the angle between $\mathbf{v}$ and $\mathbf{u}$.

Solution:

$$\mathbf{v} \cdot \mathbf{u} = 3(4) + 4(3) = 24$$

$$\cos\theta = \frac{v \cdot u}{\|v\| \cdot \|u\|} = \frac{24}{\sqrt{3^2 + 4^2} \cdot \sqrt{4^2 + 3^2}} = \frac{24}{25} = 0.96 \quad \therefore \theta = 16.26°$$

10. An object is pulled by a force of 10 units to the south and a force of 6 units at a heading (clock-wise from due north) of 60°. Find the magnitude and direction of the resultant.

Solution:

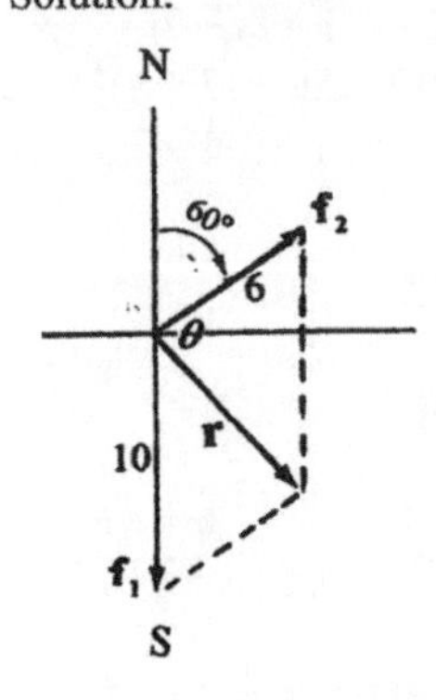

The force of 10 units to the south is $\mathbf{f}_1 = -10\mathbf{j}$

The force of 6 units at a heading of 60° is

$$\mathbf{f}_2 = (6\cos 30°)\ \mathbf{i} + (6\sin 30°)\ \mathbf{j} = 3\sqrt{3}\ \mathbf{i} + 3\ \mathbf{j}$$

The resultant $\mathbf{r} = (-10\ \mathbf{j}) + (3\sqrt{3}\ \mathbf{i} + 3\ \mathbf{j}) = 3\sqrt{3}\ \mathbf{i} - 7\ \mathbf{j}$

The magnitude of the resultant is

$$\|r\| = \sqrt{(3\sqrt{3})^2 + (-7)^2} = \sqrt{76} = 8.72$$

The direction of the resultant is given by

$$\cos\theta = \frac{f_2 \cdot r}{\|f_2\| \cdot \|r\|} = \frac{(3\sqrt{3})(3\sqrt{3}) + 3(-7)}{(6)(8.72)} = \frac{6}{52.32} = 0.1147$$

$\therefore\ \theta = 83.4°$ (it is at a heading of 143.4°)

2-44 Limits and Continuity

The concept of limits and continuity is very important to the study of calculus. Suppose we sketch the graph of the following function by direct substitution:

$$f(x)=\frac{x^2-1}{x-1}, \quad x \neq 1$$

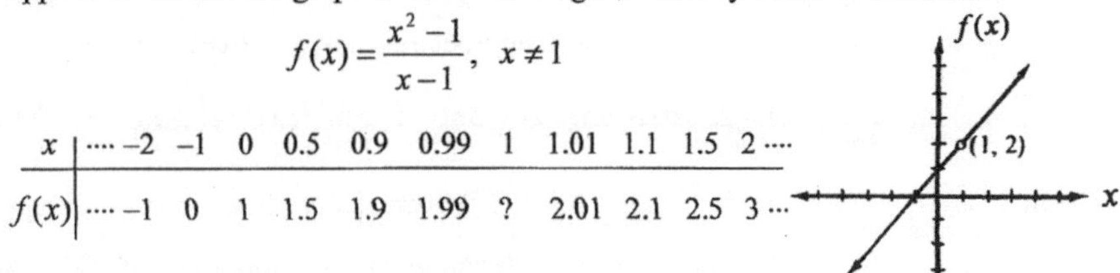

x	···· −2	−1	0	0.5	0.9	0.99	1	1.01	1.1	1.5	2 ····
$f(x)$	···· −1	0	1	1.5	1.9	1.99	?	2.01	2.1	2.5	3 ···

The value of $f(x)$ at $x=1$ is undefined.
Since x cannot equal to 1, we are not sure what the value of $f(x)$ is at $x=1$.
The graph of $f(x)$ is a straight line that has a hole at the point (1, 2).
The point (1, 2) is not part of the graph.

To find the value of $f(x)$ near the undefined point at $x=1$, we could use the values of x that approaches 1 from the left, and the values of x from the right. We estimate that the value of $f(x)$ moves close to 2 when x approaches 1 from either the right or the left. We say that " the limit of $f(x)$ is 2 as x approaches 1 ", and we write the result in limit notation: $\lim_{x\to 1} f(x)=2$

Definition of a limit: If the values of a function $f(x)$ approach or equal L as the values of x approaches c, we say that $f(x)$ has limit L as x approaches c.
We write: $\lim_{x\to c} f(x)=L$.
Read " The limit of $f(x)$ equals L as x approaches c.

Some functions do not have a limit as $x \to c$. If a limit of a function exists, it is unique. In other words, a function cannot have two different limits as $x \to c$.

By reducing the function, we have $f(x)=\dfrac{x^2-1}{x-1}=\dfrac{(x+1)\cancel{(x-1)}}{\cancel{x-1}}=x+1,\ x \neq 1$.

Therefore, for all points other than $x=1$, $\lim_{x\to 1}\dfrac{x^2-1}{x-1}=\lim_{x\to 1}(x+1)=2$.

We use the following ways to evaluate the limit of $f(x)$ as $x \to c$:

1. Evaluated by direct substitution.
2. If $f(x)$ is undefined at $x=c$, we try to reduce $f(x)$ and find a new function that agrees with $f(x)$ for all x other than c.

One-side limits: $\lim_{x \to c^+} f(x) = L$ represents the limit of $f(x)$ from the right. It is the limit of $f(x)$ as x approaches c from values greater than c.

$\lim_{x \to c^-} f(x) = L$ represents the limit of $f(x)$ from the left. It is the limit of $f(x)$ as x approaches c from values less than c.

$\lim_{x \to c} f(x) = L$ **exists only and only if** $\lim_{x \to c^+} f(x) = L$ **and** $\lim_{x \to c^-} f(x) = L$.

Infinite Limits: $\lim_{x \to c} f(x) = \infty$ means that $f(x)$ increases without bound as x approaches c.

$\lim_{x \to c} f(x) = -\infty$ means that $f(x)$ decreases without bound as x approaches c.

Since infinite " ∞ " is not a number, we may say that "the limit does not exist".

Limits of the greatest integer functions: If n is an integer, then $\lim_{x \to n} [x]$ does not exist.

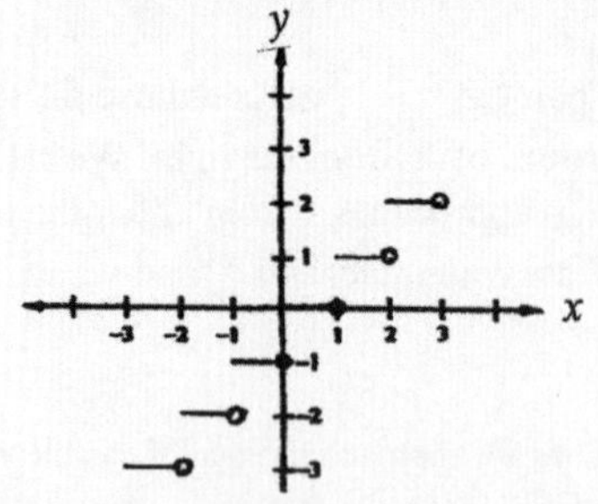

Example: $\lim_{x \to 2^+} [x] = \lim_{x \to 2^+} 2 = 2$

$\lim_{x \to 2^-} [x] = \lim_{x \to 2^-} 1 = 1$

Therefore, $\lim_{x \to 2} [x]$ does not exist.

If the degree of $p(x)$ is greater than the degree of $q(x)$, then $\lim_{x \to \infty} \frac{p(x)}{q(x)} = \pm \infty$.

If the degree of $p(x)$ is less than the degree of $q(x)$, then $\lim_{x \to \infty} \frac{p(x)}{q(x)} = 0$.

To find the limit of a quotient of polynomials, we divide the highest power of x and use the fact that $\lim_{x \to \infty} \frac{1}{x} = 0$.

$$\lim_{x \to \infty} \frac{-x^3}{2 - x^2} = \lim_{x \to \infty} \frac{-1}{\frac{2}{x^3} - \frac{1}{x}} = \frac{-1}{0 - 0} = \frac{-1}{0} = -\infty.$$

The limit does not exist in this example.

Examples

1. $\lim\limits_{x\to 2} 3 = 3$ **2.** $\lim\limits_{x\to 2} x^3 = 2^3 = 8$ **3.** $\lim\limits_{x\to 1}(x^2+x+2) = 1^2+1+2 = 4$ **4.** $\lim\limits_{x\to -2}|x| = 2$

5. $\lim\limits_{x\to 1}\dfrac{x^2+x-2}{x-1} = \lim\limits_{x\to 1}\dfrac{(x+2)(x-1)}{x-1} = \lim\limits_{x\to 1}(x+2) = 1+2 = 3$.

6. $\lim\limits_{x\to -4}\dfrac{x^2+2x-8}{x+4} = \lim\limits_{x\to -4}\dfrac{(x+4)(x-2)}{x+4} = \lim\limits_{x\to -4}(x-2) = -4-2 = -6$.

7. $\lim\limits_{x\to 2}\dfrac{x-2}{x^3-2x^2+3x-6} = \lim\limits_{x\to 2}\dfrac{x-2}{x^2(x-2)+3(x-2)} = \lim\limits_{x\to 2}\dfrac{\cancel{x-2}}{\cancel{(x-2)}(x^2+3)} = \lim\limits_{x\to 2}\dfrac{1}{x^2+3} = \dfrac{1}{7}$.

8. $\lim\limits_{x\to 0}\dfrac{x}{x^3} = \lim\limits_{x\to 0}\dfrac{1}{x^2} = \dfrac{1}{0} = \infty$ (undefined). The limit does not exist.

$f(x)$ increases without bound as x approaches 0 from either the right or the left. It has no limit as $x \to 0$.

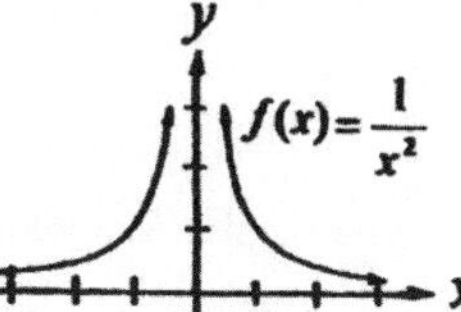

9. $\lim\limits_{x\to 0}\dfrac{\sqrt{x+4}-2}{x} = \lim\limits_{x\to 0}\dfrac{(\sqrt{x+4}-2)(\sqrt{x+4}+2)}{x(\sqrt{x+4}+2)} = \lim\limits_{x\to 0}\dfrac{(x+4)-4}{x(\sqrt{x+4}+2)} = \lim\limits_{x\to 0}\dfrac{x}{x(\sqrt{x+4}+2)}$

$= \lim\limits_{x\to 0}\dfrac{1}{\sqrt{x+4}+2} = \dfrac{1}{\sqrt{0+4}+2} = \dfrac{1}{4}$.

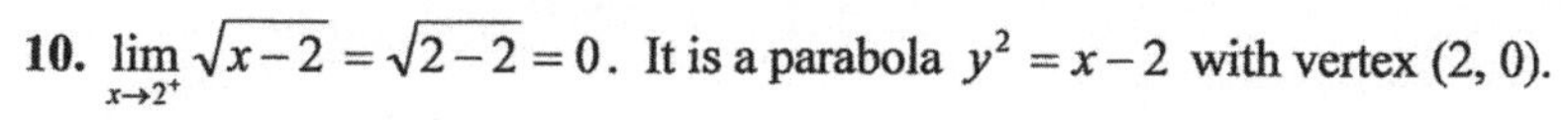

10. $\lim\limits_{x\to 2^+}\sqrt{x-2} = \sqrt{2-2} = 0$. It is a parabola $y^2 = x-2$ with vertex (2, 0).

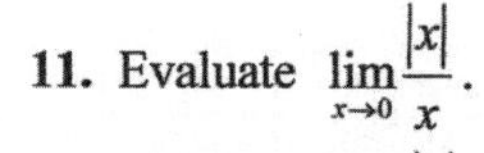

11. Evaluate $\lim\limits_{x\to 0}\dfrac{|x|}{x}$.

Solution: $\lim\limits_{x\to 0^+}\dfrac{|x|}{x} = \lim\limits_{x\to 0^+}\dfrac{x}{x} = \lim\limits_{x\to 0^+} 1 = 1$

$\lim\limits_{x\to 0-}\dfrac{|x|}{x} = \lim\limits_{x\to 0^-}\dfrac{-x}{x} = \lim\limits_{x\to 0^-}(-1) = -1$.

Therefore, the limit does not exist.

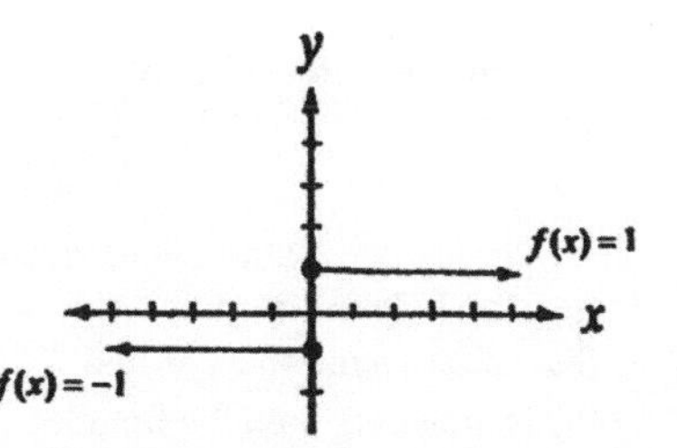

12. Evaluate $\lim\limits_{x\to 1} f(x)$, Where $f(x) = \begin{cases} 2x^2, --- -1 \le x \le 1 \\ -x+3, -- 1 \le x \le 3 \end{cases}$

Solution:

$\lim\limits_{x\to 1^+}(-x+3) = -1+3 = 2$

$\lim\limits_{x\to 1^-}(2x^2) = 2(1)^2 = 2$

Therefore, $\lim\limits_{x\to 1} f(x) = 2$.

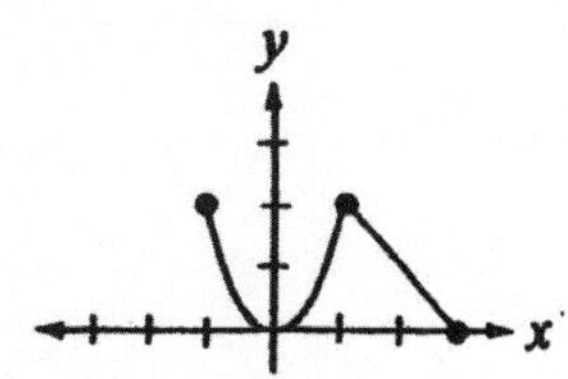

Continuity

A function $f(x)$ is said to be continuous at $x = c$ if its graph is unbroken (no holes) at $x = c$. If a function is not continuous at $x = c$, we say that it is discontinuous at $x = c$.

A function $f(x)$ is said to be continuous on the closed interval $[a, b]$ if it is continuous on the interval $a \le x \le b$.

A function $f(x)$ that is continuous on the entire real numbers $(-\infty, +\infty)$ is called a continuous function.

Definition of Continuity: A function $f(x)$ is said to be continuous at $x = c$ if the following three conditions are satisfied:

1. $f(x)$ is defined. **2.** $\lim_{x \to c} f(x)$ exists. **3.** $\lim_{x \to c} f(x) = f(c)$.

Continuity on a closed interval: If a function $f(x)$ is continuous on a closed interval $[a, b]$, we say that it is continuous at every interior point, continuous on the right at a, and continuous on the left at b.

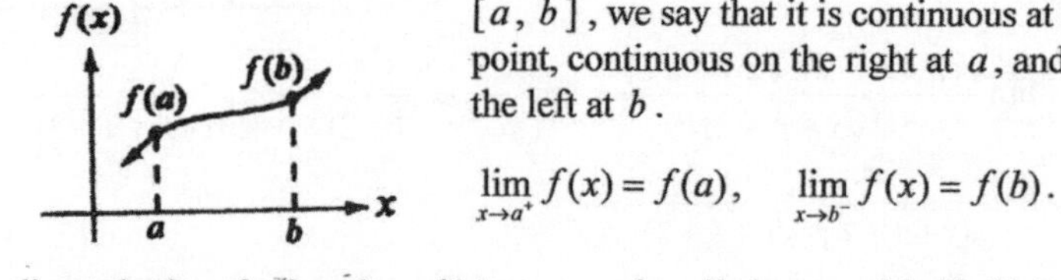

$\lim_{x \to a^+} f(x) = f(a), \quad \lim_{x \to b^-} f(x) = f(b).$

A discontinuity of a function $f(x)$ at $x = c$ is called removable if $f(x)$ can be made continuous by redefining $f(x)$ at $x = c$.

Intermediate Value Theorem: If $f(x)$ is continuous on the closed interval $[a, b]$ and k is any number between $f(a)$ and $f(b)$, then there is at least one number c between a and b such that $f(c) = k$.

The Intermediate Value Theorem states that if a function is continuous on a closed interval, there is no hole in its graph.

If there are two numbers a and b in the function $f(x)$, such that $f(a)$ is negative and $f(b)$ is positive, then the function $f(x)$ must have at least one number c between a and b such that $f(c) = 0$. Its graph must cross the $x - axis$ at $x = c$.

The equation $f(x) = 0$ has a real root at $x = c$.

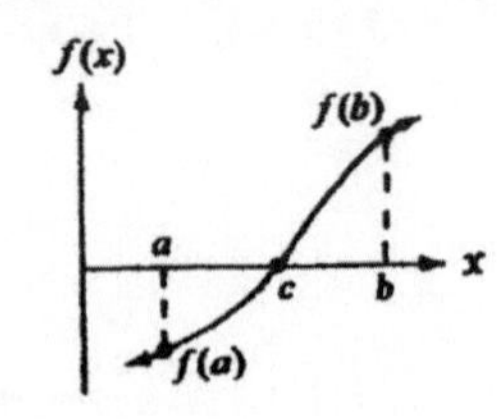

Chapter Three : Geometry

3-1 Understanding Geometry

Geometry is the study of the properties and mutual relations of points, lines, planes, and solids in space. There are many applications of geometry in science and our real life, such as archeology, navigation, astronomy, housing and gardening.
The famous ancient mathematicians who contributed in finding geometric facts and theorems are Pythagoras, Plato, Aristotle, and Euclid. The Greek mathematician Euclid (300 B.C.) is the first person to find and organize basic geometric principles.
Geometry includes **Plane (Euclidean) Geometry**,
Coordinate (Analytic) Geometry,
and **Solid (Analytic) Geometry.**
Plane (Euclidean) Geometry deals with figures in **plane** (or two-dimensions).
Coordinate (Analytic) Geometry deals with the solutions by **algebraic analysis**.
Solid (Analytic) Geometry deals with figures in **space** (or three-dimensions).
In the study of geometry, we analyze the relationships among the geometric facts. Then, we make statements and conclusions. There are many postulates and theorems in geometry. Some basic statements are presented below.
Postulates: (Statements that we assume to be true without proof.)

1. Every line contains at least two distinct points.
2. Every plane contains at least three distinct, noncollinear points.
3. Two distinct points determine one and only one straight line.
4. Three distinct noncollinear points determine one and only one plane.
5. The line containing two distinct points in a plane lies in the plane.
6. Two intersecting planes have at least two points in common.
7. The length of a segment is the shortest distance between two points.
8. A segment has one and only one midpoint.
9. An angle has one and only one bisector.

Theorems: (Statements that can be proved to be true.)

1. If two distinct lines intersect, then their intersection is a point.
2. If two distinct planes intersect, then their intersection is a line.
3. The sum of the measures of the angles of a triangle is 180°.
4. Corollary: The measure of each angle of an equilateral triangle is 60°.
5. Pythagorean Theorem: In any right triangle, the square of the length of the hypotenuse equals the sum of the squares of the lengths of the two legs.

3-2 Points, Lines, Planes, and Angles

A **point** indicates position only. It is represented by a dot " • " and is named by a single capital letter near the dot. A point has no size (length, width, or height).

Collinear points are points that lie on the same line.
Noncollinear points are points that do not lie on the same line.
Coplanar points are points that lie in the same plane.
Noncoplanar points are points that do not lie in the same plane.

A **line** is represented by a set of continuous points that extend without end in two opposite directions and is named by two capital letters of any points on the line and drawing a " — " over the letters. A line has infinite length but no width or height.

A **ray** is the part of a line that extends without end from its one endpoint.

A **line segment** of a line consists of two endpoints with all the points between them. It is customary to use " $m\overline{AB}$ " or " AB " to express the length of line segment $\overline{AB}$.

The **midpoint** on a line segment is the point that bisects (two equal halves) the line segment

A **plane** is a flat surface that has length and width but no height. The length and width of a plane can be extended without end in all directions. A plane is represented by a four-sided figure and is named by placing a capital letter at one of its corners.

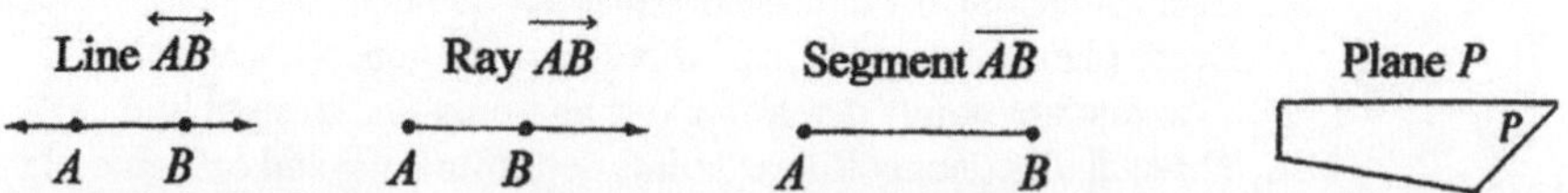

An **angle** is the figure formed by two noncollinear rays having the same endpoint and is named by the symbol " $<$ ". The rays are the sides of the angle and the endpoint is the vertex. The angle with vertex B below may be named in different ways as long as it does not cause any confusion, such as $<ABC$, $<CBA$, $<B$, or <1. We use a protractor to measure an angle. In using a protractor, we place the center of the protractor at the vertex of the angle and line up on ray of the angle with the "0" mark on the protractor.
No matter how large or small the two sides are, the size of the angle would not be changed.

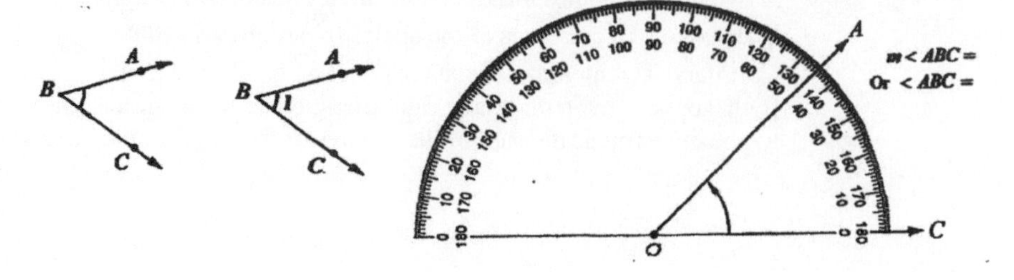

The measure of an angle is the number of degrees or radians it contains. If an angle $< B$ measures 45 degrees, we write $m < B = 45$ or $< B = 45^o$. An angle is often denoted by α **(alpha)**, β **(belta)**, θ **(theta)**, γ **(gamma)**, or ϕ **(phi)**.

An **acute angle** is an angle whose measure is less than 90°.
A **right angle** is an angle whose measure is 90° and is denoted by a square.
An **obtuse angle** is an angle whose measure is more than 90° and less than 180°.
A **straight angle** is an angle whose measure is 180°.
A **reflex angle** is an angle whose measure is more than 180° and less than 360°.

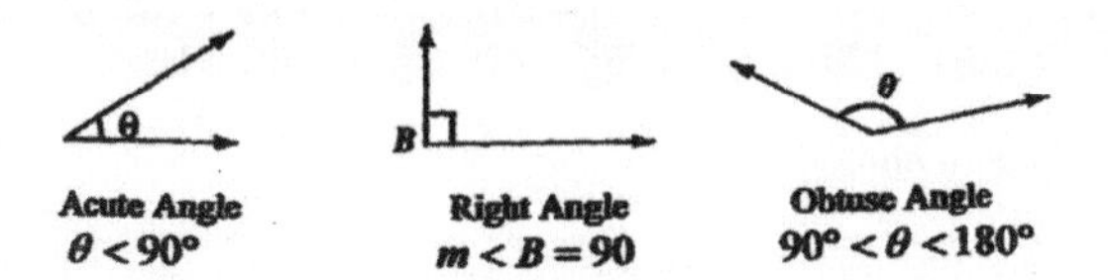

Complementary Angles are two angles whose measures have a sum of 90°.
Supplementary Angles are two angles whose measures have a sum of 180°.
Two angles form a linear pair are supplementary.

Vertical Angles are two nonadjacent angles when two lines or line segments intersect.
Vertical angles are equal.

Linear Pairs: <1 and <2, <2 and <3, <3 and <4, <1 and <4

Vertical Angles: <1 and <3, <2 and <4
$m<1 = m<3$
$m<2 = m<4$

Examples

1: Find the measure of each angle of the figure.

a. $< CBE$ **b.** $< CBF$ **c.** $< CBG$
d. $< ABH$ **e.** $< DBH$ **f.** $< PBD$
g. $< PBF$ **h.** $< ABC$

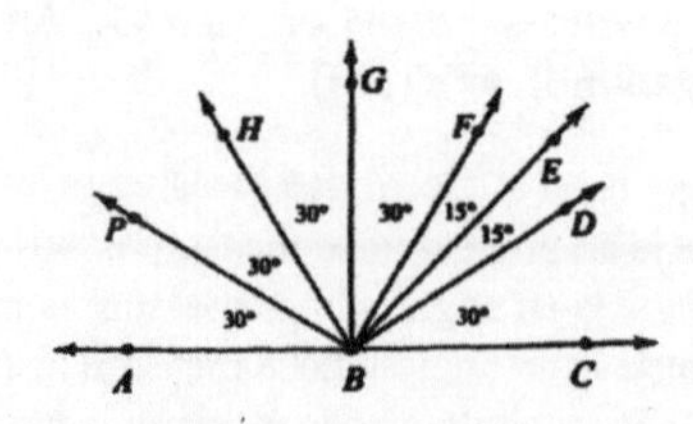

Solution:

a. $< CBE = 45°$ b. $< CBF = 60°$ c. $< CBG = 90°$ d. $< ABH = 60°$
e. $< DBH = 90°$ f. $< PBD = 120°$ g. $< PBF = 90°$ h. $< ABC = 180°$.

2: Identify each angle as acute, right, or obtuse.

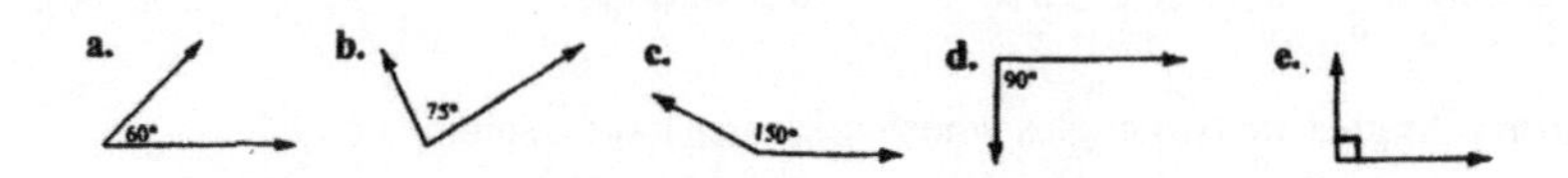

Solution: a. Acute b. Acute c. Obtuse d. Right e. Right.

3: The measure of an angle is 30^0 .

a. Find the measure of its supplement. b. Find the measure of its complement.

Solution:

a. Measure of the supplement: $180° - 30° = 150°$.
b. Measure of the complement: $90° - 30° = 60°$.

4: The measure of an angle is 30^0 larger than its supplement. Find the measure of the angle and the measure of its supplement.

Solution:

Let $x =$ the measure of the angle
$x - 30 =$ the measure of its supplement

We have the equation:

$$x + x - 30 = 180$$
$$2x = 210$$
$$\therefore x = 105^0 \rightarrow \text{The measure of the angle.}$$
$$\therefore x - 30 = 75^0 \rightarrow \text{The measure of its supplement.}$$

3-3 Parallel Lines and Angles

Perpendicular Lines: Two lines (rays, segments) are perpendicular if they intersect to form a right ($90°$) angle. The symbol is " $\perp$ ".

A line (or ray, segment) is called a **perpendicular bisector** when it is perpendicular to the segment at the midpoint.

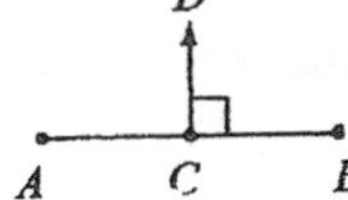

$\overrightarrow{CD} \perp \overline{AB}$ and $\overline{AC} = \overline{CB}$

$\overrightarrow{CD}$ is the perpendicular bisector of $\overline{AB}$.

Parallel Lines: Two lines are parallel if they will never intersect. The symbol is " // ".

If two parallel lines are cut by a transversal, the transversal will form eight angles (4 **interior angles** and 4 **exterior angles**).

Corresponding angles are two angles in corresponding positions relative to the two lines and the transversal. **Alternate angles** are nonadjacent angles on opposite sides of a transversal.

Among the eight angles, we have the following special relationships:

1. The **corresponding angles** are congruent.
2. The **alternate interior angles** are congruent.
3. The **alternate exterior angles** are congruent.

In geometry, **congruent figures** have the same shape and size. The symbol is " $\cong$ ".

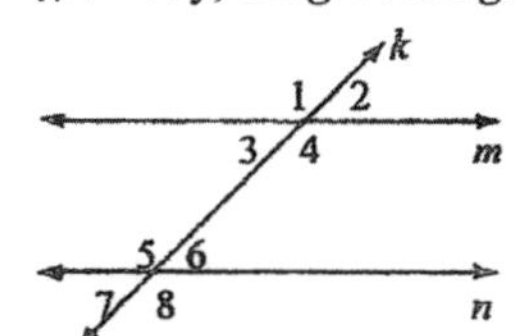

Lines m and n cut by transversal k, $m // n$, we have:

Corresponding angles: $<1 \cong <5$, $<2 \cong <6$, $<3 \cong <7$, $<4 \cong <8$

Alternate interiors angles: $<3 \cong <6$, $<4 \cong <5$

Alternate exterior angles: $<2 \cong <7$, $<1 \cong <8$

Example 1: If $<1 = 135°$ on the above figure, find each measure.

Solution:

$<2 = 45°$, $<3 = 45°$, $<4 = 135°$, $<5 = 135°$

$<6 = 45°$, $<7 = 45°$, $<8 = 135°$. Ans.

Example 2: If line f and line g are parallel, find the sum of measures of $<a$ and $<b$.

Solution:

$m<a + m<1 = 180$

$<b = <1$ (corresponding angles)

$\therefore\ m<a + m<b = 180$. Ans. (Hint: $a + b = 180°$)

Fred's Theorem: When two parallel lines are cut by a transversal, two kinds of angles are formed, little angles and big angles. All of the little angles are equal. All of the big angles are equal. Any little angle plus any big angle equals $180°$.

Examples

3: If line s // line t and $<2=62°$,
find $m<5$ and $m<8$.
Solution:

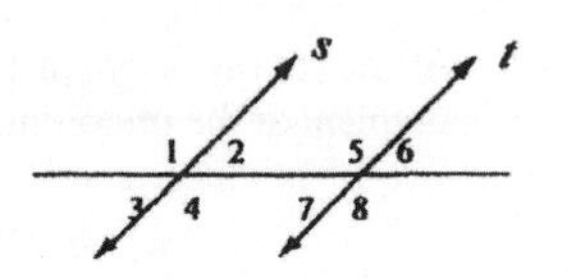

$$m<1+m<2=180 \text{ (linear pair)}$$
$$m<1+62=180 \quad \therefore <1=118°$$
$$\therefore <5 \cong <1=118°. \text{ (corresponding angles)}$$
$$\therefore <8 \cong <5=118°. \text{ (vertical angles)}$$

4. Classify the statement as true or false:
If $<4=135°$ and $<6=45°$,
then line m // line n.
Solution:

$$m<3+m<4=180 \text{ (linear pair)}$$
$$m<3+135=180 \quad \therefore <3=45°$$
$$<2 \cong <3=45° \text{ (vertical angles)}$$
$$<6=45° \text{ (given)}$$
$$<2 \cong <6 \text{ (They are parallel.)}$$

$\therefore$ The statement is **true**. (The corresponding angles are congruent.)

5. Classify the statement as true or false:
If $<4=135°$ and $<7=47°$,
then line t // line s.
Solution:

$$<1 \cong <4=135° \text{ (vertical angles)}$$
$$m<1+m<3=180 \text{ (linear pair)}$$
$$135+m<3=180 \quad \therefore <3=45°$$
$$<7=47° \text{ (given)}$$
$$m<3 \neq m<7 \text{ (They are not parallel.)}$$

$\therefore$ The statement is **false**. (The corresponding angles are not congruent.)

6. If line k and line p are parallel and $<1=140°$, find $m<5-m<6$.

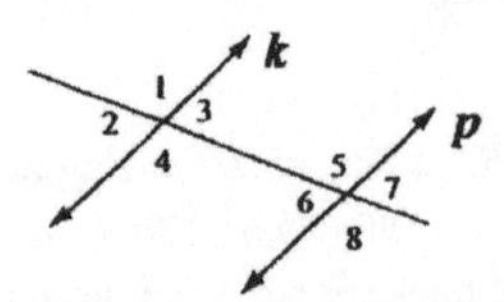

Solution:

$$<5 \cong <1=140° \text{ (corresponding angles)}$$
$$m<6+m<5=180 \text{ (linear pair)}$$
$$m<6+140=180 \quad \therefore <6=40°$$
$$\therefore m<5-m<6=140-40=100.$$

3-4 Basic Rules in Triangles

A triangle is the figure formed by connecting three noncollinear points with three line segments and is named by the symbol " Δ ". A triangle below may be named with its three letters in any order, such as Δ*ABC*, Δ*BCA*, or Δ*CAB*. Its three sides are $\overline{AB}$, $\overline{BC}$, and $\overline{AC}$. It has three vertices *A*, *B*, and *C* and three angles $< A$, $< B$, and $< C$.
An **equilateral triangle (equiangular triangle)** is a triangle having three equal sides (angles).
An **isosceles triangle** is a triangle having at least two equal sides (called legs). An equilateral triangle is also an isosceles triangle.
A **scalene triangle** is a triangle having no equal sides or no equal measure of angels.
A **right triangle** is a triangle having a right ($90°$) angle.
An **acute triangle** is a triangle having three acute angles.
An **obtuse triangle** is a triangle having an obtuse angle.

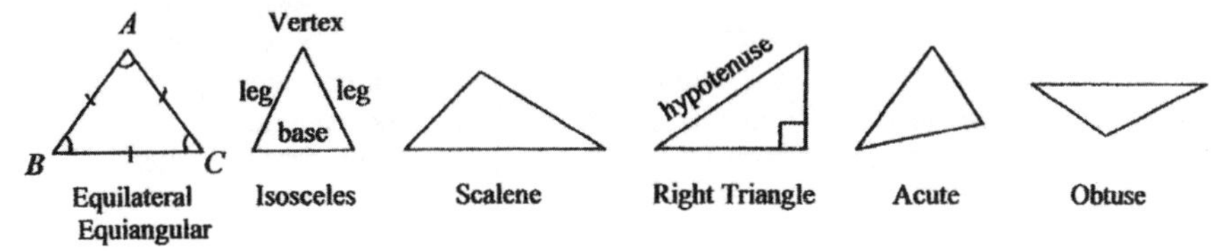

Basic Rules in Triangles

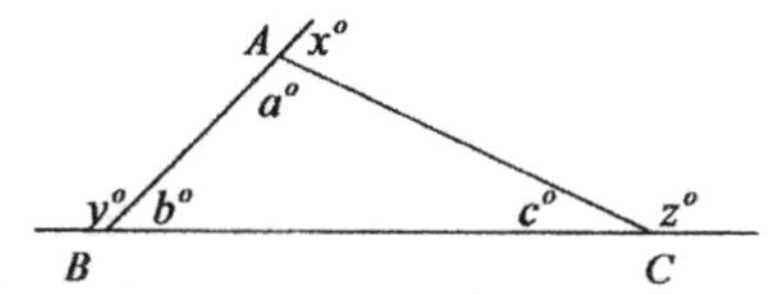

1. **Triangle-Sum Theorem:** The sum of the measures of interior angles of any triangle is $180°$. The sum of the measures of exterior angles of any triangle is $360°$.

$$a+b+c=180, \quad x+y+z=360$$

2. **Measure of an Exterior Angle:** The measure of an exterior angle of a triangle is equal to the sum of its two remote interior angles.

$$x=b+c, \quad y=a+c, \quad z=a+b$$

3. **Angle-Side (Proportionality) Relationships:** In every triangle, the larger angle is opposite the larger side and the smaller angle is opposite the smaller side.

If $a>b>c$, then $\overline{BC}>\overline{AC}>\overline{AB}$

4. **The Triangle-Inequality Theorem (Third-Side Rule):**
The sum of the lengths of any two sides is greater than the length of the third side.
The difference of the lengths of any two sides is less than the length of the third side.

$$\overline{AB}+\overline{AC}>\overline{BC}, \quad \overline{AB}+\overline{BC}>\overline{AC}, \quad \overline{AC}+\overline{BC}>\overline{AB}$$

$$\left|\overline{AB}-\overline{AC}\right|<\overline{BC}, \quad \left|\overline{AB}-\overline{BC}\right|<\overline{AC}, \quad \left|\overline{AC}-\overline{BC}\right|<\overline{AB}$$

Basic Rules in Triangles (Continued)

5. **Isosceles-Triangle Theorem:**
If two sides of a triangle are congruent, then the two base angles are congruent.

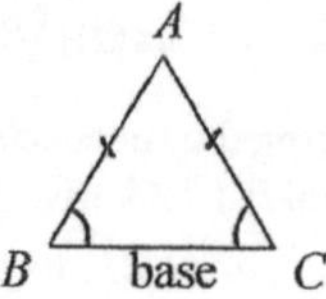

If $\overline{AB} = \overline{AC}$,
then $< B \cong < C$.

6. The **median** of a triangle is a segment from one vertex to the midpoint of the opposite side. There are three medians in a triangle. The point of concurrency of three medians is the **centroid**, or **center of gravity** of the triangle. The distance between the centroid to the vertex is $\frac{2}{3}$ of the length of the median.

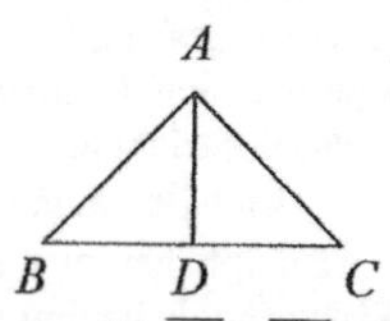

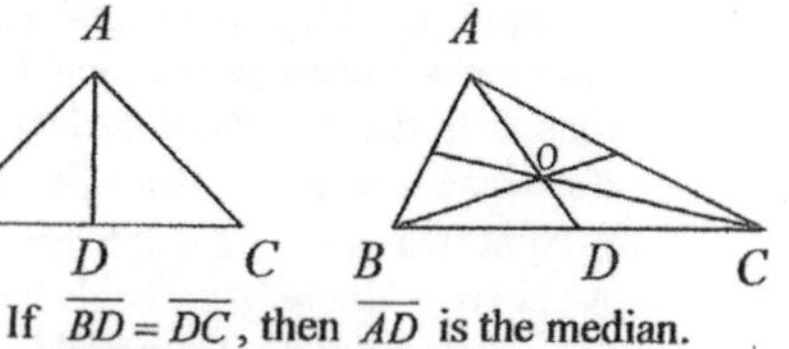

If $\overline{BD} = \overline{DC}$, then $\overline{AD}$ is the median.
Point o is the centroid.

7. The **height (altitude)** of a triangle is a segment from one vertex perpendicular to the opposite side **(the base)**, or perpendicular to the line containing the opposite side. The height of an isosceles triangle is also the median of the triangle. The point of concurrency of three heights is the **orthocenter** of the triangle.

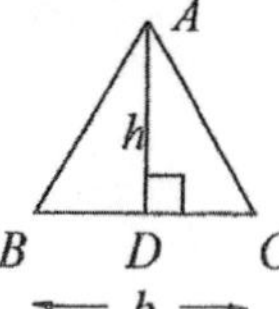
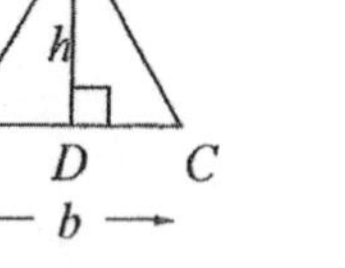

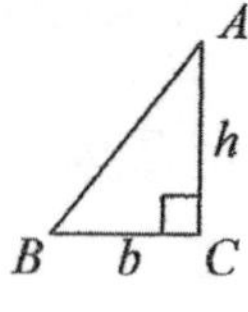

A h B b C D

$\overline{AD}$ is the height.
$\overline{BC}$ is the base.

$\overline{AC}$ is the height.
$\overline{BC}$ is the base.

$\overline{AD}$ is the height.
$\overline{BC}$ is the base.

8. The **angle bisector** of a triangle is a segment from one side of the triangle and divides an angle int two congruent angles. There are three angle bisectors in a triangle.
The angle bisector of an isosceles triangle is also a median and the height of the triangle.
The point of concurrency of three angle bisectors is the **incenter** of the triangle.
The incenter of the triangle is equidistance from the sides of the triangle. The incenter is the center of the inscribed circle.

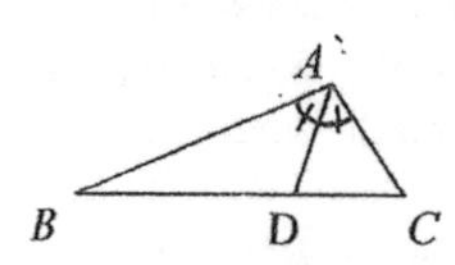

$m < BAD = m < CAD$
$\overline{AD}$ is a angle bisector of ΔABC.

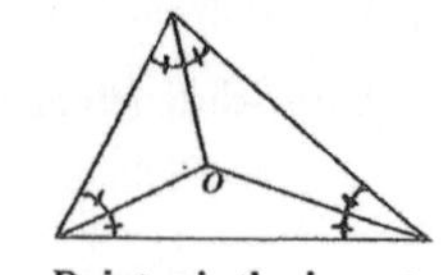

Point o is the incenter.

9. The **circumcenter** of a triangle is the point of concurrency of the perpendicular bisectors of the sides of a triangle. The circumcenter of a triangle is equidistance from the vertices of the triangle. The circumcenter is the center of the circumscribed circle.

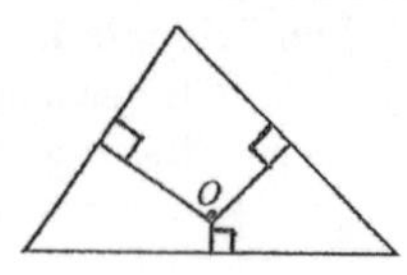

Point o is the circumcenter.

Basic Rules in Triangles (Continued)

10. Area Formula of a Triangle: The area of a triangle is one-half of (base × height). $A = \frac{1}{2}bh$.

11. Area of an Equilateral Triangle with side of length s: $A = \dfrac{s^2\sqrt{3}}{4}$

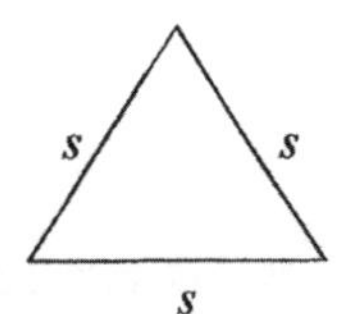

Examples

1. Find the measure of the angle x.
   Solution:
   $$x + 75 + 35 = 180$$
   $$\therefore x = 70.$$

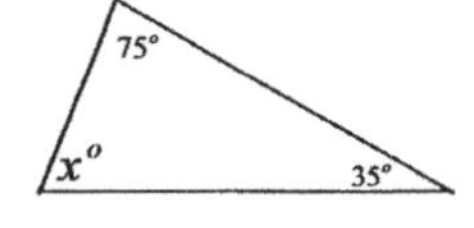

2. Find the measure of the angle a.
   Solution:
   $$a = 100 + 30 = 130.$$

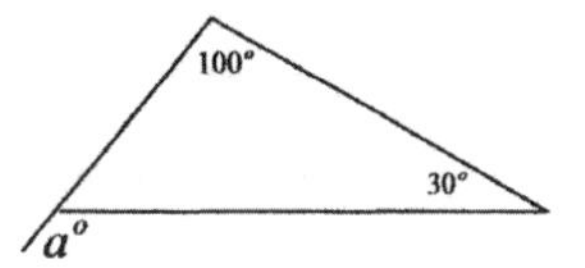

3. Identify the statement as true or false: The lengths of sides of a triangle are 2, 4, 6.
   Solution:
   False.
   (The sum of the lengths of any two sides of a triangle must be greater than the length of the third side.)

4. A triangle has the lengths of sides 7, 12, and x. What are the possible values of x ?
   Solution:
   $$12 - 7 < x < 12 + 7$$
   $$\therefore 5 < x < 19.$$

5. A triangle has the lengths of two sides 7 and 12. The perimeter is p. What are the possible values of p ?
   Solution:
   The length of the third side is $(p - 12 - 7)$
   $$12 - 7 < p - 12 - 7 < 12 + 7$$
   $$5 < p - 19 < 19$$
   $$\therefore 24 < p < 38.$$

6. An isosceles triangle has the length of sides 7, 15, and x. What are the possible values of x ?
   Solution:
   A 7- 7-15 (sides) triangle is not possible.
   A 15-15-7 (sides) triangle is the only isosceles triangle:
   $$\therefore x = 15 \text{ only}.$$

Examples

7. The sides of an equilateral triangle are each 6 inches long. Find the area.
Solution:

$$A=\frac{s^2\sqrt{3}}{4}=\frac{6^2\sqrt{3}}{4}=9\sqrt{3}\approx 15.60\ in.^2$$

8. The area of an equilateral triangle are $9\sqrt{3}$. What is the length of its perimeter ?
Solution:

$$A=\frac{s^2\sqrt{3}}{4}=9\sqrt{3},\quad s^2=36,\quad s=6,\quad \therefore p=3\times 6=18.$$

9. In the figure, what is the measure of x ?
Solution:

The sum of the measure of interior angles is 180°.
$m<1=45$, $45+84+x=180$
$\therefore x=51$.

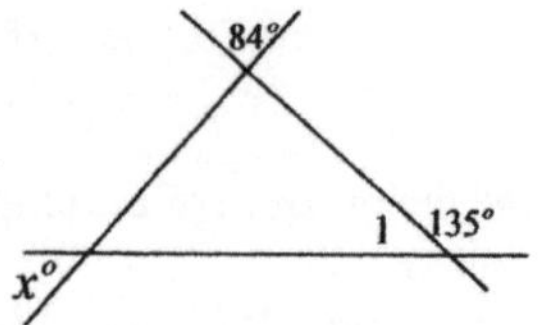

10. ΔABC is an isosceles triangle. $\overline{DB}$ and $\overline{DC}$ are two angle bisectors. $m<BDC=140$.
Find $m<A$.
Solution:

$$m<DBC+m<DCB+140=180$$
$$m<DBC=m<DCB=20$$
We have $m<B=m<C=40$
$$m<A+m<B+m<C=180$$
$$m<A+40+40=180$$
$$\therefore m<A=100.$$

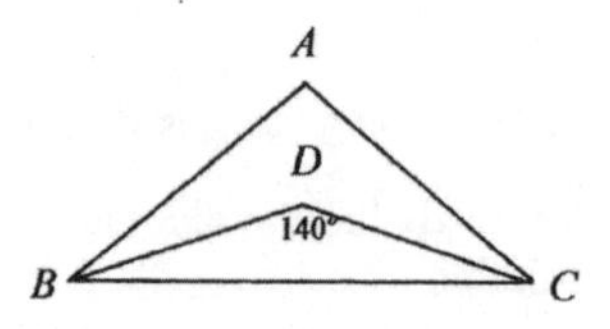

11. ΔABC is a triangle and $<A:<B:<C$ = 2:3:4. What is the measure of $<C$?
Solution:

$$m<C=180\times\frac{4}{2+3+4}=180\times\frac{4}{9}=80.$$

12. A triangle has an area of 0.5. If all of the lengths of sides in this triangle are decreased as one-half of the original lengths, what is the new area ?
Solution:

We choose the original triangle with a base of 1 and a height of 1.
We have the new area:

$$A=\tfrac{1}{2}bh=\tfrac{1}{2}\cdot\tfrac{1}{2}\cdot\tfrac{1}{2}=\tfrac{1}{8}=0.125.$$

Other Method: If two triangles are similar, the ratio of their areas is the square of the ratio of the lengths of the corresponding sides.

$$\frac{A}{0.5}=\left(\frac{0.5}{1}\right)^2\qquad \therefore A=0.5\times 0.25=0.125.$$

Notes

Notes

3-5 The Pythagorean Theorem

Right angle: It is an angle with measure 90° and indicated by a small square.

Right triangle: It is a triangle having one right (90°) angle.

The hypotenuse: It is the side opposite the right angle in a triangle.
The other two sides are the **legs.**

The Pythagorean Theorem:
In any right triangle, the square of the length of the hypotenuse equals the sum of the squares of the lengths of the two legs.

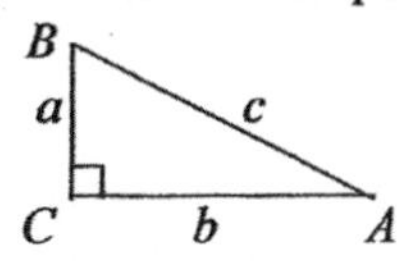

Formula: $a^2 + b^2 = c^2$

We can use the **Pythagorean Theorem** to find the length of one side of a right triangle when the lengths of the other two sides are known. To apply the Pythagorean Theorem, we need the knowledge of squares and square roots.

To determine whether or not the triangle with the given lengths of its three sides is a right triangle, we apply the Pythagorean Theorem. If the sum of the squares of the lengths of the two shorter sides is equal to the square of the length of the longest side, then it is a right triangle.

Examples

1. Find the length of the unknown side.

a)

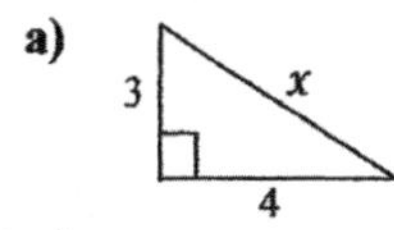

b)

c)

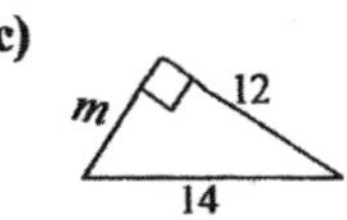

Solution:

a) $x^2 = 3^2 + 4^2$
$x^2 = 9 + 16$
$x^2 = 25$
$\therefore x = \sqrt{25} = 5.$

b) $y^2 = 2^2 + 4^2$
$y^2 = 4 + 16$
$y^2 = 20$
$\therefore y = \sqrt{20} = 4.47.$

c) $m^2 + 12^2 = 14^2$
$m^2 + 144 = 196$
$m^2 = 52$
$\therefore m = \sqrt{52} = 7.21.$

2. Determine whether or not the triangle with the given lengths of its three sides is a right triangle.

a) 6, 8, 10 **b)** 4, 5, $5\sqrt{2}$ **c)** 6, $6\sqrt{2}$, 6

Solution:

a) $6^2 + 8^2 = 100$
$10^2 = 100$
$6^2 + 8^2 = 10^2$ **Yes**

b) $4^2 + 5^2 = 41$
$(5\sqrt{2})^2 = 50$
$4^2 + 5^2 \neq (5\sqrt{2})^2$ **No**

c) $6^2 + 6^2 = 72$
$(6\sqrt{2})^2 = 72$
$6^2 + 6^2 = (6\sqrt{2})^2$ **Yes**

Examples

3. A baseball field is shaped like a square diamond. Each side is 100 feet long. Find the distance from home plate to second base.

Solution:

$$x^2 = 100^2 + 100^2$$
$$x^2 = 20000$$
$$\therefore x = \sqrt{20000} \approx 141.42 \text{ feet.}$$

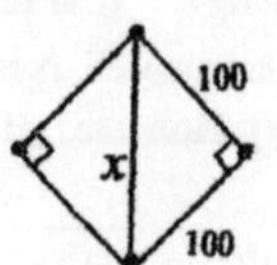

4. The sides of an equilateral triangle are each s units long. Find the altitude(h) in terms of s.

Solution:

$$s^2 = h^2 + (\tfrac{1}{2}s)^2$$
$$s^2 = h^2 + \tfrac{1}{4}s^2$$
$$h^2 = \tfrac{3}{4}s^2 \quad \therefore h = \tfrac{\sqrt{3}}{2}s.$$

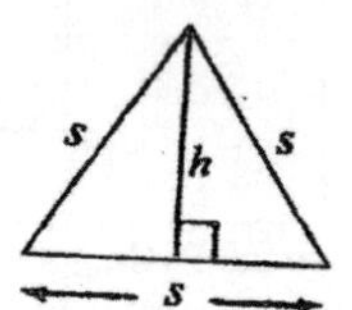

5. Find the area of the triangle.

Solution:

$$(BC)^2 = (4\sqrt{2})^2 + (4\sqrt{2})^2 = 32 + 32 = 64 \quad \therefore BC = 8 \text{ (the base)}$$
$$h^2 + 4^2 = (4\sqrt{2})^2$$
$$h^2 + 16 = 32 \quad \therefore h = 4$$

We have the area:

$$A = \tfrac{1}{2}bh = \tfrac{1}{2} \cdot 8 \cdot 4 = 16.$$

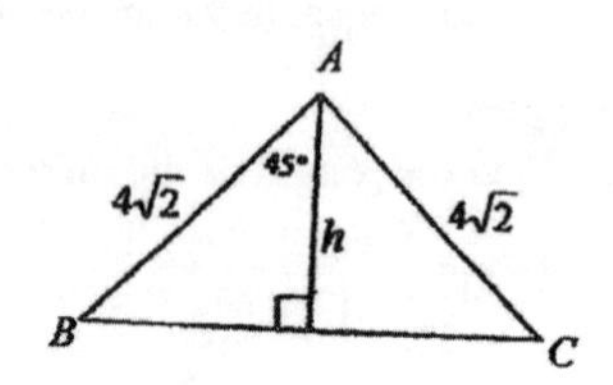

3-6 Special Right Triangles

There are two special right triangles, **30-60-90 triangle** and **45-45-90 triangle.** The lengths of sides of each special right triangle follow a special pattern.

1. A **30-60-90 Triangle** is a triangle with angles of $30°$, $60°$, and $90°$.
   The sides of every 30-60-90 triangle will follow "$1: 2 : \sqrt{3}$ ratio".
   In a 30-60-90 triangle, the length of the shorter side (opposite side of $30°$ angle) is one-half of the length of the hypotenuse. The length of the longer side is $\sqrt{3}$ times the length of the shorter side.

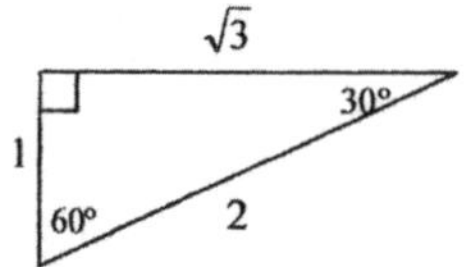

Basic 30-60-90 triangle

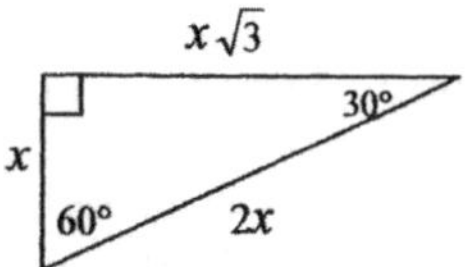

Any 30-60-90 triangle with shorter side x
(Area $= \frac{\sqrt{3}}{2}x^2$)

2. A **45-45-90 Triangle** is a triangle with angles of $45°$, $45°$, and $90°$. It is also an isosceles right triangle.
   The sides of every 45-45-90 triangle will follow "$1:1:\sqrt{2}$ ratio".
   In a 45-45-90 triangle, the two legs are always equal. The length of the hypotenuse is $\sqrt{2}$ times the length of the leg.

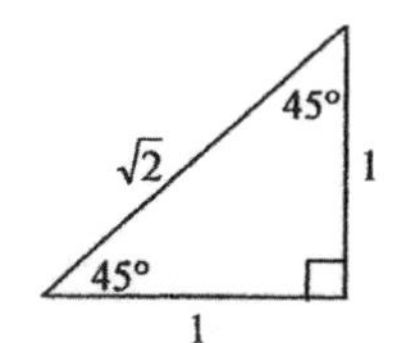

Basic 45-45-90 triangle

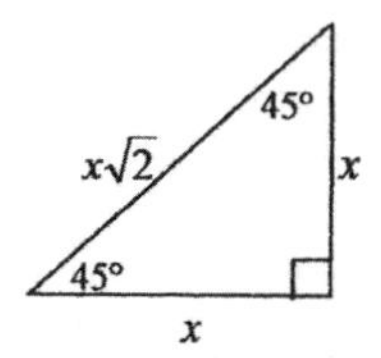

Any 45-45-90 triangle with leg x
(Area $= \frac{1}{2}x^2$)

Examples

1. The hypotenuse of a 30-60-90 triangle is 10. Find the lengths of its two sides.
   Solution:
   The shorter side is one-half of the hypotenuse $= 5$
   The longer side is $\sqrt{3}$ times the shorter side $= 5\sqrt{3}$.

2. The hypotenuse of a 45-45-90 triangle is 10. Find the length x of its leg.
   Solution:
   $x\sqrt{2} = 10 \quad \therefore x = \frac{10}{\sqrt{2}} = 5\sqrt{2}$.

Examples

3. Find the length of the diagonal of a square with sides of 20 inches.

Solution:

$d^2 = 20^2 + 20^2$

$d^2 = 800$

$\therefore d = \sqrt{800} = \sqrt{400 \times 2} = 20\sqrt{2}$.

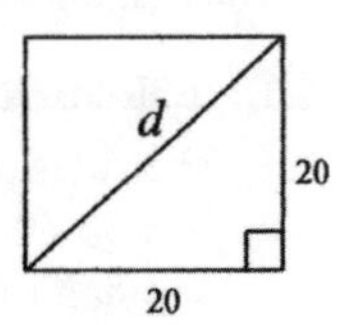

Or: It is a 45-45-90 triangle. The length of the hypotenuse is $\sqrt{2}$ times the length of the leg.

$d = 20\sqrt{2}$.

4. On the figure, Find the measures of $\overline{AD}$, $\overline{AB}$, $\overline{BD}$, and $\overline{DC}$.

Solution:

ΔADC is a 30-60-90 triangle.

$AD = \frac{1}{2}(AC) = \frac{1}{2}(20) = 10$

$\therefore DC = 10\sqrt{3}$.

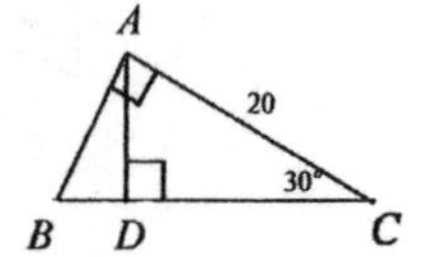

ΔABD is a 30-60-90 triangle.

$AD = \sqrt{3}(BD) \quad \therefore BD = \frac{1}{\sqrt{3}}(AD) = \frac{1}{\sqrt{3}}(10) = \frac{10\sqrt{3}}{3}$.

$\therefore AB = 2(BD) = 2 \cdot \frac{10\sqrt{3}}{3} = \frac{20\sqrt{3}}{3}$.

5. A 45-45-90 triangle has a perimeter of 30.73. Find the length of its hypotenuse and area.

Solution:

$x + x + x\sqrt{2} = 30.73$

$(2 + \sqrt{2})x = 30.73$

$x = \dfrac{30.73}{2+\sqrt{2}} = \dfrac{30.73}{3.414} = 9 \quad \therefore$ Hypotenuse $= x\sqrt{2} = 9\sqrt{2} \approx 12.73$.

$\therefore$ Area $= \frac{1}{2}x^2 = \frac{1}{2} \cdot 9^2 = 40.5$.

6. The equilateral triangle with the side length of 12 is inscribed in the circle. Find the area of the circle.

Solution:

$m < AOC = 120$, $m < AOD = 60$

ΔAOD is a 30-60-90 triangle.

The side opposite to the 60° angle is $\sqrt{3} \times$ (the side opposite to the 30° angle).

$AD = \sqrt{3}(OD)$, $OD = \frac{1}{\sqrt{3}}(AD) = \frac{1}{\sqrt{3}}(6) = 2\sqrt{3}$

$OA = 2(OD) = 2 \cdot 2\sqrt{3} = 4\sqrt{3}$

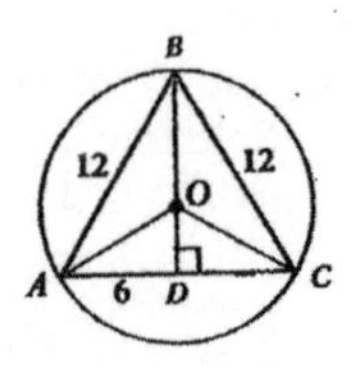

The hypotenuse $\overline{OA}$ is the radius of the circle: $r = 4\sqrt{3}$

The area of the circle: $A = \pi r^2 = \pi(4\sqrt{3})^2 = 48\pi \approx 150.72$.

Notes

Notes

3-7 Similar and Congruent Triangles

Two triangles are **similar** if the ratio of the lengths of their corresponding sides is a constant. The symbol is " $\sim$ ". The corresponding angles of two similar triangles are congruent.

AAA Similarity or AA similarity: If three angles of one triangle are congruent to the three angles of another triangle, then the triangles are similar.

If a line parallel to one side of a triangle determines the second triangle, then the second triangle is similar to the original triangle.

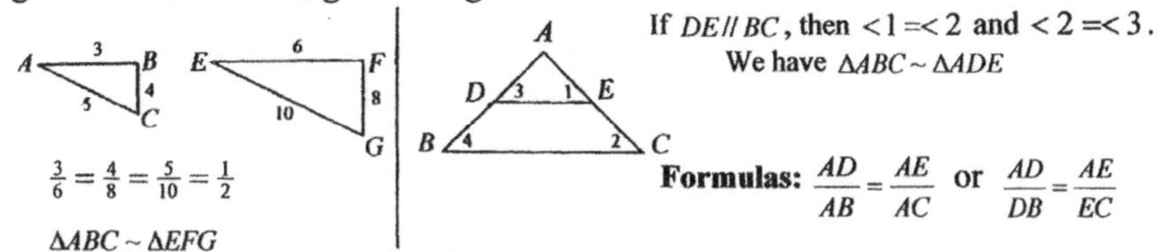

$\frac{3}{6} = \frac{4}{8} = \frac{5}{10} = \frac{1}{2}$

$\Delta ABC \sim \Delta EFG$

If $DE // BC$, then $<1 = <2$ and $<2 = <3$. We have $\Delta ABC \sim \Delta ADE$

Formulas: $\frac{AD}{AB} = \frac{AE}{AC}$ or $\frac{AD}{DB} = \frac{AE}{EC}$

Congruent Triangles: Triangles that have the same size and same shape are congruent. To prove two triangles are congruent, it is not necessary to show all six parts are congruent. The following rules (postulates) can simply determine and prove that two triangles are congruent.

1. **SSS Postulate (Side-Side-Side)**
   If three sides of one triangle are congruent to three sides of another triangle, then the triangles are congruent.
2. **SAS Postulate (Side-Angle-Side)**
   If two sides and the included angle of one triangle are congruent to the corresponding two sides and the included angle of another triangle, then the triangles are congruent. An included angle is the angle formed by the two sides.
3. **ASA Postulate (Angle-Side-Angle)**
   If two angles and the included side of one triangle are congruent to the corresponding two angles and the included side of another triangle, then the triangles are congruent. An included side is the side formed by the two angles.
   Since the sum of the measure of interior angles of a triangle is $180°$, we can conclude that **AAS** is also valid for congruent except ambiguous cases.

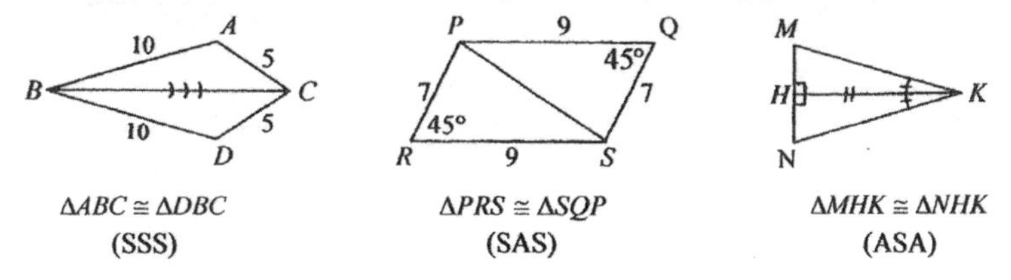

$\Delta ABC \cong \Delta DBC$ (SSS) $\Delta PRS \cong \Delta SQP$ (SAS) $\Delta MHK \cong \Delta NHK$ (ASA)

Note that **SSA is not valid for congruent:**
$\Delta ABD \neq \Delta ABC$ (SSA) (Ambiguous Case)

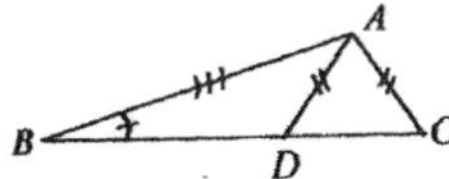

Examples

1. In the figure, $\Delta ABC \sim \Delta CDE$, find x and y.

Solution:

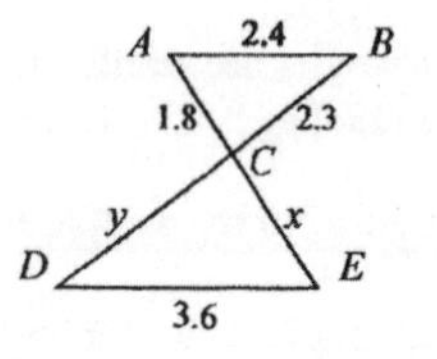

$$\frac{x}{1.8}=\frac{y}{2.3}=\frac{3.6}{2.4}=1.5$$

Every side of ΔCDE is 1.5 times the corresponding side of ΔABC.

$$\therefore x=1.8\times1.5=2.7.$$

$$\therefore y=2.3\times1.5=3.45.$$

2. Two triangles are similar. The lengths of sides of the large triangle is twice as long as the lengths of sides of the small triangle. If the area of the large triangle is 1, what is the area of the small triangle ?

Solution:

We choose the larger triangle with a base of 2 and a height of 1.

We have the area of the small triangle:

$$\therefore A=\tfrac{1}{2}bh=\tfrac{1}{2}\cdot1\cdot\tfrac{1}{2}=\tfrac{1}{4}=0.25.$$

3. In the figure, $\overline{DE}/\!/\overline{BC}$. If $\overline{DB}=\frac{1}{3}\overline{AD}$, find the ratio of the area of ΔADE to the area of ΔABC.

Solution:

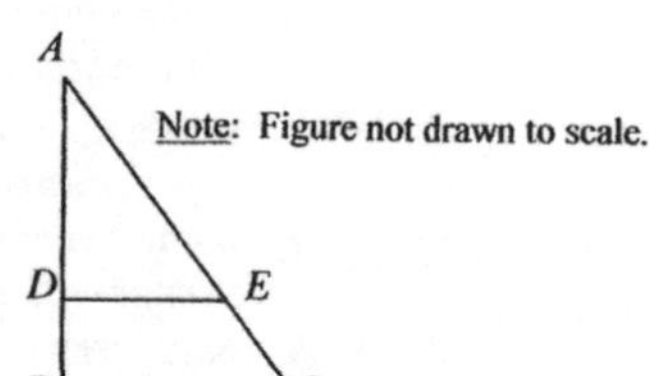

Since $\overline{DE}/\!/\overline{BC}$, $\Delta ADE \sim \Delta ABC$

We choose $AB=4$, $AD=3$.

If two triangles are similar, the ratio of their areas is the square of the ratio of the lengths of the corresponding sides.

$$\therefore \frac{\Delta ADE}{\Delta ABC}=\left(\frac{3}{4}\right)^2=\frac{9}{16}.$$

4. State whether each pair of triangles is congruent by the SAS, ASA, SSS, or None.

a.

b.

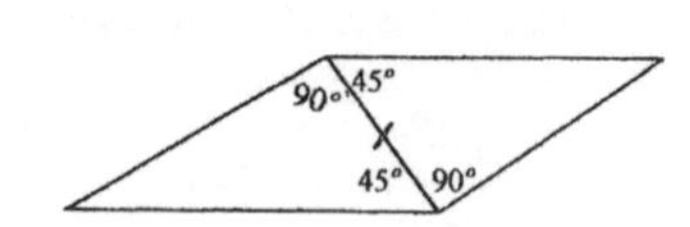

c.

d.

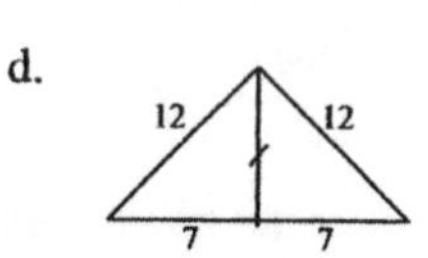

Solution:

3-8 Quadrilaterals and Polygons

Polygon: It is a flat shape formed by straight line segments. A polygon has at least three sides.
Polygon are named according to the number of sides, such as:
Triangle (3), Quadrilateral (4), Pentagon (5), Hexagon (6), Heptagon (7),
Octagon (8), Nonagon (9), Decagon (10), Dodecagon (12)

Regular Polygon: It is a polygon that has all equal sides and angles.
Diagonal of a polygon: It is a segment, other than a side, joining two vertices. A triangle has no diagonal.
Exterior angles of a polygon: They are the angles formed by extending each side in succession. The interior and exterior angle at each vertex form a linear pair.
Perimeter of a polygon is the sum of the lengths of its sides.
A polygon always means a convex polygon such that no line containing a side of the polygon also contain a point in the interior of the polygon.
A quadrilateral is a polygon having four sides. Each of the following quadrilaterals has a special name.

Parallelogram: A quadrilateral with both pairs of opposite sides parallel.
Rectangle: A parallelogram with four right angles.
Square: A parallelogram with four right angles and all four sides congruent.
Rhombus: A parallelogram with all four sides congruent and the diagonals are perpendicular to each other.
Trapezoid: A quadrilateral with exactly one pair of parallel sides. It includes right trapezoid and isosceles trapezoid.

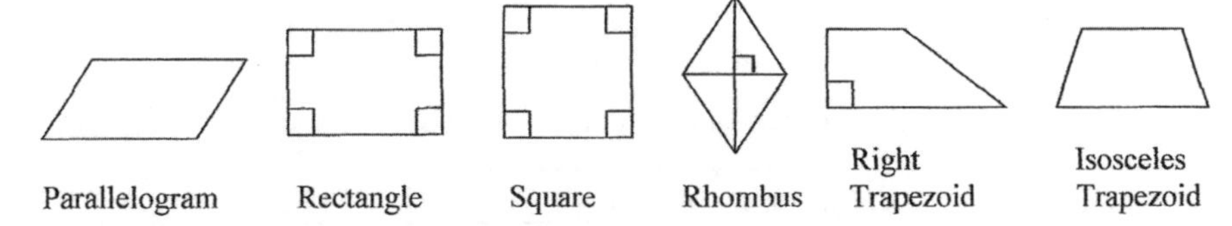

Note that rectangle, square, and rhombus are all parallelograms. Trapezoid is the only quadrilateral that is not a parallelogram.
A circle inscribed in a regular polygon is tangent to each side of the polygon.
A circle circumscribed about a regular polygon contains each vertex of the polygon.
The radius of a regular polygon is a radius of its circumscribed circle.
An **apothem** of a regular polygon is an altitude of its divided triangles. It is the radius of its inscribed circle.

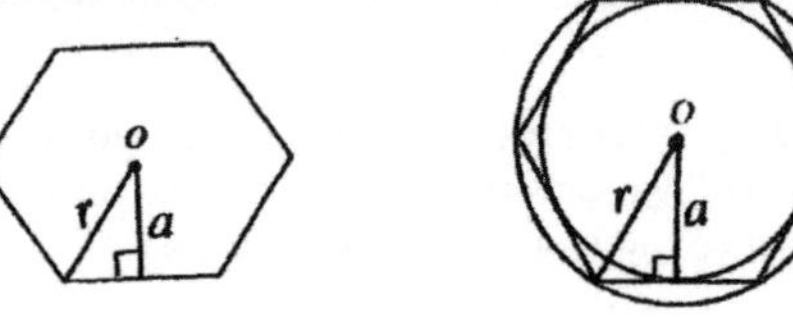

r: the radius
a: apothem

Basic Rules in Polygons

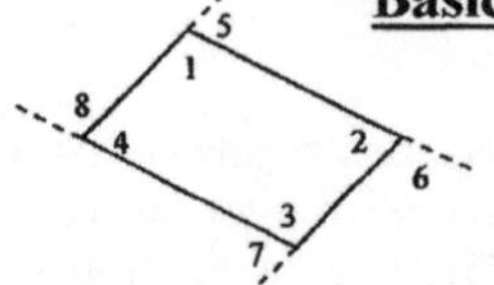

<1, <2, <3, and <4 are the interior angles.
<5, <6, <7, and <8 are the exterior angles.

1. Interior Angle-Sum Theorem: The sum of the measures of the interior angles of a polygon polygon with n sides is $180(n-2)$.

Note that a polygon with n sides can be divided into $(n-2)$ triangles.
Therefore, the sum of the measures of the interior angles is $180(n-2)$.

The measure of each interior angle of a regular polygon with n sides is $\frac{180(n-2)}{n}$.

2. Exterior Angle-Sum Theorem: The sum of the measures of the exterior angles of a polygon with n sides is 360.

Note that a polygon with n sides has n linear pairs. The sum of the measures of the interior and exterior angles is $180 \times n$. The sum of the measures of the interior angles is $180(n-2)$. Therefore, the sum of the measures of the exterior angles is $180n - 180(n-2) = 360$.

The measure of each exterior angle of a regular polygon with n sides is $\frac{360}{n}$.

3. The Properties of a Parallelogram: A parallelogram has the following properties:

a. Opposite sides are equal.
b. Opposite angles are congruent.
c. Adjacent angles are supplementary (they add up to 180°).

4. The Diagonals of a Polygon: The total number of diagonals of a polygon with n sides can be expressed as $\frac{n(n-3)}{2}$.

We can also use sequence to find the number of diagonals of a polygon.

n sides	4	5	6	7	8	9	10	11	12
diagonals	2	5	9	14	20	27	35	44	54

5. Area of a Regular Polygon:

a. A regular polygon with n sides and side length s can be divided into n equilateral triangles having a side of length s. Apply the area of the equilateral triangle, the area of a regular polygon: $A = n \cdot \frac{s^2\sqrt{3}}{4}$

b. The area of a regular polygon equals one-half of its apothem (a) times its perimeter (p): $A = \frac{1}{2}ap$

6. Similarity of Polygons: Regular polygons of the same number of sides are similar.
The perimeters (or areas) of two similar polygons have the same ratio as their corresponding sides, as their radii, or as their apothems.

Examples

1. Find the sum of the measures of the angels of a polygon with 32 sides.
   Solution:
   The sum of angels is $S = 180(n-2)$
   For $n = 32$, $S = 180(32-2) = 5400$.

2. A regular polygon has 10 sides. Find the measure of each angle.
   Solution:
   The measure of each angle of a regular polygon is $A = \frac{180(n-2)}{n}$.
   For $n = 10$, $A = \frac{180(10-2)}{10} = 144$.

3. The measure of each angle of a polygon is 144. How many sides does the polygon has ?.
   Solution:
   $144 = \frac{180(n-2)}{n}$, $144n = 180(n-2)$, $36n = 360$
   $\therefore n = \frac{360}{36} = 10$ sides.

4. How many diagonals can we draw in a decagon (10 sides) ?
   Solution:
   $\frac{n(n-3)}{2} = \frac{10(10-3)}{2} = 35$ diagonals.
   (Hint: It is easier to find the answer by using sequence.)

5. An equilateral triangle is inscribed in a circle whose radius is 18. Find the length of the apothem (a).
   Solution: $m < OBD = 30$
   ΔOBD is a 30-60-90 triangle.
   $OD = \frac{1}{2}(OB)$
   $\therefore\ a = \frac{1}{2}(18) = 9$.

6. Find the area of a regular hexagon (6 sides) if its apothem is $12\sqrt{3}$ inches.
   Solution:
   $m < AOB = \frac{360}{6} = 60$
   ΔAOC is a 30-60-90 triangle. $m < AOC = 30$
   $AC = \frac{1}{2}r$, $a = \frac{1}{2}r \cdot \sqrt{3}$
   $r = \frac{2a}{\sqrt{3}} = \frac{2 \cdot 12\sqrt{3}}{\sqrt{3}} = 24$, $AC = \frac{1}{2}r = \frac{1}{2}(24) = 12$, $AB = 24$
   $\therefore$ Perimeter: $p = 24 \times 6 = 144$
   $\therefore$ Area: $A = \frac{1}{2}ap = \frac{1}{2} \cdot 12\sqrt{3} \cdot 144 = 864\sqrt{3} \approx 1496.45\ in.^2$.

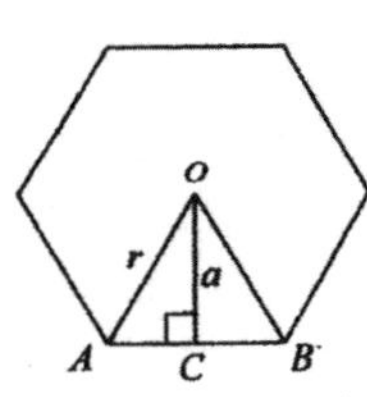

Notes

3-9 Basic Rules in Circles

Circle: It is the set of all points in a plane at a same distance (called the radius) from a fixed point (called the center).

Radius (r): It is a segment joining the center of the circle with a point on the circle.

Chord: It is a segment with endpoints on the circle.

Diameter (d): It is a chord passing through the center of the circle.

Circumference: It is the distance around a circle. $C = 2\pi r$ or $C = \pi d$

Concentric Circles: They are circles having the same center and radii of different lengths.

Central Angle: It is the angle having its vertex at the center of the circle.

Minor Arc: The interior of a central angle forms a minor arc on the circle.

Major Arc: The exterior of a central angle forms a major arc on the circle. We use three letters to name a major arc.

Basic Rules in Circles

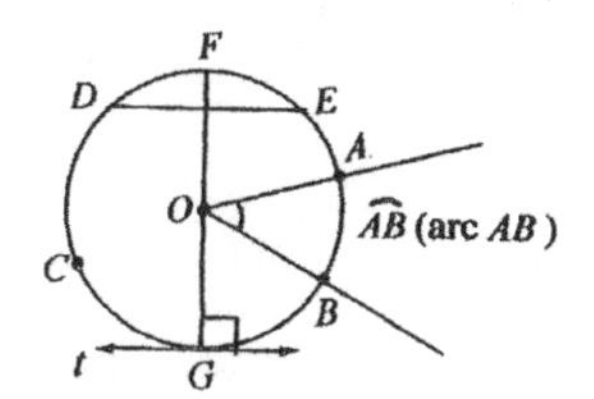

$\overline{OA}$, $\overline{OB}$, $\overline{OG}$, $\overline{OF}$ are the radii.
$\overline{FG}$ is the diameter. $\overline{DE}$ is a chord.
$< AOB$ is a central angle.

$\overparen{AB}$ is a minor arc. $\overparen{ACB}$ is a major arc.
Line t is tangent to the circle at G.

1. The measure of a minor arc equals the measure of its central angle. $m < AOB = m\overparen{AB}$

2. Two circles are congruent if their radii are congruent.

3. In the same circle, if two arcs are congruent, then their chords are congruent.
   In the same circle, if two chords are congruent, then their arcs are congruent.
   In the same circle, two chords are congruent if and only if they are equidistant from the center.

4. If a diameter of a circle is perpendicular to a chord, then the diameter bisects the chord and its arc.

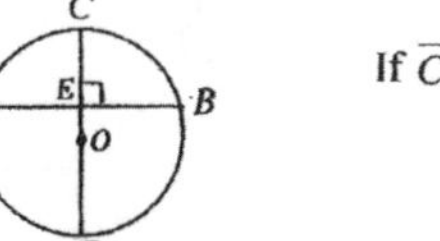

If $\overline{CD} \perp \overline{AB}$, then $\overline{AE} = \overline{EB}$, $\overparen{AC} = \overparen{CB}$.

5. If a line is tangent to a circle, then the line is perpendicular to the radius drawn to the point of tangency.
   $\overline{AB} \perp \overline{OB}$, $\overline{AC} \perp \overline{OC}$

6. The two tangent segments from the same exterior point of a circle are congruent.

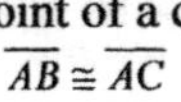

$\overline{AB} \cong \overline{AC}$

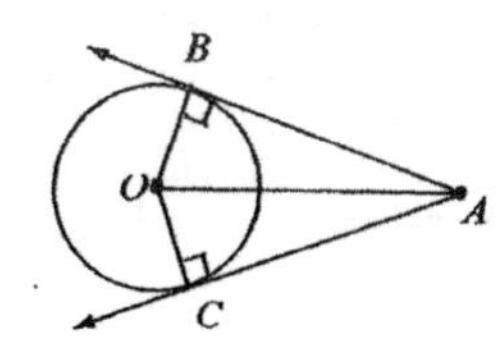

Basic Rules in Circles (Continued)

7. The measure of an inscribed angle of a circle is one-half the measure of its intercepted arc.

8. If the inscribed angles intercepts the congruent arcs, the angles are congruent.

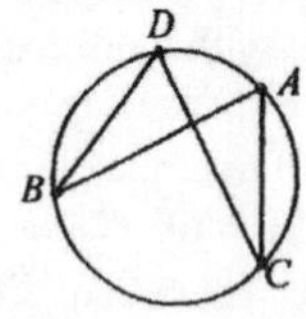

$m < BAC = \frac{1}{2} m\overset{\frown}{BC}$, $m < BDC = \frac{1}{2} m\overset{\frown}{BC}$

$m < BAC = m < BDC$

9. If an inscribed angle of a circle is inscribed in a semicircle, the inscribed angle is a right angle.

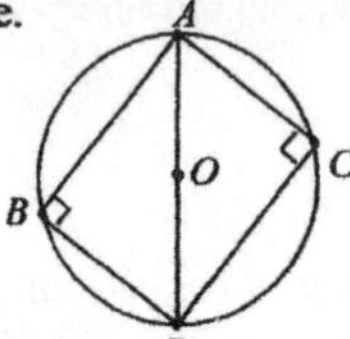

$m < ACD = 90$

$m < ABD = 90$

10. In a circle, the opposite angles of an inscribed quadrilateral are supplementary.

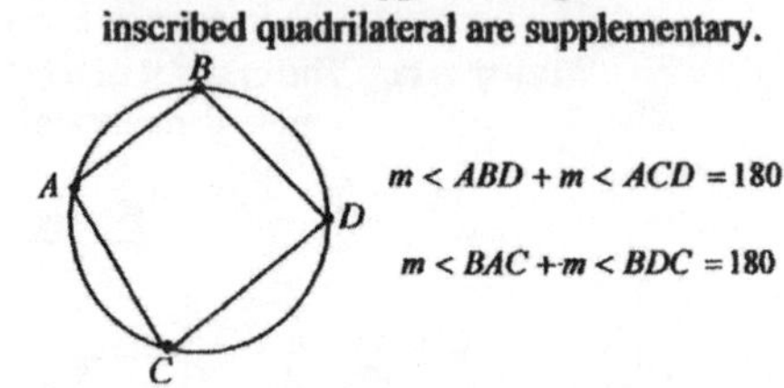

$m < ABD + m < ACD = 180$

$m < BAC + m < BDC = 180$

11. If a tangent and a secant intersect at a point on the circle, the measure of the angle formed is one-half the measure of the intercepted arc.

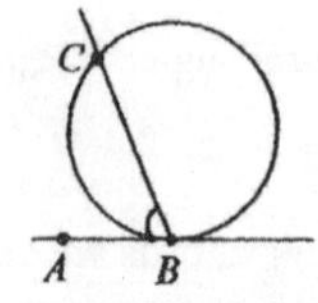

$m < ABC = \frac{1}{2} m\overset{\frown}{BC}$

12. If two secants intersect in the interior of a circle, the measure of the angle formed is one-half the sum of the measures of the intercepted arcs.

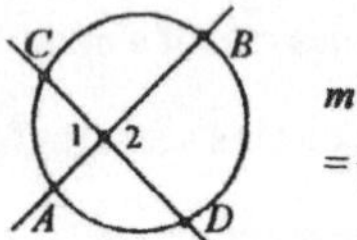

$m < 1 = m < 2$

$= \frac{1}{2}(m\overset{\frown}{AC} + m\overset{\frown}{BD})$

13. If two secants, a tangent and a secant, or two tangents intersect in the exterior of a circle, the measure of the angle formed is one-half the difference of the measures of the intercepted arcs.

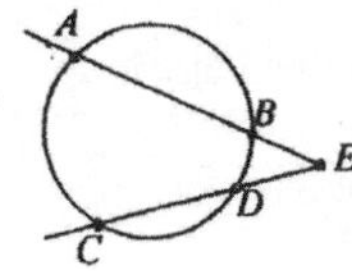

$m < E = \frac{1}{2}(m\overset{\frown}{AC} - m\overset{\frown}{BD})$

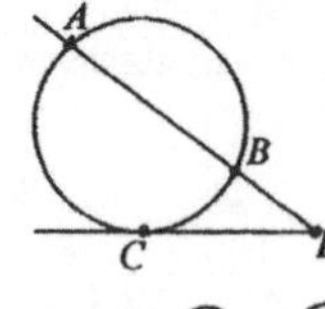

$m < D = \frac{1}{2}(m\overset{\frown}{AC} - m\overset{\frown}{BC})$

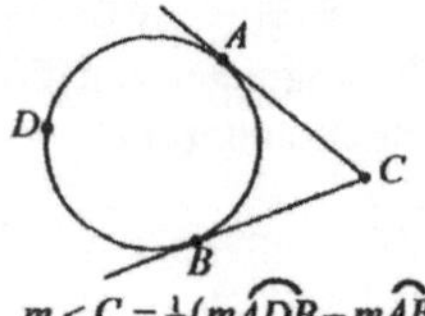

$m < C = \frac{1}{2}(m\overset{\frown}{ADB} - m\overset{\frown}{AB})$

14. If two chords intersect in the interior of a circle, the two triangles formed are similar.

15. If a tangent and a secant intersect in the exterior of a circle, the two triangles formed by the tangent and secant segments with each of the two chords are similar.

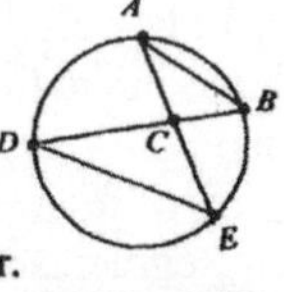

$\Delta ABC \sim \Delta CDE$

$AC \cdot CE = BC \cdot CD$

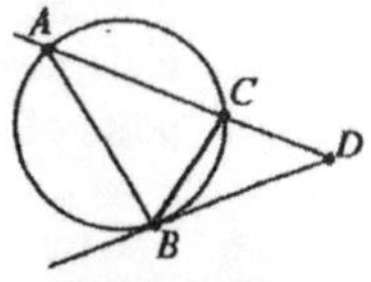

$\Delta ABD \sim \Delta CBD$

$AD : BD = BD : CD$

$(BD)^2 = AD \cdot CD$

Examples

1. In the circle, $\overline{AB}$ is the diameter, $m<C=50$, $m<D=70$. Find the measure of each numbered angle.
   Solution:

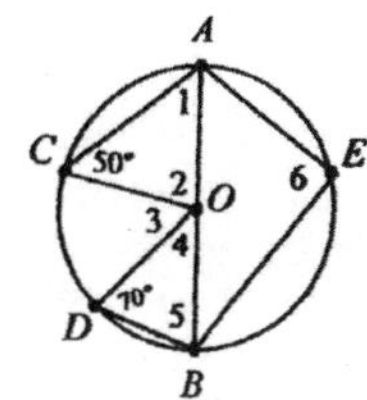

$\overline{OA}\cong\overline{OC},\ m<1=m<C=50$
$m<2=180-50-50=80$
$\overline{OD}\cong\overline{OB},\ m<5=m<D=70$
$m<4=180-70-70=40$
$m<3=180-80-40=60$
$m<6=90$ (Inscribed in a semicircle)

2. In the circle, $\overline{BC}$ is the diameter, the measures of the central angles are given, $m<1=45$, $m<2=60$. Find mAB, mAC, mAD, and $mABC$.
   Solution:

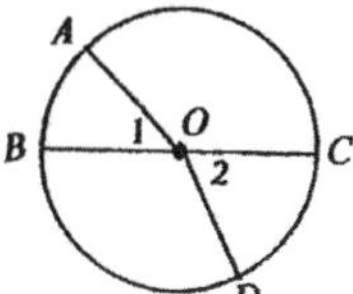

$mAB=m<1=45$
$mAC=180-45=135$
$mAD=m<1+m<BOD=45+120=165$
$mABC=360-m<AOC=360-135=225$

3. In the circle, $\overline{AB}$ is the diameter, $\overline{CD}\perp\overline{AB}$, $CD=30$, and $OE=8$. Find the length of $\overline{CO}$.
   Solution:

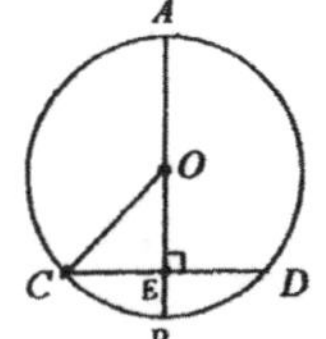

$CE=\frac{1}{2}(CD)=\frac{1}{2}(30)=15$
$(CO)^2=(CE)^2+(OE)^2=15^2+8^2=289$
$\therefore CO=17$.

4. $\overline{AB}$ and $\overline{AC}$ are two tangent segments to the circle, $BO=8, AO=17$. Find AB and AC.
   Solution:

ΔABO and ΔACO are right triangles and $\overline{AB}\cong\overline{AC}$.

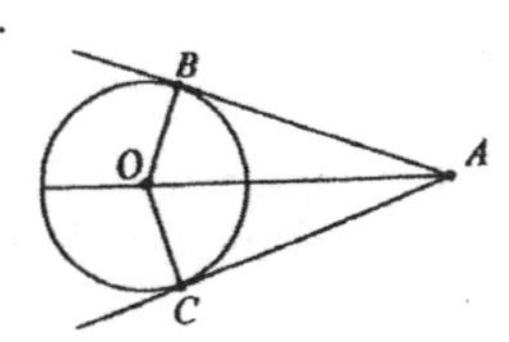

$(AB)^2+(BO)^2=(AO)^2$
$(AB)^2+8^2=17^2$
$(AB)^2+64=289$
$(AB)^2=225\ \therefore AB=15,\ AC=15$.

Examples

5. In the circle, find $m<M$, $m<MPN$, $m<1$, and $m<Q$.
Solution:

$m<M=\frac{1}{2}m\widehat{OP}=\frac{1}{2}(40)=20$

$m<MPN=\frac{1}{2}\widehat{MN}=\frac{1}{2}(100)=50$

$m<1=\frac{1}{2}(m\widehat{MN}+m\widehat{OP})=\frac{1}{2}(100+40)=70$

$m<Q=\frac{1}{2}(m\widehat{MN}-m\widehat{OP})=\frac{1}{2}(100-40)=30$

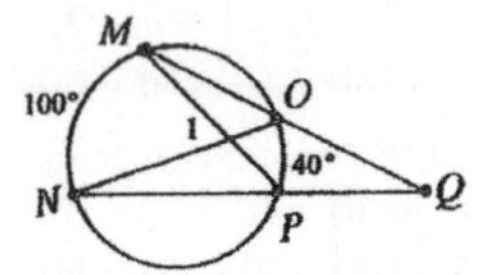

6. In the figure, quadrilateral PQSR is inscribed in the circle. Find $m<QPS$, $m<RPS$, $m<Q$, $m<R$, and $m<TPR$.
Solution:

$m<QPS=\frac{1}{2}m\widehat{QS}=\frac{1}{2}(90)=45$

$m<RPS=\frac{1}{2}m\widehat{RS}=\frac{1}{2}(100)=50$

$<Q$ and $<R$ are right angles.
$m<Q=m<R=90$

$m<TPR=\frac{1}{2}m\widehat{PR}=\frac{1}{2}(180-100)=40$

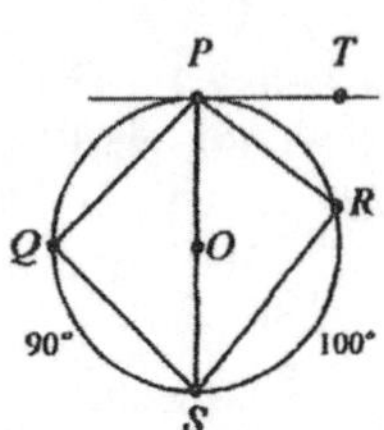

7. In the circle, chord AB and CD intersect at E. If $CE=12$, $DE=15$, and $AE=18$, find BE.
Solution:

$18\cdot BE=15\cdot 12$

$BE=\frac{180}{18}=10.$

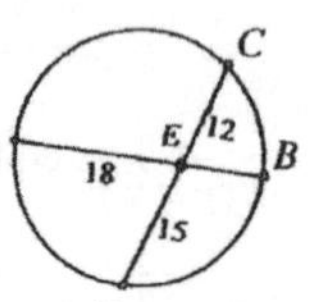

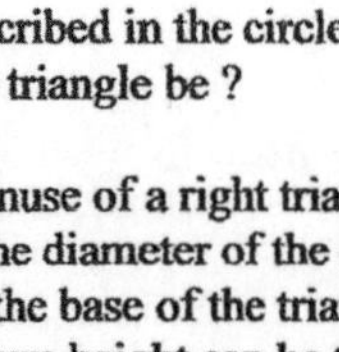

8. In the figure, $\overline{MP}$ and $\overline{QP}$ intersect at P, $MN=16$, $NP=9$, find QP.
Solution:

$(QP)^2=MP\cdot NP$
$(QP)^2=25\cdot 9=225$
$QP=15.$

9. A right triangle is inscribed in the circle. The diameter of the circle is 10 *cm*. What would the maximum area of the triangle be ?
Solution:

The hypotenuse of a right triangle inscribed a circle is the diameter of the circle.
Therefore, the base of the triangle is 10 *cm*.
The maximum height can be the radius 5 *cm*.

Max. Area $=\frac{1}{2}bh=\frac{1}{2}\cdot 10\cdot 5=25\ cm^2$.

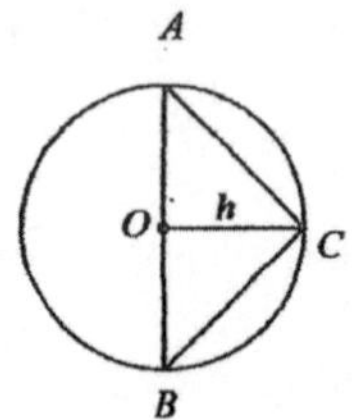

Notes

Notes

3-10 Area Formulas and Perimeters

Area is the measure of the region enclosed by a figure. Area is measured in square units, such as cm^2 (square centimeters), m^2 (square meters), $in.^2$ (square inches), and ft^2 (square feet), ... **Perimeter** of a polygon is the sum of the lengths of its sides.

1. Area of a Triangle: $A = \frac{1}{2}bh$

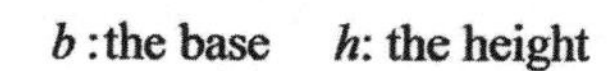

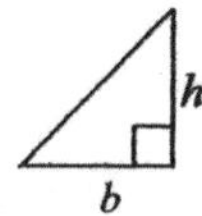

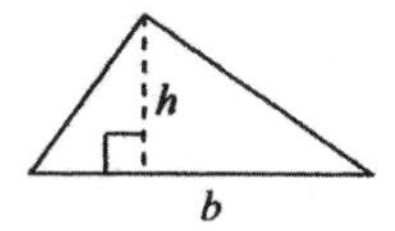

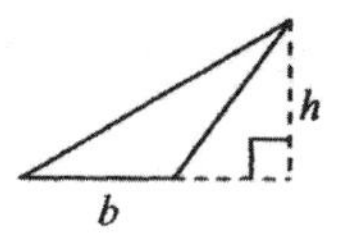

2. Area of an Equilateral Triangle: $A = \dfrac{s^2\sqrt{3}}{4}$

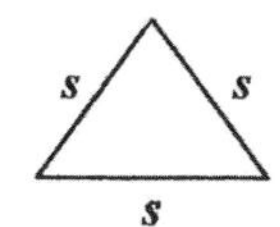

3. Area of a Square: $A = s^2$

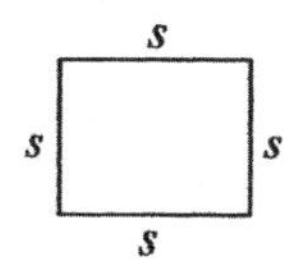

4. Area of a Rectangle: $A = \ell \cdot w$

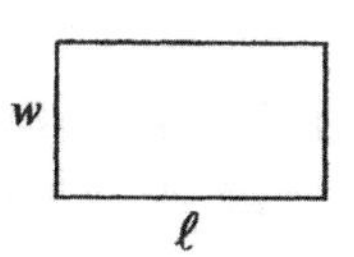

5. Area of a Parallelogram: $A = b \cdot h$

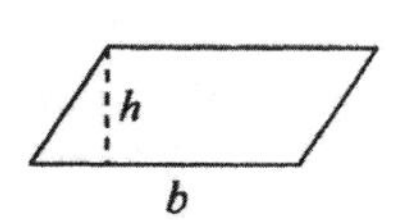

6. Area of a Trapezoid: $A = \frac{1}{2}(b_1 + b_2)h$

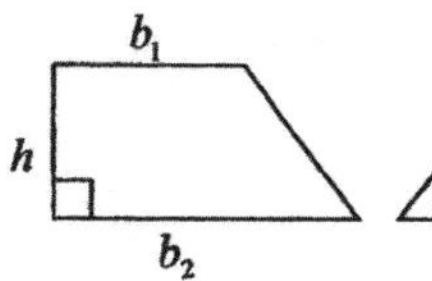

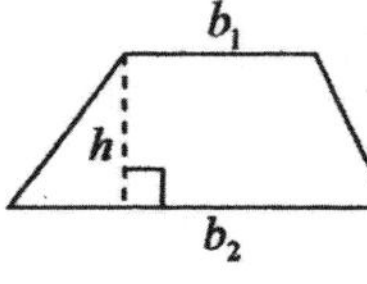

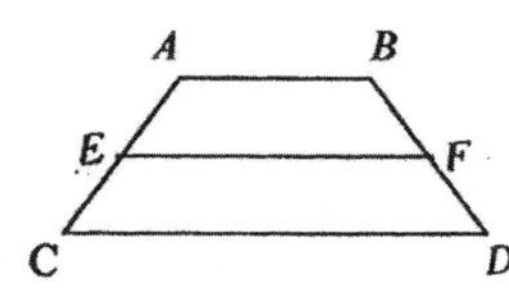

Other Formula:

$EF = \frac{1}{2}(AB + CD)$

(EF is the median.)

7. Area of a Rhombus: $A = \frac{1}{2}(d_1 d_2)$

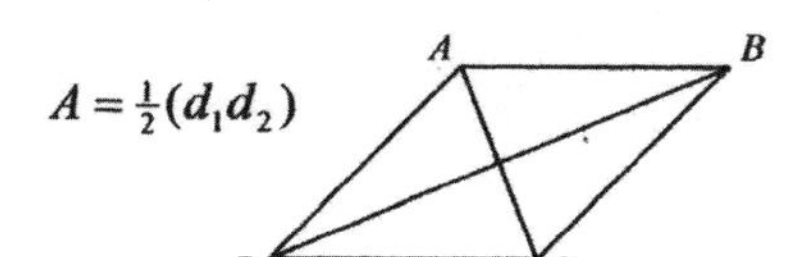

$d_1 = AC$

$d_2 = BD$

Area Formulas (Continued)

8. Area of a Circle: $A = \pi r^2$

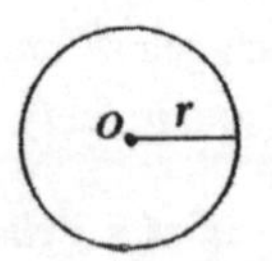

9. Area of a Sector of a Circle: $A = \pi r^2 (\frac{m}{360})$

m: measure of the arc of the sector *OAB*

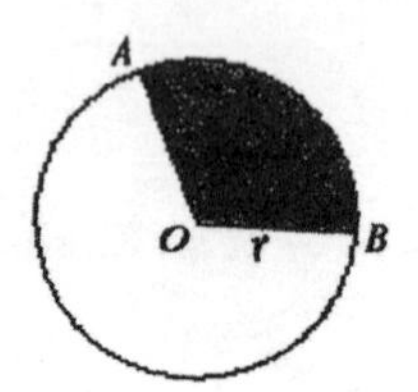

10. Area of a Regular Polygon: $A = \frac{1}{2} a p$

a: apothem
p: perimeter

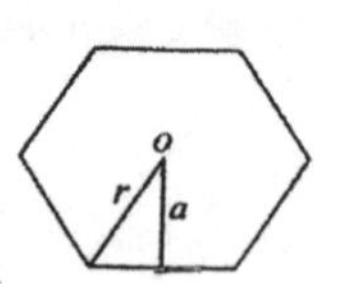

11. Area of a sector and a segment of a Circle:

A segment of a circle is the region bounded by an arc and its chord.

The area of the shaded region is:

Area of sector *OAB* – Area of ΔOAB

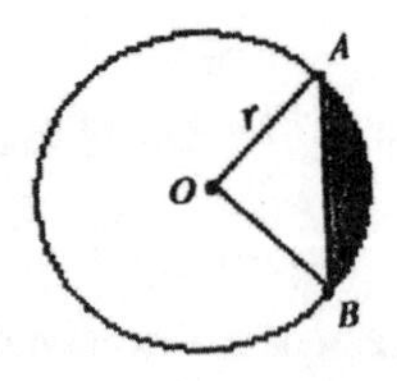

Examples

1. In the triangle, $AC = 20$, find the area of ΔABC.
   Solution:

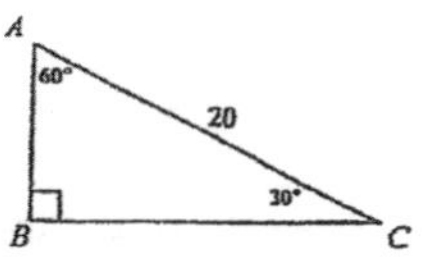

$$h = AB = \tfrac{1}{2}(AC) = \tfrac{1}{2}(20) = 10$$
$$b = BC = \sqrt{3}(AB) = \sqrt{3}(10) = 10\sqrt{3}$$
Area: $A = \tfrac{1}{2}bh = \tfrac{1}{2}(10\sqrt{3})(10) = 50\sqrt{3}$.

2. In the triangle, $AB = 20$, $BC = 50$, and $m < D = 45^o$ find the area of ΔABC.
   Solution:

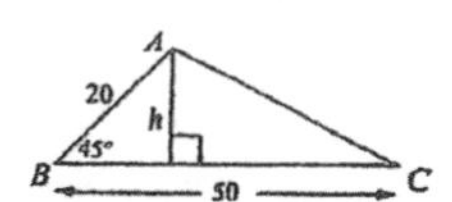

$$b = BC = 50$$
$$AB = h\sqrt{2}, \quad h = \frac{AB}{\sqrt{2}} = \frac{20}{\sqrt{2}} = 10\sqrt{2}$$
Area: $A = \tfrac{1}{2}bh = \tfrac{1}{2}(50)(10\sqrt{2}) = 250\sqrt{2}$.

3. In the figure, a square is cut out from a rectangle of length 25 and width 15. Find the area of the dashed region.
   Solution:

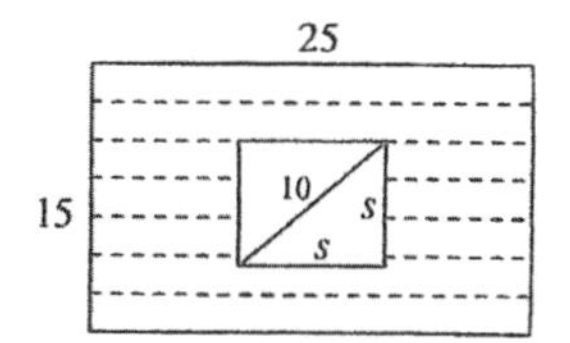

$s^2 + s^2 = 10^2$, $2s^2 = 100$ $\therefore s^2 = 50$
Area of the square $= s^2 = 50$
Area of the shaded region
$=$ Area of the rectangle $-$ Area of the square
$= 25 \cdot 15 - 50 = 325$.

4. In the figure, a rhombus is inscribed in a rectangle. Find the area of the dashed region.
   Solution:

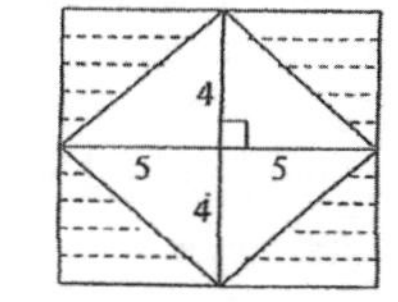

Area of the rhombus $= \tfrac{1}{2}(d_1 d_2) = \tfrac{1}{2}(10)(8) = 40$
Area of the rectangle $= l \cdot w = 10 \cdot 8 = 80$
Area of the shaded region $= 80 - 40 = 40$.

5. In the parallelogram $ABCD$, $AB = 18\,cm$, $AD = 12\,cm$, and $m < D = 60$, find the area.
   Solution:

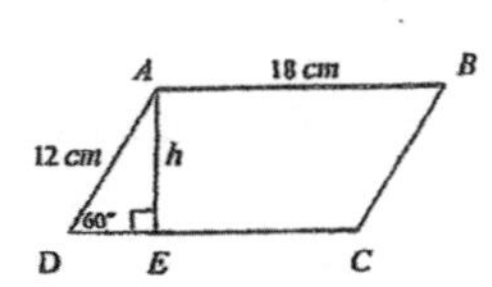

$$b = AB = 18$$
$$DE = \tfrac{1}{2}(AD) = \tfrac{1}{2}(12) = 6$$
$$h = \sqrt{3}(DE) = 6\sqrt{3}$$
Area of the parallelogram:
$A = bh = 18(6\sqrt{3}) = 108\sqrt{3}\ cm^2$.

6. In the trapezoid $MNPQ$, $MN = 18$ *in.*, $QP = 30$ *in.*, and $m < Q = 45$, find the area.

Solution:

$12 = \sqrt{2}h$, $h = \frac{12}{\sqrt{2}} = 6\sqrt{2}$

$b_1 = MN = 18$, $b_2 = QP = 30$

Area of the trapezoid:

$A = \frac{1}{2}(b_1 + b_2)h = \frac{1}{2}(18+30)(6\sqrt{2}) = 144\sqrt{2}$ $in.^2$

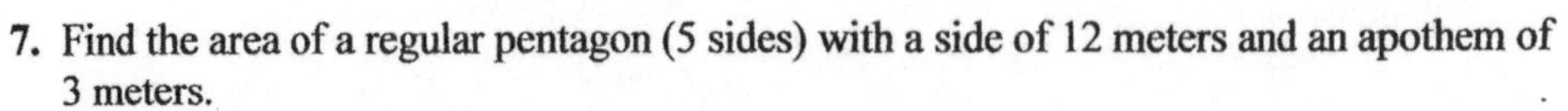

7. Find the area of a regular pentagon (5 sides) with a side of 12 meters and an apothem of 3 meters.

Solution:

Perimeter of the pentagon $p = 12 \times 5 = 60$.

Area of the pentagon $A = \frac{1}{2}ap = \frac{1}{2}(3)(60) = 90\ m^2$.

8. Find the area of a regular hexagon (6 sides) if its apothem is $12\sqrt{3}$ inches.

Solution:

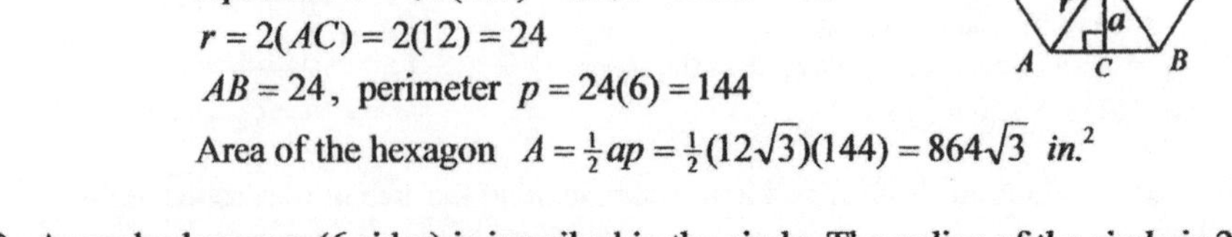

$m < AOB = \frac{360}{6} = 60$, $m < AOC = 30$

Apothem $a = \sqrt{3}(AC) = 12\sqrt{3}$ $\therefore AC = 12$

$r = 2(AC) = 2(12) = 24$

$AB = 24$, perimeter $p = 24(6) = 144$

Area of the hexagon $A = \frac{1}{2}ap = \frac{1}{2}(12\sqrt{3})(144) = 864\sqrt{3}$ $in.^2$

9. A regular hexagon (6 sides) is inscribed in the circle. The radius of the circle is 24 inches. Find the area of the shaded region

Solution:

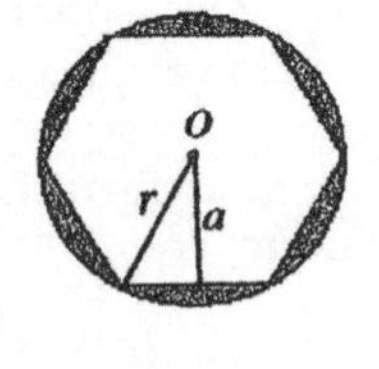

Apothem of the hexagon $a = \sqrt{3}(\frac{1}{2}r) = \sqrt{3}(12)$

Perimeter of the hexagon $p = 24(6) = 144$

Area of the hexagon $A = \frac{1}{2}ap = \frac{1}{2}(12\sqrt{3})(144) \approx 1496.45$

Area of the circle $A = \pi r^2 = \pi(24)^2 \approx 1808.64$

Area of the shaded region $= 1808.64 - 1496.45 \approx 312.19\ in.^2$

10. In the circle, find the area of the shaded area.

Solution:

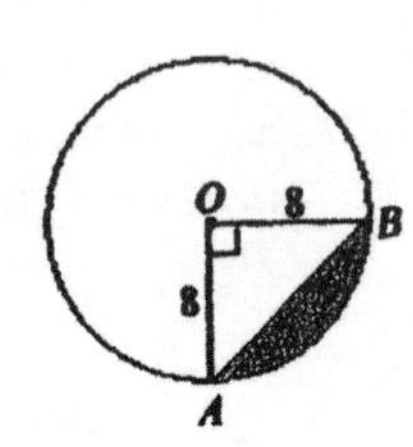

Area of sector OAB $= \pi\ (8^2)(\frac{90}{360}) = 16\pi$

Area of $\Delta OAB = \frac{1}{2}bh = \frac{1}{2}(8)(8) = 32$

Area of the shaded region $= 16\pi - 32 \approx 18.27$.

Notes

Notes

3-11 Parabolas

A conic section is a curve that can be formed by slicing a right circular cone with a plane. The figure of the intersection of the cone and the plane is a conic section. By changing the position of the plane (horizontally or slant) , we obtain four types of conic sections, **a circle, an ellipse, a parabola**, or part of **a hyperbola.**
In this section, we will learn to obtain an equation for each of the conic sections by applying its geometric definition.

Circle **Ellipse** **Parabola** **Hyperbola**

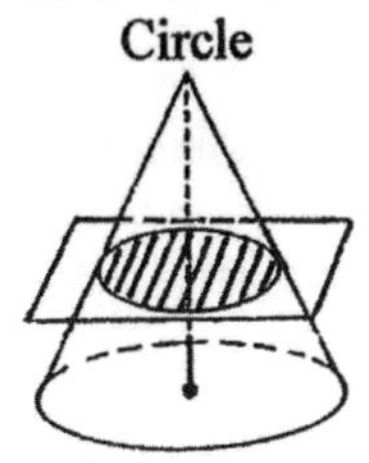
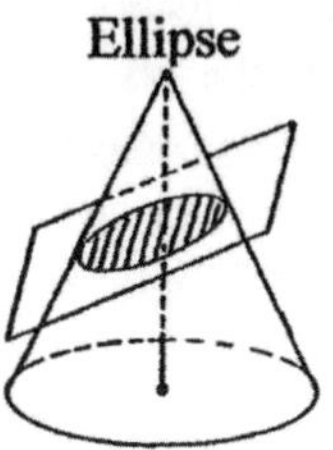
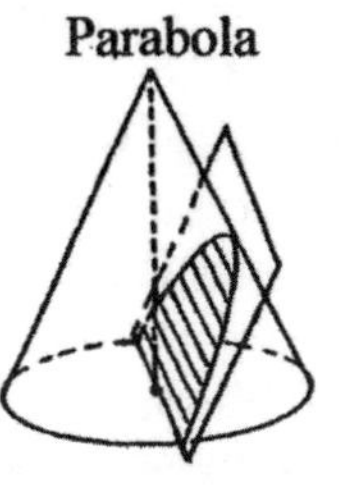
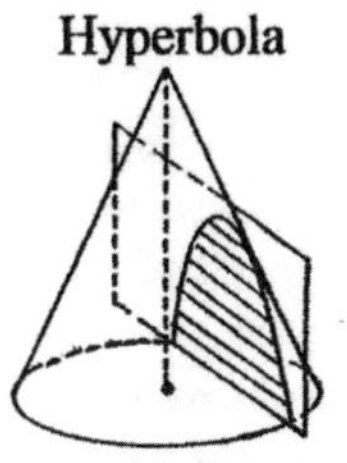

We have learned the basic equation of a parabola $y = ax^2 + bx + c \ (a \neq 0)$ and **the General Form of a Parabola**: $y - k = a(x-h)^2$. The vertex is (h, k).

If $a > 0$, the parabola open upward. If $a < 0$, the parabola open downward.

In this section, we will learn to derive equations for parabolas by applying a general definition of a parabola and learn how to graph a parabola.
Definition of a parabola: A parabola is the set of all points equidistant from a fixed line (called the directrix) and a fixed point (called the focus).

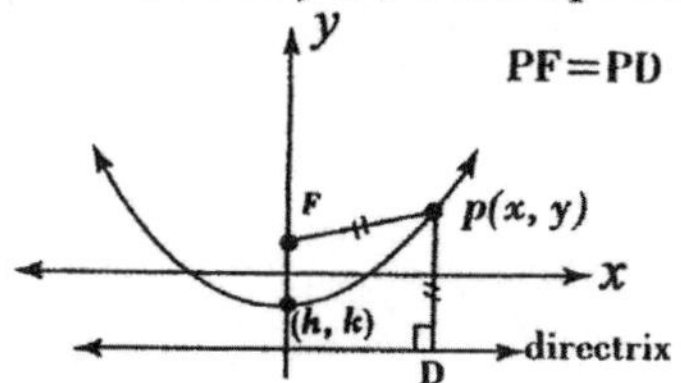

Example

1. Find the equation of a parabola whose focus is the point (0, 2) and whose directrix is the line $y = -4$.

Solution: $PF = PD$

$$\sqrt{(x-0)^2 + (y-2)^2} = \sqrt{(x-x)^2 + (y+4)^2}$$

$$x^2 + (y-2)^2 = (y+4)^2$$

$$x^2 + y^2 - 4y + 4 = y^2 + 8y + 16, \quad x^2 = 12y + 12, \quad x^2 = 12(y+1)$$

$$\therefore y + 1 = \tfrac{1}{12}x^2. \text{ (the equation)}$$

Vertex $(0, -1)$, $a = \frac{1}{12} > 0$ open upward

Example

2. Find the equation of a parabola whose focus is the point (0, 3) and whose directrix is the line $x = 4$.

Solution:

$$PF = PD$$
$$\sqrt{(x-0)^2+(y-3)^2} = \sqrt{(x-4)^2+(y-y)^2}$$
$$x^2+(y-3)^2=(x-4)^2,\ \ x^2+(y-3)^2=x^2-8x+16$$
$$(y-3)^2=-8(x-2)$$
$$x-2=-\tfrac{1}{8}(y-3)^2 \text{ (the equation)}$$

Vertex (2, 3), $a=-\frac{1}{8}<0$ open to the left.

We have learned the parabolas that have a vertical axis and open upward or downward. Now, we will also learn the parabola that have a horizontal axis and open to the right or to the left. The standard form of each parabola shown below are useful to locate the vertex, the focus, the directrix, and latus rectum.

Standard Forms of Parabolas

Latus Rectum: The line segment passing through the focus and perpendicular to the axis.

1. $y^2 = 4ax$

vertex: (0, 0), open to the right
focus : $(a, 0)$
directrix: $x=-a$
latus rectum: $4a$

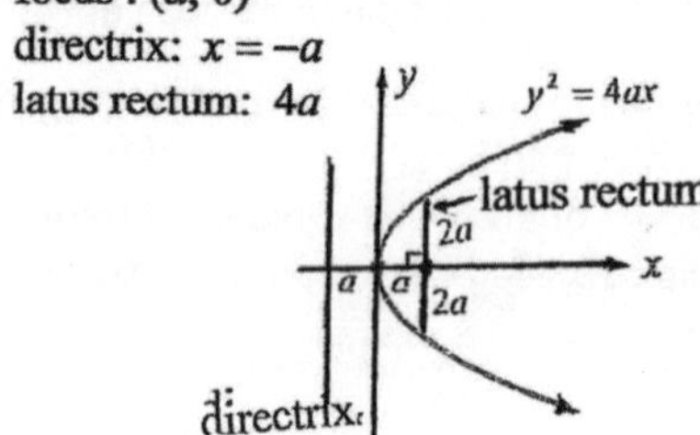

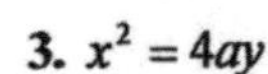

Or: $x = py^2$ where $p=\frac{1}{4a}$

2. $(y-k)^2 = 4a(x-h)$

vertex: (h, k), open to the right
focus: $(h+a, k)$
directrix: $x=h-a$
latus rectum: $4a$

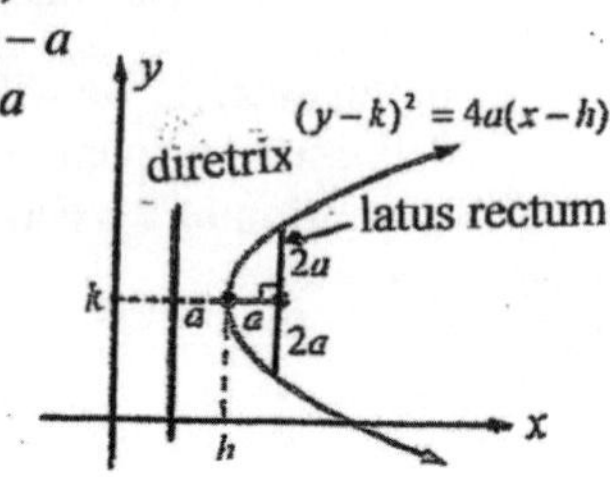

Or: $x-h=p(y-k)^2$ where $p=\frac{1}{4a}$

3. $x^2 = 4ay$

vertex: (0, 0), open upward
focus: $(0, a)$
diretrix: $y=-a$
latus rectum: $4a$

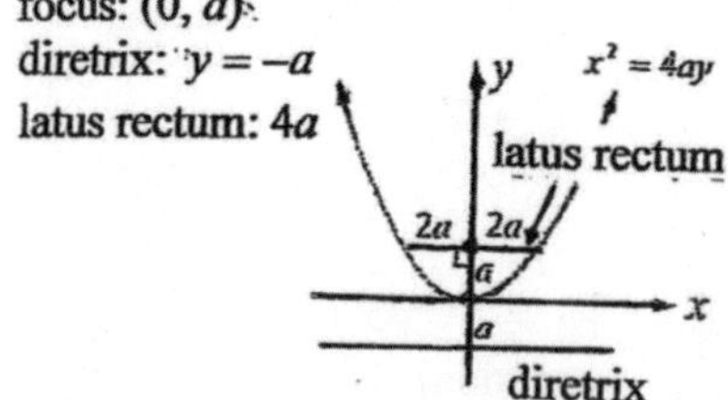

Or: $y = px^2$ where $p=\frac{1}{4a}$

4. $(x-h)^2 = 4a(y-k)$

vertex: (h, k), open upward
focus: $(h, k+a)$
diretrix: $y=k-a$
latus rectum: $4a$

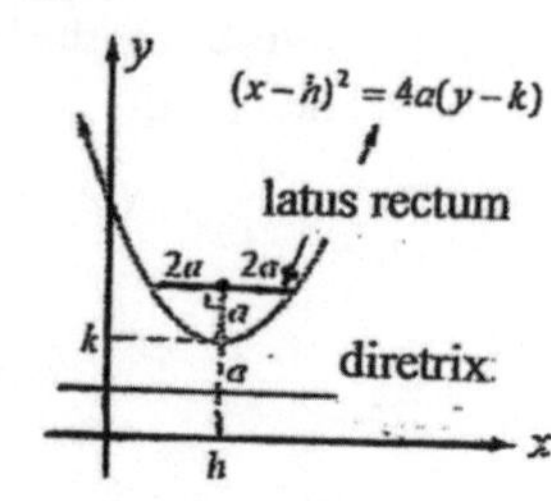

Or: $y-k=p(x-h)^2$ where $p=\frac{1}{4a}$

Standard Forms of Parabolas (Continued)

5. $y^2 = -4ax$
vertex: (0, 0), open to the left
focus: $(-a, 0)$
diretrix: $x = a$
latus rectum: $4a$

6. $(y-k)^2 = -4a(x-h)$
vertex: (h, k), open to the left
focus: $(h - a, k)$
diretrix: $x = h + a$
latus rectum: $4a$

7. $x^2 = -4ay$
vertex: (0, 0), open downward
focus: $(0, -a)$
diretrix: $y = a$
latus rectum: $4a$

8. $(x-h)^2 = -4a(y-k)$
vertex: (h, k), open downward
focus: $(h, k - a)$
diretrix: $y = k + a$
latus rectum: $4a$

Examples

1. Find an equation of the parabola with vertex (0, 0) and focus (2, 0).
Solution: Sketch and Find:
It is open to the right.
$y^2 = 4ax$ and $a = 2$
$y^2 = 4(2)x = 8x$
$\therefore y^2 = 8x$ is the equation.

2. Find an equation of the parabola with vertex (0, 0) and focus (–2, 0).
Solution: Sketch and Find:
It is open to the left.
$y^2 = -4ax$ and $a = 2$
$y^2 = -4(2)x = -8x$
$\therefore y^2 = -8x$ is the equation.

3. Find the equation of a parabola whose focus is the point (0, 4) and whose directrix is the line $y = -4$.
Solution: Sketch and find:
It is open upward. vertex (0, 0)
$x^2 = 4ay$ and $a = 4$
$x^2 = 4(4)y = 16y$
$\therefore x^2 = 16y$ is the equation.

4. Find the equation of a parabola whose focus is the point (0, 2) and whose directrix is the line $y = -4$.
Solution: Sketch and find:
It is open upward. vertex (0, –1)
$(x-h)^2 = 4a(y-k)$ and $a = 3$
$(x-0)^2 = 4(3)(y+1)$
$\therefore x^2 = 12(y+1)$ is the equation.

5. Find the vertex, focus, diretrix, and latus rectum of the parabola $x^2 + 2x + 2y - 4 = 0$.
Solution: Completing the square
$(x^2 + 2x + 4) - 4 + 2y - 4 = 0$, $(x+2)^2 + 2y - 8 = 0$
$(x+2)^2 = -2(y-4)$, open downward
$-4a = -2 \therefore a = \frac{1}{2}$
vertex: $(h, k) = (-2, 4)$; focus: $(h, k-a) = (-2, \frac{7}{2})$
directrix: $y = k + a$, $y = \frac{9}{2}$; latus rectum: $4a = 2$

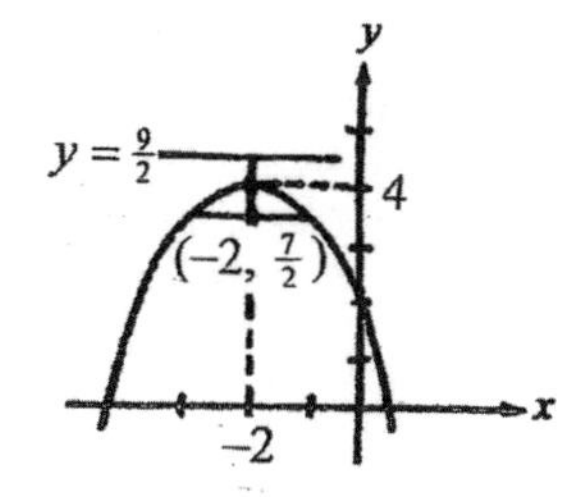

Notes

3-12 Circles

Definition of a circle: A circle is the set of all points in a plane that are a fixed distance r (called the radius) from a fixed point (called the center).

Standard Forms of Circles

1. $x^2 + y^2 = r^2$

Center (0, 0)
Radius $= r$

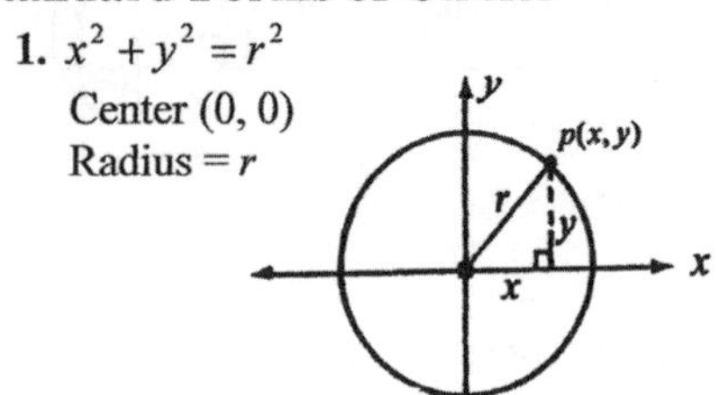

2. $(x-h)^2 + (y-k)^2 = r^2$

Center (h, k)
Radius $= r$

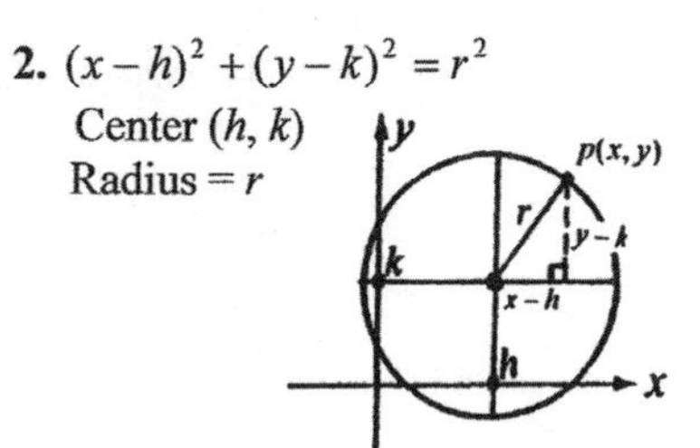

When the equation of a circle is given in the form $x^2 + y^2 + ax + by + c = 0$, we rewrite it to match the standard form by the method of completing the square. Then find its center and radius.

Examples

1. Find the center and radius of $x^2 + y^2 = 25$.

Solution:

$$x^2 + y^2 = 25$$
$$x^2 + y^2 = 5^2$$
$\therefore$ Center (0, 0)
Radius $r = 5$.

2. Find the equation of the circle with center (0, 0) and radius $3\sqrt{2}$.

Solution:

$$x^2 + y^2 = r^2$$
$$x^2 + y^2 = (3\sqrt{2})^2$$
$$\therefore x^2 + y^2 = 18.$$

3. Find the equation of the circle with center (–2, 1) and radius 3.

Solution:

$$(x-h)^2 + (y-k)^2 = r^2$$
$$(x+2)^2 + (y-1)^2 = 3^2. \text{ (Standard Form)}$$
$$x^2 + 4x + 4 + y^2 - 2y + 1 = 9$$
$$x^2 + y^2 + 4x - 2y - 4 = 0. \text{ (General Form)}$$

4. Write the equation of a circle if the endpoints of the diameter are at (5, 4) and (–1, –2).

Solution:

$$\text{Center} = \left(\frac{5+(-1)}{2}, \frac{4+(-2)}{2}\right) = (2, 1)$$

$$\text{Radius } r = \sqrt{(5-2)^2 + (4-1)^2} = \sqrt{18}$$

$$\therefore (x-2)^2 + (y-1)^2 = 18.$$

5. Graph $x^2+y^2+4x-8y+11=0$.
Solution: Completing the square

$$x^2+y^2+4x-8y+11=0$$
$$(x^2+4x)+(y^2-8y)+11=0$$
$$(x^2+4x+4)-4+(y^2-8y+16)-16+11=0$$
$$(x+2)^2+(y-4)^2=9$$
$$(x+2)^2+(y-4)^2=3^2$$
$\therefore$ Center $(-2, 4)$, Radius $=3$

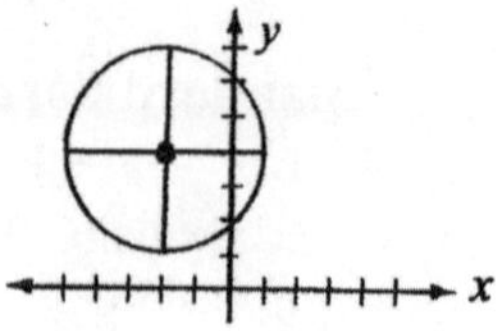

6. **Find the equation of the line that is tangent to the circle $x^2+y^2=34$ at point (3, 5).**
Solution:

The tangent line on a circle is perpendicular to the radius of the circle.
Slope of the radius $m=\frac{5-0}{3-0}=\frac{5}{3}$
Slope of the tangent line $m_1=-\frac{3}{5}$ passing the point (3, 5), $p(x, y)$
The equation of the tangent line:

$$\frac{y-5}{x-3}=-\frac{3}{5}$$
$$5y-25=-3x+9$$
$$3x+5y-34=0.$$

7. **The line $3x+5y-34=0$ is tangent to a circle whose center is at the origin (0, 0). Find the equation of the circle.**
Solution:

The tangent line on a circle is perpendicular to the radius of the circle.
Find the slope of the tangent line:

$$3x+5y-34=0$$
$$y=-\frac{3}{5}x+\frac{34}{5}, \text{ the slope } m=-\frac{3}{5}$$

Slope of the radius $m_1=\frac{5}{3}$
Equation of the radius $\frac{y-0}{x-0}=\frac{5}{3}$, $5x-3y=0$, $x=\frac{3}{5}y$
Find the point of intersection of the tangent line and the radius:

$$y=-\frac{3}{5}x+\frac{34}{5}$$
$$y=-\frac{3}{5}\left(\frac{3}{5}y\right)+\frac{34}{5}$$

$\therefore y=5$ and $x=\frac{3}{5}y=\frac{3}{5}(5)=3$, point of intersection (3, 5)
Find the length of the radius between (0, 0) and (3, 5):

$$r=\sqrt{(3-0)^2+(5-0)^2}=\sqrt{34}$$

Equation of the circle $x^2+y^2=34$.

3-13 Ellipses

Definition of an ellipse: An ellipse is the set of all points in a plane such that for each point, the sum of the distances (called the focal radii) from two fixed points (called the foci) is a constant.

Ellipse is commonly applied to the movement of the earth. The earth's orbit around the sun is shaped as an ellipse in which the sun is located and fixed at one of its two foci.

Standard Forms of Ellipses

1. $\frac{x^2}{a^2}+\frac{y^2}{b^2}=1$

   Center (0, 0)

   $c^2=a^2-b^2$

   Focal radii:

   $PF_1+PF_2=2a$

   Vertex: $(a, 0)$ and $(-a, 0)$

   Co-vertex: $(0, b)$ and $(0, -b)$

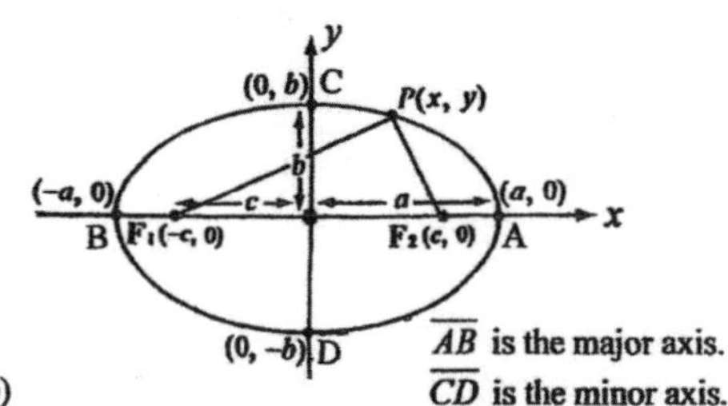

2. $\frac{(x-h)^2}{a^2}+\frac{(y-k)^2}{b^2}=1$

   Center (h, k)

   $c^2=a^2-b^2$

3. $\frac{x^2}{b^2}+\frac{y^2}{a^2}=1$

   Center (0, 0)

   $c^2=a^2-b^2$

4. $\frac{(x-h)^2}{b^2}+\frac{(y-k)^2}{a^2}=1$

   Center (h, k)

   $c^2=a^2-b^2$

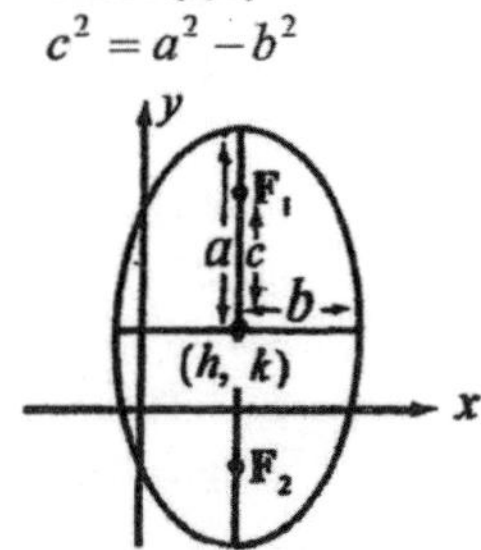

Examples

1. Find the equation of an ellipse having foci (0, 3) and (0, –3) and the sum of its focal radii is 10.

   Solution:

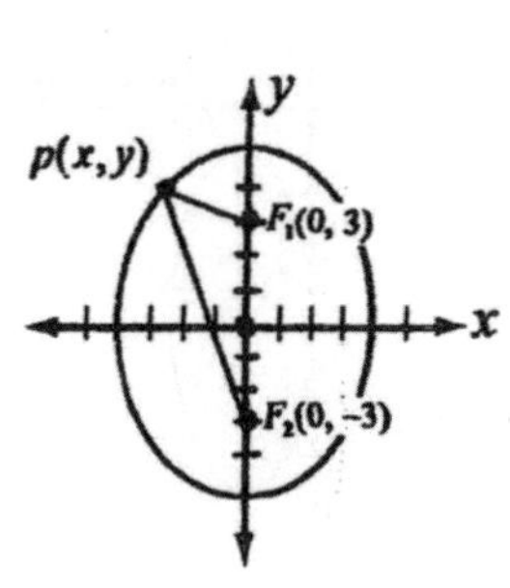

$$PF_1+PF_2=10$$

$$\sqrt{(x-0)^2+(y-3)^2}+\sqrt{(x-0)^2+(y+3)^2}=10$$

$$\sqrt{x^2+(y-3)^2}=10-\sqrt{x^2+(y+3)^2}$$

$$\cancel{x^2}+(y-3)^2=100-20\sqrt{x^2+(y+3)^2}+\cancel{x^2}+(y+3)^2$$

$$\cancel{y^2}-6y+\cancel{9}=100-20\sqrt{x^2+(y+3)^2}+\cancel{y^2}+6y+\cancel{9}$$

$$12y+100=20\sqrt{x^2+(y+3)^2}$$

$$3y+25=5\sqrt{x^2+y^2+6y+9}$$

$$9y^2+150y+625=25(x^2+y^2+6y+9)$$

$$9y^2+\cancel{150y}+625=25x^2+25y^2+\cancel{150y}+225$$

$$25x^2+16y^2=400$$

$$\frac{25x^2}{400}+\frac{16y^2}{400}=1 \quad \therefore \frac{x^2}{16}+\frac{y^2}{25}=1 \text{ is the equation. Ans.}$$

Examples

2. Graph $9x^2+25y^2=225$.

Solution:

$$9x^2+25y^2=225$$

Divided both sides by 225:

$$\frac{9x^2}{225}+\frac{25y^2}{225}=1$$

$$\frac{x^2}{25}+\frac{y^2}{9}=1$$

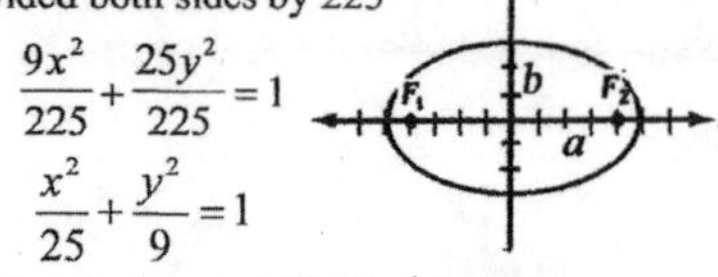

The major axis is horizontal.

Center(0, 0), $a=5$, $b=3$

$$c^2=a^2-b^2=25-9=16$$

$$\therefore c=4$$

3. Graph $25x^2+9y^2=225$.

Solution:

$$25x^2+9y^2=225$$

Divided both sides by 225:

$$\frac{25x^2}{225}+\frac{9y^2}{225}=1$$

$$\frac{x^2}{9}+\frac{y^2}{25}=1$$

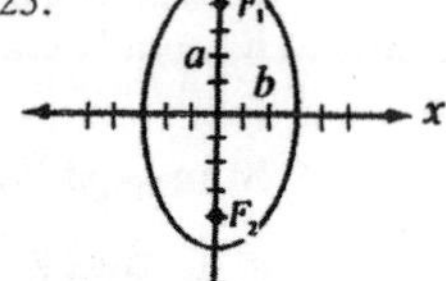

The major axis is vertical.

Center(0, 0), $a=5$, $b=3$

$$c^2=a^2-b^2=25-9=16$$

$$\therefore c=4$$

4. Find the equation of an ellipse having foci (0, 3) and (0, –3) and the sum of its focal radii is 10.

Solution:

Center (0, 0) and $c=3$

$$2a=10 \quad \therefore a=5$$

$$c^2=a^2-b^2$$

$$3^2=5^2-b^2$$

$$b^2=16 \quad \therefore b=4$$

The major axis is vertical:

$$\frac{x^2}{b^2}+\frac{y^2}{a^2}=1 \quad \therefore \frac{x^2}{16}+\frac{y^2}{25}=1. \text{ Ans.}$$

(See graph on Example 1)

5. Find the equation of an ellipse having vertices at (0, 5) and (0, –5) and foci at (0, 3) and (0, –3).

Solution:

Center (0, 0) and $a=5$, $c=3$

We find b:

$$c^2=a^2-b^2$$

$$3^2=5^2-b^2$$

$$b^2=16 \quad \therefore b=4$$

The major axis is vertical:

$$\frac{x^2}{b^2}+\frac{y^2}{a^2}=1 \quad \therefore \frac{x^2}{16}+\frac{y^2}{25}=1$$

(See graph on Example 1)

6. Graph $4x^2+y^2-16x+10y+25=0$

Solution: Completing the square:

$$4(x^2-4x)+(y^2+10y)+25=0$$

$$4[(x^2-4x+4)-4]+(y^2+10y+25)-25+25=0$$

$$4[(x-2)^2-4]+(y+5)^2=0$$

$$4(x-2)^2+(y+5)^2=16$$

Divided both sides by 16;

$\frac{(x-2)^2}{4}+\frac{(y+5)^2}{16}=1$, The major axis is vertical.

Center $(2,-5)$; $a=4$; $b=2$

$$c^2=a^2-b^2=16-4=12$$

$$\therefore c=\sqrt{12}=3.36$$

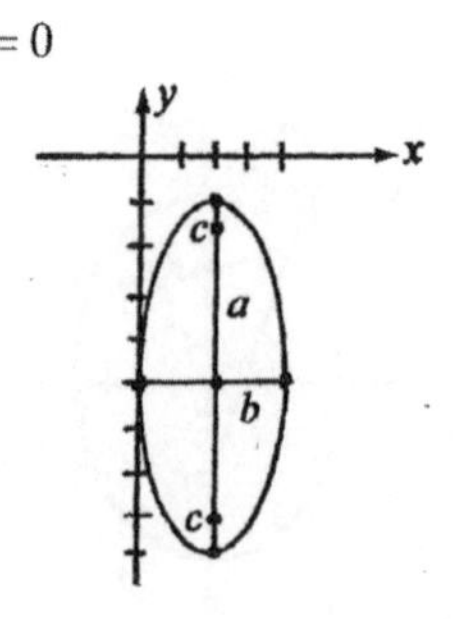

Examples

7. Find the equation of an ellipse passing through the point (3, 5) and having the foci at (3, 2) and (−1, 2).

Solution:

We graph and find: Center (1, 2) and $c=2$

The major axis is horizontal.

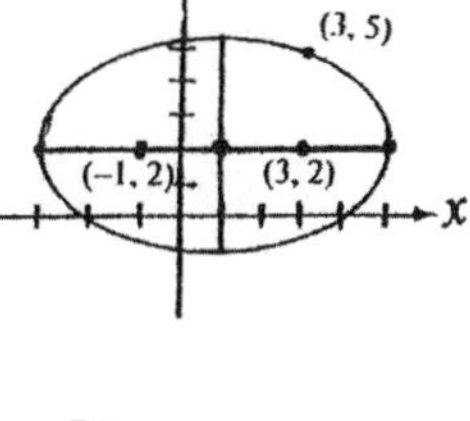

$$\frac{(x-1)^2}{a^2}+\frac{(y-2)^2}{b^2}=1$$

Passing the point (3, 5), we apply the definition of an ellipse:

$$\sqrt{(5-2)^2+(3-3)^2}+\sqrt{(5-2)^2+(3+1)^2}=2a$$

$$\sqrt{9}+\sqrt{25}=2a$$

$$3+5=2a \quad \therefore a=4$$

Since $c^2=a^2-b^2$, $c=2$ we have $4=16-b^2$ $\therefore b=\sqrt{12}$

$$\therefore \frac{(x-1)^2}{16}+\frac{(y-2)^2}{12}=1.$$

8. The earth's orbit around the sun is shaped as an ellipse in which the sun is located and fixed at one of its two foci. At its closest distance (called the perihelion), the earth is 91.45 million miles from the center of the sun. At its farthest distance (called the aphelion), the earth is 94.55 million miles from the center of the sun. Write the equation of the orbit.

Solution:

$$2a=91.45+94.55=186 \text{ million miles}$$

$$\therefore a=93 \text{ million miles}$$

We have: $c=93-91.45=1.55$ million miles

$$c^2=a^2-b^2$$

$$(1{,}55)^2=(93)^2-b^2$$

$$b^2=(93)^2-(1.55)^2=8646.60$$

$$\frac{x^2}{a^2}+\frac{y^2}{b^2}=1, \quad \frac{x^2}{(93)^2}+\frac{y^2}{8646.60}=1 \quad \therefore \frac{x^2}{8649}+\frac{y^2}{8646.60}=1 \text{ (in million miles).}$$

It is nearly circular.

Notes

3-14 Hyperbolas

Definition of a hyperbola: A hyperbola is the set of all points in a plane such that for each point, the difference of the distances (called the focal radii) from two fixed points (called the foci) is a constant.

Standard Forms of Hyperbolas

1. $\dfrac{x^2}{a^2}-\dfrac{y^2}{b^2}=1$

Center (0, 0)

$c^2=a^2+b^2$

Focal radii:

$|PF_1-PF_2|=2c$

Two asymptotes are: $y=\frac{b}{a}x$ and $y=-\frac{b}{a}x$

The length of transverse axis $=2a$

The length of conjugate axis $=2b$

2. $\dfrac{(x-h)^2}{a^2}-\dfrac{(y-k)^2}{b^2}=1$

Center (h, k)

$c^2=a^2+b^2$

Two asymptotes are :

$y-k=\pm\frac{b}{a}(x-h)$

3. $\dfrac{y^2}{a^2}-\dfrac{x^2}{b^2}=1$

Center (0, 0)

$c^2=a^2+b^2$

Two asymptotes are:

$y=\frac{a}{b}x$ and $y=-\frac{a}{b}x$

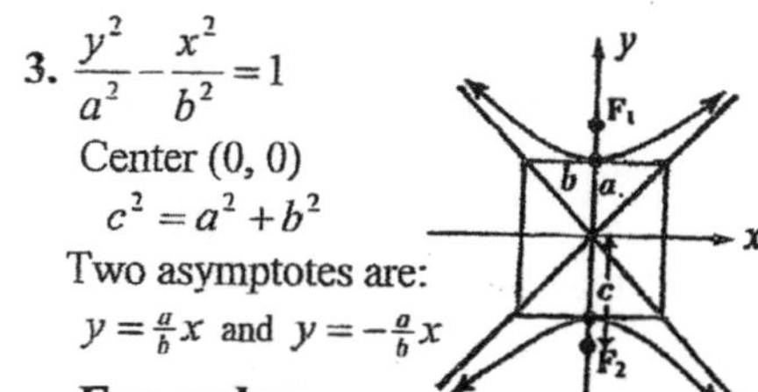

4. $\dfrac{(y-k)^2}{a^2}-\dfrac{(x-h)^2}{b^2}=1$

Center (h, k)

$c^2=a^2+b^2$

Two asymptotes are:

$y-k=\pm\frac{a}{b}(x-h)$

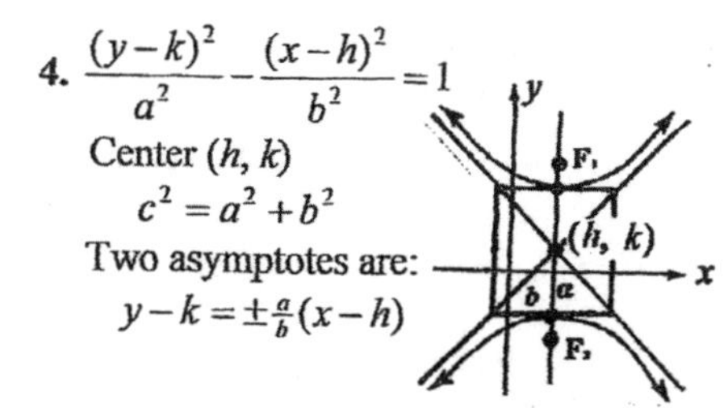

Examples

1. Find the equation of a hyperbola having foci (0, 3) and (0, –3) and the difference of its focal radii is 4.

Solution: $|PF_1=PF_2|=4$

$$\left|\sqrt{(x-0)^2+(y-3)^2}-\sqrt{(x-0)^2+(y+3)^2}\right|=4$$

$$\sqrt{x^2+(y-3)^2}-\sqrt{x^2+(y+3)^2}=\pm 4$$

$$\sqrt{x^2+(y-3)^2}=\sqrt{x^2+(y+3)^2}\pm 4$$

$$x^2+(y-3)^2=x^2+(y+3)^2\pm 8\sqrt{x^2+(y+3)^2}+16$$

$$y^2-6y+9=y^2+6y+9\pm 8\sqrt{x^2+(y+3)^2}+16$$

$$12y+16=\pm 8\sqrt{x^2+(y+3)^2}\text{ , }3y+4=\pm 2\sqrt{x^2+(y+3)^2}$$

$$9y^2+24y+16=4(x^2+y^2+6y+9)$$

$$5y^2-4x^2=20\text{ , }\frac{5y^2}{20}-\frac{4x^2}{20}=1\quad \therefore \frac{y^2}{4}-\frac{x^2}{5}=1.$$

Examples

2. Graph $16x^2 - 9y^2 = 144$.

Solution:

$$16x^2 - 9y^2 = 144$$

Divided both sides by 144:

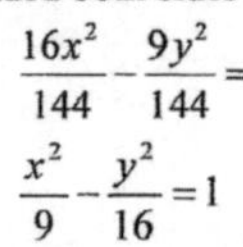

$$\frac{16x^2}{144} - \frac{9y^2}{144} = 1$$

$$\frac{x^2}{9} - \frac{y^2}{16} = 1$$

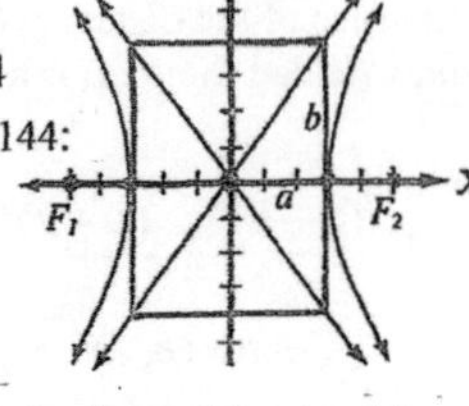

The transverse axis is horizontal (x – axis).

Center (0, 0), $a = 3$, $b = 4$

$c^2 = a^2 + b^2 = 9 + 16 = 25 \quad \therefore c = 5$.

3. Graph $4y^2 - 9x^2 = 36$.

Solution:

$$4y^2 - 9x^2 = 36$$

Divided both sides by 36:

$$\frac{4y^2}{36} - \frac{9x^2}{36} = 1$$

$$\frac{y^2}{9} - \frac{x^2}{4} = 1$$

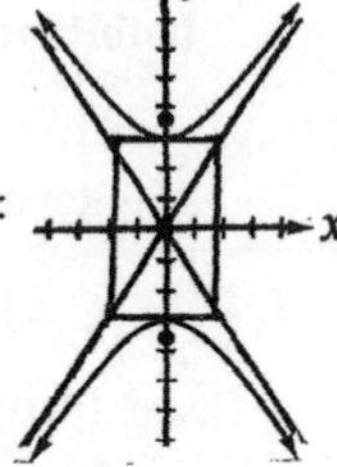

The transverse axis is vertical (y – axis).

Center (0, 0), $a = 3$, $b = 2$

$c^2 = a^2 + b^2 = 9 + 4 = 13 \quad \therefore c = \sqrt{13}$.

4. Find the equation of a hyperbola having foci (0, 3) and (0, –3) and the difference of its focal radii is 4.

Solution:

Center (0, 0), $c = 3$

$2a = 4 \quad \therefore a = 2$

We find b: $c^2 = a^2 + b^2$

$3^2 = 2^2 + b^2$

$b^2 = 5 \quad \therefore b = \sqrt{5}$

The transverse axis is vertical (y – axis).

$$\frac{y^2}{a^2} - \frac{x^2}{b^2} = 1 \quad \therefore \frac{y^2}{4} - \frac{x^2}{5} = 1.$$

(See graph on Example 1)

5. Find the equation of a hyperbola having vertices at (0, 2) and (0, –2) and foci (0. 3) and (0, –3).

Solution:

Center (0, 0) and $a = 2$, $c = 3$

We find b: $c^2 = a^2 + b^2$

$3^2 = 2^2 + b^2$

$b^2 = 5$

$\therefore b = \sqrt{5}$

The transverse axis is vertical ((y – axis).

$$\frac{y^2}{a^2} - \frac{x^2}{b^2} = 1 \quad \therefore \frac{y^2}{4} - \frac{x^2}{5} = 1.$$

(See graph on Example 1)

6. Graph $y^2 - 2x^2 + 6y + 8x - 3 = 0$.

Solution: Completing the square:

$$(y^2 + 6y) - 2(x^2 - 4x) - 3 = 0$$

$$(y^2 + 6y + 9) - 9 - 2[(x^2 - 4x + 4) - 4] - 3 = 0$$

$$(y+3)^2 - 9 - 2(x-2)^2 + 8 - 3 = 0$$

$$(y+3)^2 - 2(x-2)^2 = 4$$

Divided both sides by 4:

$$\frac{(y+3)^2}{4} - \frac{(x-2)^2}{2} = 1$$

The transverse axis is vertical ($x = 2$).

Center (2, –3), $a = 2$, $b = \sqrt{2}$

$c^2 = a^2 + b^2 = 4 + 2 = 6 \quad \therefore c = \sqrt{6}$

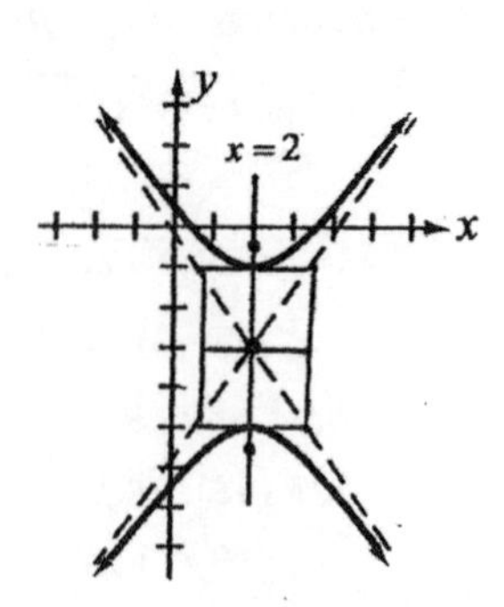

Examples

7. Find the equation of a hyperbola passing through the point (1, –5) and having the foci at (–2, –1) and (–2, –5).

Solution: We graph and find: Center (–2, –3) and $c = 2$

The transverse axis is vertical.

We have the equation: $\frac{(y+3)^2}{a^2} - \frac{(x+2)^2}{b^2} = 1$

Passing the point (1, –5), we apply the definition of a hyperbola:

$$\left|\sqrt{(1+2)^2+(-5+1)^2} - \sqrt{(1+2)^2+(-5+5)^2}\right| = 2a$$

$$\left|\sqrt{25} - \sqrt{9}\right| = 2a$$

$$|5-3| = 2a \quad \therefore a = 1$$

y
x = –2
x
(–2, –1)
(1, –5)
(–2, –5)

Since $c^2 = a^2 + b^2$, $c = 2$, we have $2^2 = 1^2 + b^2$ $\therefore b = \sqrt{3}$

$$\therefore (y+3)^2 - \frac{(x+2)^2}{3} = 1.$$

8. Find the equations of the asymptotes of the hyperbola $x^2 - 4y^2 = 4$.

Solution:

$$x^2 - 4y^2 = 4$$

Divided both sides by 4:

$$\frac{x^2}{4} - \frac{y^2}{1} = 1 \quad \therefore a = 2,\ b = 1$$

The equations of the asymptotes are:

$$y = \pm \tfrac{b}{a} x$$

$$\therefore y = \pm \tfrac{1}{2} x.$$

9. Find the equations of the asymptotes of the hyperbola $3y^2 - 4x^2 = 36$.

Solution:

$$3y^2 - 4x^2 = 36$$

Divided both sides by 36:

$$\frac{y^2}{12} - \frac{x^2}{9} = 1 \quad \therefore a = \sqrt{12},\ b = 3$$

The equations of the asymptotes are:

$$y = \pm \tfrac{a}{b} x$$

$$\therefore y = \pm \tfrac{2\sqrt{3}}{3} x.$$

10. Find the equations of the asymptotes of the hyperbola $9x^2 - 4y^2 - 36x - 24y - 36 = 0$.

Solution:

$$9x^2 - 4y^2 - 36x - 24y - 36 = 0$$

Completing the square, we have:

$$\frac{(x-2)^2}{4} - \frac{(y+3)^2}{9} = 1 \quad \therefore a = 2,\ b = 3$$

The equations of the asymptotes are:

$$y - k = \pm \tfrac{b}{a}(x - h)$$

$$\therefore y + 3 = \pm \tfrac{3}{2}(x - 2).$$

11. Find the equations of the asymptotes of the hyperbola $y^2 - x^2 - 6y - 4x + 1 = 0$.

Solution:

$$y^2 - x^2 - 6y - 4x + 1 = 0$$

Completing the square, we have:

$$\frac{(y-3)^2}{4} - \frac{(x+2)^2}{4} = 1 \quad \therefore a = b = 2$$

(Hint: It is an **equilateral hyperbola**)

The equations of the asymptotes are:

$$y - k = \pm \tfrac{a}{b}(x - h)$$

$$\therefore y - 3 = \pm(x + 2).$$

Notes

Notes

Notes

3-15 Distance and Midpoint Formulas

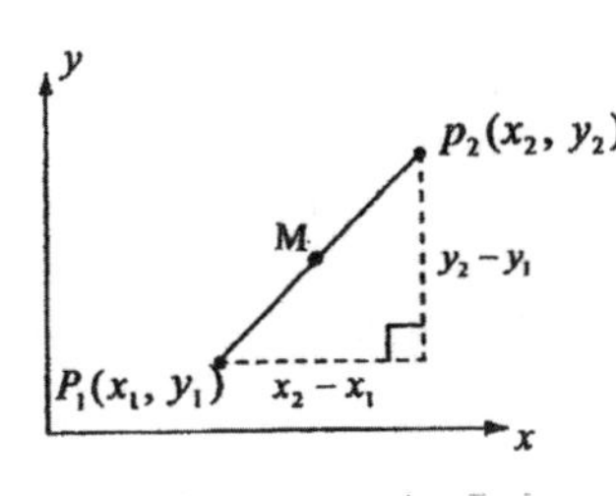

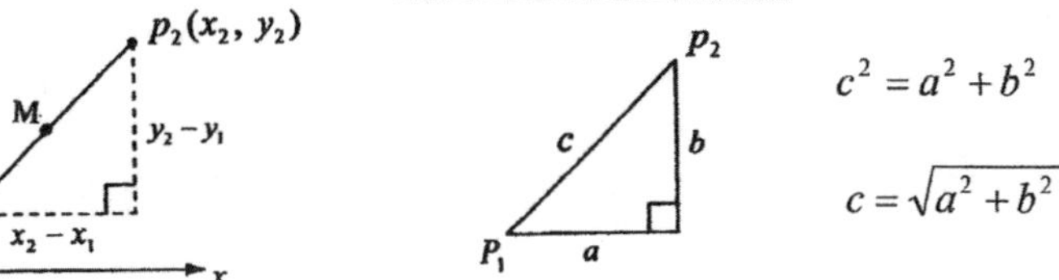

Distance Formula: The distance between two points p_1 and p_2 is:

$$d = \overline{p_1 p_2} = \sqrt{(x_2 - x_1)^2 + (y_2 - y_1)^2}$$

Midpoint Formula: The midpoint between two points p_1 and p_2 is:

$$M = \left(\frac{x_1 + x_2}{2}, \frac{y_1 + y_2}{2}\right)$$

Examples

1. Find the distance and midpoint between $A(-2, 1)$ and $B(5, 4)$.
   Solution:
   $$d = \overline{AB} = \sqrt{[5 - (-2)]^2 + (4 - 1)^2} = \sqrt{49 + 9} = \sqrt{58} \approx 7.62.$$
   $$M = \left(\frac{-2+5}{2}, \frac{1+4}{2}\right) = (\ 1.5, 2.5\ ).$$

2. Find the distance and midpoint between $(4, 3\sqrt{3})$ and $(2, -\sqrt{3})$.
   Solution:
   $$d = \sqrt{(2-4)^2 + (-\sqrt{3} - 3\sqrt{3})^2} = \sqrt{(-2)^2 + (-4\sqrt{3})^2} = \sqrt{4 + 48} = \sqrt{52} \approx 7.21.$$
   $$M = \left(\frac{4+2}{2}, \frac{3\sqrt{3} + (-\sqrt{3})}{2}\right) = (3, \sqrt{3}) \approx (\ 3, 1.73\ ).$$

3. If $M(1, 0)$ is the midpoint of the segment $\overline{AB}$ and $B(-1, -2)$ is the coordinates of point B. Find the coordinates of point A.
   Solution:
   $$A(x, y),\quad M(1, 0),\quad B(-1, -2)$$
   $$\frac{x + (-1)}{2} = 1,\quad x - 1 = 2,\ \therefore x = 3$$
   $$\frac{y + (-2)}{2} = 0,\quad y - 2 = 0,\ \therefore y = 2$$
   $\therefore$ The coordinates of point A is $A(3, 2)$.

Notes

3-16 Coordinate Geometry in Three-Dimensions

We have learned how to plot a point in a **plane** in the two-dimensional coordinate system. The plane is called the Cartesian plane determined by two perpendicular number lines. The lines are called the x-**axis** and the y-**axis**. A point in the plane is determined by an **ordered pair** (x, y).
To plot a point in a **space**, we add a third dimension by a z-**axis** perpendicular to both the x-axis and the y-axis at the origin. In a three-dimensional coordinate system, the axes determine three coordinate planes, the ***xy*-plane**, the ***xz*-plane**, and the ***yz*-plane**. A point in the space is determined by an **ordered triple** (x, y, z). The three axes divide the space into eight **octants**. In the first octant, all three coordinates are positive.

The geometry determined by the coordinate system is called the **Coordinate Geometry**, or **Analytic Geometry**. The geometry determined by the three-dimensional coordinate system is also called the **Solid Analytic Geometry**.

Coordinate System in Three-Dimensions

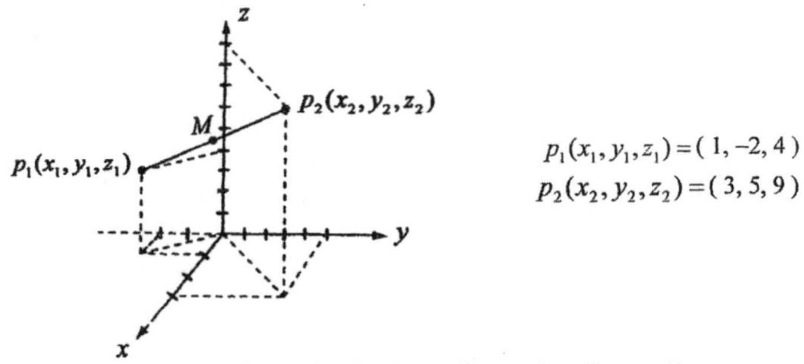

In algebra and geometry, many formulas in three-dimensional coordinate system are similar to the formulas in two-dimensional coordinate system.

Distance Formula: The distance between two points p_1 and p_2 is

$$d = \overline{p_1 p_2} = \sqrt{(x_2 - x_1)^2 + (y_2 - y_1)^2 + (z_2 - z_1)^2}$$

Midpoint Formula: The midpoint between two points p_1 and p_2 is

$$M = \left(\frac{x_1 + x_2}{2}, \frac{y_1 + y_2}{2}, \frac{z_1 + z_2}{2}\right)$$

Example

1. Find the distance and midpoint between $(1, -2, 4)$ and $(3, 5, 9)$.
   Solution:

$$d = \sqrt{(3-1)^2 + (5+2)^2 + ((9-4)^2} = \sqrt{4 + 49 + 25} = \sqrt{78} \approx 8.83.$$

$$M = \left(\frac{1+3}{2}, \frac{-2+5}{2}, \frac{4+9}{2}\right) = (2,\ 1.5,\ 6.5).$$

Linear Equations in Three-Dimensions

We have learned that the linear equation $ax+by=c$ is a straight line in two-dimensional system. If we write the equation $ax+by=c$ by solving for y in terms of x, we have a linear function $y=-\frac{a}{b}x+\frac{c}{b}$, or $f(x)=-\frac{a}{b}x+\frac{c}{b}$.

The linear equation $ax+by+cz=d$ is a plane in three-dimensional system. If we write the equation $ax+by+cz=d$ by solving for z in terms of x and y, we have a linear function

$$z=-\frac{a}{c}x-\frac{b}{c}y+\frac{d}{c}, \text{ or } f(x, y)=-\frac{a}{c}x-\frac{b}{c}y+\frac{d}{c}.$$

To graph a linear equation in three-dimensions, we find the points at which the graph intersects the three axes. Then, we connect these points of intercepts with lines to form a triangular plane that lies in the first octant and extends to all other octants.

Examples

1. Write the linear equation $2x+3y+4z=12$ as a function of x and y.
   Solution:

   $$2x+3y+4z=12$$

   Solving for z:

   $$4z=12-2x-3y$$

   $$\therefore z=\tfrac{1}{4}(12-2x-3y).$$

   Or: $f(x, y)=\tfrac{1}{4}(12-2x-3y)$.

2. Graph the linear equation $2x+3y+4z=12$ in the first octant.
   Solution:

   Find the intercepts:

   Let $x=0$, $y=0$, z-intercept $=3$.

   Let $x=0$, $z=0$, y-intercept $=4$.

   Let $y=0$, $z=0$, x-intercept $=6$.

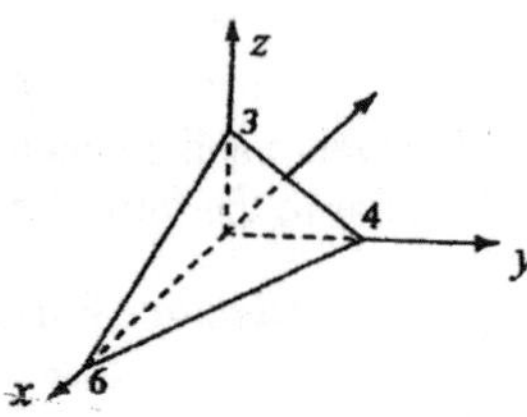

3. Find the volume of the rectangular solid with a given point (2, 6, 5).
   Solution:

   Length: $\ell=6$
   Width: $w=2$
   Height: $h=5$
   Volume: $V=\ell\times w\times h=6\times2\times5=60$.

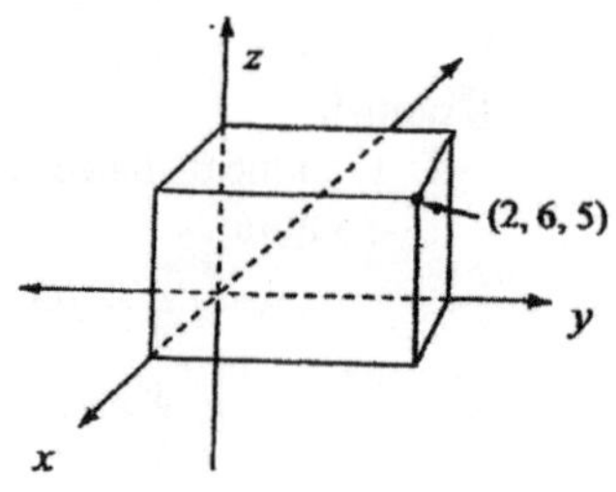

The Equation of a Sphere in Three-Dimensions

We have learned that, in two-dimensions, the equation $x^2 + y^2 = r^2$ is a circle whose center is the origin (0, 0) and whose radius is r. The equation $(x-h)^2 + (y-k)^2 = r^2$ is a circle whose center is the point (h, k) and whose radius is r.

In three-dimensions, the equation of a sphere is:

$x^2 + y^2 + z^2 = r^2$ **(Standard Form)**

Its center is the origin (0, 0, 0) and radius is r.

In three-dimensions, the equation of a sphere is:

$(x-h)^2 + (y-k)^2 + (z-j)^2 = r^2$ **(Standard Form)**

Its center is the point (h, k, j) and radius is r.

In three-dimensions, the equation $x^2 + y^2 + z^2 + ax + by + cz = 0$ is the **general form** of a sphere. To find its center and radius, we rewrite it to match the standard form by the method of **completing the square**.

Examples

1. Find the equation of the sphere whose center is (2, 5, 3) and whose radius is 3.
   Solution:

$$(x-2)^2 + (y-5)^2 + (z-3)^2 = 3^2$$

$$(x-2)^2 + (y-5)^2 + (z-3)^2 = 9.$$

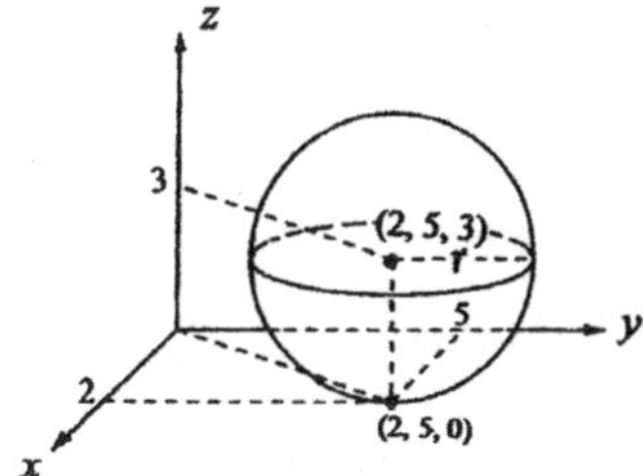

2. Find the center and radius of the sphere:

$$x^2 + y^2 + z^2 - 4x - 10y - 6z + 29 = 0$$

   Solution:

$$x^2 + y^2 + z^2 - 4x - 10y - 6z + 29 = 0$$

Completing the square:

$$(x^2 - 4x) + (y^2 - 10y) + (z^2 - 6z) + 29 = 0$$

$$(x^2 - 4x + 4) - 4 + (y^2 - 10y + 25) - 25 + (z^2 - 6z + 9) - 9 + 29 = 0$$

$$(x^2 - 4x + 4) + (y^2 - 10y + 25) + (z^2 - 6z + 9) = 9$$

$$(x-2)^2 + (y-5)^2 + (z-3)^2 = 3^2$$

$\therefore$ Center (2, 5, 3) and radius = 3.

(See the graph in Problem 1.)

Distance Formulas between Lines and Planes

1. The distance between a point (x_1, y_1) and a straight line $ax+by+c=0$ is given by the formula:

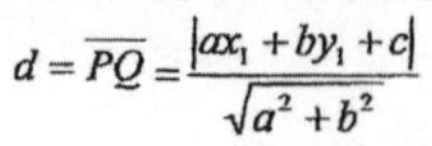

$$d=\overline{PQ}=\frac{|ax_1+by_1+c|}{\sqrt{a^2+b^2}}$$

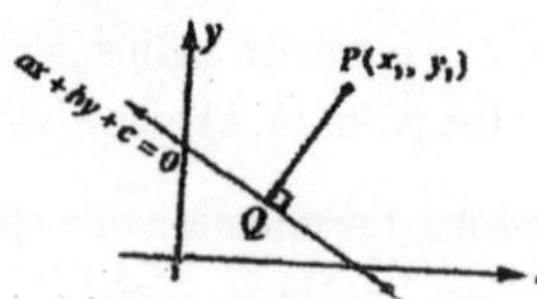

2. The distance between two parallel lines $ax+by+c_1=0$ and $ax+by+c_2=0$ is given by the formula:

$$d=\overline{PQ}=\frac{|c_1-c_2|}{\sqrt{a^2+b^2}}$$

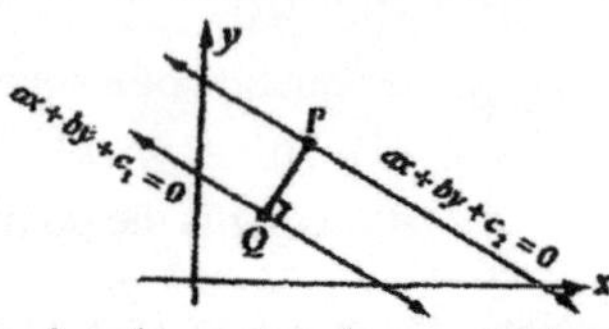

3. The distance between a point (x_1, y_1, z_1) and a plane $ax+by+cz+d=0$ is given by the formula:

$$d=\overline{PQ}=\frac{|ax_1+by_1+cz_1+d|}{\sqrt{a^2+b^2+c^2}}$$

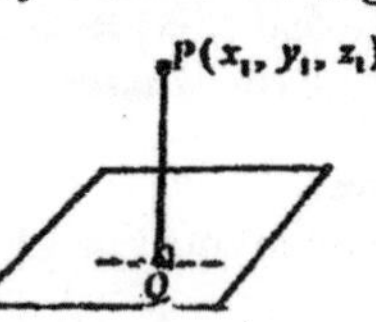

Examples

1. Find the distance between the line $2x+4y-5=0$ and the point (3, -1).

   Solution: $d=\frac{|ax_1+by_1+c|}{\sqrt{a^2+b^2}}=\frac{|2\cdot 3+4(-1)+(-5)|}{\sqrt{2^2+4^2}}=\frac{|-3|}{\sqrt{20}}=\frac{3}{4.472}=0.67.$

2. Find the distance between the two parallel lines $x-2y+3=0$ and $x-2y-5=0$.

   Solution: $d=\frac{|c_1-c_2|}{\sqrt{a^2+b^2}}=\frac{|3-(-5)|}{\sqrt{1^2+(-2)^2}}=\frac{8}{\sqrt{5}}=\frac{8}{2.236}=3.58\cdot$

3. Find the distance between the point (1, 2, 1) and the plane $2x+y-z-1=0$.

   Solution: $d=\frac{|ax_1+by_1+cz_1+d|}{\sqrt{a^2+b^2+c^2}}=\frac{|2\cdot 1+1\cdot 2-1\cdot 1-1|}{\sqrt{2^2+1^2+(-1)^2}}=\frac{2}{\sqrt{6}}=\frac{2}{2.449}=0.82\cdot$

4. Two parallel planes are $2x-3y+z-1=0$ and $2x-3y+z-5=0$. Find the distance between the planes.

   Solution: Find one point on the first plane (0, 0, 1)

   $$d=\frac{|ax_1+by_1+cz_1+d|}{\sqrt{a^2+b^2+c^2}}=\frac{|0+0+1-5|}{\sqrt{4+9+1}}=\frac{4}{\sqrt{14}}=\frac{4}{3.742}=1.07\cdot$$

5. Determine whether the point (3, 4, –2) is outside or inside the sphere $x^2+y^2+z^2=9$.

   Solution: The center of the sphere is the origin (0, 0, 0) and the radius is 3.
   A point is inside the sphere if the distance between it and the origin is smaller than the radius. $d=\sqrt{3^2+4^2+(-2)^2}=\sqrt{9+16+4}=\sqrt{29}\approx 5.34>3$

   $\therefore$ The point (3, 4, -2) is not inside the sphere.

Notes

Notes

3-17 Symmetry and Transformations

We have learned how to graph a polynomial function by transformations. In this section, we will discuss the transformations of a geometric figure in the plane. Transformations can help us to design or sketch the image of a figure in arts or constructions. In each transformation, we move the original figure by reflects, slides, or turns. Each point of the new figure corresponds (or pairs) to exactly one point of the original figure. The transformations include **Reflections, Translations, Rotations**, and **Dilations.**

Reflections, translations, and rotations are transformations that preserve size and shape. The image and the original figure are congruent figures (congruent mapping). Each of these three transformations is also called an **isometry. Dilations** are transformations that involve enlargements or reductions. Dilation is not an isometry.

1. **Symmetry & Reflections**
   In the **reflection**, each point of a figure reflects over from one side of a line to the opposite side. Reflection preserves its size and shape.
   The reflection image and the original figure has the same distance from **the line of reflection**.
   To draw the reflection image of a figure, the line of reflection is the perpendicular bisector of the line between the point and the point of its own image.

Reflect ΔABC over line k

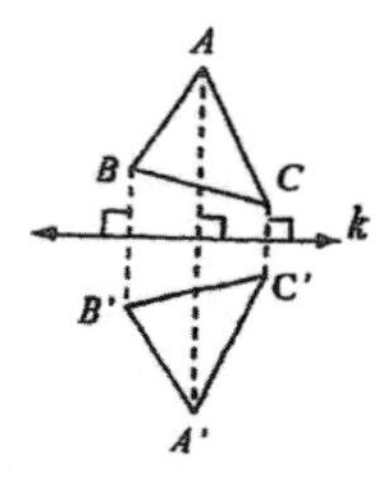

Reflect $< ABC$ over line p

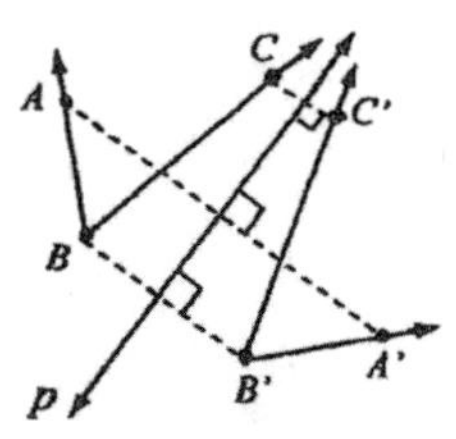

If a line can be drawn through a figure and divides the figure into two parts that are mirror images of each other, then the figure is said to have **a line of symmetry**, or **axis of symmetry**.

Polygon *ABCDEFGH* has a line k as its line of symmetry.

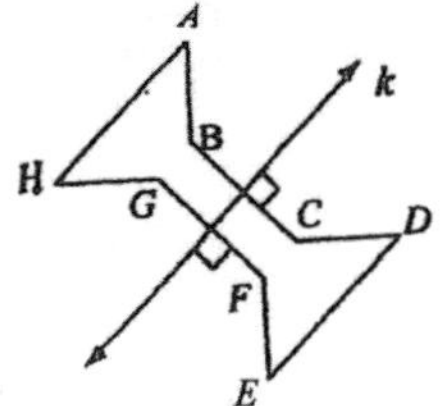

Examples

1. Reflect ΔABC over the x-axis.
   Solution:

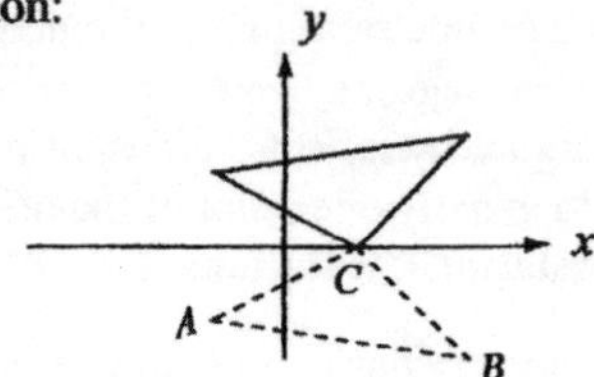

2 Reflect the quadrilateral $ABCD$ over the y-axis.
   Solution:

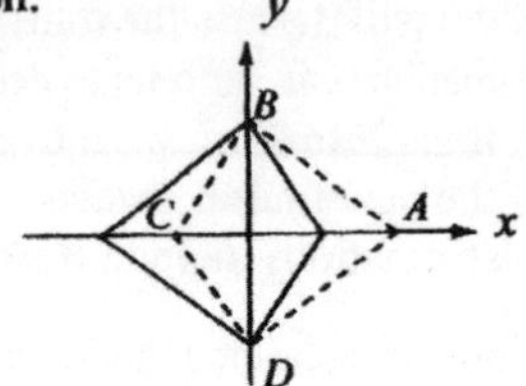

3. Draw the graph
   1. to make it symmetric with respect to the x-axis.
   2. to make it symmetric with respect to the y-axis.
   3. to make it symmetric with respect to the origin.
   4. to make it symmetric with respect to the x-axis, y-axis, and origin.

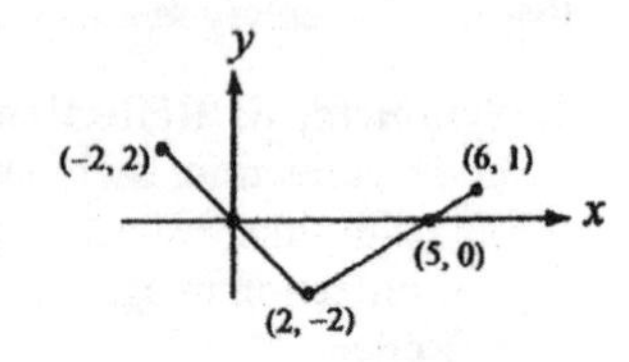

Solution:

1.

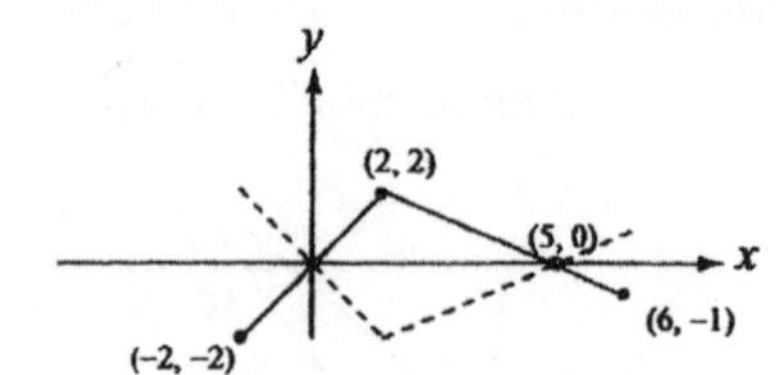

2.

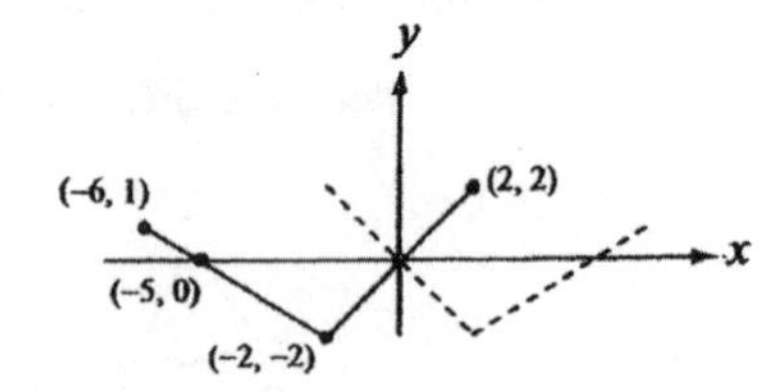

3.

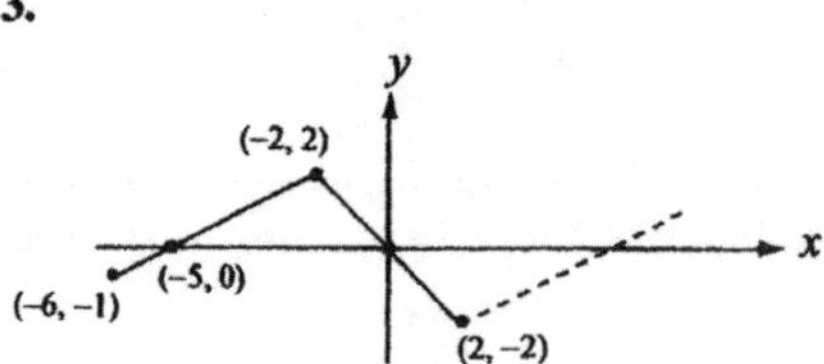

4.

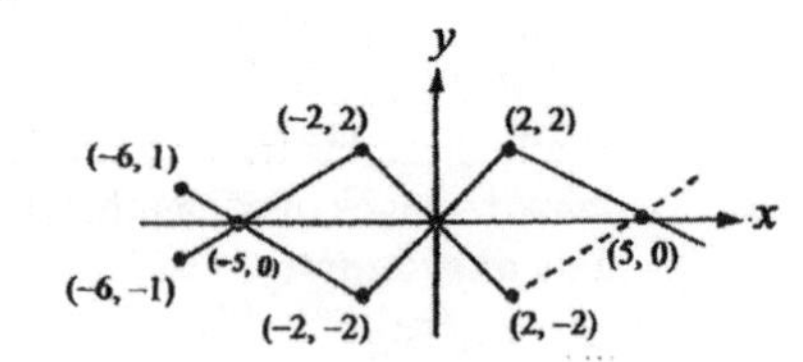

2. Translations

In the **translation**, each point of a figure slides the same distance in the same direction. Translation preserves its size and shape.

A composition of two reflections over two parallel lines is a translation. The distance between the original figure and the image after a translation is twice the distance between the parallel lines.

Examples

1. Translate $\overline{AB}$ 4 units to the left and and 2 units down, or by the translation vector $\vec{v} = (-4,-2)$.

   Solution:

2. Lines m and n are parallel. Reflect ΔABC over line m, followed by a reflection over line n.

   Solution:

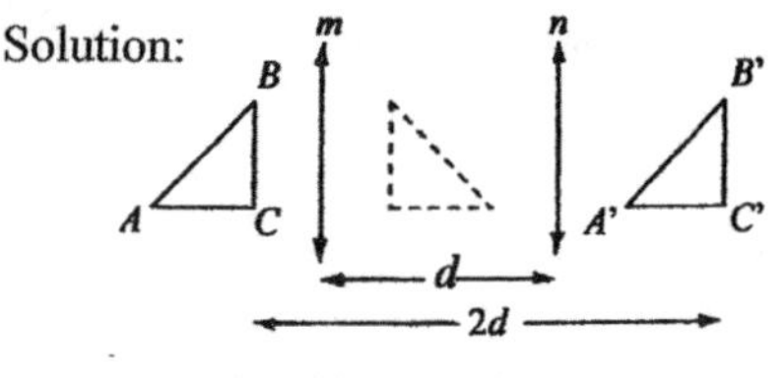

3. Rotations

In the **rotation**, each point of a figure moves along a circular path about a fixed point. Rotation preserves its size and shape. If the direction of a rotation is counterclockwise, the measure of the angle of rotation is positive. If the direction is clockwise, the measure of the angle of rotation is negative.

$\overline{AB}$ is rotated by $-90°$ (clockwise) about origin. $m < AOA' = 90$, $m < BOB' = 90$

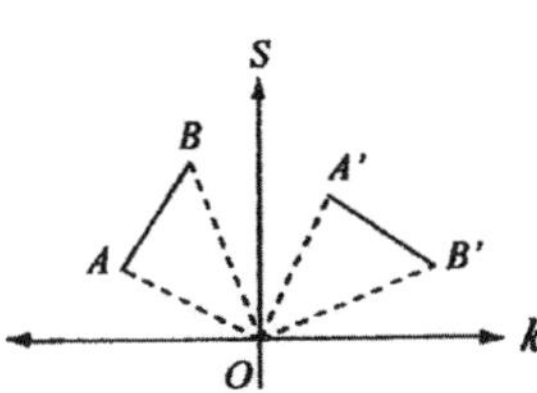

ΔABC is rotated by $-90°$ (clockwise) about the the origin.

$m < AOA' = 90$, $m < BOB'$, $m < COC' = 90$

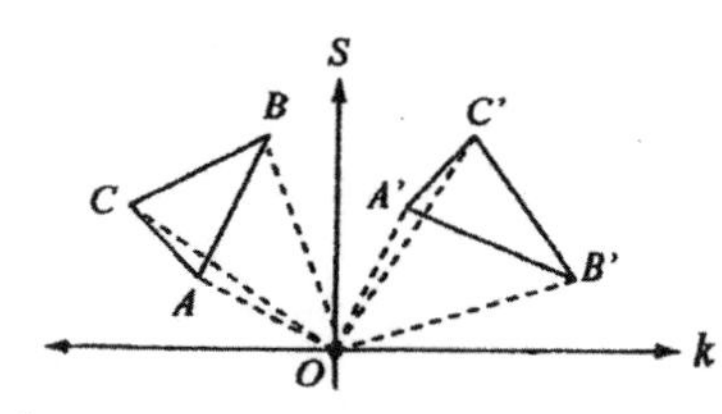

Formula for Rotation of a Point

Let $p(x, y)$ be a point in the plane, and $p'(x', y')$ be the resulting point after a rotation of angle θ with the origin as the center of rotation.

$$x = r\cos\alpha \qquad x' = r\cos(\theta + \alpha)$$

$$y = r\sin\alpha \qquad y' = r\sin(\theta + \alpha)$$

$$x' = r\cos(\theta + \alpha) = r(\cos\theta\cos\alpha - \sin\theta\sin\alpha) = x\cos\theta - y\sin\theta$$

$$y' = r\sin(\theta + \alpha) = r(\sin\theta\cos\alpha + \cos\theta\sin\alpha) = x\sin\theta + y\cos\theta$$

Formula: $x' = x\cos\theta - y\sin\theta$

$y' = x\sin\theta + y\cos\theta$

In matrix form: $\begin{bmatrix} x' \\ y' \end{bmatrix} = \begin{bmatrix} \cos\theta & -\sin\theta \\ \sin\theta & \cos\theta \end{bmatrix} \cdot \begin{bmatrix} x \\ y \end{bmatrix}$

4. Dilations

In the dilation (or dilatation), each point of a figure is multiplied by a **scale factor** to create a similar image. Dilation preserves the angle measure of the original figure while enlarging or reducing the size and shape. The center of dilation is the point from which the dilation is created. The perimeter of the image enlarges or reduces the original perimeter figure in the same **ratio** as the scale factor. Dilation is not an isometry. The image A' of point A under a dilation is usually denoted by $D_n(A)$.

In the graph, $\overline{A'B'}$ is the image of $\overline{AB}$ under a dilation of $n = \frac{2}{3}$.

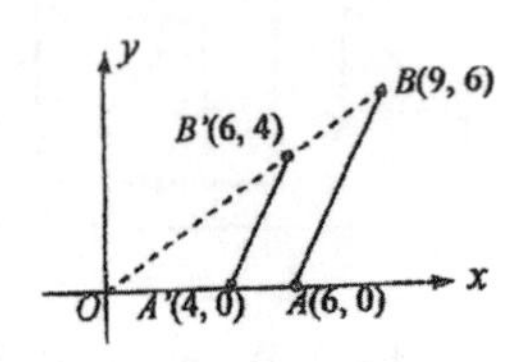

$A' = (6 \cdot \frac{2}{3},\ 0 \cdot \frac{2}{3}) = (4,\ 0)$

$B' = (9 \cdot \frac{2}{3},\ 6 \cdot \frac{2}{3}) = (6,\ 4)$

Center of dilation is (0, 0)

$\overline{A'B'} = \frac{2}{3}(\overline{AB})$

In the graph, $\Delta OA'B'$ is the image of ΔOAB under a dilation of $n = 2$.

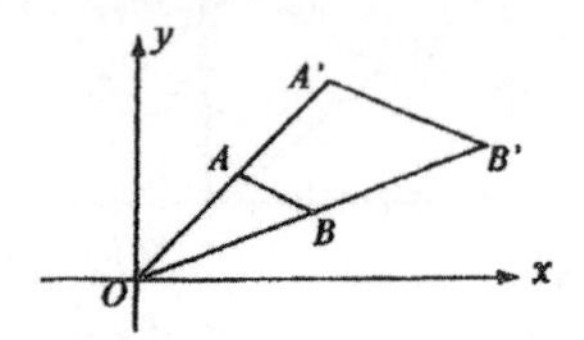

$\overline{OA'} = 2(\overline{OA})$

$\overline{OB'} = 2(\overline{OB})$

Center of dilation is (0, 0)

$\overline{OA'} + \overline{OB'} + \overline{A'B'} = 2(\overline{OA} + \overline{OB} + \overline{AB})$

In the graph, enlarge ΔABC so that the perimeter of $\Delta A'B'C'$ triples the original perimete

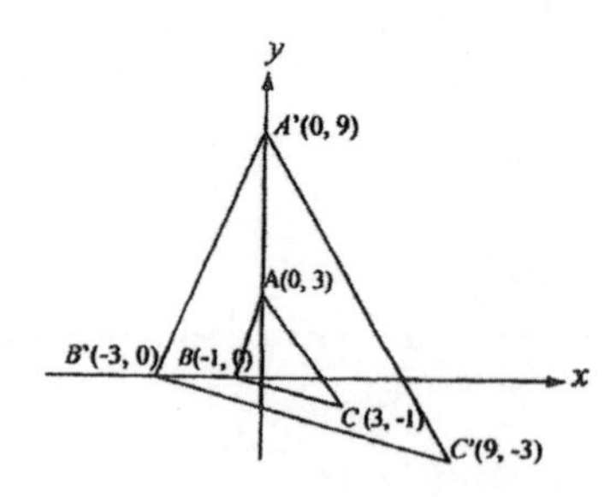

Dilation Theorem: If a closed figure F is dilated to a closed figure F' with ratio k, then the following statements are true:

1. The perimeter of F' is k times the perimeter of F.
2. The area of F' is k^2 times the perimeter of F.
3. The volume of F' is k^3 times the volume of F.

Notes

Notes

3-18 Cylinders

A **Cylinder** is a three-dimensional figure that consists of two parallel and congruent circular bases and a lateral surface. The axis of a cylinder is the segment that connects the centers of the two bases. Cylinders include **right cylinders** and **oblique cylinders**. In a right cylinder, the axis is its height (altitude).

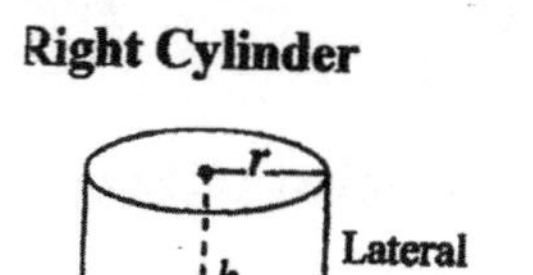

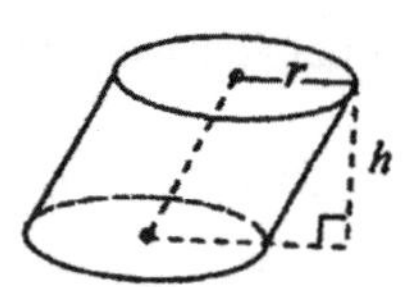

The lateral surface of a right cylinder is a rectangle. The length of the rectangle is the circumference ($C = 2\ \pi r$) of the circle (the base). The width of the rectangle is the height of the cylinder.

The **total surface area** (**SA**) of a right cylinder with base radius r and height h is the sum of the lateral surface area (LA) and the area of the two congruent circles.

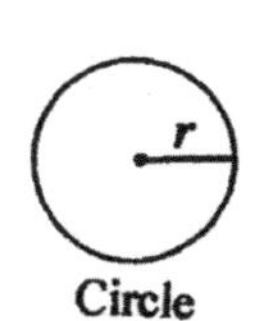

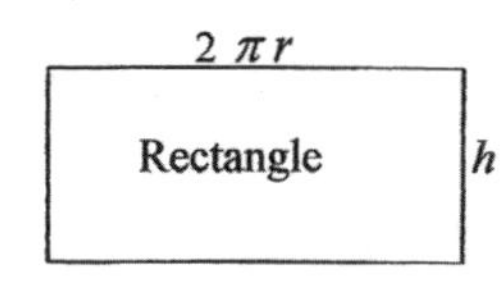

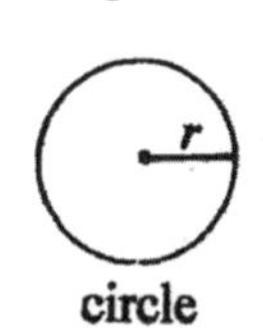

Lateral Surface Area of a Cylinder: $LA = 2\pi\ r\,h$

Total Surface Area of a Cylinder: $SA = 2(\pi\ r^2)\ +\ 2\pi\ r\,h$

The **volume** (V) of a right cylinder with base radius r and height h is the product of the base area (B) and the height h.

Volume of a Cylinder: $V = Bh = \pi\ r^2 h$

Examples

1. Find the total surface area and volume of the right cylinder of radius 5 inches and height 8 inches.

   Solution:

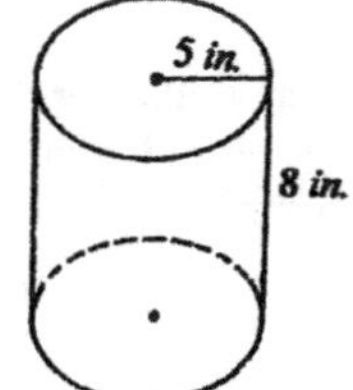

   Area of the bases = $2(\pi\ r^2) = 2\pi(5^2) = 50\pi$

   Area of the lateral surface $LA = 2\pi\ rh = 2\pi(5)(8) = 80\pi$

   Total surface area $SA = 50\pi + 80\pi = 130\pi\ \ in.^2$

   Volume $V = \pi\ r^2 h = \pi(5^2)(8) = 200\pi\ \ in.^3$

Examples

2. Find the volume of the oblique cylinder of radius 20 centimeters and height 30 centimeters
Solution:

$$V = \pi r^2 h = \pi(20^2)(30) = 12{,}000\pi \ cm^3.$$

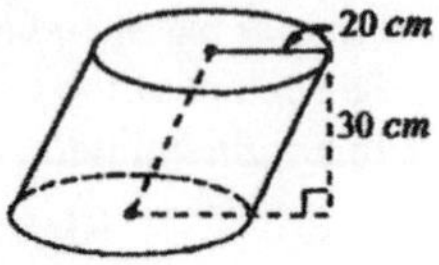

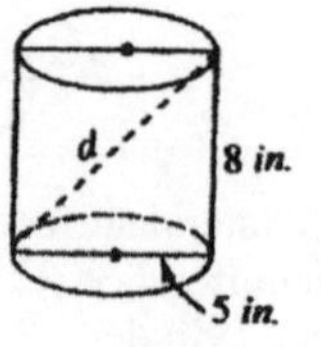

3. Find the longest line that can be drawn inside the cylinder of length 8 inches and radius 5 inche
Solution:

$$d^2 = 10^2 + 8^2 = 164 \text{ (Pythagorean Theorem)}$$

$$d = \sqrt{164} \approx 12.81 \text{ inches.}$$

4. A water tank is shaped like a right cylinder of diameter 1.6 meters and length 4 meters. Find its volume.
Solution:

$$V = \pi r^2 h = \pi(0.8)^2(4) \approx 8.04 \ m^3.$$

5. In the diagram, a concrete tunnel pipe is shaped as a hollowed-out right cylinder. How much concrete in cubic meters is needed to make this pipe ?
Solution:

$$r_1 = 2\,m, \quad r_2 = 1.7\,m, \quad h = 5\,m$$

$$V = \pi r_1^2 h - \pi r_2^2 h = \pi h(r_1^2 - r_2^2) = \pi(5)(2^2 - 1.7^2)$$
$$= \pi(5)(4 - 2.89) = \pi(5)(1.11) \approx 17.43 \ m^3.$$

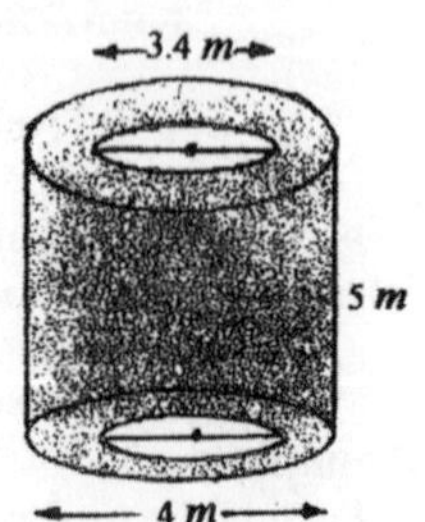

6. The volume and lateral surface area of a cylinder are equal. Find the total surface area of this cylinder with height 12 centimeters.
Solution:

$$\pi r^2 h = 2\pi r h$$

We have: $r = 2$

Total surface area: $SA = 2(\pi r^2) + 2\pi rh = 2\pi(2^2) + 2\pi(2)(12) = 56\pi \ cm^2$.

3-19 Prisms

A **prism** is a three-dimensional figure that consists of two parallel and congruent polygons as its two **faces** (called **bases**) connecting by the **lateral faces** (parallelograms). The intersections of all faces (the bases and lateral faces) are called the **edges.**

A **right prism** is a prism whose lateral edge is perpendicular to the bases.
A **regular prism** is a right prism whose bases are regular congruent polygons.
An **oblique prism** is a prism whose lateral edge is not perpendicular to the bases.

Prisms are classified by the shapes of their bases, such as **Rectangular Solids, Cubes, Triangular Prisms, Pentagonal Prisms, Hexagonal Prisms,** ....

1. **Rectangular Solids (Rectangular Prisms):** A rectangular solid is a prism whose two bases and four lateral faces are rectangles.

Volume of a Rectangular Solid:

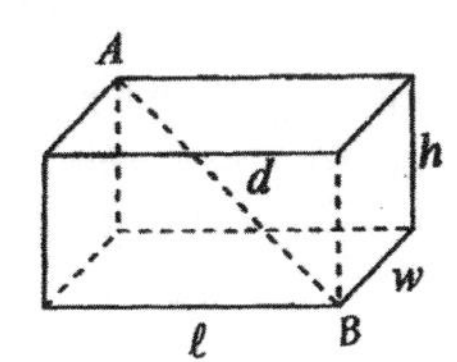

$V = Bh = \ell \times w \times h$; $B =$ the area of a base
$h =$ the height

Surface Area of a Rectangular Solid:

$SA = 2(\ell h) + 2(\ell w) + 2(wh)$

$d = \overline{AB}$

The diagonal of a rectangular solid is the segment ($d = \overline{AB}$) that connects the two opposite corners through the center of the solid.

The Length of the Diagonal of a Rectangular Solid:

$d^2 = \ell^2 + w^2 + h^2$ (This is the Pythagorean Theorem in three-dimensions.)

$d = \sqrt{\ell^2 + w^2 + h^2}$

2. **Cubes:** A cube is a rectangular solid whose two bases and four lateral faces are squares. In a cube, the length, width, and height are all equal.

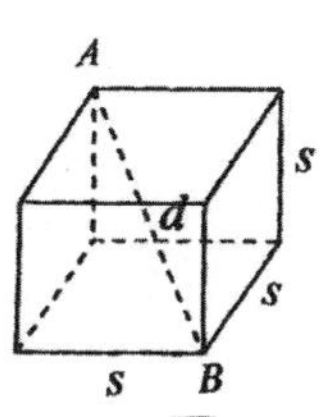

Volume of a Cube: $V = B\,h = s \times s \times s = s^3$

Surface Area of a Cube: $SA = 6(s \times s) = 6\,s^2$

$d = \overline{AB}$

The length of the diagonal of a Cube: $d = \sqrt{s^2 + s^2 + s^2} = \sqrt{3}\,s$

3. In general, the basic formulas for any right prism are given by:

Volume of a Prism: $V = B \times h$; $B =$ the area of a base
$h =$ the height

Lateral Surface Area of a Right Prism: $LA = ph$; $p =$ the perimeter of a base
$h =$ the height

Surface Area of a Right Prism: $SA =$ Area of two bases + Area of lateral faces

Examples

1. Find the volume and surface area of the rectangular solid having a length 10 inches, width 4 inches, and height 5 inches.
   Solution:
   $V = B\ h = \ell wh = (10)(4)(5) = 200\ in.^3$ Ans.
   $SA = 2(10 \times 5) + 2(10 \times 4) + 2(4 \times 5) = 100 + 80 + 40 = 220\ in.^2$

2. Find the volume and surface area of the cube whose edge length is 8 *cm*.
   Solution:
   $V = B\ h = s^3 = 8^3 = 512\ cm^3$.
   $SA = 6\ s^2 = 6(8^2) = 384\ cm^2$.

3. Find the volume and surface area of the right triangular prism shown below.
   (Hint: A right triangular prism is a prism whose two bases are triangles and lateral faces are three rectangles.)
   Solution:
   $V = B\ h = \frac{1}{2}(4)(3) \times 9 = 54\ yd.^3$
   Area of two bases $2 \times \frac{1}{2}(4)(3) = 12$
   Area of three faces $= 9(5) + 9(4) + 9(3) = 108$
   $SA = 12 + 108 = 120\ yd.^2$

4. Find the volume of the right triangular prism shown below.
   Solution:
   $V = B\ h = \frac{1}{2}(1\frac{1}{2})(1) \times 2 = 1\frac{1}{2}\ ft^3$.

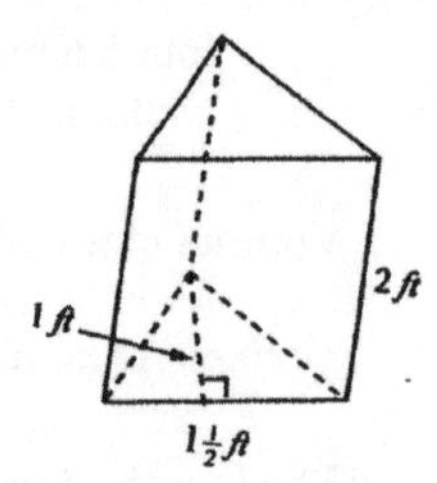

Examples

5. Find the surface area of the regular hexagonal prism shown below.
   Solution:

   Hint: The area of a regular polygon equals one-half of its apothem times its perimeter. (See page 178, Chapter 3-11)

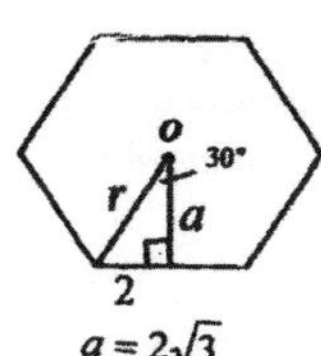

   Surface area of 6 rectangles $= 6(10 \times 4) = 240$

   Area of a base $= \frac{1}{2}ap = \frac{1}{2}(2\sqrt{3})(24) = 24\sqrt{3}$

   $SA = 240 + 2(24\sqrt{3}) = (240 + 48\sqrt{3})\ cm^2$.

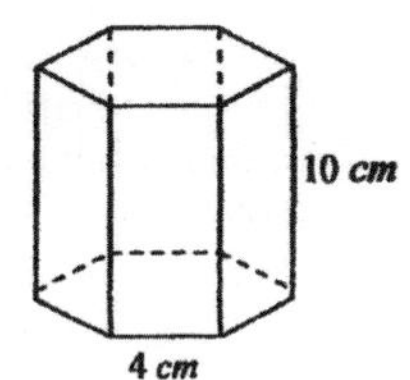

6. Find the volume of the regular hexagonal prism shown on Problem 5.
   Solution:

   Area of a base $= \frac{1}{2}ap = \frac{1}{2}(2\sqrt{3} \times 24) = 24\sqrt{3}$

   $V = B\ h = 24\sqrt{3} \times 10 = 240\sqrt{3}\ cm^3$.

7. Find the length of diagonal in the rectangular solid shown below.
   Solution:

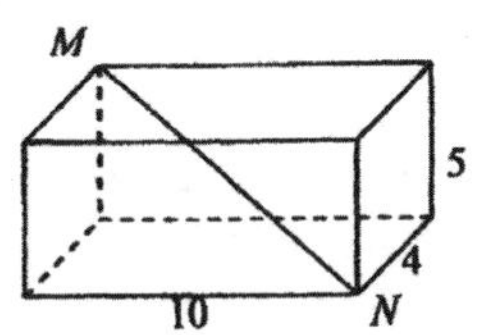

   $d = \overline{MN} = \sqrt{10^2 + 4^2 + 5^2} = \sqrt{100 + 16 + 25} = \sqrt{141} \approx 11.87$.

8. Find the volume of the solid made by rotating the rectangle $MNPQ$ 360° around side $\overline{MN}$.
   Solution:

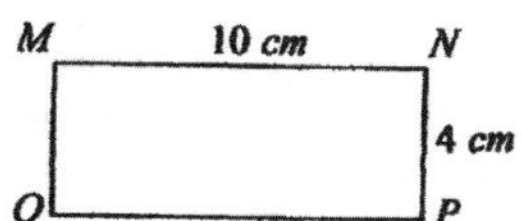

   $V = B\ h = \pi\ r^2 h = \pi(4^2)(10) = 160\pi\ cm^3$.

9. The volume and lateral surface area of a cube are equal. Find the edge length.
   Solution: $s^3 = 4s^2 \quad \therefore s = 4$.

10. The volume and total surface area of a cube are equal. Find the edge length.
    Solution: $s^3 = 6s^2 \quad \therefore s = 6$.

11. The surface area of a cube is 384 $in.^2$. Find the edge length.
    Solution:

    $6s^2 = 384, \quad s^2 = 64 \quad \therefore s = 8$ inches.

Examples

12. The surface area of a cube is 216. Find the longest line that can be drawn in the cube.
Solution:

$$6s^2 = 216, \quad s^2 = 36 \quad \therefore s = 6$$

The longest line that can be drawn in the cube is the diagonal of the cube.

$$\therefore\ d = \sqrt{6^2 + 6^2 + 6^2} = \sqrt{108} \approx 10.4\,.$$

13. The volume of a cube is 64. Find the longest line that can be drawn in the cube.
Solution:

$$s^3 = 64 \quad \therefore s = 4$$

The longest line that can be drawn in the cube is the diagonal of the cube.

$$\therefore\ d = \sqrt{4^2 + 4^2 + 4^2} = \sqrt{64} = 8\,.$$

14. If we double the side length of a cube, how much do we increase the volume of the cube ?
Solution:

$$V = s^3$$

$$\therefore\ V_1 = (2s)^3 = 8s^3 = 8V\,.$$

The volume of the cube is multiplied by a factor of 8.
We may simply work on the side length to find the scale factor.

$$(2s)^3 = 8s^3$$

15. If the side length of a cube is halved, how much does the surface area decrease ?
Solution:

$$SA = 6s^2$$

$$\therefore\ (SA)_1 = 6\left(\frac{s}{2}\right)^2 = 6\left(\frac{s^2}{4}\right) = \tfrac{1}{4}(6s^2) = \tfrac{1}{4}(SA)\,.$$

The surface area of the cube is multiplied by a factor of $\frac{1}{4}$ (divided by 4).
We may simply work on the side length to find the scale factor.

$$(\tfrac{1}{2}s)^2 = \tfrac{1}{4}s$$

Notes

Notes

3-20 Cones

A **cone** is a three-dimensional figure that consists of a circular base and a curved lateral surface tapering to a vertex so that any point on the surface is in a straight line between the circumference of the base and the vertex.

A **right (circular) cone** is a cone whose distance from the vertex to any point on the circumference of the base is a constant. An **oblique (circular) cone** is a cone whose distance from the vertex to any point on the circumference of the base is not a constant.

The **axis** of a cone is the segment that connects the vertex and the center of the base.

The **height** of a cone is the length of the segment from the vertex, perpendicular to the base plane. In a right cone, the axis and the height are the same segment.

The **slant height** (s) of a cone is the distance from the vertex to a point on the edge of the base. The **lateral surface area** (**LA**) of a cone is the area of the curved surface. In a right cone, the height, the radius, and the slant height form a right triangle.

When a plane intersects a cone, the cross section can be a circle, an ellipse, a hyperbola, or a triangle (the plane cuts through the vertex.).

Right Cone **Oblique Cone**

vertex
axis
s
h
r

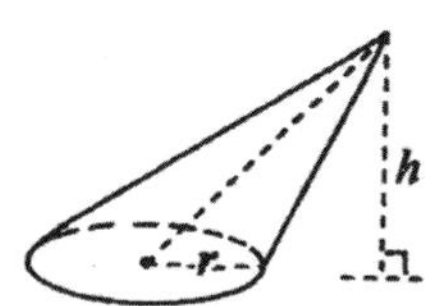

The **volume** (V) of a cone is one-third of the product of the area (B) of the base and the height h: **Volume of a Cone:** $V = \frac{1}{3}Bh = \frac{1}{3}\pi r^2 h$

(Hint: It is one-third of the volume of a cylinder with same base and h.)

The **lateral surface area** (LA) of a right cone of radius r equals one-half the product of the circumference of the base and the slant height.
(Hint: As the number of sides of the base increases, a regular pyramid approaches a cone.)

Lateral Surface Area of a Right Cone: $LA = \frac{1}{2}cs = \pi r s$ s: slant height
where $c = 2\pi r$ (circumference)

The **surface area** (SA) of a right cone with base radius r and slant height s equals the sum of the lateral area and the base area.

Surface Area of a Right Cone: $SA = \pi r s + \pi r^2 = \pi r(s + r)$; s: slant height

There is no general formula for the surface area of an oblique cone. To find the surface area of an oblique cone, we use the technique in Calculus.

Examples

1. Find the volume of the cone.

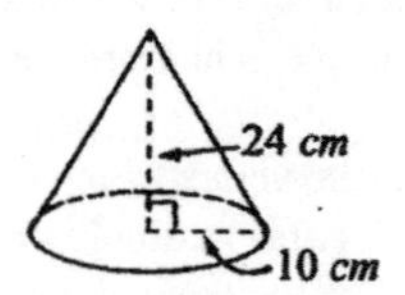

Solution:

$$r = 10, \quad h = 24$$

$$V = \tfrac{1}{3}\pi r^2 h = \tfrac{1}{3}\pi(10^2)(24)$$
$$= 800\pi \ cm^3.$$

2. Find the volume of the cone.

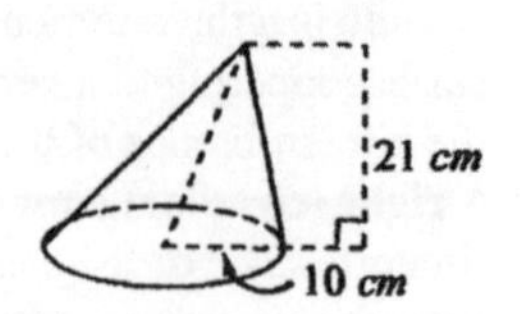

Solution:

$$r = 10, \quad h = 21$$

$$V = \tfrac{1}{3}\pi r^2 h = \tfrac{1}{3}\pi(10^2)(21)$$
$$= 700\pi \ cm^3.$$

3. Find the lateral area and total surface area of the cone shown on Problem 1.
Solution:

Find the slant height by the Pythagorean Theorem:

$$s^2 = 24^2 + 10^2$$
$$s^2 = 676 \quad \therefore s = 26$$

Lateral area: $LA = \pi r s = \pi(10)(26) = 260\pi \ cm^2$.

Surface area: $SA = \pi rs + \pi r^2 = \pi(10)(26) + \pi(10^2) = 360\pi \ cm^2$.

4. A right cone and a right cylinder have the same radius of base and volume. If the cylinder has a height of 9 centimeters, find the height of the cone.
Solution:

Volume of the cone = Volume of the cylinder

Let h = height of the cone, $h_1 = 9$ (height of the cylinder)

We have: $\tfrac{1}{3}\pi r^2 h = \pi r^2 h_1$

$$\tfrac{1}{3}h = h_1$$
$$\tfrac{1}{3}h = 9 \quad \therefore h = 27 \ cm.$$

5. Right triangle ABC is revolved about $\overline{AB}$ as an axis generating a right cone, $\overline{AB} = 7$ and $\overline{BC} = 9$. Find the volume of the cone.
Solution:

$$r = 10, \quad h = 7$$

Volume: $V = \tfrac{1}{3}\pi r^2 h = \tfrac{1}{3}\pi(9^2)(7) = 189\pi$.

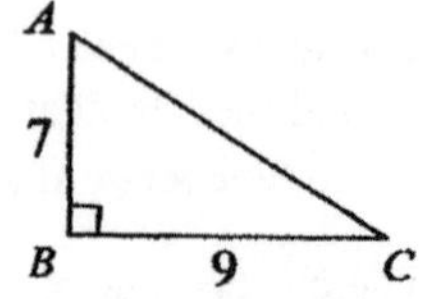

3-21 Pyramids

A **pyramid** is a three-dimensional figure that consists of a polygonal base and triangular surfaces meeting at a vertex.
A **regular pyramid** is a pyramid whose base is a regular polygon. Its height passes through the center of the base and is perpendicular to the base plane.
In a regular pyramid, the height is the **axis.**
The **slant height** (s) of a regular pyramid is the length of the height of one of its triangular surfaces. The **lateral surface area** (LA) of a pyramid is the sum of the areas of the triangular surfaces. The intersections of the lateral surfaces are the **edges.**

Pyramids are classified by the shapes of their bases, such as **Rectangular Pyramids, Triangular Pyramids, Pentagonal Pyramids, Hexagonal Pyramids,** ····· .

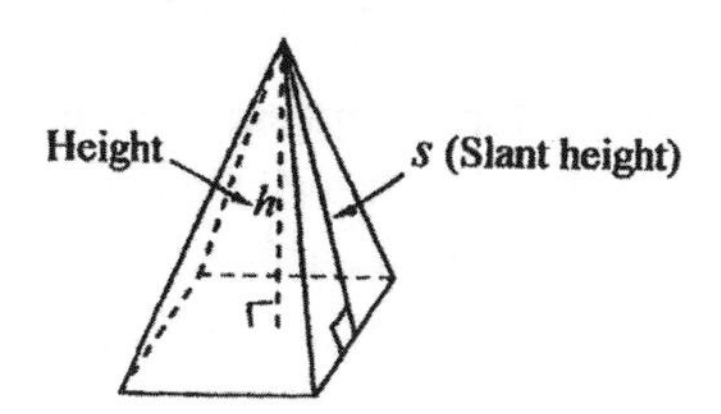

Regular Pyramid

The **volume** (V) of a pyramid is one-third of the product of the area (B) of the base and the height h.

Volume of a Pyramid: $V = \frac{1}{3}Bh$

(Hint: It is one-third of the volume of a prism with same base and h.)

The **lateral surface area** (LA) of a regular pyramid equals one-half of the product of the slant height and the perimeter of the base.

Lateral Surface Area of a Regular Pyramid: $LA = \frac{1}{2}sp$; s: slant height
p: perimeter of the base

The **surface area** (SA) of a regular pyramid with base area B and slant height s equals the sum of the lateral area and the base area.

Surface Area of a Regular Pyramid: $SA = \frac{1}{2}sp + B$; s: slant height
B: base area

There is no general formula for the surface area of a pyramid which is not regular. To find the surface area of a general pyramid, we add up the areas of all polygonal surfaces.

Examples

1. Find the volume of the pyramid whose base is a rectangle.

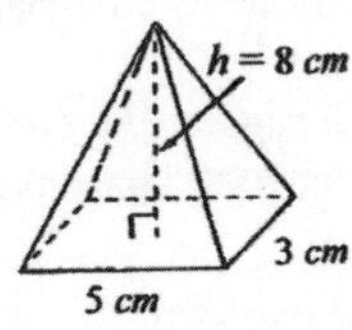

Solution:

Base area: $B = 5 \times 3 = 15$

$\therefore V = \frac{1}{3}Bh = \frac{1}{3}(15)(8) = 40\ cm^3$.

2. Find the volume of the pyramid whose base is an equilateral triangle with side length 6 inches, and height 8 inches.

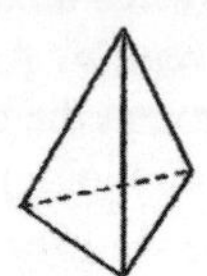

Solution:

Base area: $B = \frac{s^2\sqrt{3}}{4} = \frac{6^2\sqrt{3}}{4} = 9\sqrt{3}$

(Area of equilateral triangle)

$\therefore V = \frac{1}{3}Bh = \frac{1}{3}(9\sqrt{3})(8) = 24\sqrt{3}\ in.^3$

3. Find the lateral surface area of the regular pyramid.

Solution:

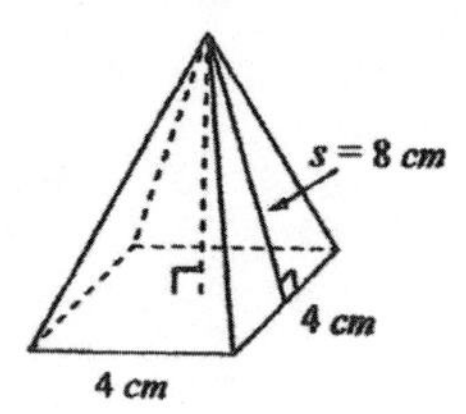

Method 1: There are four congruent triangles. $h = s$

$\therefore LA = 4(\frac{1}{2}bh) = 4(\frac{1}{2} \cdot 4 \cdot 8) = 64\ cm^2$.

Method 2: $s = 8$, $p = 4(4) = 16$

$\therefore LA = \frac{1}{2}sp = \frac{1}{2}(8)(16) = 64\ cm^2$.

4. Find the total surface area of the regular hexagonal pyramid.

Solution:

Lateral area: $LA = \frac{1}{2}sp = \frac{1}{2}(10)(4 \times 6) = 120$

Base area: $a = 2\sqrt{3}$, $p = 4(6) = 24$

$B = \frac{1}{2}ap = \frac{1}{2}(2\sqrt{3})(24) = 24\sqrt{3}$

Surface area: $SA = (120 + 24\sqrt{3})\ cm^2$.

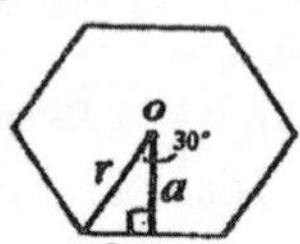

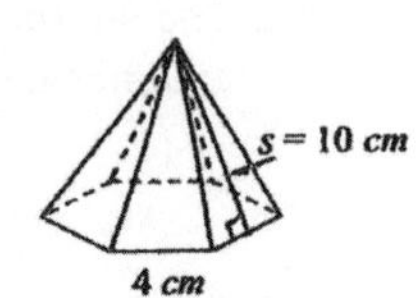

5. The base of a pyramid is a square with edge length 3, and slant height 3. Find the volume of the pyramid.

Solution:

$h^2 + (\frac{3}{2})^2 = 3^2$, $h^2 + \frac{9}{4} = 9$, $h^2 = \frac{27}{4}$, $h = \frac{3\sqrt{3}}{2}$

$\therefore V = \frac{1}{3}Bh = \frac{1}{3}(3^2)(\frac{3\sqrt{3}}{2}) = \frac{9\sqrt{3}}{2}$.

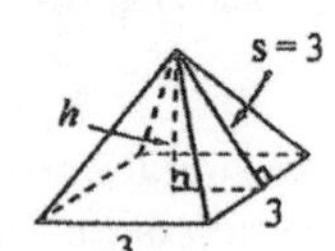

3-22 Spheres

A **sphere** is a three-dimensional figure having all points on the surface equally distance from the center. A sphere is a three-dimensional circle.
The **radius** of s sphere is the distance from the center to any point on the sphere.
If a plane intersects a sphere, the intersection is a circle. If a plane is tangent to a sphere, the intersection is point.

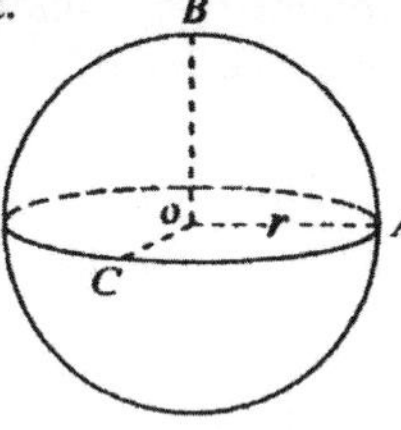

$\overline{OA} = \overline{OB} = \overline{OC} = r$ (radius)

The **volume** (V) of a sphere is given as following.

Volume of a sphere: $V = \frac{4}{3}\pi r^3$; r: radius

The **Surface Area** (SA) of a sphere is given as following.

Surface Area of a sphere: $SA = 4\pi r^2$; r: radius

Rules of Inscribed Solids

1. When a cylinder is inscribed in a sphere, the diameter of the sphere is equal to the diagonal of the rectangle formed by the heights and diameters of the cylinder.
2. When a sphere is inscribed in a cylinder, the sphere and the cylinder have the same radius.
3. When a sphere is inscribed in a cube, the diameter of the sphere is equal to the edge length of the cube.
4. When a rectangular solid or cube is inscribed in a sphere, the diagonal of the solid is equal to the diameter of the sphere.

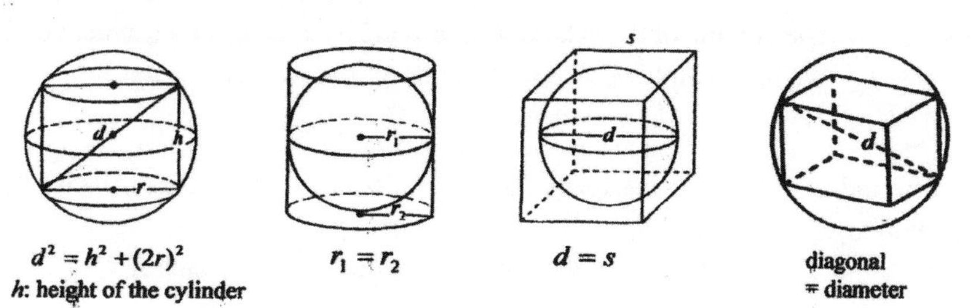

$d^2 = h^2 + (2r)^2$
h: height of the cylinder

$r_1 = r_2$

$d = s$

diagonal = diameter

Examples

1. Find the volume and surface area of a sphere with a radius 6 inches.
Solution:

$$V = \tfrac{4}{3}\pi r^3 = \tfrac{4}{3}\pi(6^3) = 288\pi \ in.^3$$
$$SA = 4\pi r^2 = 4\pi(6^2) = 144\pi \ in.^2$$

2. The surface area of a sphere and the volume of a cylinder are equal. What is the ratio of the radius of the sphere to the radius of the cylinder ?
Solution:

Let r_1 = radius of the sphere
r_2 = radius of the cylinder

We have:

$$4\pi(r_1)^2 = \pi(r_2)^2(h)$$
$$\frac{(r_1)^2}{(r_2)^2} = \frac{h}{4}$$
$$\therefore \ \frac{r_1}{r_2} = \sqrt{\frac{h}{4}} = \frac{\sqrt{h}}{2}.$$

3. If we double the radius of a sphere, how much do we increase the volume of the sphere ?
Solution:

$$V = \tfrac{4}{3}\pi r^3$$
$$V_1 = \tfrac{4}{3}\pi(2r)^3 = \tfrac{4}{3}\pi(8r^3) = 8(\tfrac{4}{3}\pi r^3) = 8V.$$

The volume of the sphere is multiplied by a factor of 8.
We may simply work on the radius to find the scale factor.

$$(2r)^3 = 8r^3$$

4. If the radius of a sphere is halved, how much does the volume decrease ?
Solution:

$$V = \tfrac{4}{3}\pi r^3$$
$$V_1 = \frac{4}{3}\pi\left(\frac{r}{2}\right)^3 = \frac{4}{3}\pi\left(\frac{r^3}{8}\right) = \frac{1}{8}\left(\frac{4}{3}\pi r^3\right) = \frac{1}{8}V.$$

The volume of the sphere is multiplied by a factor of $\frac{1}{8}$ (divided by 8).
We may simply work on the radius to find the scale factor.

$$(\tfrac{1}{2}r)^3 = \tfrac{1}{8}r^3$$

5. If the radius of a sphere is tripled, how much does the surface area increase ?
Solution: $(3r)^2 = 9r^2$

The surface area of the sphere is multiplied by a factor of 9.

Examples

6. In the figure, a sphere is inscribed in a cube. Find the ratio of the area of the sphere to the area of the cube.

Solution:

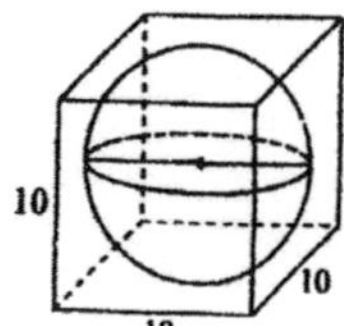

Area of the sphere: $A_1 = 4\pi\, r^2 = 4\pi(5^2) = 100\pi$

Area of the cube: $A_2 = 6s^2 = 6(10^2) = 600$

$\therefore\ \dfrac{A_1}{A_2} = \dfrac{100\pi}{600} = \dfrac{\pi}{6} \approx 0.52$.

7. In the figure, a cylinder is inscribed in the sphere. The height of the cylinder is equal to its the diameter of its circular base. Find the ratio of the volume of the cylinder to the volume of the sphere.

Solution:

Volume of the cylinder: $V_1 = \pi\, r^2 h = \pi(3^2)(6) = 54\pi$

Volume of the sphere: $d^2 = 6^2 + 6^2 = 72$

$d = \sqrt{72} = 6\sqrt{2}\,,\quad r = 3\sqrt{2}$

$V_2 = \frac{4}{3}\pi\, r^3 = \frac{4}{3}\pi(3\sqrt{2})^3 = \frac{4}{3}\pi(54\sqrt{2}) = 72\sqrt{2}\,\pi$

$\therefore\ \dfrac{V_1}{V_2} = \dfrac{54\pi}{72\sqrt{2}\pi} = \dfrac{3}{4\sqrt{2}} = \dfrac{3\sqrt{2}}{8} \approx 0.53$.

8. A cube is inscribed in a cylinder. The height of the cylinder is 12. The length of the diagonal of the cube is $8\sqrt{3}$. Find the volume of the cylinder.

Solution:

First, find the edge length s of the cube.

Length of diagonal is given: $d = 8\sqrt{3}$

$d^2 = s^2 + s^2 + s^2 = 3s^2$

$d = \sqrt{3}\, s,\quad 8\sqrt{3} = \sqrt{3}\, s\quad \therefore s = 8$

The diagonal of a face of the cube is the diameter of the circular base.

$(d_1)^2 = 8^2 + 8^2 = 128$

$d_1 = \sqrt{128} = 8\sqrt{2}$

The radius of the circular base: $r = 4\sqrt{2}$

The volume of the cylinder: $V = \pi\, r^2 h = \pi(4\sqrt{2})^2(12) = 384\pi$.

Notes

3-23 Chart of Measurements

Chart of Measurements

Length	Weight (Mass)	Volume (Capacity)
1 yard (yd) = 3 feet (ft) 1 foot (ft) = 12 inches (in.) 1 in. = 2.54 centimeters (cm) 1 mile (mi) = 5280 feet (ft) 1 mile (mi) = 1.6 km	1 ton = 2,000 pounds (lb) 1 pound (lb) = 16 ounces (oz) 1 kilogram (kg) = 2.2 lb 1 ounce (oz) = 28 grams (g)	1 gallon (gal) = 4 quarts (qt) 1 quart (qt) = 2 pints (pt) 1 pint (pt) = 2 cups (cu) 1 cup(cu) = 8 fluid ounces(fl.oz) 1 gallon (gal) = 3.8 liters (ℓ) = 128 fl.oz.
1 kilometer(km)=1000 meters (m) 1 meter(m)=100 centimeters (cm) 1 meter(m)=1000 millimeters(mm)	1 kilogram (kg)=1000 grams (g) 1 gram (g)=100 centigrams (cg) 1 gram(g)=1000 milligrams (mg)	1 kiloliter ($k\ell$)=1000 liters (ℓ) 1 liter (ℓ)=100 centiliters ($c\ell$) 1 liter(ℓ)=1000 milliliters($m\ell$)

Other Measurements

1 cc.(or c.c) = 1 cubic centimeter 1 tablespoon = 3 teaspoons = 15 cc. 1 teaspoon = $\frac{1}{3}$ tablespoon = 5 cc. 1 tablespoon = $\frac{1}{2}$ fluid ounce	1 bushel = 36 liters 1 peck = 9 liters 1 bushel = 4 pecks	" hecto " means " hundred ". " deca " means " ten ". 1 hectometer and 2 decameters = 120 meters

Area of a Rectangle: $A = \ell \times w$ where ℓ is the length and w is the width

Volume of a Rectangular Solid: $V = \ell \times w \times h$ where h is the height

Area of a Parallelogram: $A = b\,h$ where b is the base and h is the height

Area of a Triangle: $A = \frac{1}{2} b\,h$ where b is the base and h is the height

Area of a Circle: $A = \pi\, r^2$ where r is the radius

Circumference of a Circle: $C = \pi\, d$ where d is the diameter

Surface Area of a Sphere: $S = 4\pi\, r^2$ where r is the radius

Volume of a Sphere: $V = \frac{4}{3}\pi\, r^3$ Where r is the radius

Area of a Right Cylinder: $A = B\,h$ Where B is the base area and h is the height

Volume of a Right Circular Cone: $V = \frac{1}{3}\pi\, r^2 h$ where r is the radius and h is the height

Volume of a Pyramid: $V = \frac{1}{3} B h$ where B is the base area and h is the height

Lateral Area of a Right Circular Cone: $S = \frac{1}{2} c\,\ell$ where c is the circumference and ℓ is the slant height

Notes

Notes

Notes

3-24 Polar and Complex Coordinates

A point $p(x, y)$ in the rectangular coordinates x and y can be represented in the **polar coordinates** $p(r, \theta)$:

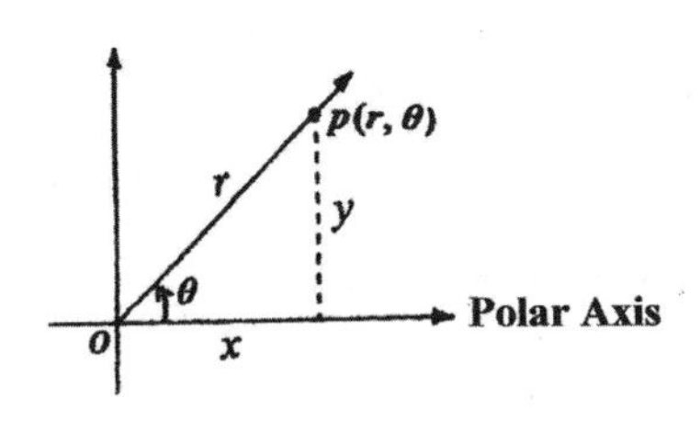

o : the **pole** or **origin**.
r : directed distance from o to p.
θ : directed angle, counterclockwise from polar axis to $\overline{op}$.

The relationship between polar and rectangular coordinates are:

$$x = r\cos\theta, \quad y = r\sin\theta$$

$$\tan\theta = \frac{y}{x}, \quad r = \sqrt{x^2 + y^2}$$

In the polar coordinates, the coordinates of the pole is $p(0, \theta)$, where θ can be any angle.
In the polar coordinates, a point $p(r, \theta)$ has infinitely many representations.
The point $p(r, \theta)$ can be represented as $p(r, \theta \pm 2n\pi)$ or $p(-r, \theta \pm (2n\pi + \pi))$, where n is an integer.

Example: $(5, \frac{\pi}{6}) = (5, \frac{\pi}{6} + 2\pi) = (5, \frac{13\pi}{6})$
$(5, \frac{\pi}{6}) = (5, \frac{\pi}{6} - 2\pi) = (5, -\frac{11\pi}{6})$
$(5, \frac{\pi}{6}) = (-5, \frac{\pi}{6} + \pi) = (-5, \frac{7\pi}{6})$
$(5, \frac{\pi}{6}) = (-5, \frac{\pi}{6} - \pi) = (-5, -\frac{5\pi}{6})$

To locate the point $(-5, \frac{7\pi}{6})$, we use the ray in the direction opposite the terminal side of $\frac{7\pi}{6}$ at a distance 5 units from the pole.

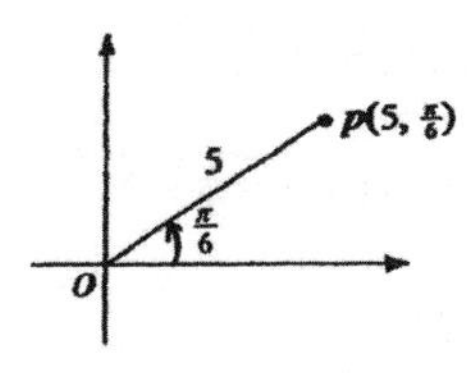

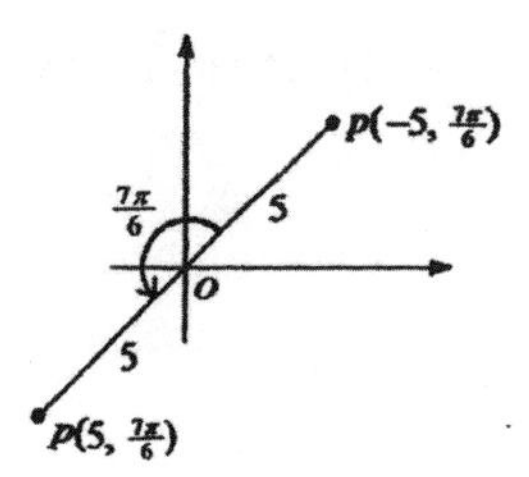

Examples

1. Express the point $(\sqrt{3}, 1)$ in terms of polar coordinates.
Solution:
$$r = \sqrt{x^2 + y^2} = \sqrt{(\sqrt{3})^2 + 1^2} = 2, \quad \tan\theta = \tfrac{y}{x} = \tfrac{1}{\sqrt{3}}, \ \theta = \tfrac{\pi}{6}.$$
$\therefore$ The polar coordinates of p are $(2, \frac{\pi}{6})$.

2. Find the polar coordinates of the point $(0, -5)$.
Solution:
The point lies on the negative $y-axis$.
$r = 5$, $\theta = \frac{3\pi}{2}$. The polar coordinates of p are $(5, \frac{3\pi}{2})$.

3. Find the polar coordinates of $p(-1, -\sqrt{3})$.
Solution:
$$r = \sqrt{x^2 + y^2} = \sqrt{(-1)^2 + (-\sqrt{3})^2} = 2. \quad \tan\alpha = \tfrac{-\sqrt{3}}{-1} = \sqrt{3}, \ \alpha = \tfrac{\pi}{3}.$$
The point lies on the 3rd quadrant. $\therefore \theta = \frac{\pi}{3} + \pi = \frac{4\pi}{3}$.
The polar coordinates of p are $(2, \frac{4\pi}{3})$.

4. Find the rectangular coordinates of the point $(2, \frac{\pi}{6})$.
Solution:
$$x = r\cos\theta = 2\cos\tfrac{\pi}{6} = 2(\tfrac{\sqrt{3}}{2}) = \sqrt{3}$$
$$y = r\sin\theta = 2\sin\tfrac{\pi}{6} = 2(\tfrac{1}{2}) = 1$$
The rectangular coordinates of the point are $(\sqrt{3}, 1)$.

5. Transform the equation $xy = 2$ to polar form.
Solution:
$$xy = 2, \ (r\cos\theta)(r\sin\theta) = 2, \ r^2\cos\theta\sin\theta = 2, \ r^2(\tfrac{1}{2}\sin 2\theta) = 2$$
The polar form of the equation is $r^2 \sin 2\theta = 4$.

6. Transform the equation $r^2 \sin 2\theta = 4$ to rectangular coordinates.
Solution:
$$r^2 \sin 2\theta = 4, \ r^2(2\sin\theta\cos\theta) = 4, \ r^2\sin\theta\cos\theta = 2$$
$$(r\cos\theta)(r\sin\theta) = 2$$
The rectangular coordinates of the equation is $xy = 2$.

Graphing Polar Equations

One method to graph a polar equation is to convert the equation to rectangular form. However, it is not always easy or helpful to graph a polar equation by converting it to rectangular form. Usually, we graph a polar equation by **checking for symmetry and plotting points.**

Checking for symmetry:

1. If we replace (r, θ) by $(r, -\theta)$ and the equation is unchanged, it is symmetry with respect to the polar axis.
2. If we replace (r, θ) by $(-r, \theta)$ and the equation is unchanged, it is symmetry with respect to the pole (the origin).
3. If we replace (r, θ) by $(r, \pi - \theta)$ and the equation is unchanged, it is symmetry with respect to the line $\theta = \frac{\pi}{2}$.

However, the above rules are not necessary conditions for symmetry. It is possible for a graph to have certain symmetry status which the above rules fail to test.

Examples:

1. Graph the polar equation $\theta = \dfrac{3\pi}{4}$.

   Solution:

   Converting it to rectangular form.

   $\tan\theta = \frac{y}{x}$, $\tan\frac{3\pi}{4} = -\tan 45° = -1$, $\frac{y}{x} = -1$,

   $\therefore x + y = 0$

   It is a line passing through the pole at an angle of $135°$ with the polar axis (regardless of the value of r.)

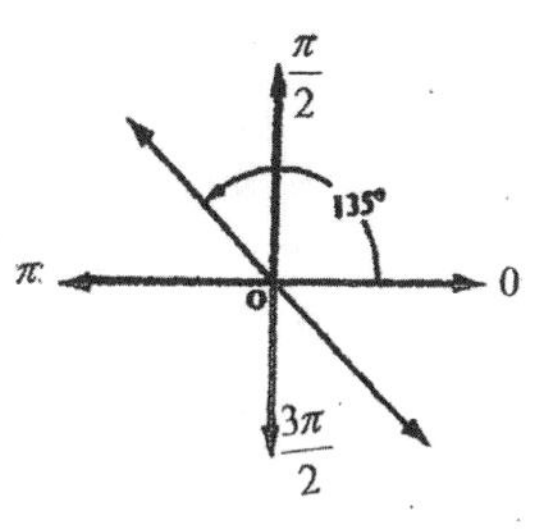

2. Graph the polar equation $r = -4$.

   Solution:

   Converting it to rectangular form.

   $r = \sqrt{x^2 + y^2} = -4$, $\therefore x^2 + y^2 = 16$

   It is a circle with center (0, 0) and radius 4.

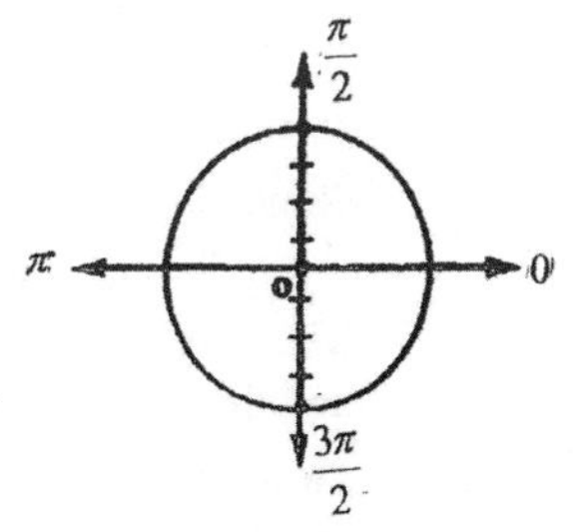

3. Graph the polar equation $r = -4\cos\theta$.

   Solution:

   Converting it to rectangular form.

   $r = -4\cos\theta$, $r^2 = -4r\cos\theta$, $x^2 + y^2 = -4x$

   $x^2 + y^2 + 4x = 0$

   Completing the square: $(x+2)^2 + y^2 = 4$.

   It is a circle with center (–2, 0) and radius 2.

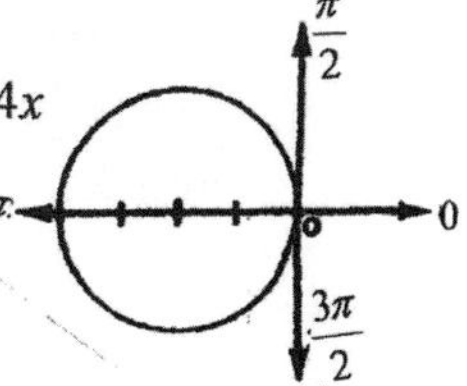

Examples

4. Graph the polar equation $r = 1 - \cos\theta$.

Solution:

Replacing θ by $(-\theta)$, $r = 1 - \cos(-\theta) = 1 - \cos\theta$.

The graph is symmetric with respect to the polar axis.

Plotting the points, we only need to select values of θ from 0 to π.

θ	0	$\frac{\pi}{6}$	$\frac{\pi}{3}$	$\frac{\pi}{2}$	$\frac{2\pi}{3}$	$\frac{5\pi}{6}$	π
$\cos\theta$	1	$\frac{\sqrt{3}}{2}$	$\frac{1}{2}$	0	$-\frac{1}{2}$	$-\frac{\sqrt{3}}{2}$	-1
$r = 1 - \cos\theta$	0	0.13	0.5	1	1.5	1.87	2

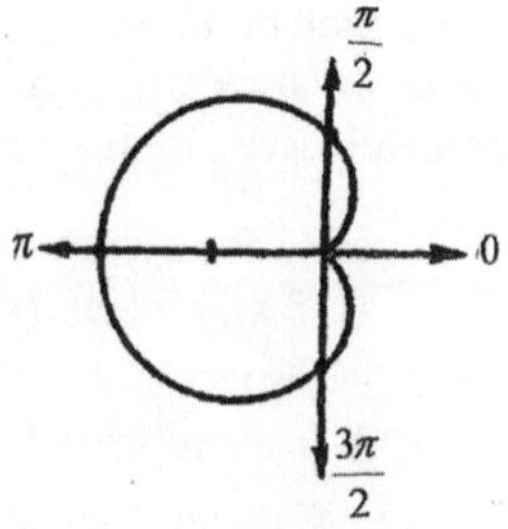

It is a **Cardioid**.
(A heart-shaped curve).

5. Graph the polar equation $r = 3 + 2\sin\theta$.

Solution:

Replacing θ by $(\pi - \theta)$, $r = 3 + 2\sin(\pi - \theta) = 3 + 2\sin\theta$.

The graph is symmetric with respect to the line $\theta = \frac{\pi}{2}$.

Plotting the points, we only need to select values of θ from $-\frac{\pi}{2}$ to $\frac{\pi}{2}$.

θ	$-\frac{\pi}{2}$	$-\frac{\pi}{3}$	$-\frac{\pi}{6}$	0	$\frac{\pi}{6}$	$\frac{\pi}{3}$	$\frac{\pi}{2}$
$\sin\theta$	-1	$-\frac{\sqrt{3}}{2}$	$-\frac{1}{2}$	0	$\frac{1}{2}$	$\frac{\sqrt{3}}{2}$	1
$r = 3 + 2\sin\theta$	1	1.27	2	3	4	4.73	5

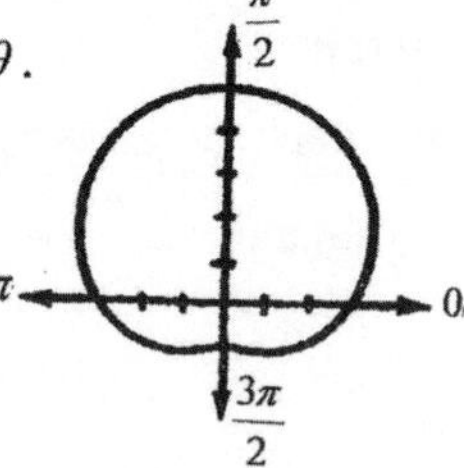

It is a **Limacon**.

6. Graph the polar equation $r = 4\cos 2\theta$.

Solution:

Replacing θ by $(-\theta)$, $r = 4\cos 2(-\theta) = 4\cos 2\theta$.

Replacing θ by $(\pi - \theta)$,

$$r = 4\cos 2(\pi - \theta) = 4\cos(2\pi - 2\theta) = 4\cos 2\theta.$$

The graph is symmetric with respect to the polar axis and the line $\theta = \frac{\pi}{2}$.

Plotting the points, we only need to select values of θ from 0 to $\frac{\pi}{2}$.

θ	0	$\frac{\pi}{6}$	$\frac{\pi}{4}$	$\frac{\pi}{3}$	$\frac{\pi}{2}$
$\cos 2\theta$	1	$\frac{1}{2}$	0	$-\frac{1}{2}$	-1
$r = 4\cos 2\theta$	4	2	0	-2	-4

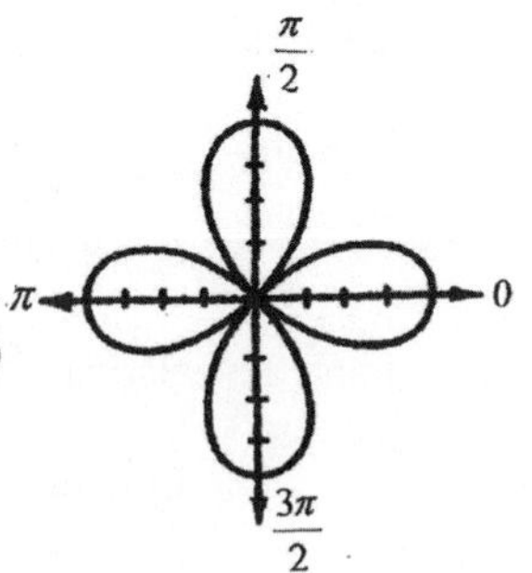

It is a **Rose** with four petals.

Example 7.

Graph the polar equation $r = 1 + 2\sin\theta$.

Solution: Replacing θ by $(\pi - \theta)$: $r = 1 + 2\sin(\pi - \theta) = 1 + 2\sin\theta$.

The graph is symmetric with respect to the line $\theta = \frac{\pi}{2}$.

Plotting the points, we only need to select values of θ from $-\frac{\pi}{2}$ to $\frac{\pi}{2}$.

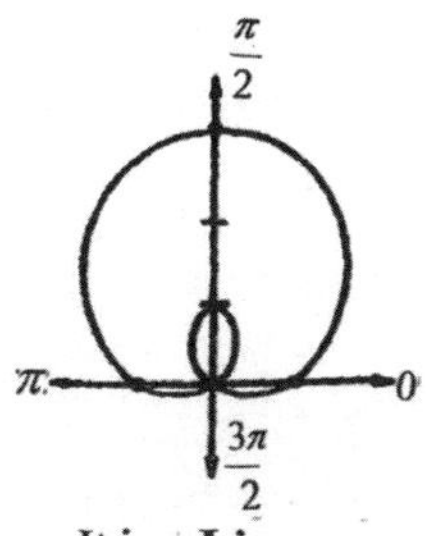

θ	$-\frac{\pi}{2}$	$-\frac{\pi}{3}$	$-\frac{\pi}{6}$	0	$\frac{\pi}{6}$	$\frac{\pi}{3}$	$\frac{\pi}{2}$
$\sin\theta$	-1	$-\frac{\sqrt{3}}{2}$	$-\frac{1}{2}$	0	$\frac{1}{2}$	$\frac{\sqrt{3}}{2}$	1
$r = 1 + 2\sin\theta$	-1	-0.73	0	1	2	2.73	3

It is a **Limacon**. with an inner loop.

Polar Forms of Conics: In order to find a convenient form to represent a conic, we locate a focus at the pole and apply the definition of a conic.

Definition of a Conic: A conic is the set of all points such that the distance from a fixed point (called the focus) is in a constant ratio to its distance from a fixed line (called the directrix). The constant ratio is the eccentricity of the conic and is denoted by e.

$$\frac{\overline{PF}}{\overline{PD}} = e \quad \therefore \overline{PF} = e \cdot \overline{PD}$$

$$r = e(p + r\cos\theta)$$

$$r = ep + er\cos\theta$$

$$r - er\cos\theta = ep$$

$$r(1 - e\cos\theta) = ep$$

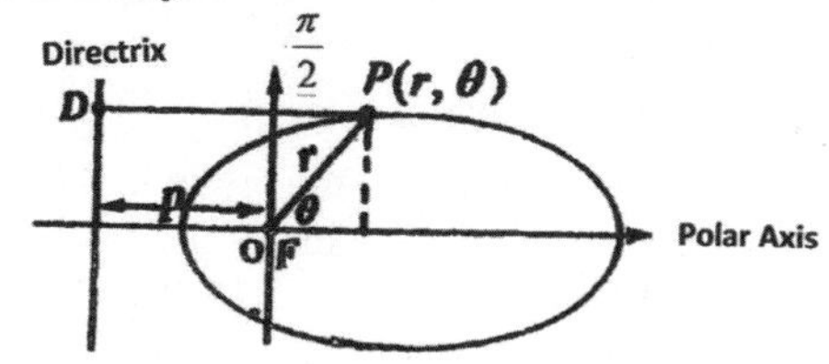

$$\therefore r = \frac{ep}{1 - e\cos\theta}$$ (Directrix is located at the left of the pole.)

Similarly, we can obtain the following polar forms of conics:

$$r = \frac{ep}{1 + e\cos\theta}$$ (Directrix is located at the right of the pole.)

$$r = \frac{ep}{1 - e\sin\theta}$$ (Directrix is located below the pole.)

$$r = \frac{ep}{1 + e\sin\theta}$$ (Directrix is located above the pole.)

Hint: $e = 1$ (parabola), $e < 1$ (ellipse), $e > 1$ (hyperbola)

Example 8.

Identify and graph the equation $r = \dfrac{9}{3 - 3\sin\theta}$.

Solution: $r = \frac{3}{1 - \sin\theta}$ and $e = 1$

It is a parabola with focus at the pole.

$ep = 3 \quad \therefore p = 3$

Directrix is 3 units below the pole.

We plot two addition points $(3, 0^\circ)$ and $(3, \pi)$ to assist in graphing.

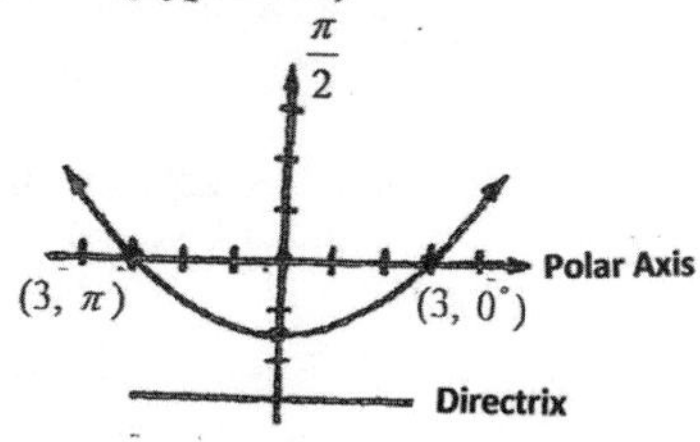

Polar Coordinates involving Complex Numbers

A **complex number** $z = a + b\,i$ is formed when we solve an algebraic equation having an imaginary root. A complex number $z = a + b\,i$ can be interpreted geometrically as the corresponding point (a, b) in the rectangular plane. The polar coordinates used to represent the complex number is called the **complex plane.**

In a complex plane, the horizontal axis is called the **real axis**, and the vertical axis is called the **imaginary axis.**

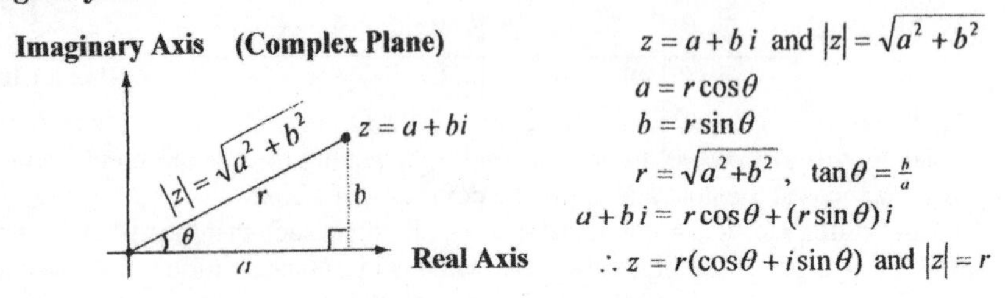

The number r is called the **modulus** (or **amplitude**) of z. The angle θ is called an **argument** of z.

The **absolute value** of a complex number $z = a + b\,i$ is defined as $|z| = \sqrt{a^2 + b^2}$. It is the distance between z and the origin in the complex plane.

Since there are infinitely many choices for θ, the polar form of a complex number is not unique. We choice the values of θ for $0 \le \theta < 2\pi$, or $\theta < 0$.

The complex number i is represented by the point (0, 1) and $i = 1(\cos 90° + i\sin 90°)$.

A complex number represented in polar form will give us an efficient and easy way to do some applications and computations (add, subtract, multiply, and divide).

To add or subtract two complex numbers, we use the **parallelogram rule.**

Example: Find the sum and difference of $z_1 = 2 + 6i$ and $z_2 = 4 - 2i$ by using the polar polar coordinates.

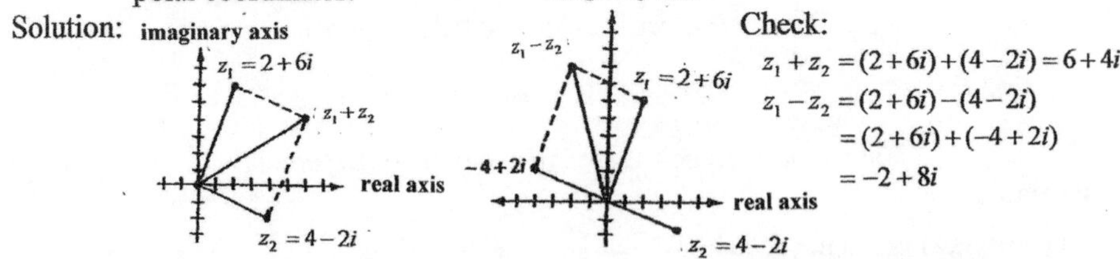

To multiply or divide two complex numbers in polar forms, we use the following two formulas: $z_1 = r_1(\cos\theta_1 + i\sin\theta_1)$, $z_2 = r_2(\cos\theta_2 + i\sin\theta_2)$

1) $z_1 \cdot z_2 = r_1 r_2[\cos(\theta_1 + \theta_2) + i\sin(\theta_1 + \theta_2)]$

2) $\dfrac{z_1}{z_2} = \dfrac{r_1}{r_2}[\cos(\theta_1 - \theta_2) + i\sin(\theta_1 - \theta_2)]$, $z_2 \ne 0$

Example 1: Given: $z_1 = r_1(\cos\theta_1 + i\sin\theta_1)$, $z_2 = r_2(\cos\theta_2 + i\sin\theta_2)$, find: 1. $z_1 z_2$ 2. $\frac{z_1}{z_2}$.

Solution:

$$\begin{aligned} 1.\ z_1 z_2 &= r_1(\cos\theta_1 + i\sin\theta_1)\cdot r_2(\cos\theta_2 + i\sin\theta_2) \\ &= r_1 r_2[(\cos\theta_1\cos\theta_2 - \sin\theta_1\sin\theta_2) + i(\sin\theta_1\cos\theta_2 + \cos\theta_1\sin\theta_2)] \\ &= r_1 r_2[\cos(\theta_1+\theta_2) + i\sin(\theta_1+\theta_2)].\ \text{Ans.} \end{aligned}$$

$$\begin{aligned} 2.\ \frac{z_1}{z_2} &= \frac{r_1(\cos\theta_1 + i\sin\theta_1)}{r_2(\cos\theta_2 + i\sin\theta_2)} = \frac{r_1(\cos\theta_1 + i\sin\theta_1)}{r_2(\cos\theta_2 + i\sin\theta_2)}\cdot\frac{\cos\theta_2 - i\sin\theta_2}{\cos\theta_2 - i\sin\theta_2} \\ &= \frac{r_1}{r_2}\cdot\frac{\cos\theta_1\cos\theta_2 + \sin\theta_1\sin\theta_2 + i(\sin\theta_1\cos\theta_2 - \cos\theta_1\sin\theta_2)}{\cos^2\theta_2 + \sin^2\theta_2} \\ &= \frac{r_1}{r_2}\left[\frac{\cos(\theta_1-\theta_2) + i\sin(\theta_1-\theta_2)}{1}\right] = \frac{r_1}{r_2}[\cos(\theta_1-\theta_2) + i\sin(\theta_1-\theta_2)]. \end{aligned}$$

Example 2: Given $z_1 = 2\sqrt{2}(\cos 45^\circ + i\sin 45^\circ)$ and $z_2 = 3\sqrt{2}(\cos 45^\circ + i\sin 45^\circ)$.

Find $z_1 z_2$ and $\frac{z_1}{z_2}$.

Solution:

$$\begin{aligned} z_1 z_2 &= 2\sqrt{2}(\cos 45^\circ + i\sin 45^\circ)\cdot 3\sqrt{2}(\cos 45^\circ + i\sin 45^\circ) \\ &= 2\sqrt{2}\cdot 3\sqrt{2}(\cos 90^\circ + i\sin 90^\circ) = 12(0+i) = 12i.\ \text{Ans.} \end{aligned}$$

$$\frac{z_1}{z_2} = \frac{2\sqrt{2}(\cos 45^\circ + i\sin 45^\circ)}{3\sqrt{2}(\cos 45^\circ + i\sin 45^\circ)} = \frac{2\sqrt{2}}{3\sqrt{2}}(\cos 0^\circ + i\sin 0^\circ) = \frac{2}{3}(1+0) = \frac{2}{3}.$$

Check: $z_1 z_2 = (2+2i)(3+3i) = 6+6i+6i-6 = 12i$

$$\frac{z_1}{z_2} = \frac{2+2i}{3+3i}\cdot\frac{3-3i}{3-3i} = \frac{6-6i+6i+6}{9+9} = \frac{12}{18} = \frac{2}{3}$$

To find the n th power of a complex number, we use De Moivre's theorem.

De Moivre's Theorem

For any positive integer n and $z = r(\cos\theta + i\sin\theta)$, then

$$z^n = r^n(\cos n\theta + i\sin n\theta)$$

To find the n th roots of a complex number, we use the Nth-root formula.

Nth root formula

For any positive integer n and $z = r(\cos\theta + i\sin\theta)$, then

$$z^{1/n} = r^{1/n}\left(\cos\frac{\theta+2k\pi}{n} + i\sin\frac{\theta+2k\pi}{n}\right),\quad k = 0, 1, 2, \cdots, n-1.$$

Examples

3. Express the complex number $z = 2\sqrt{3} - 2i$ in polar form.

Solution: $r = \sqrt{(2\sqrt{3})^2 + (-2)^2} = \sqrt{16} = 4$, $\tan\theta = \frac{b}{a} = \frac{-2}{2\sqrt{3}} = -\frac{\sqrt{3}}{3}$.

It lies in 4th quadrant, we have $\theta = 360° - 30° = 330°$.

$\therefore\ z = r(\cos\theta + i\sin\theta) = 4(\cos 330° + i\sin 330°)$.

4. Express $z = 4(\cos 330^\circ + i\sin 330^\circ)$ in standard form $a + bi$.

Solution: $z = 4(\cos 330^\circ + i\sin 330^\circ) = 4[\frac{\sqrt{3}}{2} + i(-\frac{1}{2})] = 2\sqrt{3} - 2i$.

5. Find z^3 of $z = 2(\cos 45° + i\sin 45°)$ in polar form.

Solution: $z^3 = 2^3[\cos 3(45°) + i\sin 3(45°) = 8(\cos 135° + i\sin 135°)$.

6. Find the cubic roots of $2 + 2i$ in standard form $a + bi$.

Solution: $r = \sqrt{2^2 + 2^2} = \sqrt{8}$, $\tan\theta = \frac{b}{a} = \frac{2}{2} = 1$, $\therefore \theta = 45°$

$$(2+2i)^{1/3} = [\sqrt{8}(\cos 45^\circ + i\sin 45^\circ)]^{1/3} = (\sqrt{8})^{1/3}(\cos\frac{45^\circ + 2k\pi}{3} + i\sin\frac{45^\circ + 2k\pi}{3})$$

$$= [(2)^{3/2}]^{1/3}(\cos\frac{45° + 2k\pi}{3} + i\sin\frac{45° + 2k\pi}{3})$$

$$= \sqrt{2}(\cos\frac{45° + 2k\pi}{3} + i\sin\frac{45° + 2k\pi}{3})$$

Therefore, for $k = 0, 1, 2$, we have three cubic roots:

$k = 0$, $z_0 = \sqrt{2}(\cos 15^\circ + i\sin 15^\circ) = \sqrt{2}(0.9659 + i\cdot 0.2588) = 1.3658 + 0.3659i$

$k = 1$, $z_1 = \sqrt{2}(\cos 135^\circ + i\sin 135^\circ) = \sqrt{2}(-0.707 + i\cdot 0.707) = -1 + i$

$k = 2$, $z_2 = \sqrt{2}(\cos 255^\circ + i\sin 255^\circ) = \sqrt{2}(-0.2588 + i\cdot -0.9659) = -0.3659 - 1.3658i$

7. Find all solutions to the equation $x^2 + 4 = 0$ in standard form $a + bi$.

Solution: $x^2 = -4$, $x = (-4)^{1/2}$

Convert -4 into polar form: $-4 = 4(\cos\pi + i\sin\pi)$. It is the 3rd quadrant .

$$x = (-4)^{1/2} = 4^{1/2}[\cos\frac{\pi + 2k\pi}{2} + i\sin\frac{\pi + 2k\pi}{2}]$$

Therefore, for $k = 0, 1$, we have two square roots:

$k = 0$, $z_0 = 2[\cos 90° + i\sin 90°) = 2(0 + i) = 2i$

$k = 1$, $z_1 = 2[\cos 270° + i\sin 270°) = 2(0 - i) = -2i$

Notes

Notes

Notes

Notes

Notes

Notes

Chapter Four : Trigonometry

4-1 Degree Measure of Angles

In geometry, the degree measures of an angle is between 0° and 180°.
In trigonometry, we consider an angle as being formed by the rotation of its **initial ray** about its **vertex** to its **terminal ray**.
An angle is often denoted by θ (theta), ϕ (phi), α (alpha), β (beta), or γ (gamma).

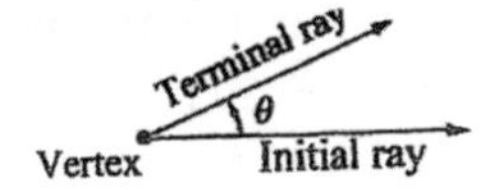

An angle is **positive** if it rotates counterclockwise. An angle is **negative** if it rotates clockwise. An angle is said to measure 360° if it is formed by rotating the initial side exactly one revolution in the counterclockwise direction until it meets with itself.
A **right angle** is an angle that measures 90°. A **straight angle** is an angle that measures 180°.

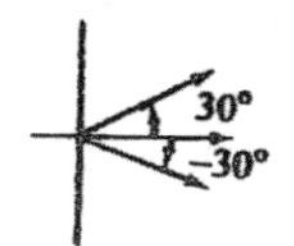

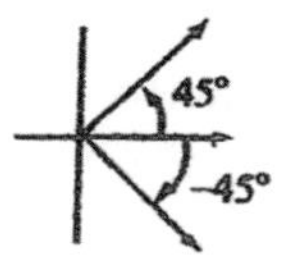

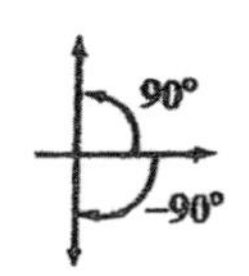

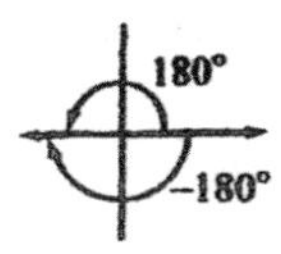

Coterminal Angles are angles which have the same initial and terminal rays. There are infinite number of angles coterminal with an angle θ.
The coterminal angles of θ are: $\pm 360^o n + \theta$ where $n =$ 0, 1, 2, 3, ····
If $\theta = 30^o$, then the coterminal angles are: $\cdots, -690^o, -330^o, 30^o, 390^o, 750^o, \cdots$
There are two different ways to express the degree measure of an angle.

1. **Decimal degrees:** 7.5944^o
2. **Degrees, minutes, seconds:** $7^o\ 35'\ 40''$ = 7 degree, 35 minutes, 40 seconds
   (1 degree = 60 minutes, 1 minute = 60 seconds, 1 degree = 3600 seconds)

Examples

1. Express 7.5944^o in degree, minutes, seconds
Solution:

$0.5944^o = 0.5944 \times 60 = 35.664$ minutes

0.664 minutes $= 0.664 \times 60 \approx 40$ seconds

$7.5944^o = 7^o\ 35'\ 40''$

2. Express $7^o 35' 40''$ in decimal degrees.
Solution:

$7^o\ 35'\ 40'' = 7^o + (\frac{35}{60})^o + (\frac{40}{3600})^o$

$= 7^o + 0.5833^o + 0.0111^o$

$= 7.5944^o$

Examples

3. Express 6.30^o in degrees, minutes, seconds.

Solution:

$$6.30^o = 6^o + (0.30 \times 60)' = 6^o 18'. \text{ Ans.}$$

4. Express -60.50^o in degrees, minutes, seconds.

Solution:

$$-60.50^o = -60^o - (0.50 \times 60) = -60^o - 30'$$
$$= -60^o 30'. \text{ Ans.}$$

5. Express 60.5030^o in degrees, minutes, seconds.

Solution:

$$60.5030^o = 60^o + (0.5030 \times 60)' = 60^o + 30.18' = 60^o + 30' + (0.18 \times 60)''$$
$$= 60^o + 30' + 0.108'' \approx 60^o 30' 11''. \text{ Ans.}$$

6. Express -120.3553^o in degrees, minutes, seconds.

Solution:

$$-120.3553^o = -120^o - (0.3553 \times 60)' = -120^o - 21.3180'$$
$$= -120^o - 21' - (0.3180 \times 60)'' = -120^o - 21' - 19.08''$$
$$\approx -120^o 21' 19''. \text{ Ans.}$$

7. Express $6^o 30'$ in decimal degrees.

Solution:

$$6^o 30' = 6^o + (\tfrac{30}{60})^o = 6^o + 0.5^o = 6.5^o. \text{ Ans.}$$

8. Express $6^o 50' 30''$ in decimal degrees.

Solution:

$$6^o 50' 30'' = 6^o + (\tfrac{50}{60})^o + (\tfrac{30}{3600})^o$$
$$= 6^o + 0.833^o + 0.0083^o$$
$$= 6.8416^o. \text{ Ans.}$$

9. Express $-120^o 35' 53''$ in decimal degrees.

Solution:

$$-120^o 35' 53'' = -120^o - (\tfrac{35}{60})^o - (\tfrac{53}{3600})^o = -120^o - 0.5833^o - 0.0147^o$$
$$= 120.598^o. \text{ Ans.}$$

4-2 Six Trigonometric Functions in a Triangle

In trigonometry, we start to find six ratios of an acute angle θ in a right triangle. Each ratio is a function of the angle. We call these six ratios " **Six Trigonometric Functions** ". If θ is an acute angle in a right (90°) triangle, the definitions of these six trigonometric functions are:

1. The **sine** of θ $= \dfrac{opposite \quad leg}{hypotenuse} = \dfrac{y}{r}$

2. The **cosine** of θ $= \dfrac{adjacent \quad leg}{hypotenuse} = \dfrac{x}{r}$

3. The **tangent** of θ $= \dfrac{opposite \quad leg}{adjacent \quad leg} = \dfrac{y}{x}$

4. The **cotangent** of θ $= \dfrac{adjacent \quad leg}{opposite \quad leg} = \dfrac{x}{y}$

5. The **secant** of θ $= \dfrac{hypotenuse}{adjacent \quad leg} = \dfrac{r}{x}$

6. The **cosecant** of θ $= \dfrac{hypotenuse}{opposite \quad leg} = \dfrac{r}{y}$

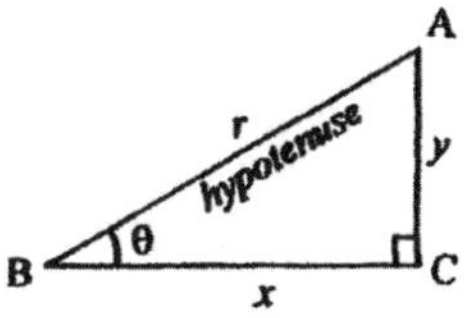

A
r
hypotenuse
y
θ
B
C
x

These trigonometry functions can be written in abbreviated forms as:

$$\sin\theta = \frac{y}{r}, \quad \cos\theta = \frac{x}{r}, \quad \tan\theta = \frac{y}{x}, \quad \cot\theta = \frac{x}{y}, \quad \sec\theta = \frac{r}{x}, \quad \csc\theta = \frac{r}{y}$$

Examples

1. If the terminal side of an angle θ passes through (3, 2), find all six basic trigonometric functions.
   Solution:
   $$r^2 = 3^2 + 2^2 = 13 \quad \therefore r = \sqrt{13}$$
   $$\sin\theta = \frac{y}{r} = \frac{2}{\sqrt{13}} = \frac{2\sqrt{13}}{13}, \quad \cos\theta = \frac{x}{r} = \frac{3}{\sqrt{13}} = \frac{3\sqrt{13}}{13}$$
   $$\tan\theta = \frac{y}{x} = \frac{2}{3}, \quad \cot\theta = \frac{x}{y} = \frac{3}{2}$$
   $$\sec\theta = \frac{r}{x} = \frac{\sqrt{13}}{3}, \quad \csc\theta = \frac{r}{y} = \frac{\sqrt{13}}{2}$$

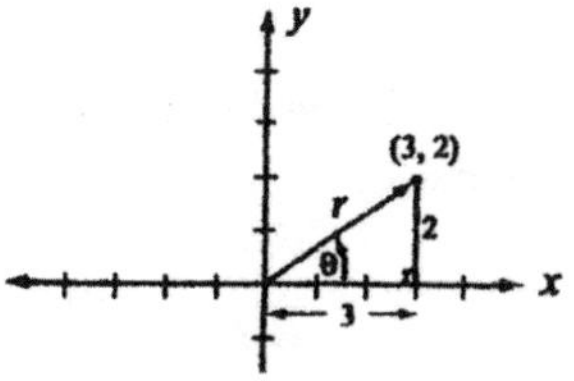

y
(3, 2)
r
2
θ
3
x

Examples

2. If the terminal side of an angle θ passes through (–1, 3), find all six basic trigonometric functions.
   Solution:

$$r^2=(-1)^2+3^2=10 \quad \therefore r=\sqrt{10}$$

$$\sin\theta=\frac{y}{r}=\frac{3}{\sqrt{10}}=\frac{3\sqrt{10}}{10}$$

$$\cos\theta=\frac{x}{r}=\frac{-1}{\sqrt{10}}=-\frac{\sqrt{10}}{10}$$

$$\tan\theta=\frac{y}{x}=\frac{3}{-1}=-3$$

$$\cot\theta=\frac{x}{y}=\frac{-1}{3}=-\frac{1}{3}$$

$$\sec\theta=\frac{r}{x}=\frac{\sqrt{10}}{-1}=-\sqrt{10}$$

$$\csc\theta=\frac{r}{y}=\frac{\sqrt{10}}{3}$$

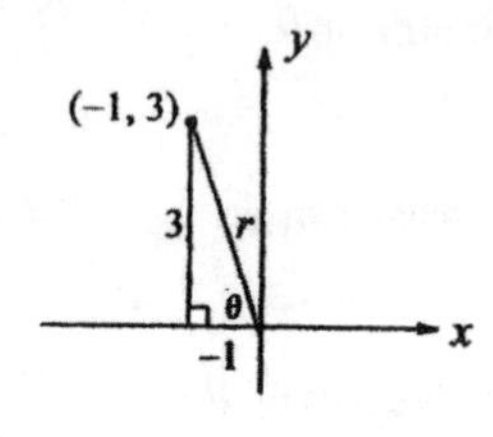

3. If the terminal side of an angle θ passes through (1, –3), find all six basic trigonometric functions
   Solution:

$$r^2=1^2+(-3)^2=10 \quad \therefore r=\sqrt{10}$$

$$\sin\theta=\frac{y}{r}=\frac{-3}{\sqrt{10}}=-\frac{3\sqrt{10}}{10}$$

$$\cos\theta=\frac{x}{r}=\frac{1}{\sqrt{10}}=\frac{\sqrt{10}}{10}$$

$$\tan\theta=\frac{y}{x}=\frac{-3}{1}=-3$$

$$\cot\theta=\frac{x}{y}=\frac{1}{-3}=-\frac{1}{3}$$

$$\sec\theta=\frac{r}{x}=\frac{\sqrt{10}}{1}=\sqrt{10}$$

$$\csc\theta=\frac{r}{y}=\frac{\sqrt{10}}{-3}=-\frac{\sqrt{10}}{3}$$

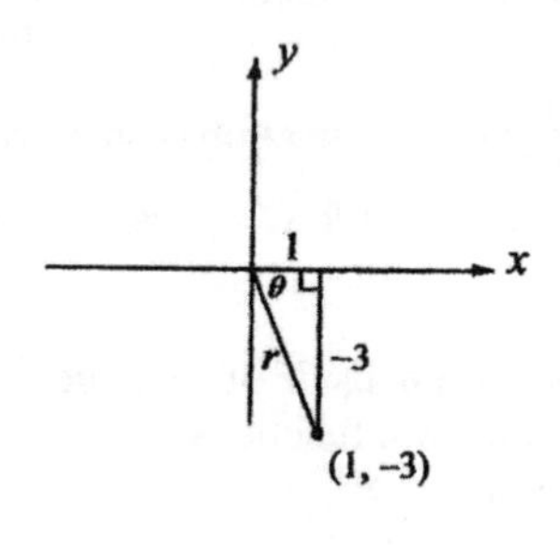

4. If $\theta<90^o$ and $\sin\theta=\frac{1}{2}$, find $\cos\theta$.
   Solution:

$$x^2+1^2=2^2,\ \therefore x=\sqrt{3}$$

$$\cos\theta=\frac{x}{r}=\frac{\sqrt{3}}{2}.\ \text{Ans.}$$

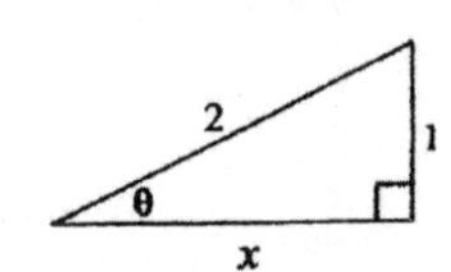

4-3 Solving a Right Triangle

We can use the basic six trigonometric functions to find the value of a specific side or angle of a right triangle if certain information of the triangle are given.

$$\sin\theta = \frac{y}{r},\ \cos\theta = \frac{x}{r},\ \tan\theta = \frac{y}{x},\ \cot\theta = \frac{x}{y},\ \sec\theta = \frac{r}{x},\ \csc\theta = \frac{r}{y}$$

Examples

1. Find a and b.

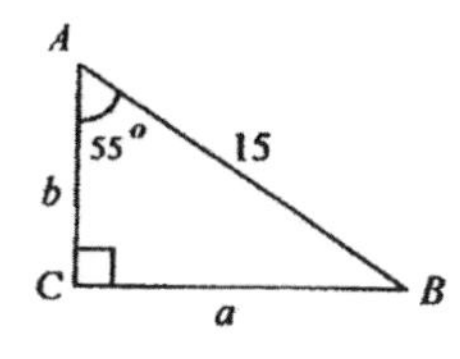

Solution:

$$\sin 55^\circ = \frac{a}{15}$$

$$\therefore a = 15\sin 55^\circ = 15\,(0.8192) = 12.3$$

$$\cos 55^\circ = \frac{b}{15}$$

$$\therefore b = 15\cos 55^\circ = 15\,(0.5736) = 8.6.$$

2. Find a.

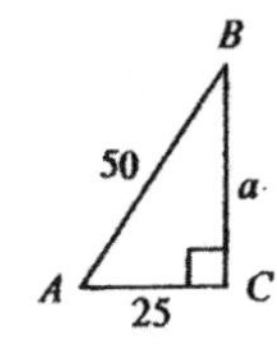

Solution:

$$\cos A = \frac{25}{50} = 0.5$$

$$\therefore A = 60^\circ$$

$$\tan A = \frac{a}{25}$$

$$\therefore a = 25\tan A = 25\tan 60^\circ = 25\,(1.732) = 43.3.$$

3. A ladder of length 20 feet leans against the side of a building by an angle of 25° with the building. Find the distance from the base of the ladder to the bottom of the building.

Solution:

$$\sin 25^\circ = \frac{x}{20}$$

$$x = 20(\sin 25^\circ) = 20(0.4226) \approx 8.45 \text{ feet.}$$

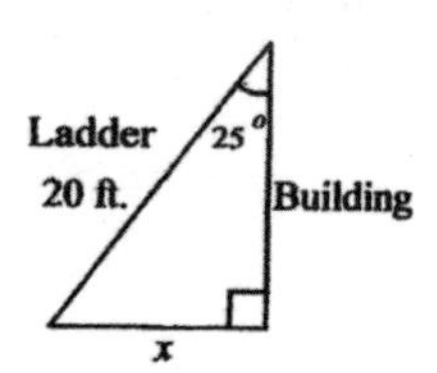

4-4 Trigonometric Identities

A **trigonometric identity** is an equation that is true for all values of the variable for which both sides of the equation are defined. We may simply say that trigonometric identities mean **trigonometric formulas** or **trigonometric relationships**.

Using trigonometric identities, we can simplify trigonometric expressions or solve trigonometric equations.

From the definition of the six trigonometric functions, we have developed the following trigonometric identities which are useful in the study of trigonometry.

1) Reciprocal Relationships

$\sin\theta = \dfrac{1}{\csc\theta}$, $\cos\theta = \dfrac{1}{\sec\theta}$, $\tan\theta = \dfrac{1}{\cot\theta}$

$\csc\theta = \dfrac{1}{\sin\theta}$, $\sec\theta = \dfrac{1}{\cos\theta}$, $\cot\theta = \dfrac{1}{\tan\theta}$

2. Basic Relationships

$\tan\theta = \dfrac{\sin\theta}{\cos\theta}$

$\cot\theta = \dfrac{\cos\theta}{\sin\theta}$

3) Cofunction Identities

$\sin\theta = \cos(90^o - \theta)$

$\cos\theta = \sin(90^o - \theta)$

$\tan\theta = \cot(90^o - \theta)$

$\cot\theta = \tan(90^o - \theta)$

$\sec\theta = \csc(90^o - \theta)$

$\csc\theta = \sec(90^o - \theta)$

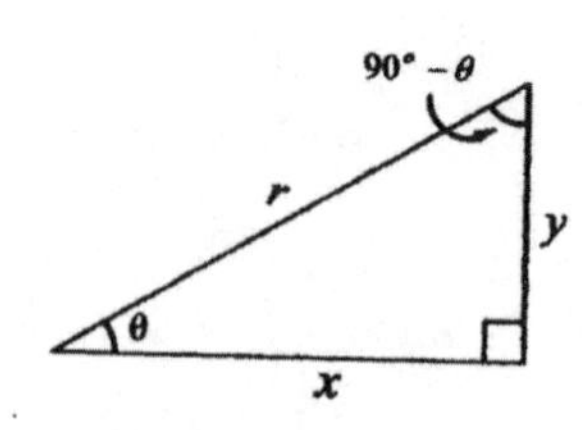

4) Pythagorean Identities

$\sin^2\theta + \cos^2\theta = 1 \rightarrow \sin^2\theta$ means $(\sin\theta)^2$

$1 + \tan^2\theta = \sec^2\theta$

$1 + \cot^2\theta = \csc^2\theta$

Examples

1. Simplify $\cos^2\theta(1+\tan^2\theta)$.

Solution:

$\cos^2\theta(1+\tan^2\theta)$

$= \cos^2\theta(\sec^2\theta)$

$= \dfrac{1}{\sec^2\theta} \cdot \sec^2\theta = 1.$

2. Simplify $\sin x + \cos x \cot x$.

Solution:

$\sin x + \cos x \cot x$

$= \sin x + \cos x \cdot \dfrac{\cos x}{\sin x}$

$= \sin x + \dfrac{\cos^2 x}{\sin x}$

$= \dfrac{\sin^2 x + \cos^2 x}{\sin x} = \dfrac{1}{\sin x} = \csc x.$

4-5 Trigonometric Functions of Special Angles

To memorize the values of the trigonometric functions of special angles 30°, 45°, 60° that appear often in the study of trigonometry, we use the following two special triangles:

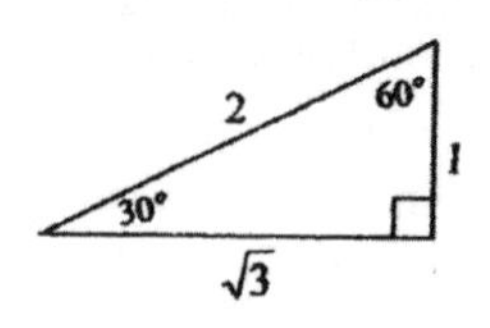

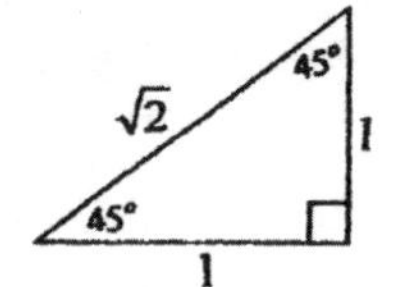

($\pi = 180^o$)

θ	$\sin\theta$	$\cos\theta$	$\tan\theta$	$\cot\theta$	$\sec\theta$	$\csc\theta$
$30^o, \frac{\pi}{6}$	$\frac{1}{2}$	$\frac{\sqrt{3}}{2}$	$\frac{\sqrt{3}}{3}$	$\sqrt{3}$	$\frac{2\sqrt{3}}{3}$	2
$45^o, \frac{\pi}{4}$	$\frac{\sqrt{2}}{2}$	$\frac{\sqrt{2}}{2}$	1	1	$\sqrt{2}$	$\sqrt{2}$
$60^o, \frac{\pi}{3}$	$\frac{\sqrt{3}}{2}$	$\frac{1}{2}$	$\sqrt{3}$	$\frac{\sqrt{3}}{3}$	2	$\frac{2\sqrt{3}}{3}$

Examples

1. Find the value of $\sin 30^o \cos 60^o$.
   Solution:
   $$\sin 30^o \cos 60^o = \frac{1}{2} \cdot \frac{1}{2} = \frac{1}{4}.$$

2. Find the value of $\tan 60^o \sec 45^o$.
   Solution:
   $$\tan 60^o \sec 45^o = \sqrt{3} \cdot \sqrt{2} = \sqrt{6} \approx 2.45.$$

3. Find the value of $\sin\frac{\pi}{4}\cos\frac{\pi}{6}$.
   Solution:
   $$\sin\frac{\pi}{4}\cos\frac{\pi}{6} = \frac{\sqrt{2}}{2} \cdot \frac{\sqrt{3}}{2} = \frac{\sqrt{6}}{4} \approx 0.612.$$

4-6 Radian Measure of Angles

We have learned the degree measure of an angle. The other way for measuring an angle is the **radian measure**.

1 radian (1^R): 1^R of an angle is the measure of the angle with counterclockwise rotation corresponding to an arc length equal to the **radius** $\boldsymbol{r}$ of the circle.

Since the circumference ($C = 2\pi r$) of a circle has a degree measure of 360°, a circle has a radian measure of 2π. Therefore, we have: ($\pi \approx 3.14$, $2\pi \approx 6.28$)

$$360^o = (2\pi)^R \approx 6.28 \text{ radians} \quad ; \quad 180^o = \pi^R \approx 3.14 \text{ radians}$$

We may simply say: $2\pi = 360^o$; $\pi = 180^o$.

Formulas: 1. $1^o = \dfrac{\pi}{180} \approx 0.0175$ radians **2.** $1^R = \dfrac{180}{\pi} \approx 57.3^o$

To change the measure of an angle from degrees to radians, or from radians to degrees, we use the following formulas:

Formulas: 3. $\theta^R = \dfrac{\pi}{180} \cdot \theta^o$ **4.** $\theta^o = \dfrac{180}{\pi} \cdot \theta^R$

For a circle of radius r and a central angle of θ in radians, the arc length s is:

Formula: 5. $s = r\theta$

Examples

1. Express 72° in radians.
Solution:

$$\theta^R = \frac{\pi}{180} \cdot \theta^o = \frac{\pi}{180} \cdot 72$$
$$= 0.4\pi \approx 1.26 \text{ radians.}$$

2. Express -75° in radians.
Solution:

$$\theta^R = \frac{\pi}{180} \cdot \theta^o = \frac{\pi}{180}(-75)$$
$$= -(\tfrac{5}{12}\pi) \approx -1.31 \text{ radians.}$$

3. Express 1.26 radians in degree.
Solution:

$$\theta^o = \frac{180}{\pi} \cdot \theta^R = \frac{180}{\pi} \cdot (1.26)$$
$$\approx 180(0.40) \approx 72^o.$$

4. Express -1.31 radians in degrees.
Solution:

$$\theta^o = \frac{180}{\pi} \cdot \theta^R = \frac{180}{\pi} \cdot (-1.31)$$
$$\approx 180(-0.417) \approx 75^o.$$

5. Express 0.4π radians in degrees.
Solution:

$$\theta^o = \frac{180}{\pi} \cdot \theta^o = \frac{180}{\pi}(0.4\pi)$$
$$= 180(0.4) = 72^o.$$

Quick Method:

$$\theta^o = 0.4\pi = 0.4(180) = 72^o.$$

4-7 Trigonometric Functions of General Angles

In the previous sections, we have learned to find the values of trigonometric functions for acute angles ($\theta < 90^o$).
Since the values of x and y have different signs (+ or –) in the various quadrants, the signs of the trigonometric function of an angle θ for $\theta > 90^o$ can be determined by the quadrant of the angle.
The following figure gives the terminal ray of an angle θ which lies in a particular quadrant, and a point (a, b) on the terminal side.

$$\sin\theta = \frac{y}{r}\ ,\quad \cos\theta = \frac{x}{r},\quad \tan\theta = \frac{y}{x},\quad \cot\theta = \frac{x}{y},\quad \sec\theta = \frac{r}{x},\quad \csc\theta = \frac{r}{y}$$

The signs (+ or –) depend on the quadrant where the angle θ lies.

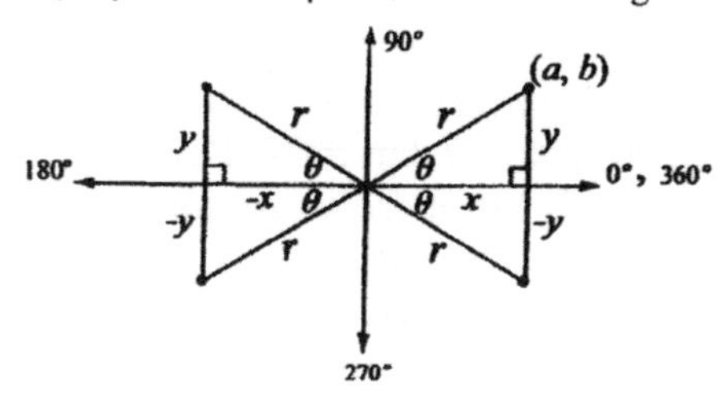

1. The value of each of the trigonometric functions at the **quadrantal angles**:

(Hint: $0^o = 0$, $90^o = \frac{\pi}{2}$, $180^o = \pi$, $270^o = \frac{3\pi}{2}$, $360^o = 2\pi$)

sin	cos	tan	cot	sec	csc
sin 0° =0	cos 0° =1	tan 0° =0	cot 0° =undefined	sec 0° =1	csc 0° =undefined
sin 90° =1	cos 90° =0	tan 90° =undefined	cot 90° =0	sec 90° =undefined	csc 90° =1
sin 180° =0	cos 180° =-1	tan 180° =0	cot 180° =undefined	sec 180° =-1	csc 180° =undefined
sin 270° =-1	cos 270° =0	tan 270° =undefined	cot 270° =0	sec 270° =undefined	csc 270° =-1
sin 360° =0	cos 360° =1	tan 360° =0	cot 360° =undefined	sec 360° =1	csc 360° = undefined

2. The value of each of the trigonometric functions at the **coterminal angles**:
Cotèrminal angles are angles which have the same initial and terminal rays.
The coterminal angles of θ are: $\pm 360^o n + \theta$ or $\pm 2\pi n + \theta$, where $n = 0, 1, 2, 3, \cdots$.

$\sin(\pm 360^o n + \theta) = \sin\theta$

$\cos(\pm 360^o n + \theta) = \cos\theta$

$\tan(\pm 360^o n + \theta) = \tan\theta$

$\cot(\pm 360^o n + \theta) = \cot\theta$

$\sec(\pm 360^o n + \theta) = \sec\theta$

$\csc(\pm 360^o n + \theta) = \csc\theta$

Examples

1. $\sin 390^o = \sin(360^o + 30^o) = \sin 30^o = \frac{1}{2}$.

2. $\cos 390^o = \cos(360^o + 30^o) = \cos 30^o = \frac{\sqrt{3}}{2}$.

3. $\tan 405^o = \tan(360^o + 45^o) = \tan 45^o = 1$.

4. $\sec\frac{5\pi}{2} = \sec(2\pi + \frac{\pi}{2}) = \sec\frac{\pi}{2} =$ undefined.

5. $\sin 3\pi = \sin(2\pi + \pi) = \sin\pi = 0$.

3. Since the six basic trigonometric functions have the reciprocal relationships and only the signs (+ or –) of x and y determine the signs of the functions in the various quadrants, we say that $\sin\theta$ and $\csc\theta$ have the same sign in each quadrant, $\cos\theta$ and $\sec\theta$ have the same sign in each quadrant, and $\tan\theta$ and $\cot\theta$ have the same sign in each quadrant.

Functions	Quadrants 1st	2nd	3rd	4th
$\sin\theta$ and $\csc\theta$	+	+	–	–
$\cos\theta$ and $\sec\theta$	+	–	–	+
$\tan\theta$ and $\cot\theta$	+	–	+	–

To find the value of the trigonometric functions of any angle θ when $\theta > 90^o$, we use the **reference angle** of θ. A reference angle is always an acute angle between 0° and 90°. The value of each of the trigonometric functions of an angle θ equals the value of the trigonometric functions of its reference angle, except for the correct sign (+ or –). The sign depends on the quadrant in which the angle θ lies.
In the figures, $\alpha = 20^o$ is called the **reference angle** of the angles $\theta = 160^o$, 200°, or 340°. We can find the value of each of the trigonometric functions of the angle θ by multiplying the same value of its reference angle by the signs ($+1$ or -1) according to the quadrant in which the angle θ lies.

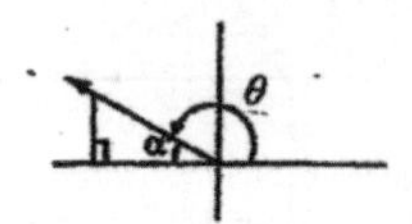

$\theta = 160^o$

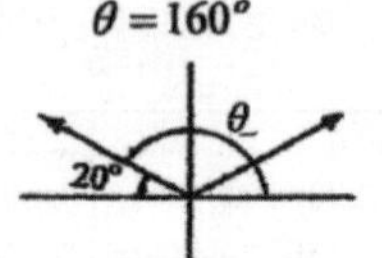

$\theta = 200^o$

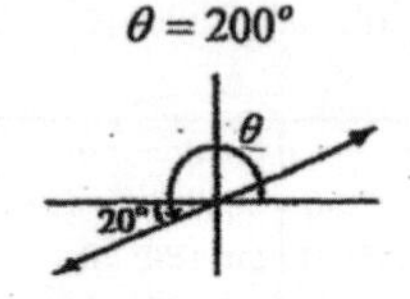

$\theta = 340^o$

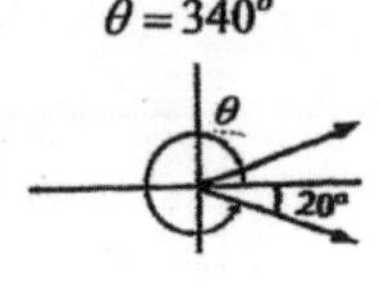

Given $\sin 20^o = 0.3420$ and $\cos 20^o = 0.9397$, we have:

$\sin 160^o = 0.3420$ (2nd qdt.) $\quad \cos 160^o = -0.9397$ (2nd qdt.)

$\sin 200^o = -0.3420$ (3rd qdt.) $\quad \cos 200^o = -0.9397$ (3rd qdt.)

$\sin 340^o = -0.3420$ (4th qdt.) $\quad \cos 340^o = 0.9397$ (4th qdt.)

We use the **coterminal angle** or the **reference angle** to find the value of a trigonometric function.

Examples

1. Find $\tan 160^o$.

Solution: $\tan 160^o = -\tan 20^o = -0.3640$ (2nd qdt.).

2. Find $\sin 4.35^R$.

Solution: reference angle $= 4.35 - 3.14 = 1.21$

$\sin 4.35^R = -\sin 1.21 = -0.9385$ (3rd qdt.).

Examples

3. Find $\cos 850^o 20'$.

Solution:

$$\cos 850^o 20' = \cos(720^o + 130^o 20') = \cos 130^o 20' = -\cos 49^o 40' = -\cos 49.67^o$$
$$= -0.647 \text{ (2nd qdt).}$$

4. Find $\tan 120^o$.

Solution: $\tan 120^o = -\tan 60^o = -\sqrt{3}$ (2nd qdt.).

5. Find $\cot 315^o$.

Solution:

$$\cot 315^o = \cot(360^o - 45^o) = -\cot 45^o = -1 \text{ (4th qdt.).}$$

6. Find $\tan(-30^o)$.

Solution:

$$\tan(-30^o) = -\tan 30^o = -\tfrac{\sqrt{3}}{3} \text{ (4th qdt.).}$$

7. Find $\sec(-225^o)$.

Solution:

$$\sec(-225^o) = \sec(-360^o + 135^o) = \sec 135^o = -\sec 45^o = -\sqrt{2} \text{ (2nd qdt).}$$

8. Find $\csc(-315^o)$.

Solution: $\csc(-315^o) = \csc(-360^o + 45^o) = \csc 45^o = \sqrt{2}$ (1st qdt.).

9. Find $\cos \dfrac{5\pi}{6}$.

Solution: $\cos\dfrac{5\pi}{6} = -\cos\dfrac{\pi}{6} = -\dfrac{\sqrt{3}}{2}$ (2nd qdt.).

10. Find $\sec\left(-\dfrac{7\pi}{4}\right)$.

Solution: $\sec\left(-\dfrac{7\pi}{4}\right) = \sec\left(-2\pi + \dfrac{\pi}{4}\right) = \sec\dfrac{\pi}{4} = \sqrt{2}$ (1st qdt).

11. Find $\tan\theta$ if $\sec\theta = -\frac{\sqrt{5}}{2}$ and θ is in the 3rd quadrant.

Solution: $\sec\theta = \dfrac{r}{x} = \dfrac{\sqrt{5}}{-2}$, $y^2 + (-2)^2 = (\sqrt{5})^2$

$y^2 + 4 = 5 \quad \therefore y = \pm 1$

x and y are negative in 3rd quadrant.

$\therefore \tan\theta = \dfrac{y}{x} = \dfrac{-1}{-2} = \dfrac{1}{2}$.

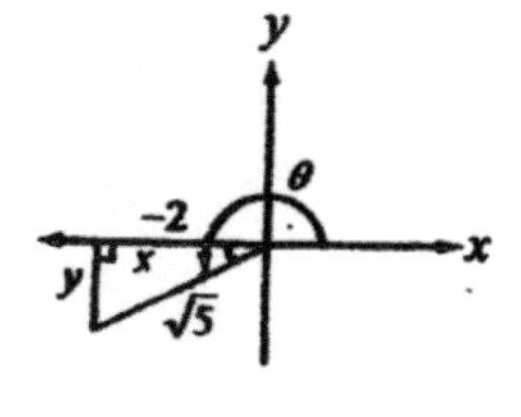

4-8 The Law of Sines

The **Law of Sines** states that for any triangle, the ratio of the sine of each angle to the length of the opposite side is a constant (proportion). We can use the Law of Sines to find the unknown sides if :

1. two sides and one of their opposite angle are given.
2. two angles and one side are given.

The Law of Sines: $\dfrac{\sin A}{a} = \dfrac{\sin B}{b} = \dfrac{\sin C}{c}$

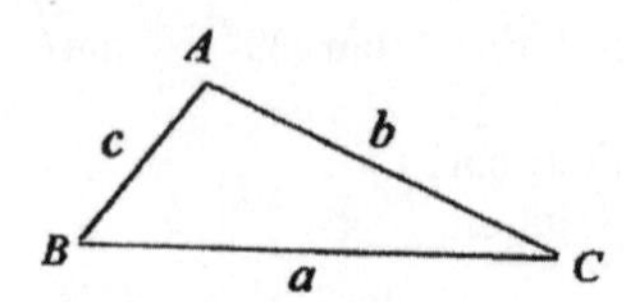

Examples

1. In ΔABC, find $< B$.
Solution:

$$\frac{\sin A}{a} = \frac{\sin B}{b}, \quad \frac{\sin 30^o}{6} = \frac{\sin B}{12}$$

$$\sin B = \frac{12\sin 30^o}{6} = 2\sin 30^o = 2(\tfrac{1}{2}) = 1$$

$$\sin 90^o = 1 \quad \therefore < B = 90^o.$$

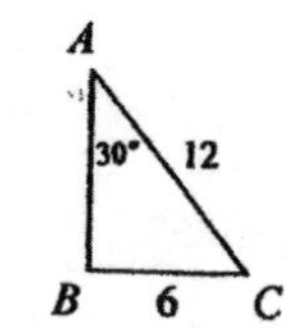

2. In ΔABC, find the length of c.
Solution:

$$< B = 180^o - 60^o - 41^o = 79^o$$

$$\frac{\sin B}{b} = \frac{\sin C}{c}, \quad \frac{\sin 79^o}{140} = \frac{\sin 60^o}{c}$$

$$\therefore \ c = \frac{140\sin 60^o}{\sin 79^o} = \frac{140(0.866)}{0.982} \approx 123.46\, cm.$$

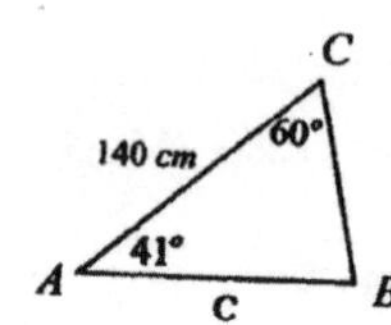

3. In ΔABC, $a = 12$, $b = 8$, $< A = 98^o$, find $< B$, $< C$, and c.

Solution:

$$\frac{\sin 98^o}{12} = \frac{\sin B}{8} = \frac{\sin C}{c}$$

$$\sin B = \frac{8(\sin 98^o)}{12} = \frac{8(0.99)}{12} = 0.66$$

$$\therefore < B \approx 41.3^o.$$

$$< C = 180^o - 98^o - 41.3^o \approx 40.7^o.$$

$$\therefore c = \frac{8(\sin C)}{\sin B} = \frac{8(\sin 40.7^o)}{\sin 41.3^o} = \frac{8(0.65)}{0.66} \approx 7.88.$$

A c B
98°
8
12
C

4-9 The Law of Cosines

The **Law of Cosines** states that for any triangle, the square of one side equals the sum of the squares of the other two sides minus twice their product times the cosine of their included angle.
We can use the Law of Cosines to find unknown dimensions of any triangle if:

1. Two sides and the included angle are given.
2. Three sides are given.

The law of Cosines:
1. $a^2 = b^2 + c^2 - 2bc\cos A$
2. $b^2 = a^2 + c^2 - 2ac\cos B$
3. $c^2 = a^2 + b^2 - 2ab\cos C$

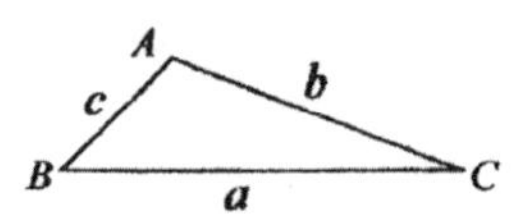

Examples

1. In ΔABC, find b.
Solution:

$$\begin{aligned} b^2 &= a^2 + c^2 - 2ac\cos 60^\circ \\ &= 25^2 + 15^2 - 2(25)(15)(\tfrac{1}{2}) \\ &= 625 + 225 - 375 \\ &= 475 \end{aligned}$$

$\therefore b = \sqrt{475} \approx 21.8$.

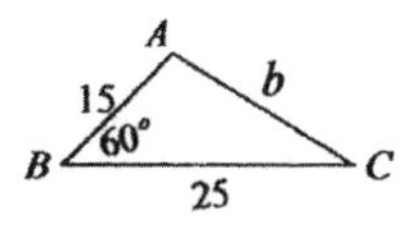

2. In ΔABC, find $< A$.
Solution:

$$\begin{aligned} a^2 &= b^2 + c^2 - 2bc\cos A \\ 10^2 &= 9^2 + 3^2 - 2(9)(3)\cos A \\ 100 &= 81 + 9 - 54\cos A \\ 10 &= -54\cos A \end{aligned}$$

$\cos A = -\dfrac{10}{54} \approx -0.1852 \quad \therefore < A \approx 100.67^\circ$.

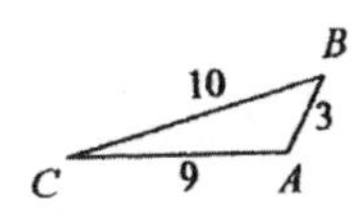

3. In the triangle, find x.
Solution:

$$\begin{aligned} x^2 &= 12^2 + (9.75)^2 - 2(12)(9.75)\cos 32^\circ \\ &= 144 + 95.1 - 234(0.8480) \\ &= 239.1 - 198.43 \\ &= 40.67 \end{aligned}$$

$\therefore x = \sqrt{40.67} \approx 6.38$.

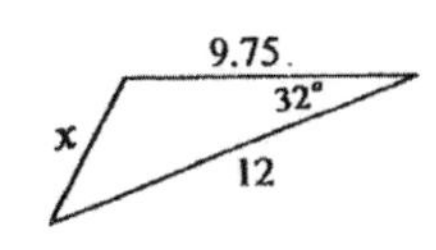

Examples

4. In a triangle, the lengths of two sides are 6 and 3. The opposite angle of side of length 3 is 70°. Find the opposite angle of side of length 6.
   Solution:
   $$\frac{\sin 70^0}{3} = \frac{\sin\theta}{6}$$
   $$\sin\theta = \frac{6(\sin 70^o)}{3} = 2(\sin 70^o) \approx 2(0.94) \approx 1.88$$
   Since there is no angle measure of A for $\sin A > 1$, there is no triangle with the given measurements.

   Ans: No solution. (It is an ambiguous case.)

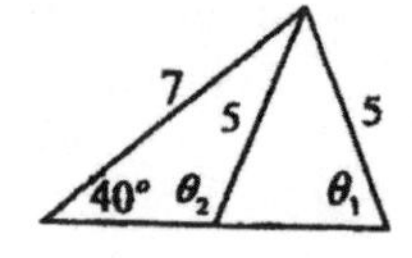

5. In the figure, find the opposite angles of side of length 7.
   Solution:
   $$\frac{\sin 40^o}{5} = \frac{\sin\theta}{7}$$
   $$\sin\theta = \frac{7(\sin 40^o)}{5} \approx \frac{7(0.643)}{5} \approx 0.9$$
   $\theta_1 \approx 64.16^o$ or $\theta_2 = 180^o - 64.16 = 115.84^o$.
   (There are two answers. It is an ambiguous case.)

6. An airplane sends the radar signals to two ground Stations A and B which are 10 miles away. The signal indicates that the bearing (direction) of the airplane from Station A is $E40^oN$ (40° north of east). The signal indicates that the bearing (direction) of the airplane from Station B is $W60^0N$ (60° north of west). How far is the airplane from each station ?
   Solution:
   The measure of the third angle $= 180^o - 40^o - 60^o = 80^o$
   $$\frac{\sin 40^o}{a} = \frac{\sin 80^o}{10} \quad \therefore a = \frac{10(\sin 40^o)}{\sin 80^o} \approx \frac{10(0.643)}{0.985} \approx 6.53 \text{ miles.}$$
   $$\frac{\sin 60^o}{b} = \frac{\sin 80^o}{10} \quad \therefore b = \frac{10(\sin 60^o)}{\sin 80^o} \approx \frac{10(0.866)}{0.985} \approx 8.79 \text{ miles.}$$

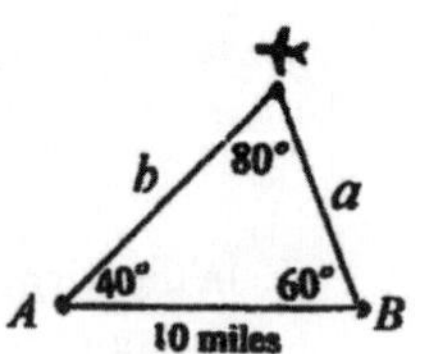

4-10 Double-Angle Formulas

We use double-Angle formulas to find the **exact values** of trigonometric function whose sine, cosine, or tangent are known exactly.

1. $\sin 2A = 2\sin A\cos A$

2. $\cos 2A = \cos^2 A - \sin^2 A$

$= 1 - 2\sin^2 A$ → **4.** $\sin^2\theta = \dfrac{1-\cos 2\theta}{2}$

$= 2\cos^2 A - 1$ → **5.** $\cos^2\theta = \dfrac{1+\cos 2\theta}{2}$

3. $\tan 2A = \dfrac{2\tan A}{1-\tan^2 A}$ **6.** $\tan^2\theta = \dfrac{1-\cos 2\theta}{1+\cos 2\theta}$

Evaluate each expression by its exact value (in fraction, finite decimal, radical, and π)

1. $2\sin 15^\circ\cos 15^\circ = \sin 2(15^\circ) = \sin 30^\circ = \dfrac{1}{2}$

2. $\cos^2 15^\circ - \sin^2 15^\circ = \cos 2(15^\circ) = \cos 30^\circ = \dfrac{\sqrt{3}}{2}$

3. $1 - 2\sin^2 105^\circ = \cos 2(105^\circ) = \cos 210^\circ = -\cos 30^\circ = -\dfrac{\sqrt{3}}{2}$

4. $2\cos^2 120^\circ - 1 = \cos 2(120^\circ) = \cos 240^\circ = -\cos 60^\circ = -\dfrac{1}{2}$

5. $\dfrac{2\tan 22.5^\circ}{1-\tan^2 22.5^\circ} = \tan 2(22.5^\circ) = \tan 45^\circ = 1$

Examples

1. If $\sin\theta = \frac{3}{5}$, find the exact value of $\sin 2\theta$ if $0^\circ < \theta < 90^\circ$.

Solution: Find $\cos\theta$: $\sin\theta = \frac{3}{5} = \frac{y}{r}$, $y = 3$, $r = 5$, $\theta \approx 37^\circ$

$x^2 + 3^2 = 5^2$, $x^2 = 16$, $x = 4$ $\therefore \cos\theta = \frac{x}{r} = \frac{4}{5}$

$\sin 2\theta = 2\sin\theta\cos\theta = 2(\frac{3}{5})(\frac{4}{5}) = \frac{24}{25}$ (2θ is in 1st qdt.).

2. If $\cos\theta = \frac{3}{5}$, find the exact value of $\cos 2\theta$ if $0^\circ < \theta < 90^\circ$.

Solution: Find $\sin\theta$: $\cos\theta = \frac{3}{5} = \frac{x}{r}$, $x = 3$, $r = 5$, $\theta \approx 53^\circ$

$3^2 + y^2 = 5^2$, $y^2 = 16$, $y = 4$ $\therefore \sin\theta = \frac{y}{r} = \frac{4}{5}$

$\cos 2\theta = 1 - 2\sin^2\theta = 1 - 2(\frac{4}{5})^2 = 1 - \frac{32}{25} = -\frac{7}{25}$ (2θ is in 2nd qdt.).

Examples

3. If $\sin\theta = \frac{2}{5}$, find the exact value of $\cos 2\theta$ if $90^o < \theta < 180^o$.

Solution:

$\sin\theta = \frac{2}{5}$, $\theta \approx 42^o$ (reference angle), $\theta \approx 138^o$

2θ lies in 4th quadrant.

$\cos 2\theta = 1 - 2\sin^2\theta = 1 - 2(\frac{2}{5})^2 = \frac{17}{25}$. .

4. If $\tan\theta = \frac{1}{2}$, find the exact value of $\sin 2\theta$ if $\pi < \theta < \dfrac{3\pi}{2}$.

Solution:

$\tan\theta = \frac{1}{2}$, $\theta \approx 27^o$ (reference angle), $\theta \approx 207^o$

$\tan\theta = \frac{y}{x}$, $y = 1$, $x = 2$

$2^2 + 1^2 = r^2$, $r = \sqrt{5}$ $\therefore \sin\theta = \frac{y}{r} = \frac{1}{\sqrt{5}} = \frac{\sqrt{5}}{5}$, $\cos\theta = \frac{x}{r} = \frac{2}{\sqrt{5}} = \frac{2\sqrt{5}}{5}$

2θ lies in 1st quadrant.

$\sin 2\theta = 2\sin\theta\cos\theta = 2(\frac{\sqrt{5}}{5})(\frac{2\sqrt{5}}{5}) = \frac{4}{5}$.

Simplify each expression

5. $\cos^2 2\theta - \sin^2 2\theta = \cos 2(2\theta) = \cos 4\theta$

6. $\cos^4\theta - \sin^4\theta = (\cos^2\theta + \sin^2\theta)(\cos^2\theta - \sin^2\theta) = \cos^2\theta - \sin^2\theta = \cos 2\theta$

4-11 Half-Angle Formulas

We use half-angle formulas to find the **exact values** of trigonometric functions whose sine or cosine are known exactly.
Using the formulas in Chapter 4-10, we obtain the following half-angle formulas:

1. $\sin\frac{A}{2}=\pm\sqrt{\frac{1-\cos A}{2}}$

2. $\cos\frac{A}{2}=\pm\sqrt{\frac{1+\cos A}{2}}$

3. $\tan\frac{A}{2}=\frac{\sin A}{1+\cos A}=\frac{1-\cos A}{\sin A}$

4. $\tan\frac{A}{2}=\pm\sqrt{\frac{1-\cos A}{1+\cos A}}$

The signs (+ or −) is determined by the quadrant of the angle $\frac{A}{2}$.

Evaluate each expression by its exact value (in fraction, finite decimal, radical, and π)

1. $\cos 22.5^o=\cos\frac{45^o}{2}=\sqrt{\frac{1+\cos 45^o}{2}}=\sqrt{\frac{1+\sqrt{2}/2}{2}}=\sqrt{\frac{2+\sqrt{2}}{4}}=\frac{\sqrt{2+\sqrt{2}}}{2}$ (1st qdt.)

2. $\sin(-15^o)=-\sin 15^o$ (4th qdt.)

$=-\sin\frac{30^o}{2}=-\sqrt{\frac{1-\cos 30^o}{2}}=-\sqrt{\frac{1-\sqrt{3}/2}{2}}=-\sqrt{\frac{2-\sqrt{3}}{4}}=-\frac{\sqrt{2-\sqrt{3}}}{2}$

3. $\sin(-67.5^o)=-\sin 67.5^o=-\sin\frac{135^o}{2}$ (4th qdt.)

$=-\sqrt{\frac{1-\cos 135^o}{2}}=-\sqrt{\frac{1-(-\cos 45^o)}{2}}=-\sqrt{\frac{1+\sqrt{2}/2}{2}}=-\sqrt{\frac{2+\sqrt{2}}{4}}=-\frac{\sqrt{2+\sqrt{2}}}{2}$

4. $\tan 157.5^o=-\tan 22.5^o=-\tan\frac{45^o}{2}$ (2nd qdt.)

$=-\frac{\sin 45^o}{1+\cos 45^o}=-\frac{\sqrt{2}/2}{1+\sqrt{2}/2}=-\frac{\sqrt{2}}{2+\sqrt{2}}\cdot\frac{2-\sqrt{2}}{2-\sqrt{2}}=-\frac{2\sqrt{2}-2}{2}=-(\sqrt{2}-1)=1-\sqrt{2}$

5. $\tan\frac{9\pi}{8}=\tan\frac{\pi}{8}=\tan 22.5^o$ (3rd qdt.)

$=\frac{\sin 45^o}{1+\cos 45^o}=\frac{\sqrt{2}/2}{1+\sqrt{2}/2}=\frac{\sqrt{2}}{2+\sqrt{2}}\cdot\frac{2-\sqrt{2}}{2-\sqrt{2}}=\frac{2\sqrt{2}-2}{2}=\sqrt{2}-1$

Examples

1. If $\sin\theta = \frac{3}{5}$ and $0^o < \theta < 90^o$, find the exact value of **a.** $\sin\frac{\theta}{2}$ **b.** $\cos\frac{\theta}{2}$ **c.** $\tan\frac{\theta}{2}$.

Solution:

Find $\cos\theta$: $\sin\theta = \frac{y}{r}$, $y = 3$, $r = 5$, $\theta \approx 37^o$

$x^2 + 3^2 = 5^2$, $x^2 = 16$, $x = 4$ $\therefore \cos\theta = \frac{x}{r} = \frac{4}{5}$

$\theta/2$ lies in 1st quadrant.

a. $\sin\frac{\theta}{2} = \sqrt{\frac{1-\cos\theta}{2}} = \sqrt{\frac{1-4/5}{2}} = \sqrt{\frac{1/5}{2}} = \sqrt{\frac{1}{10}} = \frac{\sqrt{10}}{10}$.

b. $\cos\frac{\theta}{2} = \sqrt{\frac{1+\cos\theta}{2}} = \sqrt{\frac{1+4/5}{2}} = \sqrt{\frac{9/5}{2}} = \sqrt{\frac{9}{10}} = \frac{3\sqrt{10}}{10}$.

c. $\tan\frac{\theta}{2} = \sqrt{\frac{1-\cos\theta}{1+\cos\theta}} = \sqrt{\frac{1-4/5}{1+4/5}} = \sqrt{\frac{1/5}{9/5}} = \sqrt{\frac{1}{9}} = \frac{1}{3}$.

Or: $\tan\frac{\theta}{2} = \frac{\sin\theta/2}{\cos\theta/2} = \frac{\sqrt{10}/10}{3\sqrt{10}/10} = \frac{1}{3}$.

2. If $\cos\theta = -\frac{2}{5}$ and $\pi < \theta < 3\pi/2$, find the exact value of **a.** $\sin\frac{\theta}{2}$ **b.** $\cos\frac{\theta}{2}$ **c.** $\tan\frac{\theta}{2}$.

Solution:

$\cos\theta = -\frac{2}{5}$, $\theta \approx 66^o$ (reference angle), $\theta \approx 246^o$

$\theta/2$ lies in 2nd quadrant.

a. $\sin\frac{\theta}{2} = \sqrt{\frac{1-\cos\theta}{2}} = \sqrt{\frac{1-(-2/5)}{2}} = \sqrt{\frac{7/5}{2}} = \sqrt{\frac{7}{10}} = \frac{\sqrt{70}}{10}$.

b. $\cos\frac{\theta}{2} = -\sqrt{\frac{1+\cos\theta}{2}} = -\sqrt{\frac{1+(-2/5)}{2}} = -\sqrt{\frac{3/5}{2}} = -\sqrt{\frac{3}{10}} = -\frac{\sqrt{30}}{10}$.

c. $\tan\frac{\theta}{2} = -\sqrt{\frac{1-\cos\theta}{1+\cos\theta}} = -\sqrt{\frac{1-(-2/5)}{1+(-2/5)}} = -\sqrt{\frac{7/5}{3/5}} = -\frac{\sqrt{21}}{3}$.

4-12 Sum and Difference Formulas

We use sum and difference formulas to find the **exact values** of trigonometric functions whose sine, cosine, or tangent are known exactly.

1. $\sin(A+B)=\sin A\cos B+\cos A\sin B$
2. $\sin(A-B)=\sin A\cos B-\cos A\sin B$
3. $\cos(A+B)=\cos A\cos B-\sin A\sin B$
4. $\cos(A-B)=\cos A\cos B+\sin A\sin B$
5. $\tan(A+B)=\dfrac{\tan A+\tan B}{1-\tan A\tan B}$
6. $\tan(A-B)=\dfrac{\tan A-\tan B}{1+\tan A\tan B}$

Evaluate each expression by its exact value (in fraction, finite decimal, radical, and π)

1. $\sin 75^\circ=\sin(45^\circ+30^\circ)=\sin 45^\circ\cos 30^\circ+\cos 45^\circ\sin 30^\circ=\dfrac{\sqrt{2}}{2}\cdot\dfrac{\sqrt{3}}{2}+\dfrac{\sqrt{2}}{2}\cdot\dfrac{1}{2}=\dfrac{\sqrt{6}+\sqrt{2}}{4}$

2. $\sin 15^\circ=\sin(45^\circ-30^\circ)=\sin 45^\circ\cos 30^\circ-\cos 45^\circ\sin 30^\circ=\dfrac{\sqrt{2}}{2}\cdot\dfrac{\sqrt{3}}{2}-\dfrac{\sqrt{2}}{2}\cdot\dfrac{1}{2}=\dfrac{\sqrt{6}-\sqrt{2}}{4}$

3. $\cos 120^\circ=\cos(90^\circ+30^\circ)=\cos 90^\circ\cos 30^\circ-\sin 90^\circ\sin 30^\circ=0\cdot\dfrac{\sqrt{3}}{2}-1\cdot\dfrac{1}{2}=-\dfrac{1}{2}$

4. $\cos 150^\circ=\cos(180^\circ-30^\circ)=\cos 180^\circ\cos 30^\circ+\sin 180^\circ\sin 30^\circ=-1\cdot\dfrac{\sqrt{3}}{2}+0\cdot\dfrac{1}{2}=-\dfrac{\sqrt{3}}{2}$

5. $\sin 20^\circ\cos 25^\circ+\cos 20^\circ\sin 25^\circ=\sin(20^\circ+25^\circ)=\sin 45^\circ=\dfrac{\sqrt{2}}{2}$

6. $\sin 75^\circ\cos 15^\circ-\cos 75^\circ\sin 15^\circ=\sin(75^\circ-15^\circ)=\sin 60^\circ=\dfrac{\sqrt{3}}{2}$

7. $\sin\dfrac{\pi}{4}\cos\dfrac{\pi}{12}-\cos\dfrac{\pi}{4}\sin\dfrac{\pi}{12}=\sin(\dfrac{\pi}{4}-\dfrac{\pi}{12})=\sin\dfrac{\pi}{6}=\dfrac{1}{2}$

8. $\cos 85^\circ\cos 40^\circ+\sin 85^\circ\sin 40^\circ=\cos(85^\circ-40^\circ)=\cos 45^\circ=\dfrac{\sqrt{2}}{2}$

9. $\tan 105^\circ=\tan(60^\circ+45^\circ)=\dfrac{\tan 60^\circ+\tan 45^\circ}{1-\tan 60^\circ\tan 45^\circ}=\dfrac{\sqrt{3}+1}{1-\sqrt{3}\cdot 1}=\dfrac{1+\sqrt{3}}{1-\sqrt{3}}=\dfrac{1+\sqrt{3}}{1-\sqrt{3}}\cdot\dfrac{1+\sqrt{3}}{1+\sqrt{3}}$

$=\dfrac{1+2\sqrt{3}+3}{1-3}=\dfrac{4+2\sqrt{3}}{-2}=-2-\sqrt{3}$.

Evaluate each expression by its exact value

10. $\tan 15^o = \tan(45^o - 30^o) = \dfrac{\tan 45^o - \tan 30^o}{1 + \tan 45^o \tan 30^o} = \dfrac{1 - \frac{\sqrt{3}}{3}}{1 + 1 \cdot \frac{\sqrt{3}}{3}} = \dfrac{3-\sqrt{3}/3}{3+\sqrt{3}/3} = \dfrac{3-\sqrt{3}}{3+\sqrt{3}} \cdot \dfrac{3-\sqrt{3}}{3-\sqrt{3}}$

$= \dfrac{9 - 6\sqrt{3} + 3}{9-3} = \dfrac{12 - 6\sqrt{3}}{6} = 2 - \sqrt{3}$

Simplify each expression

11. $\cos(x-y)\cos y + \sin(x-y)\sin y = \cos[(x-y)-y] = \cos(x-2y)$.

12. $\sin(x-y)\cos y + \cos(x-y)\sin y = \sin[(x-y)+y] = \sin x$.

13. $\dfrac{\tan 70^o - \tan 25^o}{1 + \tan 70^0 \tan 25^o} = \tan(70^o - 25^o) = \tan 45^o = 1$.

14. $\dfrac{\tan x + \tan 2x}{1 - \tan x \tan 2x} = \tan(x + 2x) = \tan 3x$.

15. $\dfrac{\tan\frac{\pi}{4} - \tan\frac{\pi}{12}}{1 + \tan\frac{\pi}{4}\tan\frac{\pi}{12}} = \tan(\dfrac{\pi}{4} - \dfrac{\pi}{12}) = \tan\dfrac{\pi}{6} = \dfrac{\sqrt{3}}{3}$.

16. $\sin(\dfrac{\pi}{2} + \theta) = \sin\dfrac{\pi}{2}\cos\theta + \cos\dfrac{\pi}{2}\sin\theta = 1 \cdot \cos\theta + 0 \cdot \sin\theta = \cos\theta$

17. $\tan(\pi - \theta) = \dfrac{\tan\pi - \tan\theta}{1 + \tan\pi\tan\theta} = \dfrac{0 - \tan\theta}{1 + 0 \cdot \tan\theta} = -\tan\theta$

18. Prove $\sin 3x = 3\sin x - 4\sin^3 x$.
Proof:

$\sin 3x = \sin(2x + x) = \sin 2x \cos x + \cos 2x \sin x = 2\sin x \cos x \cdot \cos x + (1 - 2\sin^2 x)\sin x$
$= 2\sin x \cos^2 x + \sin x - 2\sin^3 x = 2\sin x(1 - \sin^2 x) + \sin x - 2\sin^3 x$
$= 2\sin x - 2\sin^3 x + \sin x - 2\sin^3 x = 3\sin x - 4\sin^3 x$.

19. Prove $\cos 3x = 4\cos^3 x - 3\cos x$.
Proof:

$\cos 3x = \cos(2x + x) = \cos 2x \cos x - \sin 2x \sin x = (2\cos^2 x - 1)\cos x - (2\sin x \cos x)\sin x$
$= 2\cos^3 x - \cos x - 2\sin^2 x \cos x = 2\cos^3 x - \cos x - 2(1 - \cos^2 x)\cos x$
$= 2\cos^3 x - \cos x - 2\cos x + 2\cos^3 x = 4\cos^3 x - 3\cos x$.

4-13 The Area of a Triangle

The area of a triangle can be obtained by the following formulas if two sides and the included angle are known.

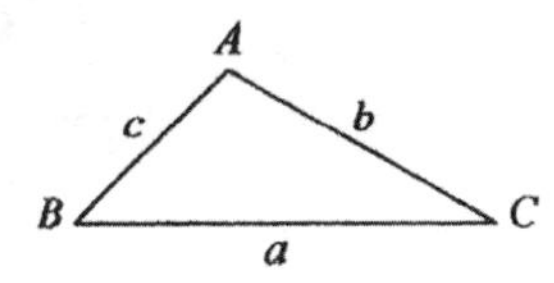

1. $Area = \frac{1}{2}ab\sin C$ 2. $Area = \frac{1}{2}bc\sin A$

3. $Area = \frac{1}{2}ac\sin B$

4. $Area = \sqrt{s(s-a)(s-b)(s-c)}$ where $s = \frac{a+b+c}{2}$

(It is called **Heron's Formula.**)

Examples

1. Find the area of ΔABC.

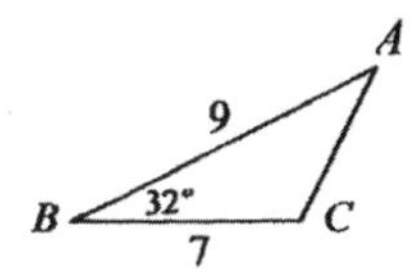

Solution:

$$Area = \frac{1}{2}ac\sin 32^o$$
$$= \frac{1}{2}(7)(9)(\sin 32^o)$$
$$\approx \frac{1}{2}(7)(9)(0.5299) \approx 16.69.$$

2. Find the area of the triangle.

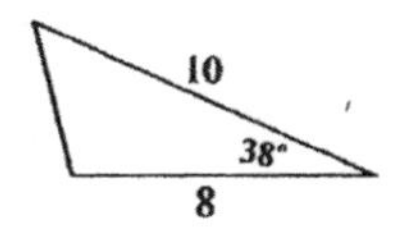

Solution:

$$Area = \frac{1}{2}(10)(8)(\sin 38^o)$$
$$\approx \frac{1}{2}(10)(8)(0.6157) \approx 24.63.$$

3. Find the area of the triangle.

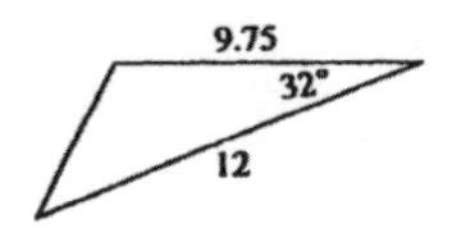

Solution:

$$Area = \frac{1}{2}(12)(9.75)(\sin 32^o)$$
$$\approx \frac{1}{2}(12)(9.75)(0.5299) \approx 31.$$

4. Find the area of a triangle with the lengths of three sides 12, 6.38, and 9.75.
Solution:

$$s = \frac{12+6.38+9.75}{2} = 14.07$$
$$Area = \sqrt{s(s-a)(s-b)(s-c)}$$
$$= \sqrt{14.07(14.07-12)(14.07-6.38)(14.07-9.75)}$$
$$= \sqrt{14.07(2.07)(7.69)(4.32)} = \sqrt{967.55} \approx 31.$$

Examples

5. Find the area of the triangle for which $a = 4$, $b = 3$, and the included angle $\alpha = 30^o$.
Solution:

$$Area = \tfrac{1}{2}ab\sin\alpha = \tfrac{1}{2}(4)(3)(\sin 30^o) = \tfrac{1}{2}(4)(3)(\tfrac{1}{2}) = 3.$$

6. Find the area of the shaded region in the circle.
Solution:

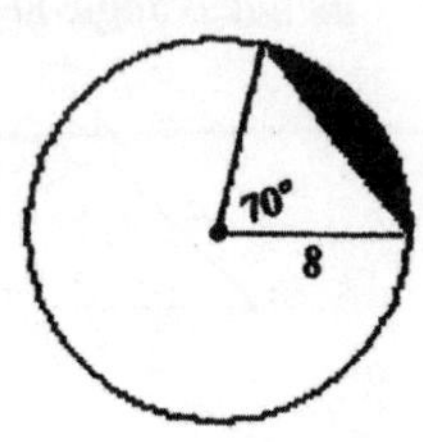

Area of the sector $= \pi\,(8^2)(\frac{70}{360}) \approx 39.095$
Area of the triangle $= \frac{1}{2}(8)(8)(\sin 70^o) \approx 30.07$
Area of the shaded region $\approx 39.095 - 30.07 \approx 9.025$.

4-14 Periodic Functions and Graphs

Trigonometric functions are called **periodic functions.** The values of a periodic funtion repeat on a regular interval. This regular interval is called the period of the funtion. The period of a trigonometric function contains a full cycle of the function.
The period of sine or cosine function is 2π radians or 360°. It means that the values of sine or cosine function repeat themselves every cycle (360°). The period of tangent or cotangent function is π radians or 180°. The period of secant or cosecant function is 2π or 360°.
Each of the six trigonometric functions can be graphed on a coordinate plane. The *x*-axis represents the value of the angle. The *y*-axis represents the value of the function of each angle.

Graphs of $y = \sin x$ and $y = \cos x$

To draw the graphs of $y = \sin x$ (or $y = \cos x$) in a coordinate plane, we begin by choosing values of *x* as an angle in radians in the interval $0 \le x \le 2\pi$. The values of $\sin x$ (or $\cos x$) repeat every 2π units with a smooth, continuous curve. These graphs are called **sine waves** and **cosine waves.** The values of $\sin x$ or $\cos x$ vary from -1 to 1. We say that both $\sin x$ and $\cos x$ have a **period** of 2π and an **amplitude** of 1.

A) Sine wave → $y = \sin x$; $-1 \le y \le 1$, $|\sin x| \le 1$

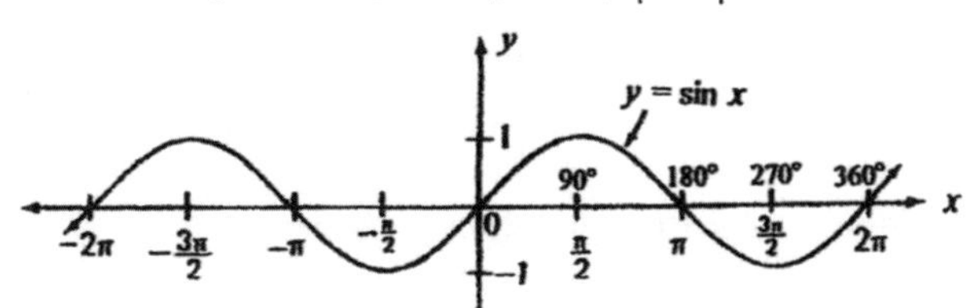

B) Cosine wave → $y = \cos x$; $-1 \le y \le 1$, $|\cos x| \le 1$

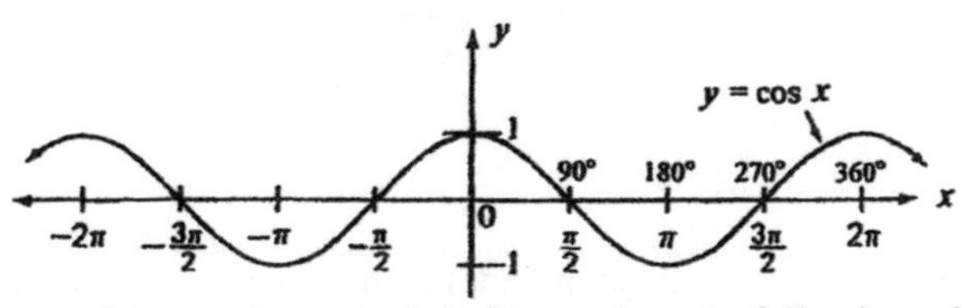

Since the sine and cosine functions have a period of 2π, we have the following rules:

$$\sin(2\pi + x) = \sin x \quad ; \quad \cos(2\pi + x) = \cos x$$

The **amplitudes** and **periods** of sine and cosine functions can be determined by the following ways: $y = A\sin kx$ or $y = A\cos kx$

$$\textbf{amplitude} = A \ , \ \textbf{period} = \frac{2\pi}{k} .$$

Graphing Variations of Trigonometric Functions using Transformations

The graphing transformations introduced in chapter 2-21 can be used to graph trigonometric functions that are variations of the functions. Example: $y = a\sin(kx \pm c)$

1. $y = -\sin x$ It reflects the graph over the x-axis.
2. $y = \sin x \pm c$ It moves (shifts) the graph by c units upward (downward).
3. $y = \sin(x \pm c)$ It moves (shifts) the graph by c units to the left (right).
4. $y = a\sin x$ It stretches the graph vertically if $a > 1$ and compresses the graph vertically if $0 < a < 1$.
5. $y = \sin ax$ It compresses the graph horizontally if $a > 1$ and stretches the graph horizontally if $0 < a < 1$.

Examples

1. Find the amplitude and period of the function $y = 2\sin x$.

   Solution: $A = 2$ (amplitude) ; The function varies from –2 to 2.

   $k = 1$ $\quad \therefore \text{period} = \dfrac{2\pi}{k} = \dfrac{2\pi}{1} = 2\pi$.

2. Find the amplitude and period of the function $y = \dfrac{2}{3}\cos 2x$.

   Solution: $A = \dfrac{2}{3}$ (amplitude) ; The function varies from $-\dfrac{2}{3}$ to $\dfrac{2}{3}$.

   $k = 2$ $\quad \therefore \text{period} = \dfrac{2\pi}{k} = \dfrac{2\pi}{2} = \pi$.

3. Graph $y = \sin 2x$.

   Solution: $A = 1$ (amplitude), period $= \dfrac{2\pi}{k} = \dfrac{2\pi}{2} = \pi$.

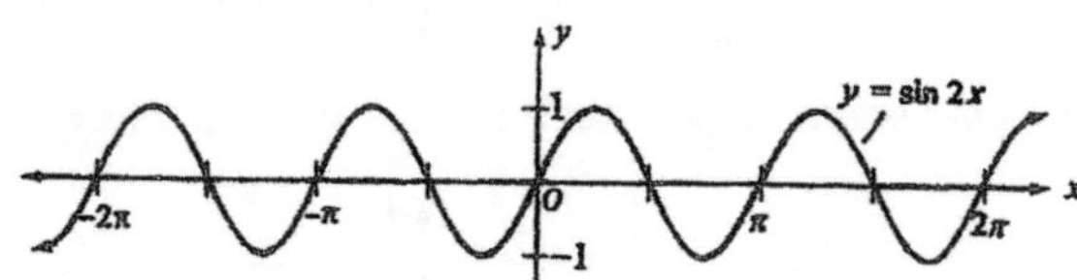

4. Graph $y = 2\cos\dfrac{1}{2}x$.

   Solution: $A = 2$ (amplitude) ; period $= \dfrac{2\pi}{k} = \dfrac{2\pi}{½} = 4\pi$.

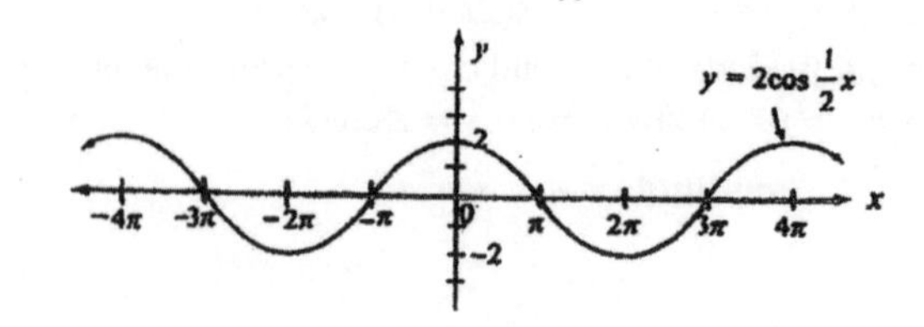

Graphs of $y = \tan x$ and $y = \cot x$

To draw the graphs of $y = \tan x$ (or $y = \cot x$) in a coordinate plane, we begin by choosing values of x as an angle in radians. The graphs of tangent and cotangent functions have no amplitude. Their graphs are drawn between successive vertical asymptotes and repeat the pattern to the right and to the left.

A) $y = \tan x$

	$-\frac{\pi}{2}$	$-\frac{\pi}{3}$	$-\frac{\pi}{4}$	$-\frac{\pi}{6}$	0	$\frac{\pi}{6}$	$\frac{\pi}{4}$	$\frac{\pi}{3}$	$\frac{\pi}{2}$
y	undefined	$-\sqrt{3}$	-1	$-\frac{\sqrt{3}}{3}$	0	$\frac{\sqrt{3}}{3}$	1	$\sqrt{3}$	undefined

1. y is undefined when $x = n\pi + \frac{\pi}{2}$, where n is an integer.
2. y increases as x increases.
3. The equations of the asymptotes are: $x = n\pi + \frac{\pi}{2}$, n is an integer.
4. A period of π

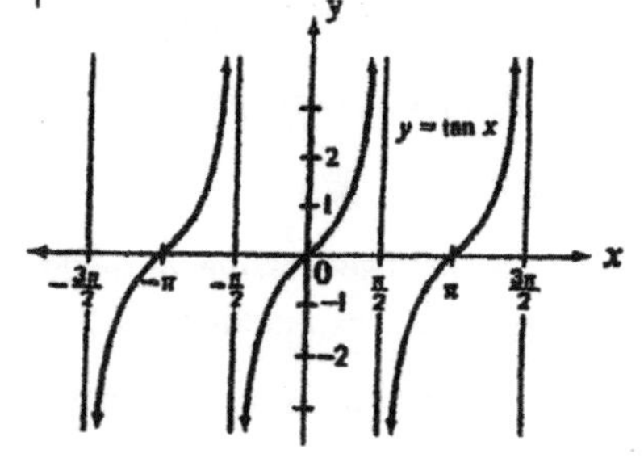

B) $y = \cot x$

	0	$\frac{\pi}{6}$	$\frac{\pi}{4}$	$\frac{\pi}{3}$	$\frac{\pi}{2}$	$\frac{2\pi}{3}$	$\frac{3\pi}{4}$	$\frac{5\pi}{6}$	π
y	undefined	$\sqrt{3}$	1	$\frac{\sqrt{3}}{3}$	0	$-\frac{\sqrt{3}}{3}$	-1	$-\sqrt{3}$	undefined

1. y is undefined when $x = n\pi$, where n is an integer.
2. y decreases as x increases.
3. The equations of asymptotes are: $x = n\pi$, n is an integer.
4. A period of π

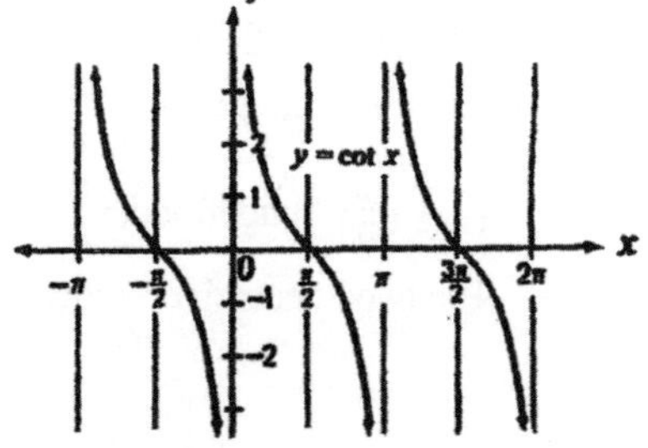

The periods of tangent and cotangent functions can be determined by the following way:

$$\text{If } y = \tan kx \text{ or } y = \cot kx, \text{ then } \textbf{period} = \frac{\pi}{k}.$$

The asymptotes occur every π/k units along the x-axis.

Examples

1. Graph the function $y = 3\tan 2x$.

Solution: $k = 2$, period $= \frac{\pi}{k} = \frac{\pi}{2}$

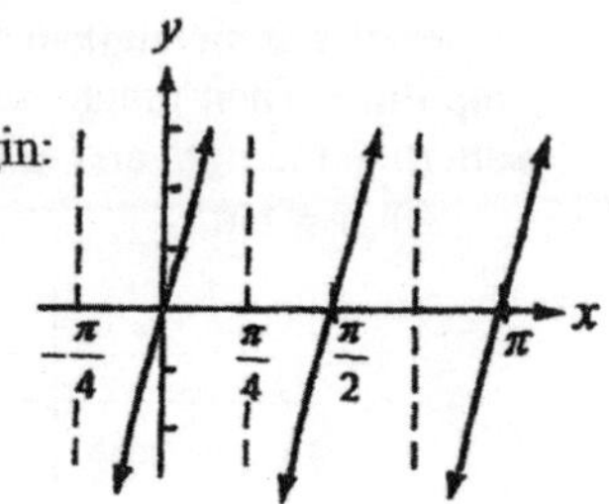

Find successive vertical asymptotes near the origin:

$\tan 2x =$ undefined, $2x = -\frac{\pi}{2}$ and $2x = \frac{\pi}{2}$

Vertical asymptotes are $x = -\frac{\pi}{4}$ and $x = \frac{\pi}{4}$

Graph the function between these asymptotes.
The same pattern occurs to the right and to the left.

2. Graph the function $y = \cot 2x$.

Solution: $k = 2$, period $= \frac{\pi}{k} = \frac{\pi}{2}$

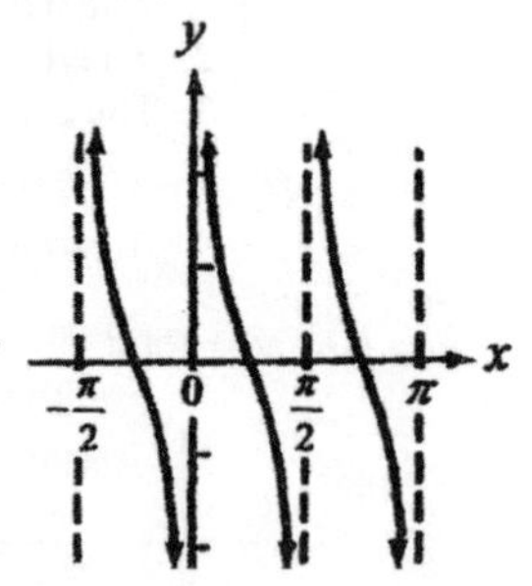

Find successive vertical asymptotes near the origin:
$\cot 2x =$ undefined, $2x = -\pi$, $2x = 0$ and $2x = \pi$

Vertical asymptotes are $x = -\frac{\pi}{2}$, $x = 0$ and $x = \frac{\pi}{2}$.

Graph the function between these asymptotes.
The same pattern occurs to the right and to the left.

3. Find the period of $y = \tan(x + \frac{\pi}{4})$ and two successive vertical asymptotes near the origin, one to the left, one to the right.

Solution: $k = 1$, period $= \frac{\pi}{k} = \frac{\pi}{1} = \pi$.

$\tan(x + \frac{\pi}{4}) =$ undefined, $x + \frac{\pi}{4} = -\frac{\pi}{2}$ and $x + \frac{\pi}{4} = \frac{\pi}{2}$

The two successive vertical asymptotes near the origin are:

$$x = -\frac{3\pi}{4} \quad \text{and} \quad x = \frac{\pi}{4}.$$

4. Find the period of $y = \cot(2x - \frac{\pi}{2})$ and two successive vertical asymptotes to the right of the origin

Solution: $k = 2$, period $= \frac{\pi}{k} = \frac{\pi}{2}$.

$\cot(2x - \frac{\pi}{2}) =$ undefined, $2x - \frac{\pi}{2} = 0$ and $2x - \frac{\pi}{2} = \pi$

The two successive vertical asymptotes near the origin are:

$$x = \frac{\pi}{4} \quad \text{and} \quad x = \frac{3\pi}{4}.$$

Graphs of $y = \sec x$ and $y = \csc x$

To draw the graphs of $y = \sec x$ (or $y = \csc x$) in a coordinate plane, we begin by choosing values of x as an angle in radians. The graphs of secant and cosecant functions have no amplitude.

Their graphs of secant and cosecant functions can be drawn by making use of the reciprocals of corresponding cosine and sine functions: $\sec x = \dfrac{1}{\cos x}$ and $\csc x = \dfrac{1}{\sin x}$

For example, if $x = 0$, $\sin 0 = 0$, then $\csc = \frac{1}{0} =$ undefined.

Their graphs can also be drawn between successive vertical asymptotes and repeat the pattern to the right and to the left.

A) $y = \sec x$ $\quad y \le -1$ or $y \ge 1$, $|\sec| \ge 1$

x	0	$\frac{\pi}{4}$	$\frac{\pi}{2}$	$\frac{3\pi}{4}$	π	$\frac{5\pi}{4}$	$\frac{3\pi}{2}$	$\frac{7\pi}{4}$	2π
y	1	$\sqrt{2}$	undefined	$-\sqrt{2}$	-1	$-\sqrt{2}$	undefined	$\sqrt{2}$	1

1. y is undefined when $x = n\pi + \frac{\pi}{2}$, where n is an integer.
2. Intersects with y-axis at (0, 1)
3. No graph between $y = 1$ and $y = -1$
4. Graph discontinues at the asymptotes: $x = n\pi + \frac{\pi}{2}$, n is an integer.
5. Symmetric about y-axis: $\sec(-x) = \sec x$
6. A period of 2π

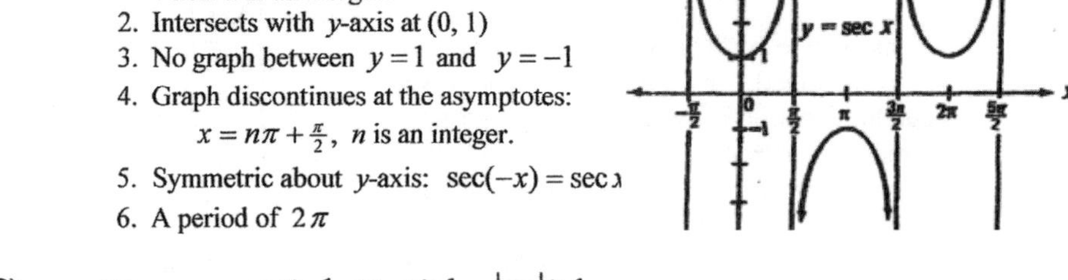

B) $y = \csc x$ $\quad y \le -1$ or $y \ge 1$, $|\csc| \ge 1$

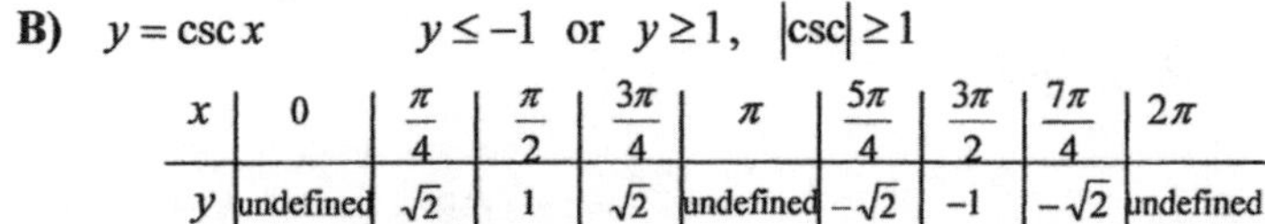

x	0	$\frac{\pi}{4}$	$\frac{\pi}{2}$	$\frac{3\pi}{4}$	π	$\frac{5\pi}{4}$	$\frac{3\pi}{2}$	$\frac{7\pi}{4}$	2π
y	undefined	$\sqrt{2}$	1	$\sqrt{2}$	undefined	$-\sqrt{2}$	-1	$-\sqrt{2}$	undefined

1. y is undefined when $x = n\pi$, where n is an integer.
2. No intersection with y-axis
3. No graph between $y = 1$ and $y = -1$
4. Graph discontinues at the asymptotes: $x = n\pi$, n is an integer.
5. Symmetric through origin: $\csc(-x) = -\csc x$
6. A period of 2π

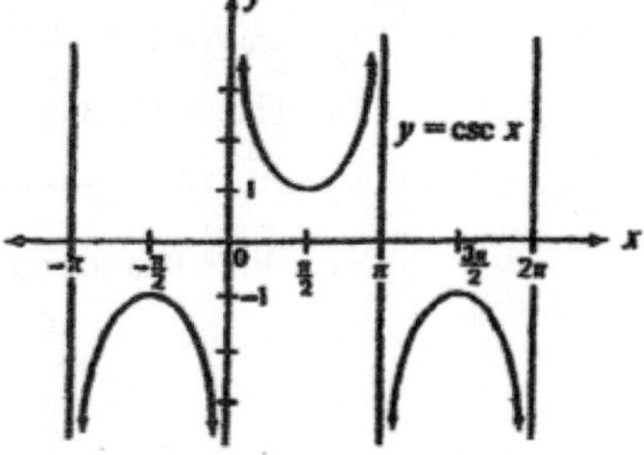

The periods of secant and cosecant functions can be determined by the following way:

If $y = \sec kx$ or $y = \csc kx$, then **period** $= \dfrac{2\pi}{k}$

Examples

1. Graph the function $y = \sec 2x$.

Solution:

$k = 2$, period $= \frac{2\pi}{2} = \pi$

Find successive vertical asymptotes near the origin:

$\sec 2x =$ undefined, $2x = -\frac{\pi}{2}$, $2x = \frac{\pi}{2}$, and $2x = \frac{3\pi}{2}$

Vertical asymptotes are $x = -\frac{\pi}{4}$, $x = \frac{\pi}{4}$, and $\frac{3\pi}{4}$

Graph the function between these asymptotes.

The same pattern occurs to the right and to the left.

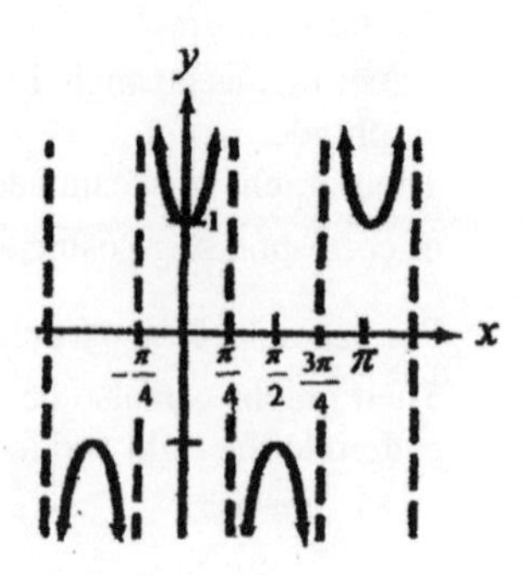

2. Graph the function $y = 2\csc x$.

Solution:

$k = 1$, period $= \frac{2\pi}{k} = \frac{2\pi}{1} = 2\pi$

Find successive vertical asymptotes near the origin:

$\csc x =$ undefined, $x = -\pi$, $x = 0$, and $x = \pi$

Graph the function between these asymptotes.

The same pattern occurs to the right and to the left.

The graph is vertical stretched by multiplying 2 units.

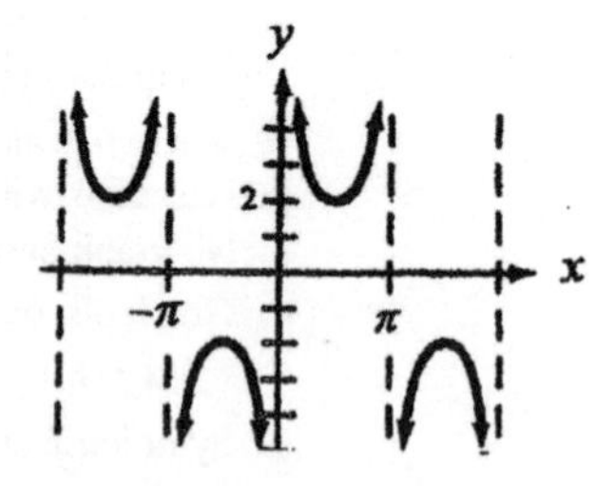

3. Find the period of $y = 2\sec(x - \frac{\pi}{4})$ and three successive vertical asymptotes near the origin, one to the left, two to the right.

Solution:

$$k = 1,\ \text{period} = \frac{2\pi}{k} = \frac{2\pi}{1} = 2\pi.\ \text{Ans.}$$

$$\sec(x - \frac{\pi}{4}) = \text{undefined},\quad x - \frac{\pi}{4} = -\frac{\pi}{2},\ x - \frac{\pi}{4} = \frac{\pi}{2}\ \text{and}\ x - \frac{\pi}{4} = \frac{3\pi}{2}$$

The three successive vertical asymptotes near the origin are:

$$x = -\frac{\pi}{4},\ x = \frac{3\pi}{4},\ \text{and}\ x = \frac{7\pi}{4}.\ \text{Ans.}$$

Hint: It shifts the graph of $y = 2\sec x$ to the right by $\frac{\pi}{4}$ units.

4-15 Trigonometric Equations

A **trigonometric equation** is an equation involving trigonometric functions. We solve a trigonometric equation just as we solve any algebraic equation. To solve trigonometric equations, we use algebraic transformations and trigonometric formulas.

Examples

1. Determine whether $\theta = 30^o$ is a solution of the equation $\sin\theta = \frac{1}{2}$.

Solution:

$$\sin 30^o = \tfrac{1}{2} \quad \therefore \theta = 30^o \text{ is a solution of the equation.}$$

2. Determine whether $\theta = \pi/4$ is a solution of the equation $\cos\theta = \frac{\sqrt{3}}{2}$.

Solution:

$$\cos \pi/4 = \tfrac{\sqrt{2}}{2} \neq \tfrac{\sqrt{3}}{2} \quad \therefore \theta = \pi/4 \text{ is not a solution of the equation.}$$

3. Determine whether $x = 150^o$ is a solution of the equation $\sin x = \frac{1}{2}$.

Solution:

$$\sin 150^o = \sin 30^o = \tfrac{1}{2} \text{ (2nd qdt)}$$

$$\therefore x = 150^o \text{ is a solution of the equation.}$$

4. Solve $2\sin x = 1$ for $0 \le x \le 360^o$.

Solution:

$$2\sin x = 1, \quad \sin x = \tfrac{1}{2}$$

$$\therefore x = 30^o \text{ and } 150^o. \quad \text{or } x = \pi/6 \text{ and } 5\pi/6.$$

5. Solve $2\sin^2 x - \sin x = 1$ for $0 \le x \le 2\pi$.

Solution:

$$2\sin^2 x - \sin x = 1$$

$$2\sin^2 x - \sin x - 1 = 0$$

$$(2\sin x + 1)(\sin x - 1) = 0$$

$2\sin x + 1 = 0$	$\sin x - 1 = 0$
$\sin x = -\frac{1}{2}$	$\sin x = 1$
$x = 210^o,\ 330^o$	$x = 90^o$

Ans: $x = 90^o, 210^o, 330^o$.

or $x = \pi/2,\ 7\pi/6,\ 11\pi/6$

Examples

6. Solve $\cos x + 2\sin^2 x = 1$ for $0 \le x \le 360^o$.

Solution:

$$\cos x + 2\sin^2 x = 1$$
$$\cos x + 2(1-\cos^2 x) = 1$$
$$\cos x + 2 - 2\cos^2 x = 1$$
$$-2\cos^2 x + \cos x + 1 = 0$$
$$2\cos^2 x - \cos x - 1 = 0$$
$$(2\cos x + 1)(\cos x - 1) = 0$$

$2\cos x + 1 = 0$	$\cos x - 1 = 0$
$\cos x = -\frac{1}{2}$	$\cos x = 1$
$x = 120^o,\ 240^o$	$x = 0^o,\ 360^o$

Ans: $x = 0^o,\ 120^o,\ 240^o,\ 360^o$

or $x = 0,\ {}^{2\pi}/_{3},\ {}^{4\pi}/_{3},\ 2\pi$

7. Solve $\tan\theta = \cot\theta$ for $0 \le \theta \le 360^o$.

Solution:

$$\tan\theta = \cot\theta,\ \tan\theta = \frac{1}{\tan\theta},\ \tan^2\theta = 1,\ \therefore \tan\theta = \pm 1$$
$$\tan\theta = 1,\quad \theta = 45^o,\ 225^o$$
$$\tan\theta = -1,\quad \theta = 135^o,\ 315^o$$

Ans: $\theta = 45^o,\ 135^o,\ 225^o,\ 315^o$

or $\theta = {}^{\pi}/_{4},\ {}^{3\pi}/_{4},\ {}^{5\pi}/_{4},\ {}^{7\pi}/_{4}$

8. Find the general solution of $\cos 2x = 5\cos x + 2$.

Solution:

$$\cos 2x = 5\cos x + 2$$
$$2\cos^2 x - 1 = 5\cos x + 2$$
$$2\cos^2 x - 5\cos x - 3 = 0$$
$$(2\cos x + 1)(\cos x - 3) = 0$$

$2\cos x + 1 = 0$	$\cos x - 3 = 0$
$\cos x = -\frac{1}{2}$	$\cos x = 3$
$x = {}^{2\pi}/_{3},\ {}^{4\pi}/_{3}$	$\lvert\cos\rvert \le 1$, no solution

The cosine function has a period of 2π.

$\therefore$ The general solution is $x = {}^{2\pi}/_{3} + 2\pi \cdot n$ and $x = {}^{4\pi}/_{3} + 2\pi \cdot n$, where $n = 1 \cdot 2 \cdot 3 \cdots\cdot$.

4-16 Inverse Trigonometric Functions

We have defined the trigonometric functions for an acute angle.

Function: $y = \sin x$

Example: $y = \sin 30^o = \frac{1}{2}$

In order to represent an angle, we restrict the domain of $y = \sin x$ to the interval of $-\frac{\pi}{2} \le x \le \frac{\pi}{2}$ to define a one-to-one inverse function $x = \sin y$ by interchanging x and y and using the symbol of **inverse trigonometric function**.

Function: $y = \sin x$ where $-1 \le y \le 1$ and $-\frac{\pi}{2} \le x \le \frac{\pi}{2}$

Inverse Function: $y = \sin^{-1} x$ where $-1 \le x \le 1$ and $-\frac{\pi}{2} \le y \le \frac{\pi}{2}$

Example: $y = \sin^{-1}(\frac{1}{2}) = 30^o$

$\sin^{-1} x$ (or $\arcsin x$) is called " inverse sine of x " or " $\arcsin x$ ".

$y = \sin^{-1} x$ and $x = \sin y$ are equivalent and read " y is the angle whose sine equals x.".

$y = \sin^{-1} x$ and $y = \sin x$ are **inverses**. We say that $y = \sin^{-1} x$ means " y is the inverse sine of x "

The graphs of $y = \sin x$ and its inverse $y = \sin^{-1} x$ are reflection of each other about the line $y = x$.

$y = \sin x$ and $-1 \le y \le 1$, $-\frac{\pi}{2} \le x \le \frac{\pi}{2}$

$y = \sin^{-1} x$ and $-1 \le x \le 1$, $-\frac{\pi}{2} \le y \le \frac{\pi}{2}$

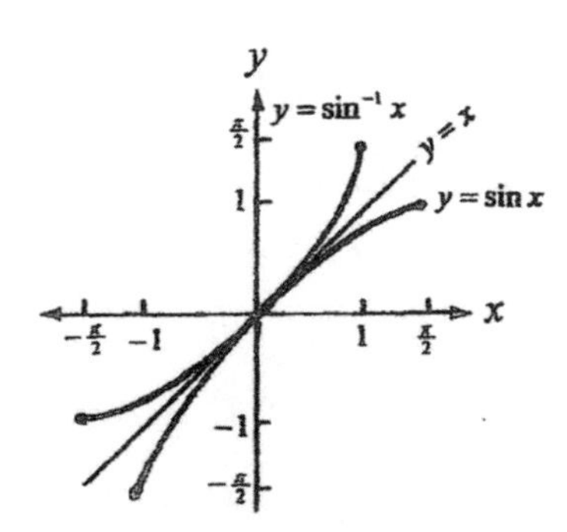

Notice that: $y = \sin^{-1} x$ and $x = \sin y$ are not inverses each other. They are only one same function in different forms. They are **equivalent**.

$y = \sin x$ and $y = \sin^{-1} x$ are inverses each other.

Similarly, an inverse function can be defined for each of the other five trigonometric functions.

$y = \cos x$ and $-1 \le y \le 1,\ 0 \le x \le \pi$

$y = \cos^{-1} x$ and $-1 \le x \le 1,\ 0 \le y \le \pi$

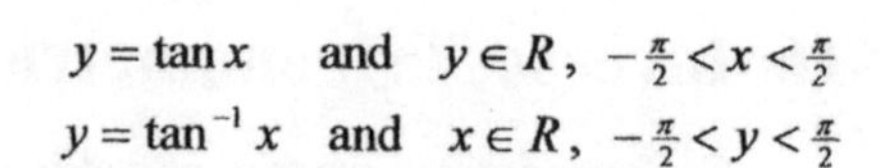

$y = \tan x$ and $y \in R,\ -\frac{\pi}{2} < x < \frac{\pi}{2}$

$y = \tan^{-1} x$ and $x \in R,\ -\frac{\pi}{2} < y < \frac{\pi}{2}$

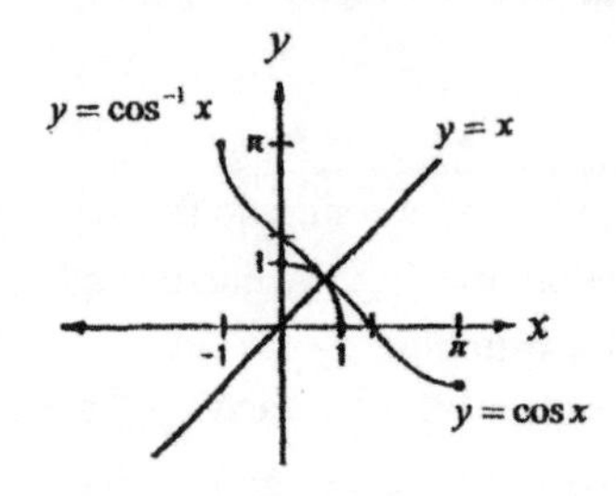

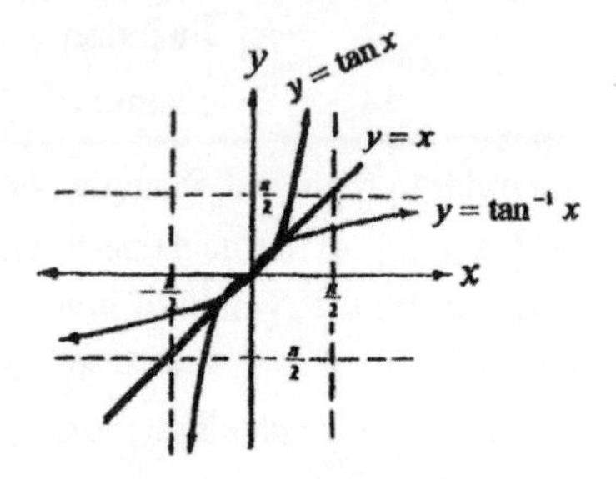

$y = \cot x$ and $y \in R,\ 0 < x < \pi$

$y = \cot^{-1} x$ and $x \in R,\ 0 < y < \pi$

$y = \sec x$ and $|y| \ge 1,\ 0 \le x \le \pi$

$y = \sec^{-1} x$ and $|x| \ge 1,\ 0 \le y \le \pi$

$\sec^{-1} x \ne \frac{\pi}{2}$

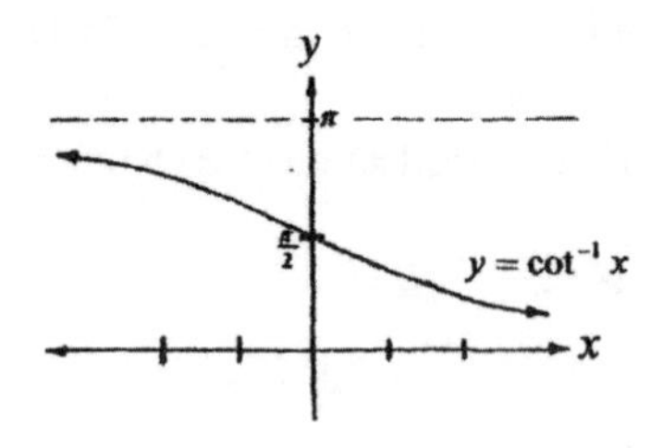

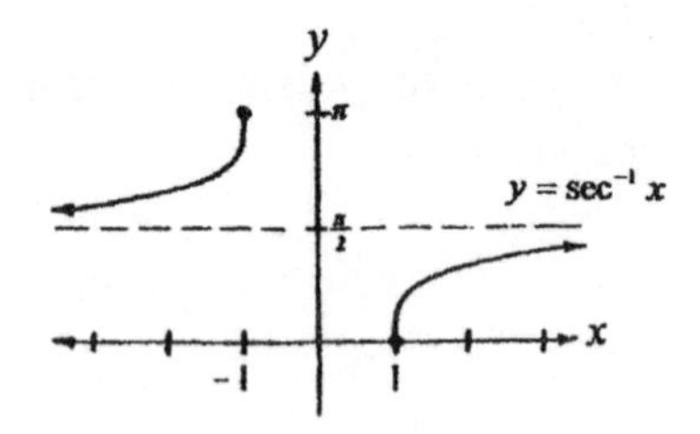

$y = \csc x$ and $|y| \ge 1,\ -\frac{\pi}{2} \le x \le \frac{\pi}{2}$

$y = \csc^{-1} x$ and $|x| \ge 1,\ -\frac{\pi}{2} \le y \le \frac{\pi}{2}$

$\csc^{-1} x \ne 0$

Notice that: 1. $\cot^{-1} x = \tan^{-1} \frac{1}{x}$, $\sec^{-1} = \cos^{-1} \frac{1}{x}$, $\csc^{-1} x = \sin^{-1} \frac{1}{x}$

2. Since $f^{-1}(f(x)) = x$ and $f(f^{-1}(x)) = x$, we have:

$\sin^{-1}(\sin x) = x$ and $\sin(\sin^{-1} x) = x$

$\cos^{-1}(\cos x) = x$ and $\cos(\cos^{-1} x) = x$

$\tan^{-1}(\tan x) = x$ and $\tan(\tan^{-1} x) = x$

The above rules apply to other 3 trigonometric functions as well.

Examples

Find the exact value of each expression

(Hint: Exact value means value with fraction, finite decimal, radical, and π .)

1. $\sin^{-1}(\frac{1}{2})$ **2.** $\sin^{-1}(-\frac{1}{2})$ **3.** $\cos^{-1}(-\frac{\sqrt{2}}{2})$ **4.** $\tan^{-1}(-1)$

5. $\sec^{-1}2$ **6.** $\csc^{-1}(-2)$

Solution: **1.** Let $y=\sin^{-1}(\frac{1}{2})$, we have

$\sin y=\frac{1}{2}$ and $-\frac{\pi}{2}\le y\le\frac{\pi}{2}$

$y=\frac{\pi}{6}$ is the only angle in the interval.

$\therefore \sin^{-1}(\frac{1}{2})=\frac{\pi}{6}$.

2. Let $y=\sin^{-1}(-\frac{1}{2})$, we have

$\sin y=-\frac{1}{2}$ and $-\frac{\pi}{2}\le y\le\frac{\pi}{2}$

$y=-\frac{\pi}{6}$ is the only angle in the interval.

$\therefore \sin^{-1}(-\frac{1}{2})=-\frac{\pi}{6}$.

3. Let $y=\cos^{-1}(-\frac{\sqrt{2}}{2})$, we have

$\cos y=-\frac{\sqrt{2}}{2}$ and $0\le y\le\pi$

$y=\frac{3\pi}{4}$ is the only angle in the interval.

$\therefore \cos^{-1}(-\frac{\sqrt{2}}{2})=\frac{3\pi}{4}$.

4. Let $y=\tan^{-1}(-1)$, we have

$\tan y=-1$ and $-\frac{\pi}{2}<y<\frac{\pi}{2}$

$y=-\frac{\pi}{4}$ is the only angle in the interval.

$\therefore \tan^{-1}(-1)=-\frac{\pi}{4}$.

5. Let $y=\sec^{-1}2$, we have

$\sec y=2$ and $0\le y\le\pi$

$y=\frac{\pi}{3}$ is the only angle in the interval.

$\therefore \sec^{-1}2=\frac{\pi}{3}$.

6. Let $y=\csc^{-1}(-2)$, we have

$\csc y=-2$ and $-\frac{\pi}{2}\le y\le\frac{\pi}{2}$

$y=-\frac{\pi}{6}$ is the only angle in the interval.

$\therefore \csc^{-1}(-2)=-\frac{\pi}{6}$.

7. Find the exact value of $\cos\left(\sin^{-1}(\frac{3}{5})\right)$.

Solution:

Let $\theta=\sin^{-1}(\frac{3}{5})$, we have

$\sin\theta=\frac{3}{5}$ and $-\frac{\pi}{2}\le\theta\le\frac{\pi}{2}$

$\theta\approx 37^{o}$ is the only angle in the interval.

By Pythagorean Theorem, we have

$x^2+y^2=r^2$

$x^2+3^2=5^2$

$x^2=16$

$\therefore x=4$

$\therefore \cos(\sin^{-1}(\frac{3}{5}))=\cos\theta=\frac{4}{5}$.

8. Find the exact value of $\sin(\sin^{-1}\frac{1}{2})$.

Solution:

$\sin(\sin^{-1}\frac{1}{2})=\sin\frac{\pi}{6}=\frac{1}{2}$.

9. Find the exact value of $\sin^{-1}(\sin\frac{7\pi}{4})$.

Solution:

$\sin^{-1}(\sin\frac{7\pi}{4})=\sin^{-1}(\sin(2\pi-\frac{\pi}{4}))$

$=\sin^{-1}(-\sin\frac{\pi}{4})=-\frac{\pi}{4}$.

10. Find the exact value of $\cos(\cos^{-1}\pi)$.

Solution: $y=\cos^{-1}x,\ -1\le x\le 1$

$\cos^{-1}\pi$ is not defined.

Therefore, $\cos(\cos^{-1}\pi)$ is not defined.

Examples (compare with Problems 1~6)

11. Find the approximate value of $\sin^{-1}(\frac{1}{2})$.
Solution:

Using a calculator: $\sin^{-1}(\frac{1}{2}) = 30^o \approx 0.52$ radians.

12. Find the approximate value of $\sin^{-1}(-\frac{1}{2})$.
Solution:

Using a calculator: $\sin^{-1}(-\frac{1}{2}) = -30^o \approx -0.52$ radians.

13. Find the approximate value of $\cos^{-1}(-\frac{\sqrt{2}}{2})$.
Solution:

Using a calculator: $\cos^{-1}(-\frac{\sqrt{2}}{2}) = 135^o \approx 2.36$ radians.

14. Find the approximate value of $\tan^{-1}(-1)$.
Solution:

Using a calculator: $\tan^{-1}(-1) = -45^o \approx -0.79$ radians.

15. Find the approximate value of $\sec^{-1} 2$.
Solution:

Using a calculator: $\sec^{-1} 2 = \cos^{-1}(\frac{1}{2}) = 60^o \approx 1.05$ radians.

16. Find the approximate value of $\csc^{-1}(-2)$.
Solution:

Using a calculator: $\csc^{-1}(-2) = \sin^{-1}(-\frac{1}{2}) = -30^o \approx -0.52$ radians.

17. Write the trigonometric expression as an algebraic expression.

$$\sin(\sin^{-1} x + \cos^{-1} x)$$

Solution: The formula: $\sin(u+v) = \sin u \cos v + \cos u \sin v$

Let $u = \sin^{-1} x$ and $v = \cos^{-1} x$
$\sin u = x$ $\cos v = x$

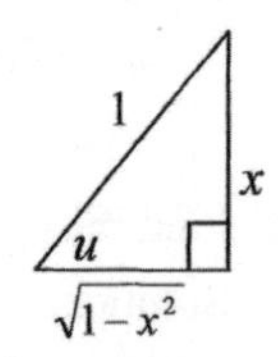

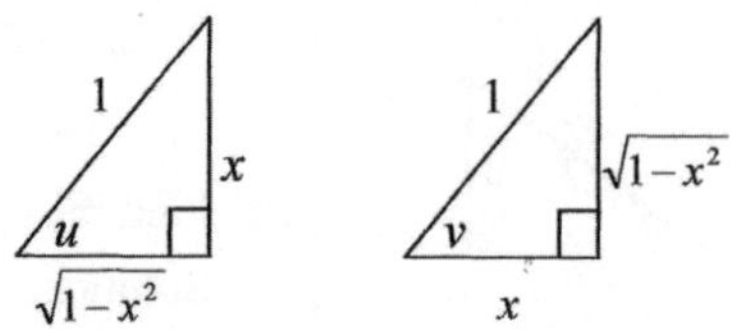

$$\sin(\sin^{-1} x + \cos^{-1} x) = \sin(\sin^{-1} x)\cos(\cos^{-1} x) + \cos(\sin^{-1} x)\sin(\cos^{-1} x)$$
$$= x \cdot x + \sqrt{1-x^2} \cdot \sqrt{1-x^2} = x^2 + (1-x^2)$$
$$=1.$$

PRACTICE TEST 1

(GRE Revised General Test)

SECTION 1

Quantitative Reasoning

Time—40 minutes

25 Questions

For each question, indicate the best answer, using the directions given.

Notes: All numbers used are real numbers.

All figures are assumed to lie in a plane unless otherwise indicated.

Geometric figures, such as lines, circles, triangles, and quadrilaterals, **are not necessarily** drawn to scale. That is, you should **not** assume that quantities such as lengths and angle measures are as they appear in a figure. You should assume, however, that lines shown as straight are actually straight, points on a line are in the order shown, and more generally, all geometric objects are in the relative positions shown. For questions with geometric figures, you should base your answers on geometric reasoning, not on estimating or comparing quantities by sight or by measurement.

Coordinate system, such as *xy*-planes and number lines, **are** drawn to scale; therefore, you can read, estimate, or compare quantities in such figures by sight or by measurement.

Graphical data presentations, such as bar graphs, circles graphs, and line graphs, **are** drawn to scale; therefore, you can read, estimate, or compare data values by sight or by measurement.

For each of Questions 1 to 9, compare Quantity A and Quantity B, using additional information centered above the two quantities if such information is given. Select one of the following four answer choices and fill in the corresponding oval to the right to the question.

Ⓐ Quantity A is greater.

Ⓑ Quantity B is greater.

Ⓒ The two quantities are equal.

Ⓓ The relationship cannot be determined from the information given.

A symbol that appears more than once in a question has the same meaning throughout the question.

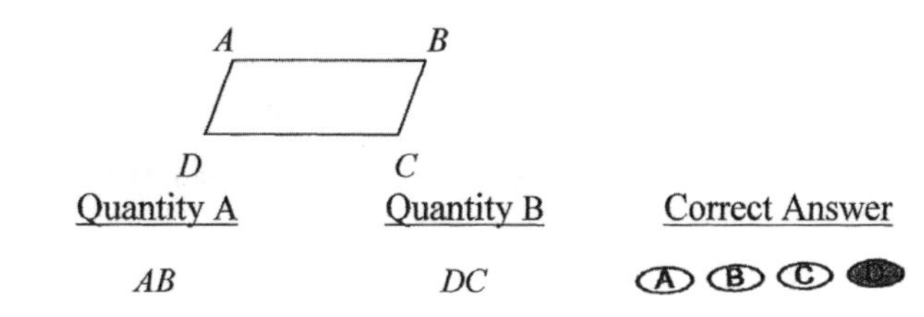

(A)	**Quantity A is greater.**
(B)	**Quantity B is greater.**
(C)	**The two quantities are equal.**
(D)	**The relationship cannot be determined from the information given.**

$$2a = b$$

	Quantity A	Quantity B	
1.	$\dfrac{a+1}{2}$	$\dfrac{b+2}{4}$	(A) (B) (C) (D)

	Quantity A	Quantity B	
2.	The perimeter of a square field with area 16 square meters.	The perimeter of a rectangular field with area 25 square meters.	(A) (B) (C) (D)

A rectangular field has an area 16.

	Quantity A	Quantity B	
3.	The perimeter of the field.	24	

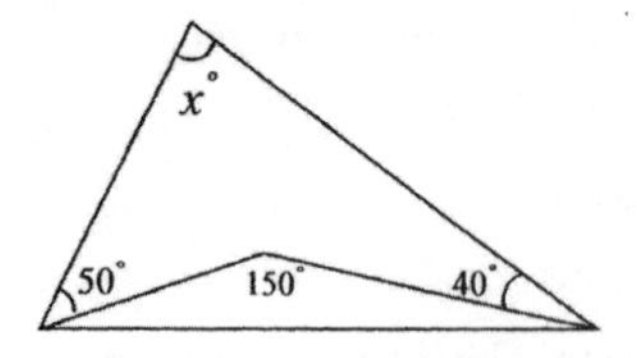

	Quantity A	Quantity B	
4.	x	65	(A) (B) (C) (D)

The average of 7, 8, and x is 14.

	Quantity A	Quantity B	
5.	$\dfrac{x+3}{2}$	15	(A) (B) (C) (D)

Ⓐ	**Quantity A is greater.**
Ⓑ	**Quantity B is greater.**
Ⓒ	**The two quantities are equal.**
Ⓓ	**The relationship cannot be determined from the information given.**

$$|x+1| = 8$$

	Quantity A	Quantity B	
6.	x	-10	 

In a sequence of 12 numbers, the first number is 2. The common difference is 1.

	Quantity A	Quantity B	
7.	The sum of the 12 numbers in the sequence	90	

In the figure, $\overline{AB}$ and $\overline{AC}$ are tangent to the circle at B and C, $OA = 5$, $OB = 3$.

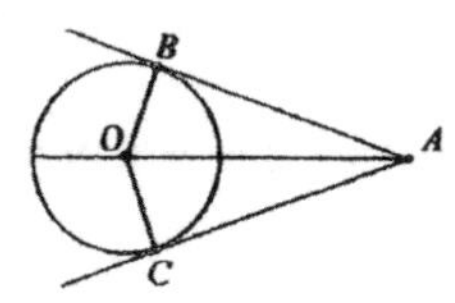

	Quantity A	Quantity B	
8.	$AB + AC$	10	  

In the figure, $\overline{DE}$ is five-sevenths the length of $\overline{DC}$.

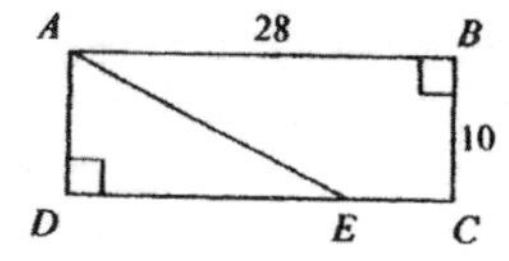

	Quantity A	Quantity B	
9.	The area of the quadrilateral $ABCE$	180	 

GO ON TO THE NEXT PAGE.

Equations 10 to 25 have several different formats. Unless otherwise directed, select a single answer choice. For Numeric Entry questions, follow the instructions below.

Numeric Entry Questions:
Enter your answer in the answer box(es) below the question.
1. **Your answer may be an integer, a decimal, or a fraction, and it may be negative.**
2. **If a question asks for a fraction, there will be two boxes —— one for the numerator and one for the denominator.**
3. **Equivalent forms of the correct answer, such as 2.5 and 2.50, are all correct. Fractions do not need to be reduced to lowest terms.**
4. **Enter the exact answer unless the equation asks you to round your answer.**

10. If $\frac{3}{5}$ of n is 27, what is $\frac{2}{5}$ of n ?

Ⓐ 15 Ⓑ 18 Ⓒ 21 Ⓓ 24 Ⓔ 25

11. If $\frac{y-4}{x+5}=0$, which of the following is true ?

Ⓐ $y=4$ and $x=-5$ Ⓑ $y=0$ and $x=-5$ Ⓒ $y=-4$ and $x\neq-5$

Ⓓ $y=0$ and $x=0$ Ⓔ $y=4$ and $x\neq-5$

12. In the figure below, $<A=90^\circ$, what is the area of ΔABC ?

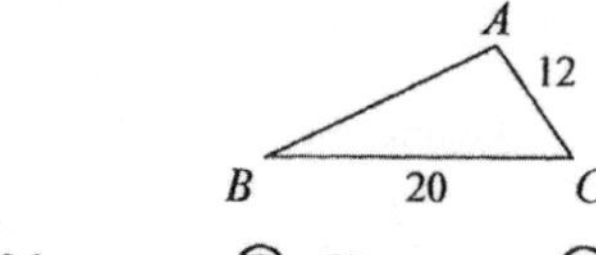

Ⓐ 96 Ⓑ 72 Ⓒ 48 Ⓓ 36 Ⓔ 24

For the following question, enter your answer in the box.

13. Roger withdrew two-fifths of the money in his bank account. Each of his three girl friends withdrew one-thirds of the balance. If Roger and one of his girl friends withdrew a total of 2,880 dollars from the bank account, what was the amount in dollars of the money in the bank account ?

[ ] dollars

14. There are 60 students in a class. 24 of them took Algebra. 33 of them took Geometry, and 6 of them took both courses. How many students took neither courses ?

Ⓐ 9 Ⓑ 7 Ⓒ 5 Ⓓ 4 Ⓔ 3

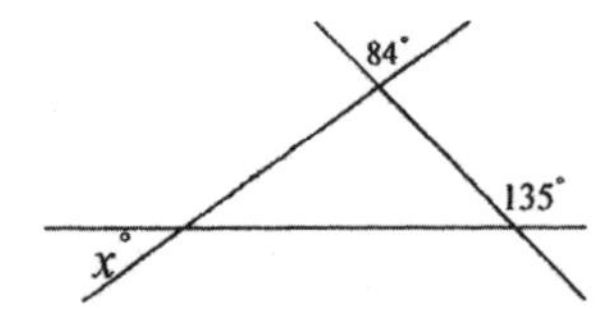

15. In the figure above, what is the measure of x ?

Ⓐ 42 Ⓑ 47 Ⓒ 51 Ⓓ 55 Ⓔ 59

16. If $x \neq 4$, then $\frac{x^2-16}{4} \div \frac{8-2x}{8} =$

Ⓐ $x+4$ Ⓑ $4-x$ Ⓒ $x-4$ Ⓓ $8-x$ Ⓔ $-4-x$

For the following two questions 17 and 18, select all the answer choices that apply.

$$|a+b| = |a| + |b|$$

17. Which of the following statements individually provide(s) sufficient additional information to determine that the absolute value of the sum of two real numbers a and b equals the sum of their absolute values as shown above ?

Indicate all such statements.

[A] $a > 0$ and $b > 0$ [B] $a < 0$ and $b < 0$
[C] $a > 0$ and $b < 0$ [D] $(a+b) > 0$

18. In a math class, there are more than three times as many boys as girls. The average (arithmetic mean) score for the boy is 75 and is 95 for the girls. Which of the following scores could be the average score for all of the students in the class ?

Indicate all such scores.

[A] 75 [B] 76 [C] 78 [D] 79 [E] 82 [F] 86

19. If $(a+b)^2 = 18$ and $ab = 4$, what is the value of $a^2 + b^2$?

Ⓐ 6 Ⓑ 8 Ⓒ 10 Ⓓ 12 Ⓔ 14

20. What is the length of the diameter of the circle whose equation is $45x^2 + 45y^2 = 180$?

Ⓐ 2 Ⓑ 4 Ⓒ 6 Ⓓ 8 Ⓔ 10

GO ON TO THE NEXT PAGE

21. A shepherd has two circular fields. The larger field has radius 4 times the radius of the smaller field. If the smaller field has an area M, then the area of the larger field has what amount greater than the area of the smaller field ?

Ⓐ 15 M Ⓑ 16 M Ⓒ 18 M Ⓓ 20 M Ⓔ 22 M

For the following question, enter your answer in the boxes.

22. A water current flows into an empty 36-gallon tank at a rate 3 gallons per hour. How many gallons per hour must flow out at the faucet on its bottom so that the tank is full in exactly 54 hours ?

Give your answer as a fraction. $\frac{\square}{\square}$

23. In three dimensional system, what is the distance between the points $(1, 3, -1)$ and $(4, 7, -6)$?

Ⓐ $4\sqrt{2}$ Ⓑ $5\sqrt{2}$ Ⓒ $6\sqrt{2}$ Ⓓ $7\sqrt{2}$ Ⓔ $8\sqrt{2}$

For the following question, enter your answer in the box.

24. What is the units digit of 3^{71} ?

Unit digit = $\square$

A-Plus Book Company
(Annual Sales of Books)

Years	Sold
2005	4,000 copies
2006	4,600
2007	5,200
2008	6,000
2009	6,900
2010	7,900

25. In the table above, which yearly period had the largest percent increase in book sold ?

Ⓐ 2005 ~ 2006 Ⓑ 2006 ~ 2007 Ⓒ 2007 ~ 2008

Ⓓ 2008 ~ 2009 Ⓔ 2009 ~ 2010

STOP: This is the end of Section 1.

PRACTICE TEST 1

(GRE Revised General Test)

SECTION 2

Quantitative Reasoning

Time—40 minutes

25 Questions

For each question, indicate the best answer, using the directions given.

Notes: All numbers used are real numbers.

All figures are assumed to lie in a plane unless otherwise indicated.

Geometric figures, such as lines, circles, triangles, and quadrilaterals, **are not necessarily** drawn to scale. That is, you should **not** assume that quantities such as lengths and angle measures are as they appear in a figure. You should assume, however, that lines shown as straight are actually straight, points on a line are in the order shown, and more generally, all geometric objects are in the relative positions shown. For questions with geometric figures, you should base your answers on geometric reasoning, not on estimating or comparing quantities by sight or by measurement.

Coordinate system, such as *xy*-planes and number lines, **are** drawn to scale; therefore, you can read, estimate, or compare quantities in such figures by sight or by measurement.

Graphical data presentations, such as bar graphs, circles graphs, and line graphs, **are** drawn to scale; therefore, you can read, estimate, or compare data values by sight or by measurement.

For each of Questions 1 to 9, compare Quantity A and Quantity B, using additional information centered above the two quantities if such information is given. Select one of the following four answer choices and fill in the corresponding oval to the right to the question.

Ⓐ Quantity A is greater.

Ⓑ Quantity B is greater.

Ⓒ The two quantities are equal.

Ⓓ The relationship cannot be determined from the information given.

A symbol that appears more than once in a question has the same meaning throughout the question.

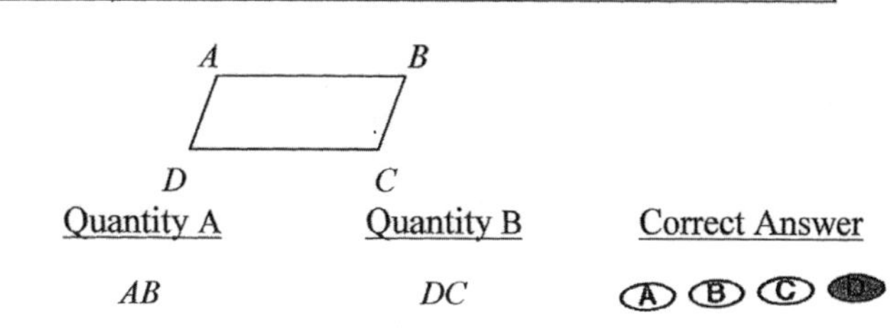

Example:

Ⓐ **Quantity A is greater.**
Ⓑ **Quantity B is greater.**
Ⓒ **The two quantities are equal.**
Ⓓ **The relationship cannot be determined from the information given.**

$x > 0$
$y > 0$

	Quantity A	Quantity B	
1.	$\frac{x+y}{y}$	$\frac{y}{x+y}$	Ⓐ Ⓑ Ⓒ Ⓓ

An equilateral triangle and a square have equal areas.

	Quantity A	Quantity B	
2.	The length of one side of the equilateral triangle	The length of one side of the square	Ⓐ Ⓑ Ⓒ Ⓓ

$ab > 0$

	Quantity A	Quantity B	
3.	$a+b$	ab	

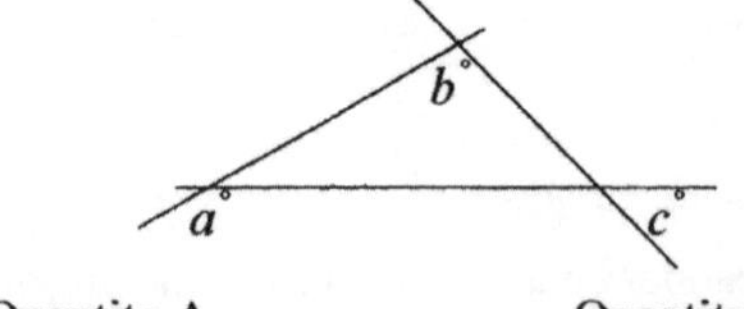

	Quantity A	Quantity B	
4.	c	$a-b$	

$0 < x < 1$
$0 < y < 1$

	Quantity A	Quantity B	
5.	$x^3 + y^3$	$x^2 + y^2$	Ⓐ Ⓑ Ⓒ Ⓓ

(A)	**Quantity A is greater.**
(B)	**Quantity B is greater.**
(C)	**The two quantities are equal.**
(D)	**The relationship cannot be determined from the information given.**

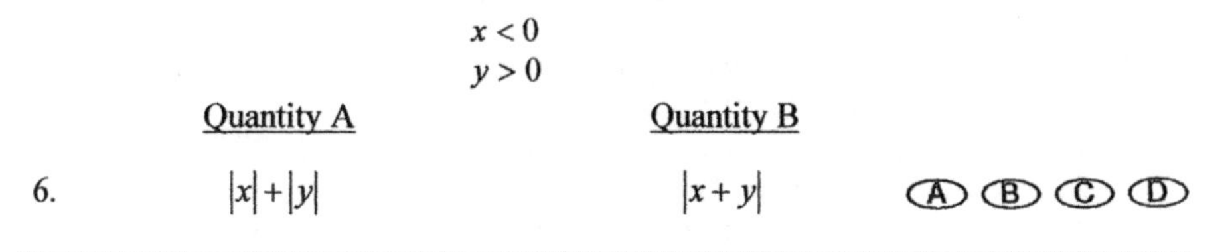

$x < 0$
$y > 0$

	Quantity A	Quantity B	
6.	$\lvert x\rvert + \lvert y\rvert$	$\lvert x + y\rvert$	(A) (B) (C) (D)

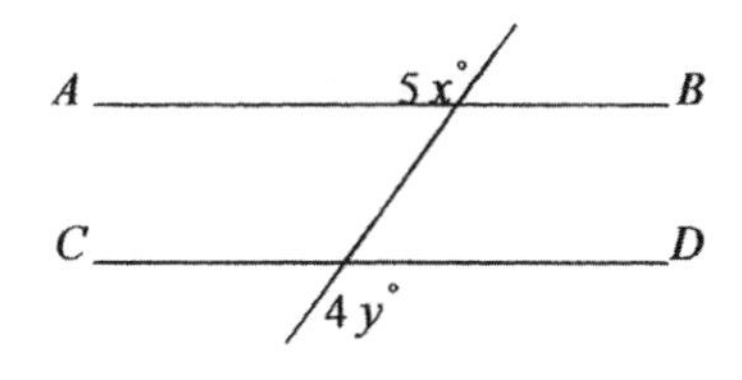

$AB // CD$

	Quantity A	Quantity B	
7.	x	y	(A) (B) (C) (D)

	Quantity A	Quantity B	
8.	$\dfrac{3^{45} - 3^{44}}{18}$	3^{42}	(A) (B) (C) (D)

In an experiment, the probability of operating life of a new brand battery is normally distributed with a mean of 22 hours and a standard deviation of 3 hours.

	Quantity A	Quantity B	
9.	The probability that a battery will last more than 28 hours	7 %	(A) (B) (C) (D)

GO ON TO THE NEXT PAGE

Equations 10 to 25 have several different formats. Unless otherwise directed, select a single answer choice. For Numeric Entry questions, follow the instructions below.

Numeric Entry Questions:
Enter your answer in the answer box(es) below the question.
1. **Your answer may be an integer, a decimal, or a fraction, and it may be negative.**
2. **If a question asks for a fraction, there will be two boxes —— one for the numerator and one for the denominator.**
3. **Equivalent forms of the correct answer, such as 2.5 and 2.50, are all correct. Fractions do not need to be reduced to lowest terms.**
4. **Enter the exact answer unless the equation asks you to round your answer.**

10. A car traveled at an average of 70 miles per hour for a trip in 7 hours. If it had traveled at an average of 50 miles per hour, how many minutes longer the trip would have taken ?

Ⓐ 65 Ⓑ 80 Ⓒ 124 Ⓓ 156 Ⓔ 168

11. In a computer store, you want to buy one of the five monitors, one of the three keyboards, and one of the four computers. If all of the choices are compatible, how many possible choices can you have ?

Ⓐ 20 Ⓑ 30 Ⓒ 40 Ⓓ 60 Ⓔ 80

12. If the following numeric expression is expressed as a scientific notation, how many nonzero digits does the decimal have ?

$$(2^{15})(5^{20})$$

Ⓐ One Ⓑ Two Ⓒ Three Ⓓ Four Ⓔ Five

For the following question, enter your answer in the box.

13. There are 8 students in a game. Each student plays exactly one game with each of the others. How many games are played ?

[ ] games

For the following question, select all the answer choices that apply.

14. If a, b, c, and d are different positive integers, a is a factor of b, b is a factor of c, c is a factor of d, which of the following statements must be true ?

Indicate all such statements.

[A] a is a factor of d. [B] b is a factor of $(d-c)$. [C] c is a factor of $(b+d)$.
[D] b is a factor of cd. [E] ad is a factor of $(c-b)$. [F] ac is a factor of d^2.

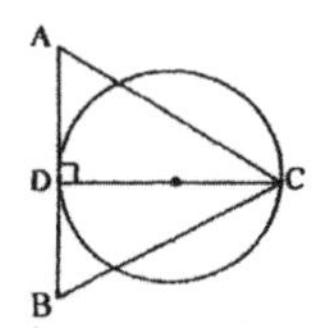

15. In the figure above, ΔABC is equilateral with side length 6. CD is the diameter of the circle. What is the area of the circle ?

Ⓐ $\frac{13}{2}\pi$ Ⓑ $\frac{27}{4}\pi$ Ⓒ 27π Ⓓ $\frac{15}{2}\pi$ Ⓔ 30π

16. Scientist tag 40 salmons in a river. Later, they caught 350 salmon. Of these, 5 have tags. Estimate how many salmons are in the river ?

Ⓐ 1,500 Ⓑ 2,000 Ⓒ 2,800 Ⓓ 3,000 Ⓔ 3,500

For the following question, enter your answer in the box.

17. Which of the following could be the reminder of $(4^{110} \div 7)$?

Reminder = ☐

For the following question, select all the answer choices that apply.

18. A plane intersects a rectangular solid. Which of the following could be the intersection ? Indicate all such intersections.

[A] A line [B] A triangle [C] A rectangle
[D] A parallelogram [E] A point

19. A line intersects a circle at points A and B. The radius of the circle is 4. What is the maximum possible length of the arc between A and B ?

Ⓐ 2 Ⓑ π Ⓒ 8 Ⓓ 4π Ⓔ 8π

For the following question, enter your answer in the boxes.

20. If $3^x + 3^x + 3^x = 27^x$, what is the value of x ?

Write your answer as a fraction. $\frac{☐}{☐}$

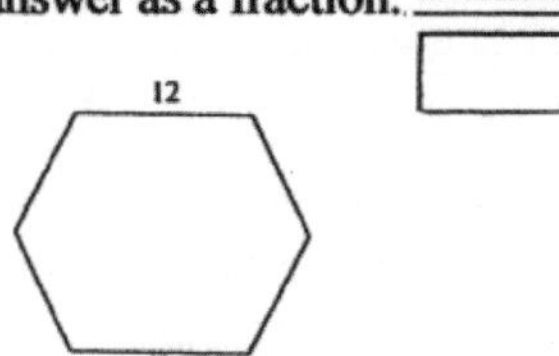

21. What is the area of the regular hexagon (6 sides) shown above ?

Ⓐ $27\sqrt{3}$ Ⓑ 54 Ⓒ $54\sqrt{3}$ Ⓓ $108\sqrt{3}$ Ⓔ $216\sqrt{3}$

GO ON TO THE NEXT PAGE.

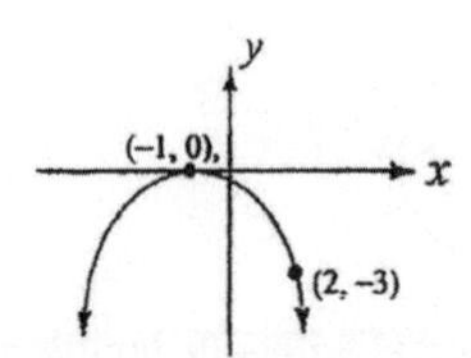

22. Which of the following is the graph of the equation of parabola above ?

Ⓐ $y=3(x-1)^2$ Ⓑ $y=-3(x-1)^2$ Ⓒ $y=\frac{1}{3}(x-1)^2$

Ⓓ $y=\frac{1}{3}(x+1)^2$ Ⓔ $y=-\frac{1}{3}(x+1)^2$

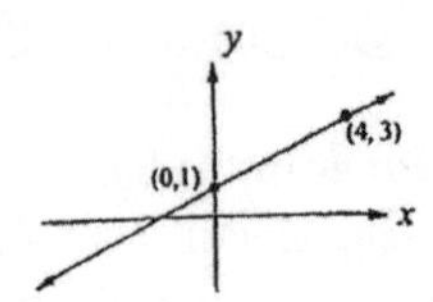

23. The figure above shows the graph of a straight line. If each point on the line is translated horizontally 3 units to the left, what is the value of y-intercept of the translating line ?

Ⓐ $\frac{3}{2}$ Ⓑ $-\frac{1}{2}$ Ⓒ 3 Ⓓ $-\frac{7}{2}$ Ⓔ 4

For the following question, select all the answer choices that apply.

24. Which of the following integers are multiples of 2, 3, and 9 ?

Indicate <u>all</u> such numbers.

[A] 18 [B] 27 [C] 36 [D] 54 [E] 72 [F] 84 [G] 90

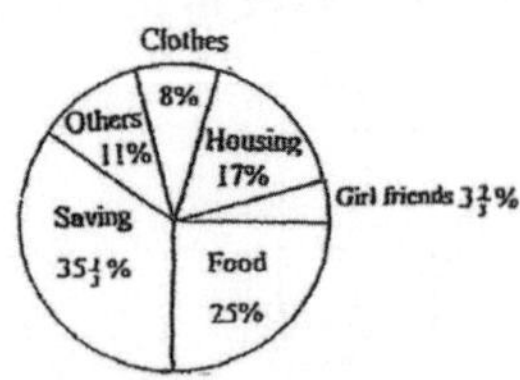

25. The graph above shows the Roger's budget to spend his money monthly. The total budget monthly is $4,800. How many items each comprised more than $720 monthly ?

Ⓐ One Ⓑ Two Ⓒ Three Ⓓ Four Ⓔ Five

STOP: This is the end of Section 1.

PRACTICE TEST 1
(GRE Revised General Test)

Page 349 ~ 360
Answer Key

Section 1: **1.** C **2.** B **3.** D **4.** B **5.** C **6.** A **7.** C **8.** B **9.** C **10.** B

11. E **12.** A **13.** 4800 **14.** A **15.** C **16.** E **17.** A and B **18.** B, C, and D

19. C **20.** B **21.** A **22.** $\frac{7}{3}$ **23.** B **24.** 7 **25.** C

Section 2: **1.** A **2.** A **3.** D **4.** C **5.** B **6.** A **7.** B **8.** C **9.** B **10.** E

11. D **12.** D **13.** 28 **14.** A, B, D, and F **15.** B **16.** C **17.** 2 **18.** B, C, D

19. D **20.** $\frac{1}{2}$ **21.** E **22.** E **23.** B **24.** A, C, D, E, and G **25.** C

Add the number of correct answers to obtain your raw math score.
Partially correct answers are considered as incorrect.
Using the table below to convert your raw score to scaled score. The scaled scores reflect degrees of difficult and is comparable with scores of different tests at different dates. Note that any test may be somewhat more or less difficult than the test shown in this table year by year.
The scaled scores for GRE revised General Test will be 130 ~ 170.

GRE REVISED GENERAL MATH TEST CONVERSION TABLE

Raw Score	Scaled Score	Raw Score	Scaled Score	Raw Score	Scaled Score
0	130	18	143	36	158
1	130	19	144	37	159
2	130	20	144	38	160
3	131	21	145	39	161
4	131	22	146	40	161
5	132	23	146	41	162
6	132	24	147	42	163
7	133	25	148	43	164
8	134	26	149	44	165
9	135	27	150	45	166
10	136	28	151	46	167
11	137	29	152	47	168
12	138	30	153	48	169
13	139	31	153	49	170
14	140	32	154	50	170
15	141	33	155		
16	142	34	156		
17	142	35	157		

Answer Explanations
(PRACTICE TEST 1)

Section 1:

1. C $\frac{a+1}{2}=\frac{a}{2}+\frac{1}{2}$ and $\frac{b+2}{4}=\frac{2a+2}{4}=\frac{a}{2}+\frac{1}{2}$. — They are equal.

2. B The perimeter of a square with area 16 must be 16. The perimeter of a rectangle with area 25 : $20 \le p < \infty$, such as $4(5)=20$, $2(25+1)=52$. — Quantity B is greater.

3. D The perimeter of a rectangular with area 16 must be $16 \le p < \infty$, such as $4(4)=16$, $2(16+1)=34$. — Cannot be determined.

4. B $180^o-150^o=30^o$, $x=180-50-40-30=60$. — Quantity B is greater.

5. C $\frac{7+8+x}{3}=14$, $x=27$. $\frac{x+3}{2}=\frac{27+3}{2}=15$. — They are equal.

6. A $|x+1|=8$, $x+1=\pm 8$, $x=7$ or -9. — Quantity A is greater.

7. C $a_{12}=a_1+(n-1)d=2+(12-1)(1)=13$. $S_{12}=\frac{n}{2}(a_1+a_{12})=\frac{12}{2}(2+13)=90$. — They are equal.

8. B $<ABO$ is a right angle. Apply Pythagorean Theorem. $(AB)^2+3^2=5^2$, $(AB)^2=16$, $AB=AC=4$, $AB+AC=8$. — Quantity B is greater.

9. C $EC=28\times\frac{2}{7}=8$, Area of $ABCE=\frac{1}{2}(28+8)(10)=180$. — They are equal.

10. B $\frac{3}{5}n=27$, $n=27\times\frac{5}{3}=45$, $45\times\frac{2}{5}=18$.

11. E $y-4=0$ and $x+5\neq 0$, therefore $y=4$, $x\neq -5$.

12. A Solve $\overline{AB}^2+12^2=20^2$, we have $AB=16$(base), Area $=\frac{1}{2}bh=\frac{1}{2}(16))(12)=96$.

13. 4800 Solve $\frac{2}{5}n+\frac{1}{3}(\frac{3}{5}n)=2880$, we have $n=4800$.

14. A Draw a Venn Diagram (see page 61). $(24+33)-6=51$, $60-51=9$.

15. C $180^\circ-135^\circ=45^\circ$, $x=180-45-84=51$.

16. E $\frac{x^2-16}{4}\div\frac{8-2x}{8}=\frac{(x+4)(x-4)}{4}\cdot\frac{8}{2(4-x)}=\frac{(x+4)(x-4)}{1}\cdot\frac{1}{-(x-4)}=-x-4$.

17. A and B Plug in real numbers to test the results.

18. B, C, and D Since there are more boys with lower average score, the average score for all of the students in the class must be greater than 75. The greatest possible score of the class is : $\frac{75(3)+95(1)}{4}=\frac{225+95}{4}=80$, therefore $75<$ (average score) <80.

19. C $(a+b)^2=a^2+2ab+b^2=a^2+2(4)+b^2=18$, $a^2+b^2=18-8=10$.

20. B $45x^2+45y^2=180$, $x^2+y^2=2^2$, $r=2$ (radius), $D=4$ (diameter).

21. A $\pi(4r)^2-\pi(r^2)=15\pi r^2=15M$.

22. $\frac{7}{3}$ There are $3\times 54=162$ gallons flow in. $162-36=126$ gallons must flow out in 54 hours. $126\div 54=\frac{126}{54}=\frac{7}{3}$ gallons per hour must flow out.

23. B $d=\sqrt{(4-1)^2+(7-3)^2+(-6+1)^2}=\sqrt{9+16+25}=\sqrt{50}=5\sqrt{2}$.

Answer Explanations
(PRACTICE TEST 1)

Section 1: 24. 7 The units digits of 3^1, 3^2, 3^3, and 3^4 are 3, 9, 7, 1.
The units digits of 3^5, 3^6, 3^7, and 3^8 are 3, 9, 7, 1.
18th row: 3^{69}, 3^{70}, 3^{71}, and 3^{72} are 3, 9, 7, 1.
The pattern of 4 units digits 3, 9, 7, 1, repeats without end.
The exponents of the 1st. number of 18th row is: $1+4(18-1)=69$.

25. C Find all of the percent increases, the largest is $\frac{6000-5200}{5200}=15.38$ %.

Section 2: 1. A $\frac{x+y}{y}=\frac{x}{y}+1>1$, $\frac{y}{x+y}=\frac{1}{\frac{x}{y}+1}<1$. Quantity A is greater.

2. A Area of an equilateral triangle = Area of a square Quantity A is greater.
$\frac{\sqrt{3}}{4}S_1^{\,2}=S_2^{\,2}$, $S_1>S_2$

3. D $ab>0$ gives a and b are both positive, or negative. Cannot be determined.
$a+b$ could be positive or negative. ab is positive.
Such as: $1+1>(1)(1)$, $-1-1<(-1)(-1)$, $2+2=(2)(2)$.

4. C $a=b+c$, $a-b=c$. They are equal.

5. B Plug in decimals to test the result. $(0.9)^3<(0.9)^2$. Quantity B is greater.

6. A Plug in real numbers to test the result. $|-5|+|4|>|-5+4|$. Quantity A is greater.

7. B $5x^\circ=4y^\circ$, $y^\circ>x^\circ$. Quantity B is greater.

8. C $\frac{3^{45}-3^{44}}{18}=\frac{3^{44}(3-1)}{18}=\frac{3^{44}}{9}=\frac{3^{44}}{3^2}=3^{42}$. They are equal.

9. B (See normal distribution on page 72.) Quantity B is greater.
The probability of an event having the value of
1 standard deviation above the mean is about 16 %,
2 standard deviation above the mean is about 2 %,
3 standard deviation above the mean is about 0.5 %.
28 hours = 22 + 2 (3), it is 2 standard deviations above the mean.

10. E $\frac{70\times7}{50}=9.8$, $9.8-7=2.8$ hours = 168 minutes.

11. D $5\times3\times4=60$.

12. D $(2^{15})(5^{20})=(2^{15})(5^{15})(5^5)=10^{15}\times3125=10^{15}\times3.125\times10^3=3.125\times10^{18}$.

13. 28

Students	1	2	3	4	5	6	7	8
Games	0	1	3	6	10	15	21	28
		+1	+2	+3	+4	+5	+6	+7

14. A, B, D, and F Plug in four different positive integers to test the results, such as a, b, c, d = 2, 4, 8, 16.

Answer Explanations
(PRACTICE TEST 1)

Section 2: 15. B diameter $d = CD$, $d^2 + 3^2 = 6^2$, $d = \sqrt{27} = 3\sqrt{3}$, radius $r = \frac{3\sqrt{3}}{2}$.

Area $= \pi \cdot \left(\frac{3\sqrt{3}}{2}\right)^2 = \frac{27}{4}\pi$.

16. C $40 \times \frac{350}{5} = 2800$

17. 2 The reminders of $4^2 \div 7$, $4^3 \div 7$, $4^4 \div 7$ are 2, 1, 4.
The reminders of $4^5 \div 7$, $4^6 \div 7$, $4^7 \div 7$ are 2, 1, 4.
The reminders of $4^{110} \div 7$, $4^{111} \div 7$, $4^{112} \div 7$ are 2, 1, 4. → 37th row
The pattern of 3 reminders 2, 1, 4, repeats without end.
The exponent of the 1st number of last row is $2 + 3(37 - 1) = 110$.

18. B, C, and D They are a rectangle, a parallelogram, and a triangle. If the plane slices through three adjacent faces at a corner of the rectangular solid, the intersection is a triangle.

19. D If the line between A and B is the diameter of the circle, the arc between A and B has the maximum length. Arc $AB = \frac{1}{2} \times 2\pi(4) = 4\pi$.

20. ½ $3^x + 3^x + 3^x = 27^x$, $3(3^x) = 3^{3x}$, $3^{1+x} = 3^{3x}$, $1 + x = 3x$, $x = \frac{1}{2}$.

21. E The sum of the interior angles $= 180^\circ(6-2) = 720^\circ$.
$< OAB = \frac{1}{2}(\frac{720^\circ}{6}) = 60^\circ$, ΔOAC is a 30-60-90 triangle.
$AC = 6$. The apothem $a = 6\sqrt{3}$.
Area $= \frac{1}{2}ap = \frac{1}{2}(6\sqrt{3})(72) = 216\sqrt{3}$.

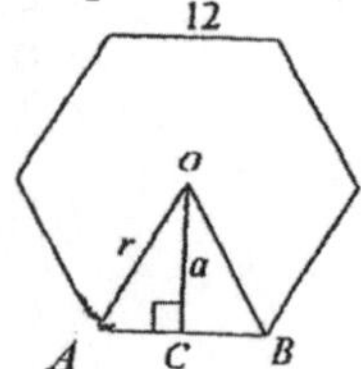

22. E The parabola $y - k = a(x-h)^2$ has vertex (−1, 0), open downward $a < 0$.
Therefore : $y = a(x+1)^2$, passing (2, -3), $-3 = a(2+1)^2$, $-3 = 9a$, $a = -\frac{1}{3}$.

23. B Slope $m = \frac{3-1}{4-0} = \frac{1}{2}$, y-intercept = 1, the equation of the line is $y = \frac{1}{2}x + 1$.
The translating line is $y = \frac{1}{2}(x-3) + 1 = \frac{1}{2}x - \frac{1}{2}$. y-intercept = $-\frac{1}{2}$.

24. A, C, D, E, and G
The least common multiple of 2, 3, and 9 is 18.
18, 36, 54, 72, and 90 are the multiples of 18.

25. C $\frac{720}{4800} = 0.15 = 15\%$. 17%, 25%, and $35\frac{1}{3}\%$ are larger than 15%.

PRACTICE TEST 2
(GRE Revised General Test)

SECTION 1
Quantitative Reasoning
Time—40 minutes
25 Questions

For each question, indicate the best answer, using the directions given.

Notes: All numbers used are real numbers.

All figures are assumed to lie in a plane unless otherwise indicated.

Geometric figures, such as lines, circles, triangles, and quadrilaterals, **are not necessarily** drawn to scale. That is, you should **not** assume that quantities such as lengths and angle measures are as they appear in a figure. You should assume, however, that lines shown as straight are actually straight, points on a line are in the order shown, and more generally, all geometric objects are in the relative positions shown. For questions with geometric figures, you should base your answers on geometric reasoning, not on estimating or comparing quantities by sight or by measurement.

Coordinate system, such as *xy*-planes and number lines, **are** drawn to scale; therefore, you can read, estimate, or compare quantities in such figures by sight or by measurement.

Graphical data presentations, such as bar graphs, circles graphs, and line graphs, **are** drawn to scale; therefore, you can read, estimate, or compare data values by sight or by measurement.

For each of Questions 1 to 9, compare Quantity A and Quantity B, using additional information centered above the two quantities if such information is given. Select one of the following four answer choices and fill in the corresponding oval to the right to the question.

(A) Quantity A is greater.

(B) Quantity B is greater.

(C) The two quantities are equal.

(D) The relationship cannot be determined from the information given.

A symbol that appears more than once in a question has the same meaning throughout the question.

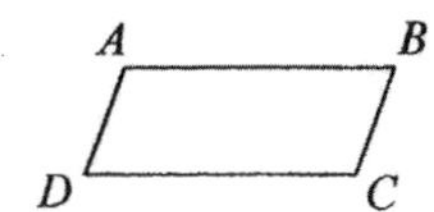

	Quantity A	Quantity B	Correct Answer
Example:	*AB*	*DC*	  

Ⓐ	**Quantity A is greater.**
Ⓑ	**Quantity B is greater.**
Ⓒ	**The two quantities are equal.**
Ⓓ	**The relationship cannot be determined from the information given.**

In the coordinate plane, the distance between the point $p(x, y)$ is exactly 10 units from the origin.

	Quantity A	Quantity B	
1.	x	y	Ⓐ Ⓑ Ⓒ Ⓓ

$$|3x-2|<10$$

	Quantity A	Quantity B	
2.	x	$-\frac{10}{3}$	Ⓐ Ⓑ Ⓒ Ⓓ

$$|3x-2|>10$$

	Quantity A	Quantity B	
3.	x	$-\frac{10}{3}$	Ⓐ Ⓑ Ⓒ Ⓓ

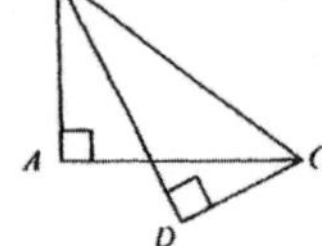

	Quantity A	Quantity B	
4.	$(AB)^2+(AC)^2$	$(DB)^2+(DC)^2$	Ⓐ Ⓑ Ⓒ Ⓓ

$$x=1+\frac{1}{1+\frac{1}{2}}$$

	Quantity A	Quantity B	
5.	x	$\frac{5}{3}$	Ⓐ Ⓑ Ⓒ Ⓓ

Ⓐ	**Quantity A is greater.**
Ⓑ	**Quantity B is greater.**
Ⓒ	**The two quantities are equal.**
Ⓓ	**The relationship cannot be determined from the information given.**

A right circular cylinder has a base radius 5 feet and has a volume of 200π cubic feet.

	Quantity A	Quantity B	
6.	The height of the cylinder.	8 feet	 Ⓐ Ⓑ Ⓒ Ⓓ

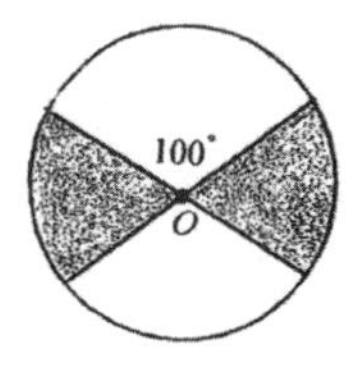

The circle above has a radius 6.

	Quantity A	Quantity B	
7.	The area of the two shaded regions	18π	Ⓐ Ⓑ Ⓒ Ⓓ

$a > 3$

	Quantity A	Quantity B	
8.	$8a-4$	$5a+7$	Ⓐ Ⓑ Ⓒ Ⓓ

	Quantity A	Quantity B	
9.	The maximum area of a field that can be built by the 48-feet fence.	$145\ ft^2$	Ⓐ Ⓑ Ⓒ Ⓓ

GO ON TO THE NEXT PAGE.

Equations 10 to 25 have several different formats. Unless otherwise directed, select a single answer choice. For Numeric Entry questions, follow the instructions below.

Numeric Entry Questions:
Enter your answer in the answer box(es) below the question.
1. **Your answer may be an integer, a decimal, or a fraction, and it may be negative.**
2. **If a question asks for a fraction, there will be two boxes —— one for the numerator and one for the denominator.**
3. **Equivalent forms of the correct answer, such as 2.5 and 2.50, are all correct. Fractions do not need to be reduced to lowest terms.**
4. **Enter the exact answer unless the equation asks you to round your answer.**

10. x is a number on the number line between 8 and 18 that is twice as far from 8 as from 18. What is the number ?

    Ⓐ 10　Ⓑ $10\frac{2}{3}$　Ⓒ 11　Ⓓ $13\frac{1}{3}$　Ⓔ $14\frac{2}{3}$

For the following question, enter your answer in the box.

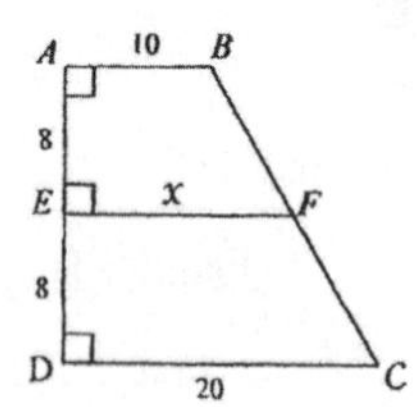

11. In the figure above, what is the length of x ?

    $x =$ ☐

12. If five students sit in a row, how many different seat arrangements are possible if two of them want to sit next to each other ?

    Ⓐ 24　Ⓑ 30　Ⓒ 36　Ⓓ 40　Ⓔ 48

13. Which of the following integers is not a factor of 28! ?

    Ⓐ 30　Ⓑ 54　Ⓒ 62　Ⓓ 69　Ⓔ 96

For the following question, select all the answer choices that apply.

14. Which of the following could be the units digit of 98^n, where n is a positive integer ?

    Indicate <u>all</u> such digits.

    [A] 2　[B] 3　[C] 4　[D] 6　[E] 7　[F] 8　[G] 9

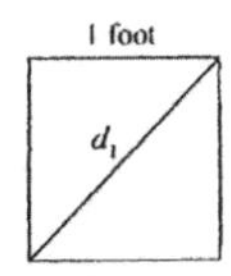

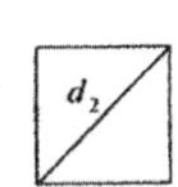

15. In the figure above, the area of the larger square is twice the area of the smaller square. The diagonal of the larger square is how many feet longer than the diagonal of the smaller square ?

Ⓐ $\sqrt{2}-\frac{1}{2}$ Ⓑ $\sqrt{2}-1$ Ⓒ $\sqrt{2}+\frac{1}{2}$ Ⓓ $\sqrt{2}+1$ Ⓔ $\frac{\sqrt{2}}{2}$

For the following question, select all the answer choices that apply.

16. Which of the following x-values could be used to show that the statement "$3^x > 2^x$" is not true ?
Indicates all such values.

[A] $x=-1$ [B] $x=-0.5$ [C] $x=0$ [D] $x=0.5$ [E] $x=1$

For the following question, enter your answer in the box.

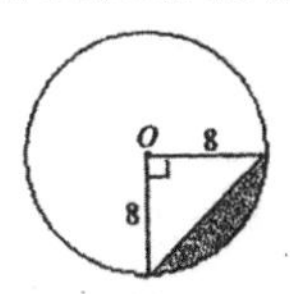

17. In the circle above, what is the area of the shaded region ? ($\pi \approx 3.14159$)

Give your answer to the nearest 0.01 of a decimal.

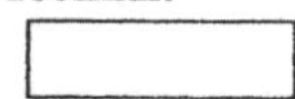

18. A bag contains 5 red balls and 4 white balls. Two balls are drawn at random from the bag. The first ball drawn is not put back into the bag before the second ball is drawn. What is the probability that the two balls drawn are both red ?

Ⓐ $\frac{4}{8}$ Ⓑ $\frac{4}{9}$ Ⓒ $\frac{5}{9}$ Ⓓ $\frac{5}{18}$ Ⓔ $\frac{25}{81}$

19. A bag contains 5 red balls and 4 white balls. Two balls are drawn together at random from the bag. What is the probability that the two balls drawn are of the same color ?

Ⓐ $\frac{4}{8}$ Ⓑ $\frac{4}{9}$ Ⓒ $\frac{5}{9}$ Ⓓ $\frac{5}{18}$ Ⓔ $\frac{25}{81}$

For the following question, enter your answer in the box.

20. There are 20 bulbs in a box and 5 are defective. You select 3 bulbs from the box. What is the probability that all three are defective ?
Give your answer to the nearest 0.1 percent.

[ ] %

GO ON TO THE NEXT PAGE.

21. John completes two-fifths of the questions in 28 minutes. At the same rate, how many minutes does he need to complete all of the questions ?

Ⓐ 52 Ⓑ 64 Ⓒ 70 Ⓓ 73 Ⓔ 78

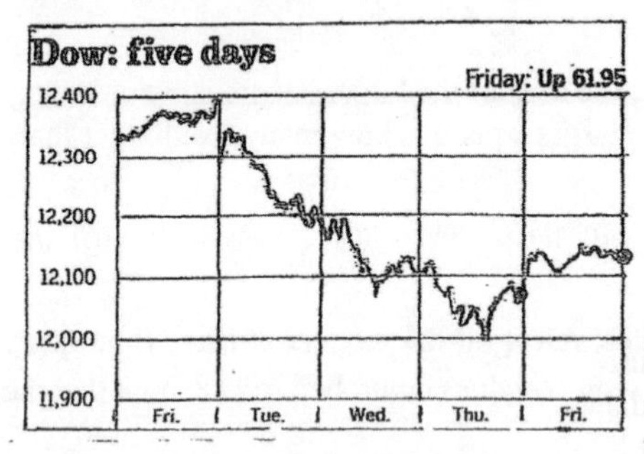

For the following question, enter your answer in the box.

22. The graph above shows that the Dow Jones industrial average of stocks rose from Thursday to close at 12,130.45 on Friday. What is the percent increase from Thursday to Friday ?

Give your answer to the nearest 0.01 percent.

 %

23. If a line passes through the origin and the point (a, 6), and is perpendicular to the line $y=-\frac{1}{2}x+4$. What is the value of a ?

Ⓐ 1 Ⓑ 2 Ⓒ 3 Ⓓ 4 Ⓔ 5

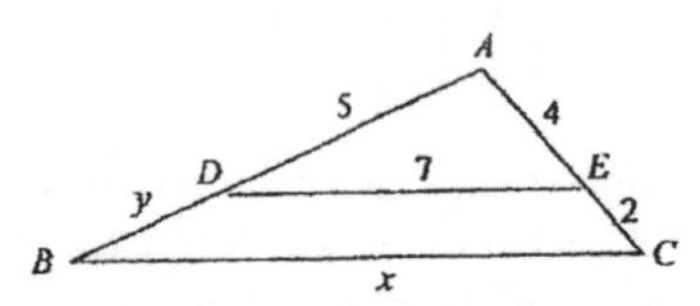

24. In the figure about, $DE//BC$. What is the value of $x+y$?

Ⓐ 12 Ⓑ 13 Ⓒ 14 Ⓓ 15 Ⓔ 16

For the following question, enter your answer in the boxes.

25. Rectangular solid M has length 12, width 6, and height 9. Rectangular solid N has length 8, width 4, and height 6. The volume of N is what fraction of the volume of M?

$$\frac{\square}{\square}$$

STOP: This is the end of Section 1.

PRACTICE TEST 2

(GRE Revised General Test)

SECTION 2

Quantitative Reasoning

Time—40 minutes

25 Questions

For each question, indicate the best answer, using the directions given.

Notes: All numbers used are real numbers.

All figures are assumed to lie in a plane unless otherwise indicated.

Geometric figures, such as lines, circles, triangles, and quadrilaterals, **are not necessarily** drawn to scale. That is, you should **not** assume that quantities such as lengths and angle measures are as they appear in a figure. You should assume, however, that lines shown as straight are actually straight, points on a line are in the order shown, and more generally, all geometric objects are in the relative positions shown. For questions with geometric figures, you should base your answers on geometric reasoning, not on estimating or comparing quantities by sight or by measurement.

Coordinate system, such as xy-planes and number lines, **are** drawn to scale; therefore, you can read, estimate, or compare quantities in such figures by sight or by measurement.

Graphical data presentations, such as bar graphs, circles graphs, and line graphs, **are** drawn to scale; therefore, you can read, estimate, or compare data values by sight or by measurement.

For each of Questions 1 to 9, compare Quantity A and Quantity B, using additional information centered above the two quantities if such information is given. Select one of the following four answer choices and fill in the corresponding oval to the right to the question.

Ⓐ Quantity A is greater.

Ⓑ Quantity B is greater.

Ⓒ The two quantities are equal.

Ⓓ The relationship cannot be determined from the information given.

A symbol that appears more than once in a question has the same meaning throughout the question.

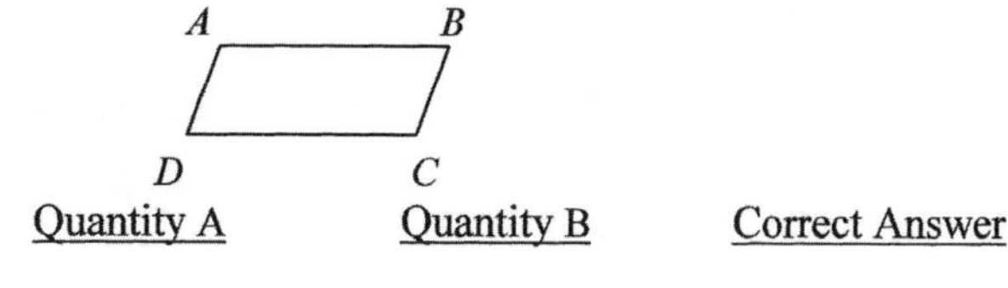

	Quantity A	Quantity B	Correct Answer
Example:	AB	DC	

(A) Quantity A is greater.
(B) Quantity B is greater.
(C) The two quantities are equal.
(D) The relationship cannot be determined from the information given.

$n > 0$

	Quantity A	Quantity B	
1.	n increased by 300 %.	$4n$	(A) (B) (C) (D)

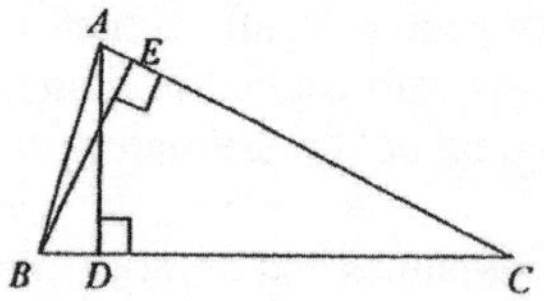

$AC > BC$

	Quantity A	Quantity B	
2.	AD	BE	(A) (B) (C) (D)

	Quantity A	Quantity B	
3.	$(4x-1)(4x+1)$	$16x^2$	(A) (B) (C) (D)

$0 < c < b < a$

	Quantity A	Quantity B	
4.	$a-b$	$b-c$	(A) (B) (C) (D)

$0 < c < b < a$

	Quantity A	Quantity B	
5.	$\frac{a}{b}$	$\frac{b}{c}$	(A) (B) (C) (D)

(A)	**Quantity A is greater.**
(B)	**Quantity B is greater.**
(C)	**The two quantities are equal.**
(D)	**The relationship cannot be determined from the information given.**

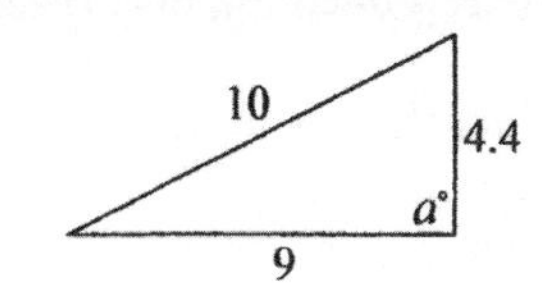

	Quantity A	Quantity B	
6.	a	90	(A) (B) (C) (D)

An equilateral polygon *ABCDE* is inscribed in a circle.

	Quantity A	Quantity B	
7.	The length of arc *ABC*	The length of arc *BCD*	(A) (B) (C) (D)

$$|n| = 18$$
$$|n-1| = 19$$

	Quantity A	Quantity B	
8.	n	18	(A) (B) (C) (D)

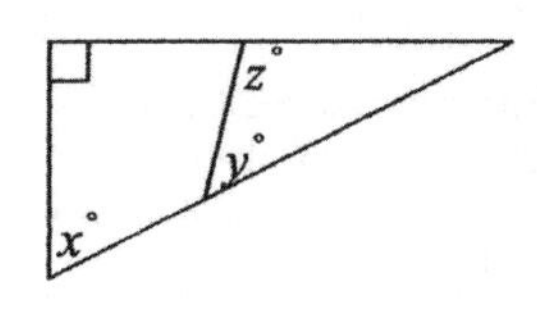

	Quantity A	Quantity B	
9.	$y+z$	$90+x$	(A) (B) (C) (D)

GO ON TO THE NEXT PAGE.

Equations 10 to 25 have several different formats. Unless otherwise directed, select a single answer choice. For Numeric Entry questions, follow the instructions below.

Numeric Entry Questions:
Enter your answer in the answer box(es) below the question.
1. **Your answer may be an integer, a decimal, or a fraction, and it may be negative.**
2. **If a question asks for a fraction, there will be two boxes —— one for the numerator and one for the denominator.**
3. **Equivalent forms of the correct answer, such as 2.5 and 2.50, are all correct. Fractions do not need to be reduced to lowest terms.**
4. **Enter the exact answer unless the equation asks you to round your answer.**

For the following question, enter your answer in the box.

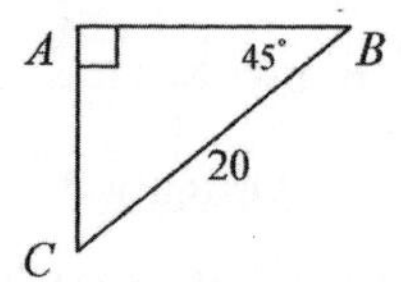

10. In the figure above, what is the area of ΔABC ?

11. In an alcohol-water solution, the ratio by volume of alcohol to water is 2 to 8. How many liters of pure alcohol will be in 45 liters of the solution ?

Ⓐ 9 Ⓑ 10 Ⓒ 11 Ⓓ 12 Ⓔ 13

12. A 30-gallon alcohol-water solution contains 20 % alcohol. How much alcohol in gallons should be added to produce the solution to 25 % alcohol ?

Ⓐ 5 Ⓑ 4 Ⓒ 3 Ⓓ 2 Ⓔ 1

For the following question, enter your answer in box.

13. If $4^8 \times 16^{20} = 4^{2x}$, what is the value of x ?

$x =$ ☐

For the following question, select all the answer choices that apply.

14. In the xy-plane, the linear function $f(x) = 2x + 3$ is defined for all numbers x. Which of the following functions must its graph intersect the graph of $f(x)$?
Indicate all such functions.

[A] $g(x) = 2x - 3$ [B] $h(x) = 2x - 4$ [C] $k(x) = -\frac{1}{2}x + 3$

[D] Line $q(x)$ passes through the points (a, b) and (c, d), where $(a-c)(b-d) < 0$.

[E] Line $r(x)$ passes through the points (a, b) and (c, d), where $(a-c)(b-d) > 0$.

15. What is the sum of $1+2+3+\cdots\cdots+100$?

Ⓐ 3,050 Ⓑ 4,050 Ⓒ 5.050 Ⓓ 6,050 Ⓔ 7.050

16. $(100001)^2 - 2(100000)^2 + (99999)^2 =$

Ⓐ 4 Ⓑ 3 Ⓒ 2 Ⓓ 1 Ⓔ 0

17. What is the maximum number of 4-digit codes can be made from 0 through 9 if no 0 is used for the first digit ?

Ⓐ 8,000 Ⓑ 9,000 Ⓒ 10,000 Ⓓ 11,000 Ⓔ 12,000

For the following question, enter your answer in the boxes.

18. The population of males of a city increased 11% and females decreased 20% this year. The ratio of males to females at the end of this year was how many times the ratio at the beginning of the year ?

Give your answer as a fraction. 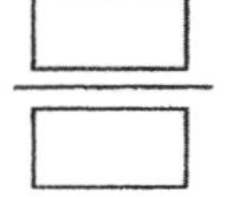

For the following question, enter your answer in the box.

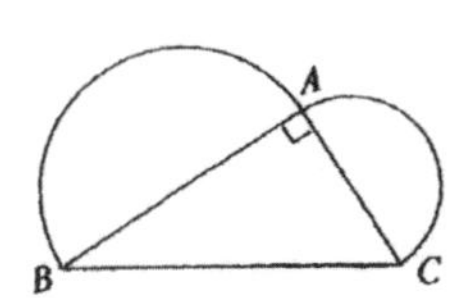

19. In the figure above, the arc AB of the semicircle has length 4π and the arc AC of the semicircle has length 3π. What is the area of ΔABC ?

Area = ☐

For the following question, select all the answer choices that apply.

20. The number a is a positive integer. If a is the square of an integer, which of the following numbers must be the square of an integer ?

Indicate all such numbers.

[A] $16a$ [B] $\sqrt{a}$ [C] a^2+2a+1 [D] a^2+4a+9

[E] $16a+8\sqrt{a}+1$ [F] $4a^2+12\sqrt{a}+9$

GO ON TO THE NEXT PAGE.

21. The probability of a person being left-handed is $\frac{1}{30}$. The probability of a person being colorblind is $\frac{1}{50}$. There are 4,500 students in a high school. How many students would probably be both left-handed and colorblind ?

Ⓐ 3 Ⓑ 4 Ⓒ 5 Ⓓ 6 Ⓔ 7

For the following question, enter your answer in the box.

22. Roger can finish a job in 20 hours. Steve can finish the same job in 15 hours. If they work together, how many hours approximately will it take them to finish the job ?

Give your answer as a decimal to the nearest 0.01.

[ ] hours

Questions 23 to 25 are based on the following graph.

ENROLLMENT AT ON-THE-MOON UNIVERSITY FOR SELECTED YEARS

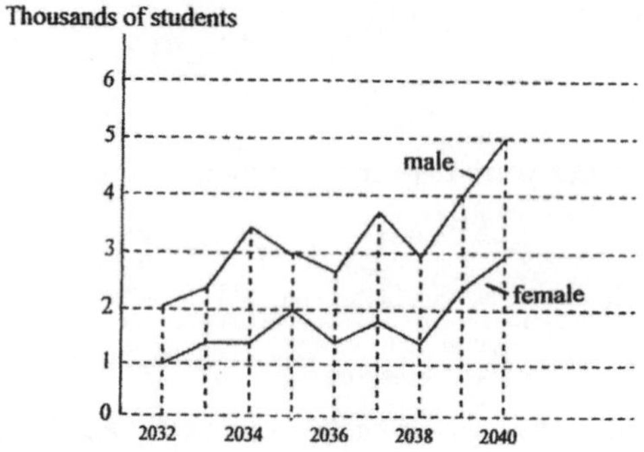

Thousands of students

0 1 2 3 4 5 6

2032 2034 2036 2038 2040

male

female

Note: graphs drawn to scale.

23. According to the information in the graph, which of the following could be the actual number of male students enrolled in 2033 ?

Ⓐ 1,420 Ⓑ 2,380 Ⓒ 2,920 Ⓓ 3,020 Ⓔ 3,420

24. Which of the following years is there a decrease over the previous year in the number of male students enrolled, but an increase in the number of female students enrolled ?

Ⓐ 2033 Ⓑ 2034 Ⓒ 2035 Ⓓ 2037 Ⓔ 2038

25. From 2038 to 2039, the percent of increase in the female students enrolled is closest to

Ⓐ 30 % Ⓑ 40 % Ⓒ 50 % Ⓓ 60 % Ⓔ 70 %

STOP: This is the end of Section 2.

PRACTICE TEST 2

(GRE Revised General Test)

Page 365 ~ 376

Answer Key

Section 1: **1.** D **2.** A **3.** D **4.** C **5.** C **6.** C **7.** B **8.** D **9.** B **10.** E

11. 15 **12.** E **13.** C **14.** A, C, D, and F **15.** B **16.** A, B, and C **17.** 18.27

18. D **19.** B **20.** 0.9 **21.** C **22.** 0.51 **23.** C **24.** B **25.** $\frac{192}{648}$, $\frac{96}{324}$, $\frac{48}{162}$, $\frac{24}{81}$, $\frac{8}{27}$

Section 2: **1.** C **2.** A **3.** B **4.** D **5.** D **6.** A **7.** C **8.** B **9.** C **10.** 100

11. A **12.** D **13.** 24 **14.** C and D **15.** C **16.** C **17.** B **18.** $\frac{111}{80}$ **19.** 24

20. A, C, and E **21.** A **22.** 8.57 **23.** B **24.** C **25.** E

Add the number of correct answers to obtain your raw math score. Partially correct answers are considered as incorrect. Using the table below to convert your raw score to scaled score. The scaled scores reflect degrees of difficult and is comparable with scores of different tests at different dates. Note that any test may be somewhat more or less difficult than the test shown in this table year by year. The scaled scores for GRE revised General Test will be 130 ~ 170.

GRE REVISED GENERAL MATH TEST CONVERSION TABLE

Raw Score	Scaled Score	Raw Score	Scaled Score	Raw Score	Scaled Score
0	130	18	143	36	158
1	130	19	144	37	159
2	130	20	144	38	160
3	131	21	145	39	161
4	131	22	146	40	161
5	132	23	146	41	162
6	132	24	147	42	163
7	133	25	148	43	164
8	134	26	149	44	165
9	135	27	150	45	166
10	136	28	151	46	167
11	137	29	152	47	168
12	138	30	153	48	169
13	139	31	153	49	170
14	140	32	154	50	170
15	141	33	155		
16	142	34	156		
17	142	35	157		

Answer Explanations
(PRACTICE TEST 2)

Section 1:

1. D $p(x, y)$ could be $p(6, 8)$ or $p(8, 6)$. Cannot be determined.

2. A $|3x-2|<10$; $-10<3x-2<10$, $-8<3x<12$, $-\frac{8}{3}<x<4$. Quantity A is greater.

3. D $|3x-2|>10$; $3x-2<-10$ or $3x-2>10$ Cannot be determined.

   $3x<-8$, $x<-\frac{8}{3}$. ; $3x>12$, $x>4$.

4. C Apply Pythagorean Theory. They are equal.

5. C $x=1+\dfrac{1}{1+\frac{1}{2}}=1+\dfrac{1}{3/2}=1+\dfrac{2}{3}=\dfrac{5}{3}$. They are equal.

6. C $V=\pi r^2 h$, $\pi(5)^2(h)=200\pi$, $h=8$. They are equal.

7. B It is $\frac{160}{360}$ of the circle. Area $=\pi(6^2)\cdot\dfrac{160}{360}=16\pi$. Quantity B is greater.

8. D $(8a-4)$ [?] $(5a+7)$, $3a$ [?] 11, a [?] $\frac{11}{3}$. Cannot be determined.

   If $a=3.1$, $a<\frac{11}{3}$; If $a=4$, $a>\frac{11}{3}$.

   We still don't know how a compares to $\frac{11}{3}$.

9. B (see page 138) The maximum area can be built by a perimeter of length p is a square. Max. area $=12^2=144$. Quantity B is greater.

10. E

    0 —— 8 —— x —— 18

    We have: $x-8=2(18-x)$; Solve for x: $x=14\frac{2}{3}$.

11. 15 x is the median of the trapezoid. $x=\dfrac{10+20}{2}=15$.

12. E $(2)(x)(x)(x) \rightarrow 4!=4\times3\times2\times1=24$. $24\times2=48$.

13. C $28!=28\times27\times26\times\cdots\times3\times2\times1$.

    Any integer that can be expressed as the product of different positive integers less than 28 is a factor of 28!.

14. A, C, D, and F The units digit of 98^n is the same as the units digit of 8^n.

    The units digits of 8^1, 8^2, 8^3, and 8^4 are 8, 4, 2, and 6.

    The pattern of 4 units digits 8, 4, 2, 6, repeats without end.

15. B The area of the larger square $A_1=1\times1=1$, $d_1=\sqrt{1^2+1^2}=\sqrt{2}$.

    The area of the smaller square $A_2=\frac{1}{2}$, side $=\sqrt{\frac{1}{2}}=\frac{\sqrt{2}}{2}$, $d_2=\sqrt{(\frac{\sqrt{2}}{2})^2+(\frac{\sqrt{2}}{2})^2}=1$.

    $d_1-d_2=\sqrt{2}-1$.

16. A, B, and C Plug in each value to test the result.

17. 18.27 Area of the shaded region = area of the sector – area of the triangle

    $=\pi(8^2)(\frac{90}{360})-\frac{1}{2}(8)(8)=16\pi-32\approx18.27$.

18. D $p=\frac{5}{9}\times\frac{4}{8}=\frac{20}{72}=\frac{5}{18}$. Or: $p=\dfrac{{}_5C_2}{{}_9C_2}=\dfrac{10}{36}=\dfrac{5}{18}$.

19. B $p=\dfrac{{}_5C_2}{{}_9C_2}+\dfrac{{}_4C_2}{{}_9C_2}=\dfrac{10}{36}+\dfrac{6}{36}=\dfrac{16}{36}=\dfrac{4}{9}$. Or: $p=\dfrac{5}{9}\cdot\dfrac{4}{8}+\dfrac{4}{9}\cdot\dfrac{3}{8}=\dfrac{4}{9}$.

Answer Explanations

(PRACTICE TEST 2)

Section 1:

20. 0.9 $p = \frac{{}_5C_3}{{}_{20}C_3} = \frac{10}{1140} \approx 0.00877 = 0.877\% \approx 0.9\%$.

Or: $p = \frac{5}{20} \cdot \frac{4}{19} \cdot \frac{3}{18} = \frac{60}{6840} \approx 0.00877 = 0.877\% \approx 0.9\%$.

21. C $28 \div \frac{2}{5} = 28 \times \frac{5}{2} = 70$.

22. 0.51 Dow on Thursday is $12{,}130.45 - 61.95 = 12068.50$.

$p(increase) = \frac{61.95}{12068.50} \approx 0.00513 = 0.513\% \approx 0.51\%$.

23. C The equation of the line passing (0, 0) and perpendicular to the given line is $y = 2x$. It passes the point (a, 6). We have $6 = 2(a)$, $a = 3$.

24. B Solve $\frac{4}{6} = \frac{7}{x}$, $x = 10.5$; Solve $\frac{4}{6} = \frac{5}{5+y}$, $y = 2.5$; $x + y = 13$.

25. $\frac{192}{648}$ (or any equivalent fraction: $\frac{V_1}{V_2} = \frac{8(4)(6)}{12(6)(9)} = \frac{192}{648}$, $\frac{96}{324}$, $\frac{48}{162}$, $\frac{24}{81}$, or $\frac{8}{27}$).

Section 2:

1. C $n + 3n = 4n$. They are equal.

2. A ΔADC and ΔBEC are similar. Quantity A is greater.
If $AC > BC$, then $AD > BE$.

3. B $(4x-1)(4x+1) = 16x^2 - 1$. Quantity B is greater.

4. D b can be any number between c and a. Cannot be determined.

5. D Plug in some numbers (b) that are close to, or far apart from c to test the result. Cannot be determined.

$\frac{9}{2} > \frac{2}{1}$, $\frac{9}{3} = \frac{3}{1}$, $\frac{9}{8} < \frac{8}{1}$.

6. A Apply Pythagorean Theorem: Quantity A is greater.
$9^2 + (4.4)^2 \approx 100.4 > 10^2$, $a > 90$.
We need two shorter sides to have 10 on the third side.

$\sqrt{100.4}$ $a°$ 4.4 9

7. C Equilateral polygon has equal sides. They are equal.
arc ABC = arc BCD

8. B $|n| = 18$, $n = \pm 18$; $|n-1| = 19$, $n - 1 = \pm 19$, $n = 20, -18$. Quantity B is greater.
$n = -18$ only.

9. C $180 - (y+z) = 180 - (90 + x)$, $y + z = 90 + x$. They are equal.

10. 100 Let $AB = AC = S$ and apply Pythagorean Theorem.
$S^2 + S^2 = 20^2$, $S^2 = 200$. Area $= \frac{1}{2}S^2 = \frac{1}{2}(200) = 100$.

11. A Pure alcohol $= 45 \times \frac{2}{10} = 9$ liters.

12. D $30 \times 20\% = 6$ gallons of pure alcohol (original).
Let x = alcohol added, we have $\frac{6+x}{30+x} = 0.25$.
Solve for x: $x = 2$ gallons of alcohol should be added.

Answer Explanations
(PRACTICE TEST 2)

Section 2: 13. 24 $4^8 \times 16^{20} = 4^{2x}$, $4^8 \times (4^2)^{20} = 4^{2x}$, $4^8 \times 4^{40} = 4^{2x}$, $4^{48} = 4^{2x}$, $2x = 48$, $x = 24$.

14. C and D The linear function $f(x) = 2x + 3$ has a slope of 2.

Evaluate each function in the answer choices.

Choice A and Choice B are lines with slope 2. They are parallel to the graph of $f(x)$.

Choice C is a line with slope $-\frac{1}{2}$. It is perpendicular to the graph of $f(x)$.

Choice D and E are lines with slope $\dfrac{b-d}{a-c}$.

D: If $(a-c)(b-d) < 0$, $(b-d)$ and $(a-c)$ have opposite signs, the slope of $q(x)$ must be negative. It intersects the graph of $f(x)$.

E: If $(a-c)(b-d) > 0$, $(b-d)$ and $(a-c)$ have same signs, the slope of $r(x)$ is positive. The slope could be 2 (parallel) or any other positive number (intersect). It cannot be determined.

15. C $100 + 1 = 101$, $99 + 2 = 101$, $98 + 3 = 101$, ······.
There are 50 pairs from 1 to 100. Each pair has a sum of 101.
The sum $= 101 \times 50 = 5{,}050$.

16. C Let $x = 100000$,
$(100001)^2 - 2(100000)^2 + (99999)^2 = (x+1)^2 - 2x^2 + (x-1)^2$
$= (x^2 + 2x + 1) - 2x^2 + (x^2 - 2x + 1) = 2$.

17. B $9 \times 10 \times 10 \times 10 = 9{,}000$.

18. $\frac{111}{80}$ Let $\frac{M}{F}$ = the ratio of males to females at the beginning of the year.
At the end of the year, we have $\frac{1.11M}{0.80F} = \frac{111}{80}\left(\frac{M}{F}\right)$.

19. 24 Circumference $= \pi D$, $8\pi = \pi D$, $D = 8 = \overline{AB}$.
$6\pi = \pi D$, $D = 6 = \overline{AC}$.
Area of the triangle $= \frac{1}{2}(8)(6) = 24$.

20. A, C, and E The number a is a perfect square.
Evaluate each number in the answer choices.

A. $16a$ is a perfect square. B. $\sqrt{a}$ is not a perfect square if a is 4, 9, 25, ······.

C. $a^2 + 2a + 1 = (a+1)^2$. It is a perfect square.

D. $a^2 + 4a + 9$ is not a perfect square.

E. $16a + 8\sqrt{a} + 1 = (4\sqrt{a} + 1)^2$. It is a perfect square.

F. $4a^2 + 12\sqrt{a} + 9$ is not a perfect square.

21. A $4500 \times \frac{1}{30} \times \frac{1}{50} = 3$.

22. 8.57 If they work together, they can finish $\frac{1}{20} + \frac{1}{15} = \frac{7}{60}$ of the job in 1 hour.
They can finish the job in $1 \div \frac{7}{60} = \frac{60}{7} \approx 8.57$ hours.

23. B

24. C 25. E Percent of increase $= \frac{2.4-1.4}{1.4} \approx 0.714 = 71.4\ \%$.

PRACTICE TEST 3
(GRE Revised General Test)
SECTION 1
Quantitative Reasoning
Time—40 minutes
25 Questions

For each question, indicate the best answer, using the directions given.

Notes: All numbers used are real numbers.

All figures are assumed to lie in a plane unless otherwise indicated.

Geometric figures, such as lines, circles, triangles, and quadrilaterals, **are not necessarily** drawn to scale. That is, you should **not** assume that quantities such as lengths and angle measures are as they appear in a figure. You should assume, however, that lines shown as straight are actually straight, points on a line are in the order shown, and more generally, all geometric objects are in the relative positions shown. For questions with geometric figures, you should base your answers on geometric reasoning, not on estimating or comparing quantities by sight or by measurement.

Coordinate system, such as xy-planes and number lines, **are** drawn to scale; therefore, you can read, estimate, or compare quantities in such figures by sight or by measurement.

Graphical data presentations, such as bar graphs, circles graphs, and line graphs, **are** drawn to scale; therefore, you can read, estimate, or compare data values by sight or by measurement.

For each of Questions 1 to 9, compare Quantity A and Quantity B, using additional information centered above the two quantities if such information is given. Select one of the following four answer choices and fill in the corresponding oval to the right to the question.

Ⓐ Quantity A is greater.

Ⓑ Quantity B is greater.

Ⓒ The two quantities are equal.

Ⓓ The relationship cannot be determined from the information given.

A symbol that appears more than once in a question has the same meaning throughout the question.

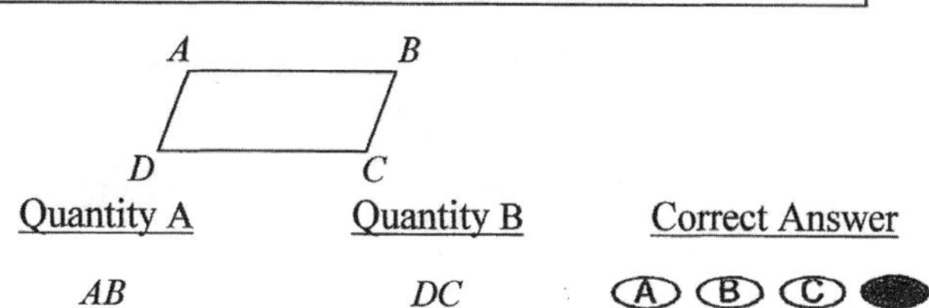

	Quantity A	Quantity B	Correct Answer
Example:	AB	DC	

(A) **Quantity A is greater.**
(B) **Quantity B is greater.**
(C) **The two quantities are equal.**
(D) **The relationship cannot be determined from the information given.**

$x > 1$

	Quantity A	Quantity B	
1.	$\frac{1}{2}$	$\frac{1}{x}-\frac{1}{2}$	

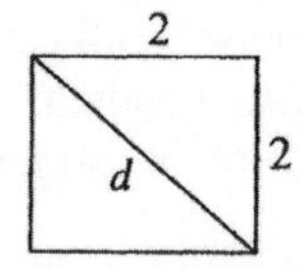

The square above has side length 2 and diagonal length d.

	Quantity A	Quantity B	
2.	d	$2\sqrt{2}$	(A) (B) (C) (D)

$(x+4)(x-5) < 0$

	Quantity A	Quantity B	
3.	x	5	(A) (B) (C) (D)

$(x+4)(x-5) > 0$

	Quantity A	Quantity B	
4.	x	0	(A) (B) (C) (D)

The average (arithmetic mean) of 11 consecutive positive integers is 40.

	Quantity A	Quantity B	
5.	The median of these 11 integers.	40	(A) (B) (C) (D)

Ⓐ	**Quantity A is greater.**
Ⓑ	**Quantity B is greater.**
Ⓒ	**The two quantities are equal.**
Ⓓ	**The relationship cannot be determined from the information given.**

$x > 0$

	Quantity A	Quantity B
6.	$x - 4$	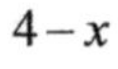 $4 - x$

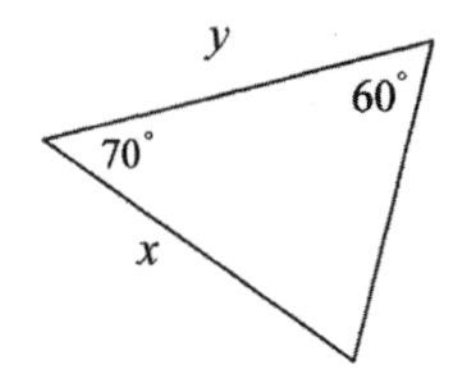

	Quantity A	Quantity B
7.	$\frac{y}{x}$	1

	Quantity A	Quantity B
8.	$4x^2 - 5$	$4x - 7$

p is the probability that the experiment will succeed. q is the probability that the experiment will fail.

	Quantity A	Quantity B
9.	$p + q$	$p \times q$

GO ON TO THE NEXT PAGE.

Equations 10 to 25 have several different formats. Unless otherwise directed, select a single answer choice. For Numeric Entry questions, follow the instructions below.

Numeric Entry Questions:
Enter your answer in the answer box(es) below the question.
1. **Your answer may be an integer, a decimal, or a fraction, and it may be negative.**
2. **If a question asks for a fraction, there will be two boxes —— one for the numerator and one for the denominator.**
3. **Equivalent forms of the correct answer, such as 2.5 and 2.50, are all correct. Fractions do not need to be reduced to lowest terms.**
4. **Enter the exact answer unless the equation asks you to round your answer.**

10. How many multiples of 7 are there between 1 and 900 ?

Ⓐ 124 Ⓑ 125 Ⓒ 126 Ⓓ 127 Ⓔ 128

For the following question, select all the answer choices that apply.

11. a, b, and c are points on a number line in that order. The origin is at the middle of a and b. b is at the middle of the origin and c. a, b, c, and the origin are equally spaced. Which of the following statements must be true ?

Indicate all such numbers.

[A] $a+b+c<0$ [B] $abc<0$ [C] $a+c=b$ [D] $a(b-c)>0$

For the following question, enter your answer in the box.

12. A rocket is fired upward with an initial speed of 160 feet per second. The height (h) of the rocket is given by the formula $h=160t-16t^2$, where t is the time in seconds. How many seconds will the rocket reach the ground again ?

[ ] seconds

13. Which of the following equals the ratio of $2\frac{1}{3}$ to $3\frac{2}{3}$?

Ⓐ 5 to 3 Ⓑ 3 to 2 Ⓒ 2 to 3 Ⓓ 11 to 7 Ⓔ 7 to 11

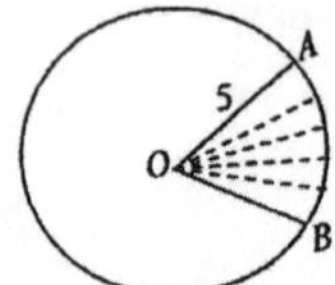

14. The area of the dashed line region of the above circle is 5π. O is the center of the circle. What is the degree measure of $<AOB$?

Ⓐ 70 Ⓑ 72 Ⓒ 74 Ⓓ 76 Ⓔ 78

For the following question, select all the answer choices that apply.

15. If a and b are both even integers, which of the following expressions must be an even number ?
Indicate all such expressions.

[A] b^a [B] $(b+1)^a$ [C] b^{a+1}

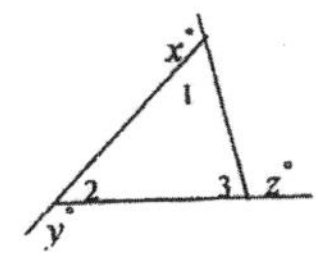

16. In the figure above, $x+y+z=$

Ⓐ 90 Ⓑ 180 Ⓒ 270 Ⓓ 360 Ⓔ 450

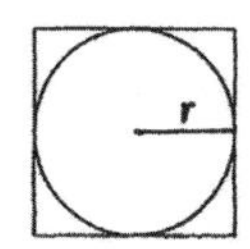

17. A circle is to be cut from a square as shown above. What percent approximately is not to be used for the circle ?

Ⓐ 18.5 % Ⓑ 20.7 % Ⓒ 21.5 % Ⓓ 24.3 % Ⓔ 25.6

For the following question, enter your answer in the boxes.

18. In a college, there are 100 students from Asia. Of these students, 67 are from China, 20 are from Japan, and the rest are from Korea. If two students are selected at random from the 100 students, what is the probability that both students selected will be students from China ?

Give your answer as a fraction. $\frac{\square}{\square}$

19. If $x=\frac{1}{2}$ and $y=\frac{1}{3}$, what is the value of $\frac{x+y}{x-y}$?

Ⓐ 6 Ⓑ 5 Ⓒ 4 Ⓓ 3 Ⓔ 2

20. If $a^2+b^2=40$ and $ab=8$, what is the value of $(a-b)^2$?

Ⓐ 24 Ⓑ 23 Ⓒ 22 Ⓓ 21 Ⓔ 20

GO ON TO THE NEXT PAGE.

21. In the basketball game, the probability that you can make the basket on free throws is 0.25. What is the probability that you can expect to make at least one basket in next three free throws ?

Ⓐ $\frac{3}{64}$ Ⓑ $\frac{9}{64}$ Ⓒ $\frac{27}{64}$ Ⓓ $\frac{37}{64}$ Ⓔ $\frac{39}{64}$

Questions 22 to 25 are based on the following graphs.

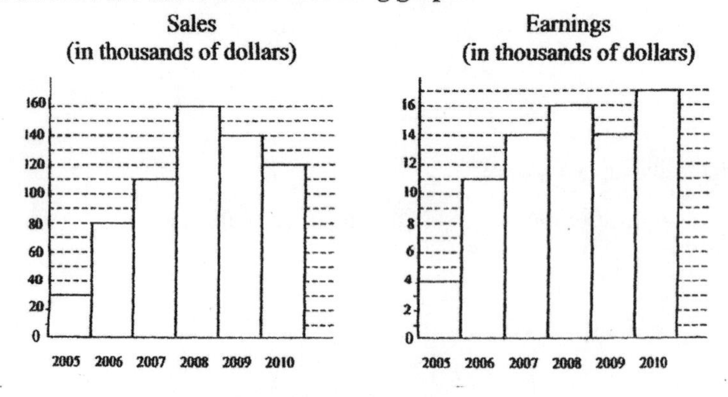

Note: Drawn to scale

22. For the year from 2006 to 2007, what was the percent of increase in sales ?

Ⓐ 27.3 % Ⓑ 35.4 % Ⓒ 37.5 % Ⓓ 39.8 % Ⓔ 42.2 %

23. For the year from 2005 to 2010 inclusive, what were the average (arithmetic mean) sales per year ? Round it to thousands.

Ⓐ \$95,000 Ⓑ \$102,000 Ⓒ \$107,000 Ⓓ \$112,000 Ⓔ \$114,000

24. In which of the year from 2006 to 2010 inclusive, did earnings change by the smallest percent over the previous years ?

Ⓐ 2007 Ⓑ 2008 Ⓒ 2009 Ⓓ 2010

Ⓔ It cannot be determined from the information given.

25. How many of the years in which earnings were at least 12 % of sales ?

Ⓐ None Ⓑ One Ⓒ Two Ⓓ Three Ⓔ Four

STOP: This is the end of Section 1.

PRACTICE TEST 3

(GRE Revised General Test)

SECTION 2

Quantitative Reasoning

Time—40 minutes

25 Questions

For each question, indicate the best answer, using the directions given.

Notes: All numbers used are real numbers.

All figures are assumed to lie in a plane unless otherwise indicated.

Geometric figures, such as lines, circles, triangles, and quadrilaterals, **are not necessarily** drawn to scale. That is, you should **not** assume that quantities such as lengths and angle measures are as they appear in a figure. You should assume, however, that lines shown as straight are actually straight, points on a line are in the order shown, and more generally, all geometric objects are in the relative positions shown. For questions with geometric figures, you should base your answers on geometric reasoning, not on estimating or comparing quantities by sight or by measurement.

Coordinate system, such as *xy*-planes and number lines, **are** drawn to scale; therefore, you can read, estimate, or compare quantities in such figures by sight or by measurement.

Graphical data presentations, such as bar graphs, circles graphs, and line graphs, **are** drawn to scale; therefore, you can read, estimate, or compare data values by sight or by measurement.

For each of Questions 1 to 9, compare Quantity A and Quantity B, using additional information centered above the two quantities if such information is given. Select one of the following four answer choices and fill in the corresponding oval to the right to the question.

Ⓐ Quantity A is greater.

Ⓑ Quantity B is greater.

Ⓒ The two quantities are equal.

Ⓓ The relationship cannot be determined from the information given.

A symbol that appears more than once in a question has the same meaning throughout the question.

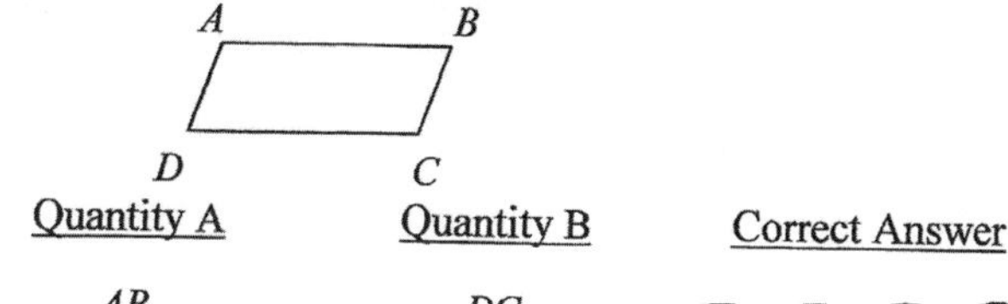

	Quantity A	Quantity B	Correct Answer
Example:	AB	DC	

Ⓐ **Quantity A is greater.**
Ⓑ **Quantity B is greater.**
Ⓒ **The two quantities are equal.**
Ⓓ **The relationship cannot be determined from the information given.**

a, *b*, *c*, and *d* are consecutive integers.

	Quantity A	Quantity B	
1.	$a \times d$	$b \times c$	Ⓐ Ⓑ Ⓒ Ⓓ

$$\frac{5^{20}}{25^{2n}} = \frac{5^{n}}{5^{7}}$$

	Quantity A	Quantity B	
2.	n	5	Ⓐ Ⓑ Ⓒ Ⓓ

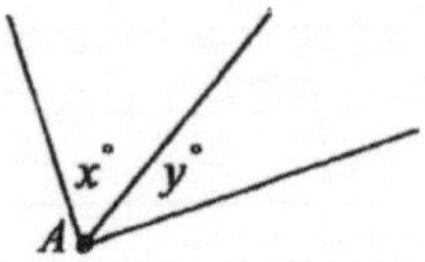

x° y° A

	Quantity A	Quantity B	
3.	x	$90 - y$	Ⓐ Ⓑ Ⓒ Ⓓ

The volume of a cube is 27.

	Quantity A	Quantity B	
4.	The total surface area of the cube	54	Ⓐ Ⓑ Ⓒ Ⓓ

$$x = 1 + 2 + 3 + \cdots\cdots + 39 + 40$$
$$y = 1 + 2 + 3 + \cdots\cdots + 29 + 30$$

	Quantity A	Quantity B	
5.	$x - y$	350	Ⓐ Ⓑ Ⓒ Ⓓ

GO ON TO THE NEXT PAGE.

Ⓐ	**Quantity A is greater.**
Ⓑ	**Quantity B is greater.**
Ⓒ	**The two quantities are equal.**
Ⓓ	**The relationship cannot be determined from the information given.**

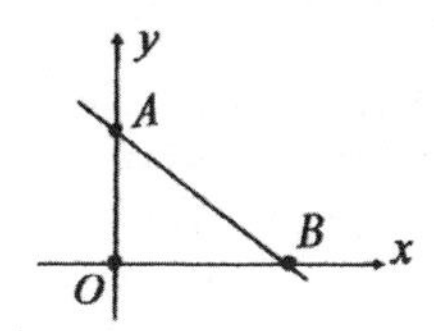

In the rectangular coordinate system above, the equation of the line is given by $y = -\frac{4}{3}x + 6$.

	Quantity A	Quantity B	
6.	$\overline{OA}$	$\overline{OB}$	Ⓐ Ⓑ Ⓒ Ⓓ

A tank contains 40 gallons of gasoline and is $\frac{2}{5}$ full.

	Quantity A	Quantity B	
7.	The number of additional gallons of gasoline needed to fill the tank in full	60	Ⓐ Ⓑ Ⓒ Ⓓ

	Quantity A	Quantity B	
8.	$x^2 + 4$	$4x + 10$	Ⓐ Ⓑ Ⓒ Ⓓ

Four consecutive integers have a sum of -118.

	Quantity A	Quantity B	
9.	The greatest of the four integers	–27	Ⓐ Ⓑ Ⓒ Ⓓ

GO ON TO THE NEXT PAGE.

Equations 10 to 25 have several different formats. Unless otherwise directed, select a single answer choice. For Numeric Entry questions, follow the instructions below.

Numeric Entry Questions:
Enter your answer in the answer box(es) below the question.
1. **Your answer may be an integer, a decimal, or a fraction, and it may be negative.**
2. **If a question asks for a fraction, there will be two boxes — one for the numerator and one for the denominator.**
3. **Equivalent forms of the correct answer, such as 2.5 and 2.50, are all correct. Fractions do not need to be reduced to lowest terms.**
4. **Enter the exact answer unless the equation asks you to round your answer.**

For the following question, enter your answer in the box.

10. David can paint a warehouse in 12 hours. Maria can paint the same warehouse in 16 hours. If they work together, how many hours approximately will it take them to paint the warehouse ?

    Give your answer as a decimal to the nearest 0.01.

    [ ] hours

11. How many different numbers that can be formed by arranging the digits in the number 2,3456 ?

    Ⓐ 120 Ⓑ 100 Ⓒ 80 Ⓓ 40 Ⓔ 20

12. How many different numbers that can be formed by arranging the digits in the number 23,226 ?

    Ⓐ 120 Ⓑ 100 Ⓒ 80 Ⓓ 40 Ⓔ 20

$$x+4y-2z=0$$
$$3x-4y+5z=0$$

13. In the system of equations above, $x \neq 0$, what is the ratio of z to x ?

    Ⓐ $-\frac{1}{2}$ Ⓑ $-\frac{4}{3}$ Ⓒ $-\frac{3}{4}$ Ⓓ $\frac{4}{3}$ Ⓔ $\frac{3}{4}$

14. If $n > 0$, which of the following must be greater than n ?

    I. $10n$
    II. n^2
    III. $10-n$

    Ⓐ I only Ⓑ II only Ⓒ III only Ⓓ I and III only Ⓔ None

15. The function $c(n) = 200 + 20n$ is used to represent the total cost c in dollars to produce calculators, n is the number of calculator produced. Based on this function, if the total cost was $5,000, how many calculators were produced ?

Ⓐ 160 Ⓑ 180 Ⓒ 200 Ⓓ 220 Ⓔ 240

16. Which of the following equations is perpendicular to the line $y = -3x + 7$?

Ⓐ $y = 3x - 7$ Ⓑ $y = -\frac{1}{3}x + 7$ Ⓒ $y = \frac{1}{3}x + 7$ Ⓓ $y = -3x - 7$ Ⓔ $y = 3x + 7$

For the following question, enter your answer in the boxes.

17. There are 20 bulbs in a box and 6 are defective. You select 2 bulbs from the box. What is the probability that neither of the bulbs selected will be defective ?

Give your answer as a fraction. $\frac{\square}{\square}$

18. A line intersects a circle at point A and B. The radius of the circle is 4. What is the maximum possible distance between A and B ?

Ⓐ 2 Ⓑ π Ⓒ 8 Ⓓ 3π Ⓔ 12

For the following question, enter your answer in the box.

19. A 25-gallon salt-water solution contains 16 % pure salt. How much salt in gallons should be added to produce the solution to 40 % salt ?

$\square$ gallons

20. There are 126 senior students in a high school. 98 took trigonometry and 43 took calculus. x represents the number of students that could have taken both trigonometry and calculus. Which of the following expressions must be true ?

Ⓐ $55 \le x \le 142$ Ⓑ $43 \le x \le 55$ Ⓒ $28 \le x \ge 83$ Ⓓ $20 \le x \le 55$ Ⓔ $15 \le x \le 43$

21. A triangle has lengths of sides 8, 15, and x, respectively. What is the possible value of x ?

Ⓐ $8 < x < 15$ Ⓑ $0 < x < 15$ Ⓒ $0 < x < 8$ Ⓓ $7 < x < 23$ Ⓔ $7 < x < 15$

For the following question, select all the answer choices that apply.

22. A triangle has lengths of two sides 12 and 7, respectively. Which of the following could be the length of the third side ?
Indicate <u>all</u> such lengths.
[A] 4 [B] 5 [C] 7 [D] 12 [E] 11 [F] 19 [G] 20

GO ON TO THE NEXT PAGE.

Questions 23 to 25 are based on the following graph.

SALES OF A-PLUS EDUCATIONAL SUPPLIES COMPANY

Math Books Sold 2006 ~ 2009 in U.S.A
(value in thousands of U.S. dollars)

	2006		2007		2008		2009	
Area	Value	% of Total	Value	% of Total	Value	% of Total	Value	% of Total
East	$50	14.3%	$52	14.6%	$48	13.4%	$54	14.4%
West	68	19.4%	63	17.7%	69	19.2%	72	19.2%
South	75	21.4%	79	22.2%	80	22.3%	78	20.8%
North	77	22.0 %	78	21.9%	80	22.3%	84	22.4%
Central	80	22.9%	84	23.6%	82	22.8%	87	23.2%
Total	$350	100.0	$356	100.0	$359	100.0	$375	100.0

SALES IN THE EAST AREA IN U.S.A

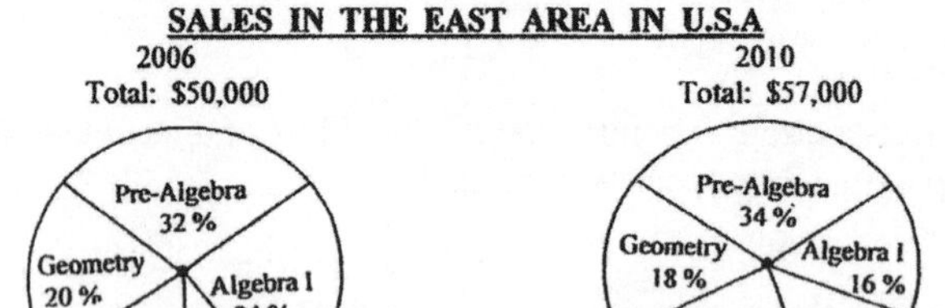

23. In 2010 what was the value, in thousands of dollars, of geometry books sold in the East Area ?

Ⓐ 10.26 Ⓑ 9.64 Ⓒ 8.48 Ⓓ 6.18 Ⓔ 5.24

24. From 2006 to 2009, the value of math books sold in the Central Area increased by what percent approximately ?

Ⓐ 6.7 % Ⓑ 7.1 % Ⓒ 8.0 % Ⓓ 8.8 % Ⓔ 9.6 %

25. In 2006 the value of pre-algebra books sold in the East Area was approximately what percent of the value of total math books sold in U.S.A ?

Ⓐ 1.24 % Ⓑ 2.10 % Ⓒ 2.94 % Ⓓ 3.50 % Ⓔ 4.58 %

STOP: This is the end of Section 2.

PRACTICE TEST 3

(GRE Revised General Test)

Page 381 ~ 392

Answer Key

Section 1: **1.** A **2.** C **3.** B **4.** D **5.** C **6.** D **7.** B **8.** A **9.** A **10.** E

11. B, C, and D **12.** 10 **13.** E **14.** B **15.** A and C **16.** D **17.** C

18. $\frac{67}{150}$ (or any equivalent fraction) **19.** B **20.** A **21.**D **22.**C **23.** C **24.** C **25.** E

Section 2: **1.** B **2.** A **3.** D **4.** C **5.** A **6.** A **7.** C **8.** D **9.** B **10.** 6.86

11. A **12.** E **13.** B **14.** A **15.** E **16.** C **17.** $\frac{91}{150}$ (or any equivalent fraction)

18. C **19.** 10 **20.** E **21.** D **22.** C, D, and E **23.** A **24.** D **25.** E

Add the number of correct answers to obtain your raw math score.
Partially correct answers are considered as incorrect.
Using the table below to convert your raw score to scaled score. The scaled scores reflect degrees of difficult and is comparable with scores of different tests at different dates. Note that any test may be somewhat more or less difficult than the test shown in this table year by year.
The scaled scores for GRE revised General Test will be 130 ~ 170.

GRE REVISED GENERAL MATH TEST CONVERSION TABLE

Raw Score	Scaled Score	Raw Score	Scaled Score	Raw Score	Scaled Score
0	130	18	143	36	158
1	130	19	144	37	159
2	130	20	144	38	160
3	131	21	145	39	161
4	131	22	146	40	161
5	132	23	146	41	162
6	132	24	147	42	163
7	133	25	148	43	164
8	134	26	149	44	165
9	135	27	150	45	166
10	136	28	151	46	167
11	137	29	152	47	168
12	138	30	153	48	169
13	139	31	153	49	170
14	140	32	154	50	170
15	141	33	155		
16	142	34	156		
17	142	35	157		

Answer Explanations
(PRACTICE TEST 3)

Section 1:

1. A $\frac{1}{2}$ [?] $\frac{1}{x}-\frac{1}{2}$, $\frac{1}{2}+\frac{1}{2}$ [?] $\frac{1}{x}$, $1>\frac{1}{x}$ if $x>1$. Quantity A is greater.

2. C Apply Pythagorean Theory. They are equal.
$d^2=2^2+2^2=8$, $d=\sqrt{8}=2\sqrt{2}$.

3. B (see page 167) $(x+4)(x-5)<0$. Quantity B is greater.
The critical points are $x=-4$ and 5.
Test each interval, we have: $-4<x<5$.

4. D (see page 167) $(x+4)(x-5)>0$. Cannot be determined.
The critical points are $x=-4$ and 5.
Test each interval, we have: $x<-4$ or $x>5$.

5. C 11 (odd) consecutive integers, the mean is the median. They are equal.

6. D $x-4$ [?] $4-x$, $2x$ [?] 8, x [?] 4. Cannot be determined.
For $x>0$, we still don't know how x compares to 4.

7. B In every triangle, the larger angle is opposite the larger side, the smaller angle is opposite the smaller side. Quantity B is greater.
The angle which is opposite the side y is a 50° angle.
We have $x>y$. Therefore $\frac{y}{x}<1$.

8. A $(4x^2-5)$ [?] $(4x-7)$, $4x^2-4x+1$ [?] -1, Quantity A is greater.
$(2x-1)^2$ [?] -1. Since $(2x-1)^2\geq 0$, $(2x-1)^2>-1$.

9. A The value of $p+q$ is always 1. $p+q=1$. Quantity A is greater.
Since $0\leq p\leq 1$, $0\leq q\leq 1$, and $p+q=1$,
we have $0\leq pq<1$.

10. E $1\times 7=7$, $2\times 7=14$, ……, $128\times 7=896$.

11. B, C, and D Draw a number line: a 0 b c
a is negative. b and c are positive.
$a=-b$, $c=2b$, and $a<b<c$. Check each choice.

12. 10 $h=0$, we have $160t-16t^2=0$, $16t(10-t)=0$, $t=10$.

13. E $\dfrac{2\frac{1}{3}}{3\frac{2}{3}}=\dfrac{\frac{7}{3}}{\frac{11}{3}}=\dfrac{7}{3}\times\dfrac{3}{11}=\dfrac{7}{11}$.

14. B Let $m=$ the degree measure of $\angle AOB$.
We have: $\pi\times 5^2\times\frac{m}{360}=5\pi$, $m=\frac{360}{5}=72$.

15. A and C The product of two or more even integers is always even. b^a and b^{a+1} are even.
The product of two or more odd integers is always odd.
$(b+1)$ is odd. Therefore $(b+1)^a$ is an odd integer.

16. D $x^\circ=\angle 2+\angle 3$, $y^\circ=\angle 1+\angle 3$, $z^\circ=\angle 1+\angle 2$.
$x^\circ+y^\circ+z^\circ=2(\angle 1+\angle 2+\angle 3)=2(180^\circ)=360^\circ$.

17. C $\dfrac{(2r)^2-\pi r^2}{(2r)^2}=\dfrac{4r^2-\pi r^2}{4r^2}=\dfrac{r^2(4-\pi)}{4r^2}=\dfrac{4-\pi}{4}\approx 0.2146\approx 21.5\%$

Answer Explanations
(PRACTICE TEST 3)

Section 1: 18. $\frac{67}{150}$ (or any equivalent fraction).

$$p=\frac{{}_{67}C_2}{{}_{100}C_2}=\frac{2211}{4950}=\frac{737}{1650}=\frac{67}{150}, \text{ or } p=\frac{67}{100}\times\frac{66}{99}=\frac{4422}{9900}=\frac{2211}{4950}=\frac{737}{1650}=\frac{67}{150}.$$

19. B $\frac{x+y}{x-y}=\frac{\frac{1}{2}+\frac{1}{3}}{\frac{1}{2}-\frac{1}{3}}=\frac{\frac{3}{6}+\frac{2}{6}}{\frac{3}{6}-\frac{2}{6}}=\frac{5/6}{1/6}=\frac{5}{6}\div\frac{1}{6}=\frac{5}{\not 6}\cdot\frac{\not 6}{1}=5.$

20. A $(a-b)^2=a^2-2ab+b^2=(a^2+b^2)-2ab=40-2(8)=24.$

21. D In three free throws, the probability that you will miss the basket is:

$$\tfrac{3}{4}\times\tfrac{3}{4}\times\tfrac{3}{4}=\tfrac{27}{64}.$$

The probability that you can make at least one basket in three throws is:

$$p=1-\tfrac{27}{64}=\tfrac{37}{64}.$$

22. C $\frac{110-80}{80}=\frac{30}{80}=0.375=37.5\ \%.$

23. C $\frac{30000+80000+110000+160000+140000+120000}{6}=\frac{640000}{6}\approx 106{,}667\approx 107{,}000.$

24. C $2006\sim2007: \frac{3}{11}\approx 27.3\ \%$, $2007\sim2008: \frac{2}{14}\approx 14.3\ \%$.

$2008\sim2009: \frac{2}{16}=12.5\ \%$, $2009\sim2010: \frac{3}{14}\approx 21.4\ \%$.

25. E $\frac{4000}{30000}\approx 13.3\%$, $\frac{11000}{80000}\approx 13.8\%$, $\frac{14000}{110000}\approx 12.7\%$,

$\frac{16000}{160000}=10\%$, $\frac{14000}{140000}=10\%$, $\frac{17000}{120000}\approx 14.2\%$.

Section 2:

1. B The consecutive integers a, b, c, and d are: — Quantity B is greater.

$$n,\ n+1,\ n+2,\ n+3.$$
$$a\times d=n(n+3)=n^2+3n,$$
$$b\times c=(n+1)(n+2)=n^2+3n+2.$$

Therefore $b\times c>a\times d$.

2. A $\frac{5^{20}}{25^{2n}}=\frac{5^n}{5^7}$, $5^{20}\cdot 5^7=25^{2n}\cdot 5^n$, $5^{27}=(5^2)^{2n}\cdot 5^n$ — Quantity A is greater.

$5^{27}=5^{4n}\cdot 5^n$, $5^{27}=5^{5n}$, $27=5n$, $n=5.4$.

3. D No information is given about the degree measure of angle A. — Cannot be determined.

4. C The side length of the cube is 3. — They are equal.

The total surface area = $(3\times 3)\times 6=54$.

5. A x has 20 pairs of $(1+40)$, $x=(1+40)\times 20=820$. — Quantity A is greater.

y has 15 pairs of $(1+30)$, $y=(1+30)\times 15=465$.

$$x-y=820-465=355.$$

6. A Let $x=0$, y-intercept $=6=OA$. — Quantity A is greater.

Let $y=0$, x-intercept $=4.5=OB$.

Answer Explanations
(PRACTICE TEST 3)

Section 2:

7. C $40 \div \frac{2}{5} = 100$ gallons for full tank, $100 - 40 = 60$ needed. They are equal.

8. D $x^2 + 4$ [?] $4x+10$, Cannot be determined.
$x^2 - 4x + 4$ [?] 10, $(x-2)^2$ [?] 10.
$(x-2)^2 \ge 0$, we still don't know how $(x-2)^2$ compares to 10.

9. B Let x = the least of the four consecutive integers. Quantity B is greater.
Solve $x + (x+1) + (x+2) + (x+3) = -118$ for x,
We have $x = -31$. The greatest integer $= -31 + 3 = -28$.

10. 6.86 They can paint $\frac{1}{12} + \frac{1}{16} = \frac{7}{48}$ of the warehouse in I hour,
$1 \div \frac{7}{48} = 1 \times \frac{48}{7} \approx 6.86$ hours needed to paint the warehouse..

11. A $5! = 5 \cdot 4 \cdot 3 \cdot 2 \cdot 1 = 120$ different numbers.

12. E $\frac{5!}{3!} = \frac{120}{6} = 20$ different numbers.

13. B Add the two equations, we have $4x + 3z = 0$, $3z = -4x$, $\frac{z}{x} = -\frac{4}{3}$.

14. A $n > 0$, A. $10n$ [?] n, we have $10 > 1$.
B. n^2 [?] n, n [?] 1, n could be greater or less than 1.
C. $10 - n$ [?] n, 10 [?] $2n$, 10 could be greater or less than $2n$.

15. E $c(n) = 200 + 20n$, solve $200 + 20n = 5000$ for n. $n = 240$.

16. C The slope of $y = -3x + 7$ is -3. The slope of $y = \frac{1}{3}x + 7$ is $\frac{1}{3}$.
Two lines are perpendicular if the product of their slopes is -1.

17. $\frac{91}{190}$ (or any equivalent fraction)
Choose 2 from 14 that are not defective. $p = \frac{{}_{14}C_2}{{}_{20}C_2} = \frac{91}{190}$, or $p = \frac{14}{20} \cdot \frac{13}{19} = \frac{91}{190}$.

18. C The maximum distance of the intersection line is the diameter D = 8.

19. 10 There are $25 \times 16\% = 4$ gallons of pure salt (original).
Let x = pure salts added, we have the equation:
$\frac{4+x}{25+x} = 0.4$, solve for x, $x = 10$ gallons of pure salts needed.

20. E Draw a Venn Diagrams.
There are $(43 - x)$ students taking calculus only.
$(43 - x) \ge 0$, we have $x \le 43$.
If all of the 126 students could have taken one or both courses, there are $(98 + 43) - 126 = 15$ students taking both courses. $x \ge 15$.

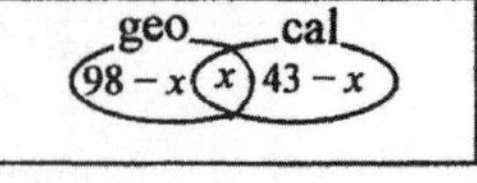

21. D The sum of the lengths of any two sides is greater than the length of the third side.
The difference of the lengths of any two sides is less than the length of the third side.

22. C, D, and E (Same as 21 above)

23. A $\$57000 \times 0.18 = \$10260$.

24. D $\frac{87-80}{80} = \frac{7}{80} = 0.0875 \approx 8.8\%$.

25. E $14.3\% \times 32\% = 0.143 \times 0.32 = 0.04576 \approx 4.58\%$.

PRACTICE TEST 4

(GRE Revised General Test)

SECTION 1

Quantitative Reasoning

Time— 40 minutes

25 Questions

For each question, indicate the best answer, using the directions given.

Notes: All numbers used are real numbers.

All figures are assumed to lie in a plane unless otherwise indicated.

Geometric figures, such as lines, circles, triangles, and quadrilaterals, **are not necessarily** drawn to scale. That is, you should **not** assume that quantities such as lengths and angle measures are as they appear in a figure. You should assume, however, that lines shown as straight are actually straight, points on a line are in the order shown, and more generally, all geometric objects are in the relative positions shown. For questions with geometric figures, you should base your answers on geometric reasoning, not on estimating or comparing quantities by sight or by measurement.

Coordinate system, such as *xy*-planes and number lines, **are** drawn to scale; therefore, you can read, estimate, or compare quantities in such figures by sight or by measurement.

Graphical data presentations, such as bar graphs, circles graphs, and line graphs, **are** drawn to scale; therefore, you can read, estimate, or compare data values by sight or by measurement.

For each of Questions 1 to 9, compare Quantity A and Quantity B, using additional information centered above the two quantities if such information is given. Select one of the following four answer choices and fill in the corresponding oval to the right to the question.

Ⓐ Quantity A is greater.

Ⓑ Quantity B is greater.

Ⓒ The two quantities are equal.

Ⓓ The relationship cannot be determined from the information given.

A symbol that appears more than once in a question has the same meaning throughout the question.

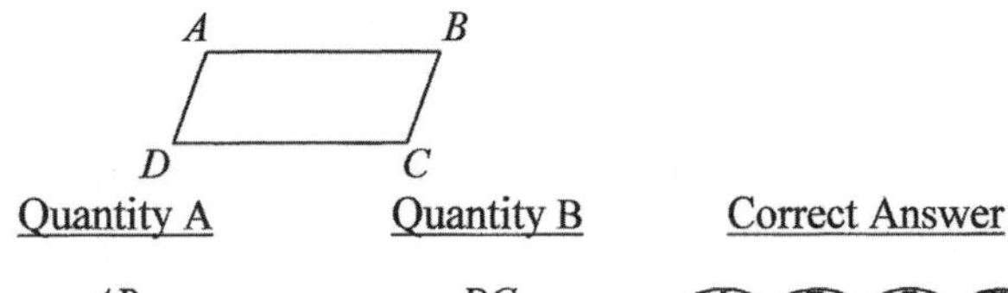

	Quantity A	Quantity B	Correct Answer
Example:	AB	DC	

Ⓐ	**Quantity A is greater.**
Ⓑ	**Quantity B is greater.**
Ⓒ	**The two quantities are equal.**
Ⓓ	**The relationship cannot be determined from the information given.**

$$2^x = 3$$

	Quantity A	Quantity B	
1.	2^{3x}	28	Ⓐ Ⓑ Ⓒ Ⓓ

$$\begin{array}{r} 0.AB \\ -\ 0.BA \\ \hline 0.72 \end{array}$$

In the correctly solved subtraction problem above, A and B represent digits.

	Quantity A	Quantity B	
2.	A	7	Ⓐ Ⓑ Ⓒ Ⓓ

The angles of a pentagon (5 sides) have measures 150°, 150°, 90°, 90°, and x°.

	Quantity A	Quantity B	
3.	x	60	Ⓐ Ⓑ Ⓒ Ⓓ

	Quantity A	Quantity B	
4.	$3^{25} + 3^{25} + 3^{25}$	3^{26}	Ⓐ Ⓑ Ⓒ Ⓓ

$$a^3b^2 > 0$$
$$a^2b^3 < 0$$

	Quantity A	Quantity B	
5.	$a + b$	0	Ⓐ Ⓑ Ⓒ Ⓓ

Ⓐ	**Quantity A is greater.**
Ⓑ	**Quantity B is greater.**
Ⓒ	**The two quantities are equal.**
Ⓓ	**The relationship cannot be determined from the information given.**

Frequency Distribution of test scores for 86 students

Test scores	50	60	70	80	90	100
Frequency	3	13	21	27	15	7

Quantity A | Quantity B

6. The average (arithmetic mean) of the 86 scores | The median of the 86 scores

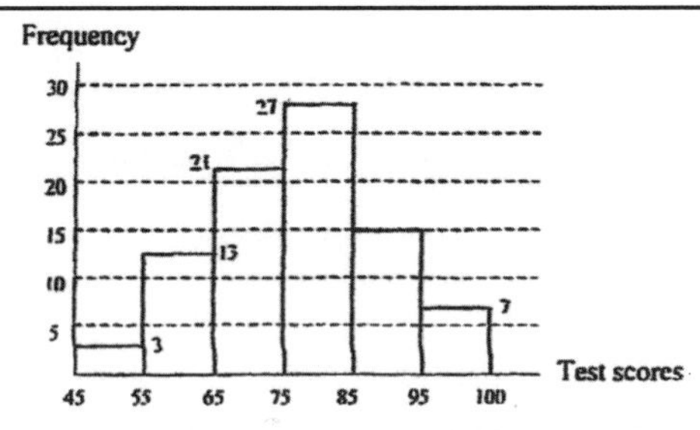

In a high school, the test scores of 86 students were recorded. The test scores are grouped into 6 convenient intervals. The histogram above shows the frequency distribution of the 86 scores by score intervals.

Quantity A | Quantity B

7. The average (arithmetic mean) of the 86 scores | The median of the 86 scores

$$a+b=-5$$

Quantity A | Quantity B

8. a | b

Quantity A | Quantity B

9. The total number of diagonals of a polygon with 9 sides | 25

GO ON TO THE NEXT PAGE.

Equations 10 to 25 have several different formats. Unless otherwise directed, select a single answer choice. For Numeric Entry questions, follow the instructions below.

Numeric Entry Questions:
Enter your answer in the answer box(es) below the question.
1. **Your answer may be an integer, a decimal, or a fraction, and it may be negative.**
2. **If a question asks for a fraction, there will be two boxes —— one for the numerator and one for the denominator.**
3. **Equivalent forms of the correct answer, such as 2.5 and 2.50, are all correct. Fractions do not need to be reduced to lowest terms.**
4. **Enter the exact answer unless the equation asks you to round your answer.**

10. A 5,000 cubic inches of water is poured into a cylindrical container with base radius 10 inches. What is the height in inches of the water ?

Ⓐ $50/\pi$ Ⓑ $40/\pi$ Ⓒ $30/\pi$ Ⓓ $20/\pi$ Ⓔ $10/\pi$

For the following question, enter your answer in the box.

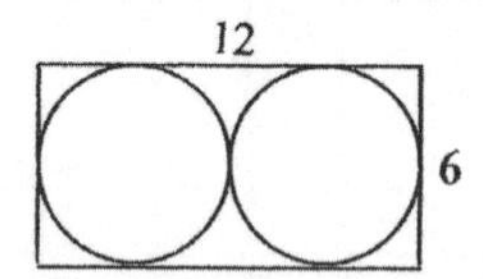

11. Suppose you throw a dart at the rectangle as target shown above. Assume the dart is equally likely to hit any point inside the target. What is the probability that the dart hits the region outside the circles ?

Give your answer as a decimal to the nearest 0.001. [ ]

For the following question, select all the answer choices that apply.

12. The range of the heart beats per minutes of the male students is 19. The range of the heart beats per minutes of the female students is 15. Which of the following statements individually provide(s) sufficient information to determine the range of the heart beats per minute of all the students ?

Indicate all such numbers.

[A] The highest heart beats per minutes of the male students is 6 heart beats higher than the highest heart beats per minute of the female students.

[B] The lowest heart beats per minutes of the female students is 2 heart beats lower than the lowest heart beats of the male students.

[C] The highest heart beats per minutes of the female students is 13 heart beats higher than the lowest heart beats per minutes of the male students.

[D] The average (arithmetic mean) heart beats is 72 for the male students, and 70 for the female students.

13. In the rectangular coordinate plane, what is the length of the circumference of the circle $x^2+y^2=1$?

Ⓐ 2 Ⓑ 3 Ⓒ 2π Ⓓ 3π Ⓔ 4π

For the following question, enter your answer in the boxes.

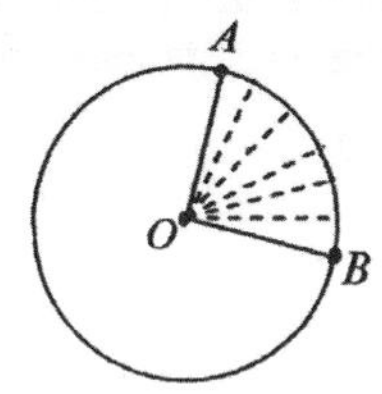

14. In the figure above, $<AOB=100^\circ$, what fraction of the circle is the region dashed ?

Give your answer as a fraction. 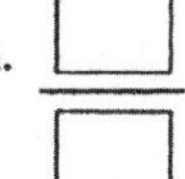

15. How many seat arrangements can be made for 6 students in a row if two of them want to sit next to each other ?

Ⓐ 60 Ⓑ 120 Ⓒ 180 Ⓓ 220 Ⓔ 240

16. A sequence is defined by $a_n=\frac{1}{n+1}-\frac{1}{n+3}$, where $n=1, 2, 3, \cdots, n$. What is the sum of 30 terms of this sequence ?

Ⓐ $\frac{1}{30}-\frac{1}{32}$ Ⓑ $\left(\frac{1}{2}+\frac{1}{3}\right)-\frac{1}{32}$ Ⓒ $\frac{1}{2}-\left(\frac{1}{30}+\frac{1}{32}\right)$

Ⓓ $\left(\frac{1}{2}+\frac{1}{3}\right)-\left(\frac{1}{32}+\frac{1}{33}\right)$ Ⓔ $\frac{1}{3}-\frac{1}{32}$

17. How many points of intersection are there with the parabola $y^2=x-5$ and the circle $x^2+y^2=36$?

Ⓐ one Ⓑ two Ⓒ three Ⓓ four Ⓔ none

For the following question, enter your answer in the box.

18. If $f(x)=3x+12$, what is the value of y for the equation $f(3y)=21$?

$y=$ ☐

GO ON TO THE NEXT PAGE.

19. A 25-gallon salt-water solution contains 40% pure salt. How much water in gallons should be added to produce the solution to 16% salt ?

Ⓐ 20.5 gallons Ⓑ 28.5 gallons Ⓒ 30.5 gallons Ⓓ 34.5 gallons Ⓔ 37.5 gallons

20. The scores of 10 games of a school sports team has a mean of 22 and a standard deviation of 3.5. What is the score of one game that is 2 standard deviations below the mean ?

Ⓐ 15 Ⓑ 18.5 Ⓒ 22 Ⓓ 22 Ⓔ 25.2

For the following question, select all the answer choices that apply.

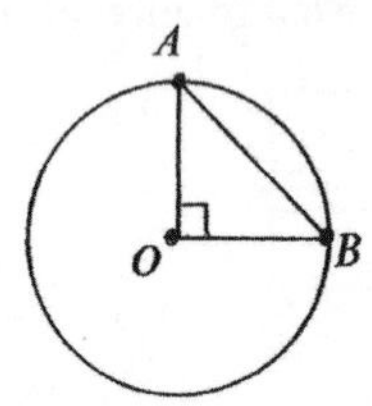

21. In the figure above, which of the following statements individually provide(s) sufficient information to determine the area of the circle with center O.

Indicate all such statements.

[A] The length of the chord AB is $\sqrt{2}$.
[B] The length of the minor arc AB is $\frac{1}{2}\pi$.
[C] The degree measure of the minor arc AB is 90.
[D] The circumference is 2π.
[E] The area of ΔAOB is $\frac{1}{2}$.

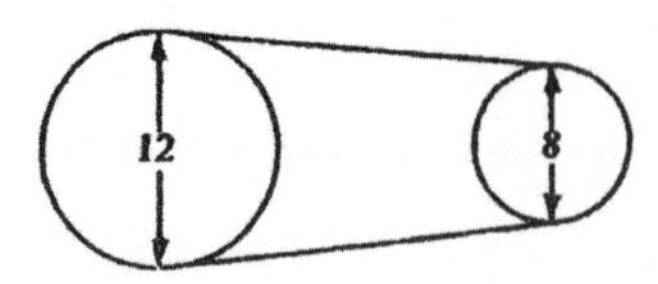

22. In the figure above, a large pulley is belted to a small pulley. If the larger pulley runs 80 revolutions per minutes, what is the speed of the small pulley, in revolutions per minute ?

Ⓐ 100 Ⓑ 110 Ⓒ 120 Ⓓ 130 Ⓔ 140

Questions 23 to 25 are based on the following Table.

ENROLLMENT AND TUITION AT DELTA UNIVERSITY FOR SELECTED YEARS

Years	2000	2005	2010
Number of Students Enrolled	5,080	5,390	5,740
Tuition per Student	$1,600	$1,800	$2,400
Income from Tuition	$8,128,000	$9,702,000	$13,776,000

23. At Delta University, the fraction of the number of students enrolled in 2000 to the number enrolled in 2010 is closest to

Ⓐ $\frac{5}{6}$ Ⓑ $\frac{6}{7}$ Ⓒ $\frac{7}{8}$ Ⓓ $\frac{8}{9}$ Ⓔ $\frac{9}{10}$

24. The increase in tuition per student from 2005 to 2010 was approximately how many times as high as the increase from 2000 to 2005 ?

Ⓐ $\frac{1}{3}$ Ⓑ $\frac{4}{5}$ Ⓒ $1\frac{1}{2}$ Ⓓ $2\frac{2}{3}$ Ⓔ 3

25. If the increase in the number of students enrolled from 2010 to 2015 is estimated half the increase from 2005 to 2010, what will be approximately the student enrollment in 2015 ?

Ⓐ 5,320 Ⓑ 5,910 Ⓒ 6,300 Ⓓ 6,420 Ⓔ 6,500

STOP: This is the end of Section 1.

NO TEST MATERIAL ON THIS PAGE

PRACTICE TEST 4

(GRE Revised General Test)
SECTION 2
Quantitative Reasoning
Time—40 minutes
25 Questions

For each question, indicate the best answer, using the directions given.

Notes: All numbers used are real numbers.

All figures are assumed to lie in a plane unless otherwise indicated.

Geometric figures, such as lines, circles, triangles, and quadrilaterals, **are not necessarily** drawn to scale. That is, you should **not** assume that quantities such as lengths and angle measures are as they appear in a figure. You should assume, however, that lines shown as straight are actually straight, points on a line are in the order shown, and more generally, all geometric objects are in the relative positions shown. For questions with geometric figures, you should base your answers on geometric reasoning, not on estimating or comparing quantities by sight or by measurement.

Coordinate system, such as *xy*-planes and number lines, **are** drawn to scale; therefore, you can read, estimate, or compare quantities in such figures by sight or by measurement.

Graphical data presentations, such as bar graphs, circles graphs, and line graphs, **are** drawn to scale; therefore, you can read, estimate, or compare data values by sight or by measurement.

For each of Questions 1 to 9, compare Quantity A and Quantity B, using additional information centered above the two quantities if such information is given. Select one of the following four answer choices and fill in the corresponding oval to the right to the question.

Ⓐ Quantity A is greater.

Ⓑ Quantity B is greater.

Ⓒ The two quantities are equal.

Ⓓ The relationship cannot be determined from the information given.

A symbol that appears more than once in a question has the same meaning throughout the question.

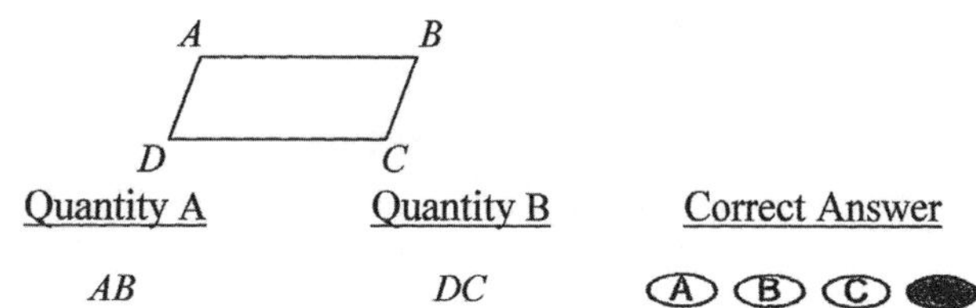

Example:

(A) Quantity A is greater.
(B) Quantity B is greater.
(C) The two quantities are equal.
(D) The relationship cannot be determined from the information given.

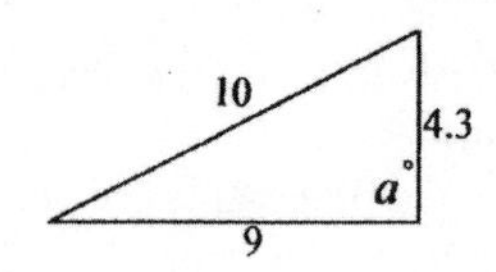

	Quantity A	Quantity B	
1.	a	90	

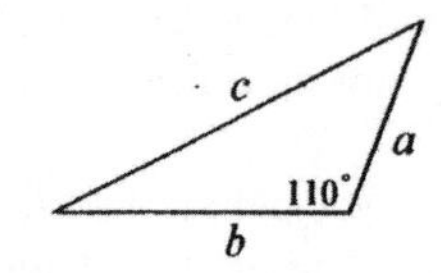

	Quantity A	Quantity B	
2.	$a^2 + b^2$	c^2	(A) (B) (C) (D)

	Quantity A	Quantity B	
3.	$\dfrac{1}{3+\dfrac{1}{4+\frac{1}{5}}}$	$\dfrac{1}{4}$	

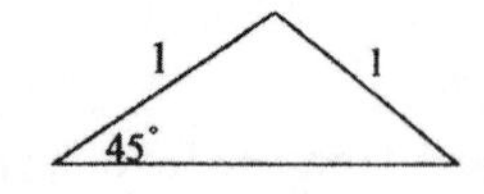

	Quantity A	Quantity B	
4.	The perimeter of the triangle above	$2+\sqrt{2}$	

n is a number.

	Quantity A	Quantity B	
5.	n^2	n^3	(A) (B) (C) (D)

Ⓐ	**Quantity A is greater.**
Ⓑ	**Quantity B is greater.**
Ⓒ	**The two quantities are equal.**
Ⓓ	**The relationship cannot be determined from the information given.**

The side length of a cube M is 4. The side length of a cube N is 3.

	Quantity A	Quantity B	
6.	The ratio of the volume of the cube M to the volume of the cube N	The ratio of the surface of the cube M to the surface area of the cube N	Ⓐ Ⓑ Ⓒ Ⓓ

The operating life of a new brand battery is normally distributed with a mean of 22 hours and a standard deviation of 3 hours.

	Quantity A	Quantity B	
7.	The probability of the event that a battery will last more than 25 hours.	$\frac{1}{10}$	Ⓐ Ⓑ Ⓒ Ⓓ

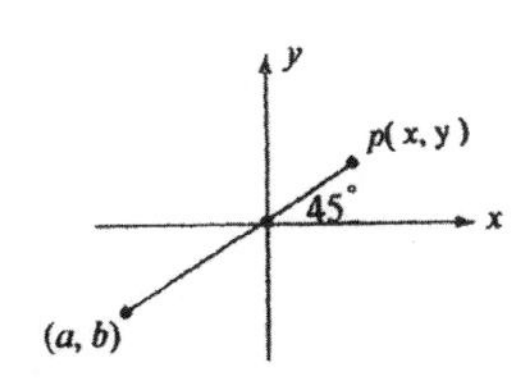

	Quantity A	Quantity B	
8.	a	b	Ⓐ Ⓑ Ⓒ Ⓓ

ΔABC is isosceles and $< ABC = 50^{\circ}$

	Quantity A	Quantity B	
9.	The sum of the measures of the two angles having equal measure	120	Ⓐ Ⓑ Ⓒ Ⓓ

GO ON TO THE NEXT PAGE.

Equations 10 to 25 have several different formats. Unless otherwise directed, select a single answer choice. For Numeric Entry questions, follow the instructions below.

Numeric Entry Questions:
Enter your answer in the answer box(es) below the question.
1. **Your answer may be an integer, a decimal, or a fraction, and it may be negative.**
2. **If a question asks for a fraction, there will be two boxes —— one for the numerator and one for the denominator.**
3. **Equivalent forms of the correct answer, such as 2.5 and 2.50, are all correct. Fractions do not need to be reduced to lowest terms.**
4. **Enter the exact answer unless the equation asks you to round your answer.**

For the following question, enter your answer in the box.

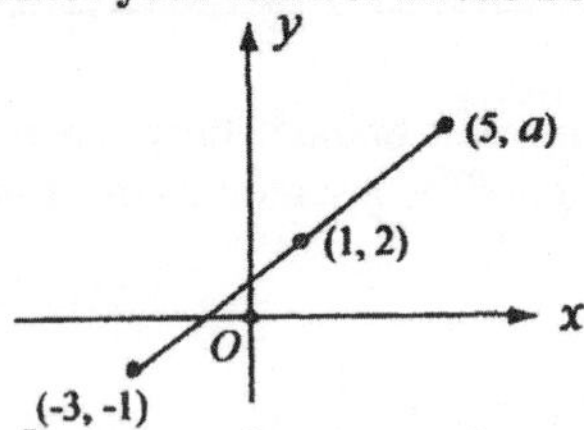

10. The rectangular coordinate plane above shows three points on the straight line. What is the value of a ?

$a =$ ☐

11. What is the units digit in the expansion of the number 7^{26} ?

Ⓐ 7 Ⓑ 9 Ⓒ 5 Ⓓ 3 Ⓔ 1

For the following question, select all the answer choices that apply.

12. If $|a+b| = |a| + |b|$, which of the following statements must be true ?
Indicate <u>all</u> such statements.
[A] $a > 0$ and $b > 0$
[B] $a > 0$ and $b < 0$
[C] $a < 0$ and $b < 0$
[D] The product of a and b are positive.

13. An alarm system uses a three-letter code and no letter can be used more than once. How many possible codes are there ?

Ⓐ 17,576 Ⓑ 16,900 Ⓒ 16,372 Ⓓ 15,900 Ⓔ 15,600

14. The average (arithmetic mean) of two numbers is $3m$. If one of the number is n, What is the other number ?

Ⓐ $3m - n$ Ⓑ $3m + n$ Ⓒ $6m - n$ Ⓓ $6m + n$ Ⓔ $6m + 3n$

For the following question, enter your answer in the box.

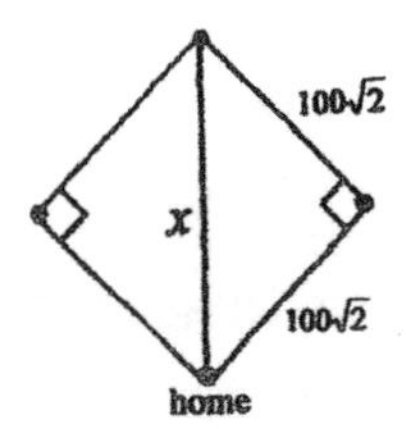

15. A baseball field is shaped as above. Each side is $100\sqrt{2}$ feet. What is the distance x in feet from the home plate to the second base ?

$x =$ ☐ feet

16. If $\frac{4}{5}$ of $\frac{3}{8}$ $=$ $\frac{1}{7}$ of $\frac{x}{10}$, what is the value of x ?

Ⓐ 25 Ⓑ 24 Ⓒ 23 Ⓓ 22 Ⓔ 21

Day	Number of computers sold
Monday	🖳🖳🖳🖳🖳🖳🖳🖳🖳
Tuesday	🖳🖳🖳🖳🖳🖳🖳
Wednesday	🖳🖳🖳🖳🖳
Thursday	🖳🖳🖳🖳
Friday	🖳🖳🖳🖳🖳🖳
Saturday	🖳🖳🖳🖳
Sunday	🖳🖳🖳🖳🖳🖳🖳

🖳= 10 computers

17. The pictograph above shows the data of computers sold within a week. What is the median of the number of computers sold during the week ?

Ⓐ 40 Ⓑ 50 Ⓒ 60 Ⓓ 70 Ⓔ 90

For the following question, enter your answer in the boxes.

18. If $(9^{3x+1})(3^{-x}) = 81$, what is the value of x ?

Give your answer as a fraction. $x = \frac{☐}{☐}$

19. If $\frac{3x-1}{6} - \frac{x-2}{12} = \frac{x-2}{6} - \frac{3x+1}{12}$, what is the value of x ?

Ⓐ $\frac{6}{5}$ Ⓑ $-\frac{6}{5}$ Ⓒ $\frac{5}{6}$ Ⓓ $-\frac{5}{6}$ Ⓔ $\frac{5}{12}$

GO ON TO THE NEXT PAGE.

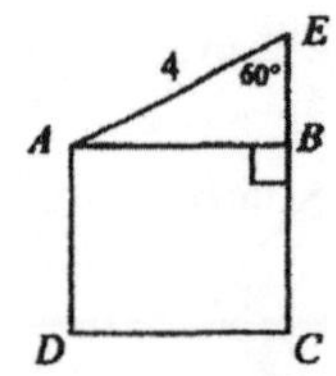

20. In the figure above, *ABCD* is a square. What is the area of the square ?

Ⓐ 10 Ⓑ 12 Ⓒ 14 Ⓓ 16 Ⓔ 18

21. A group of students has an average weight 170 pounds and a standard deviation 12. Which of the following values in pounds is more than 2.4 standard deviations from the mean (average weight) ?

Ⓐ 140 Ⓑ 150 Ⓒ 165 Ⓓ 172 Ⓔ 190

22. If $-1 < x < 0$, which of the following must be true ?

[A] $x^3 < x$ [B] $x^5 + x^3 < x^4 + x^2$ [C] $x^4 - x^3 < x^2 - x$

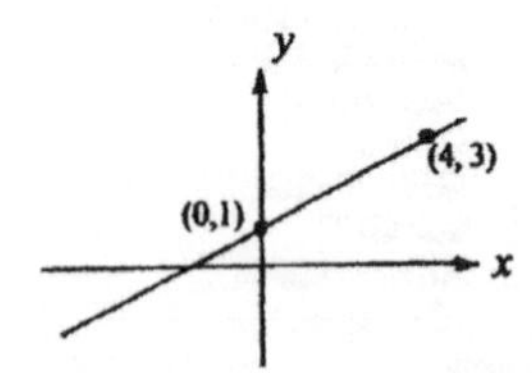

23. The figure above shows the graph of a straight line. If the line is translated horizontally 5 units to the left, what is the value of y-intercept of the translating line ?

Ⓐ $\frac{1}{2}$ Ⓑ $\frac{3}{2}$ Ⓒ 2 Ⓓ $\frac{5}{2}$ Ⓔ $\frac{7}{2}$

24. In a high school, the number of boys increased 10 % and the girls decreased by 12 % this year. The ratio of boys to girls at the end of this year was how many times the ratio at the beginning of this year ?

Ⓐ 1.25 Ⓑ 1.5 Ⓒ 1.75 Ⓓ 2 Ⓔ 2.25

For the following question, enter your answer in the box.

25. A club consists 20 boys and 30 girls. They want to select a girl as president, a boy as vice president, and a treasure who may be either sex. How many choices are possible ?

☐ choices

STOP: This is the end of Section 2.

PRACTICE TEST 4

(GRE Revised General Test)

Page 397 ~ 410

Answer Key

Section 1: **1.** B **2.** A **3.** C **4.** C **5.** D **6.** B **7.** D **8.** D **9.** A **10.** A

11. 0.215 **12.** A, B, and C **13.** C **14.** $\frac{5}{18}$ **15.** E **16.** D **17.** B

18. 1 **19.** E **20.** A **21.** A, B, D, and E **22.** C **23.** D **24.** E **25.** B

Section 2: **1.** B **2.** B **3.** A **4.** C **5.** D **6.** A **7.** A **8.** C **9.** D **10.** 5

11. B **12.** A, C, and D **13.** E **14.** C **15.** 200 **16.** E **17.** C **18.** $\frac{2}{5}$

19. D **20.** B **21.** A **22.** B and C **23.** E **24.** A **25.** 28,800

Add the number of correct answers to obtain your raw math score.
Partially correct answers are considered as incorrect.
Using the table below to convert your raw score to scaled score. The scaled scores reflect degrees of difficult and is comparable with scores of different tests at different dates. Note that any test may be somewhat more or less difficult than the test shown in this table year by year.
The scaled scores for GRE revised General Test will be 130 ~ 170.

GRE REVISED GENERAL MATH TEST CONVERSION TABLE

Raw Score	Scaled Score	Raw Score	Scaled Score	Raw Score	Scaled Score
0	130	18	143	36	158
1	130	19	144	37	159
2	130	20	144	38	160
3	131	21	145	39	161
4	131	22	146	40	161
5	132	23	146	41	162
6	132	24	147	42	163
7	133	25	148	43	164
8	134	26	149	44	165
9	135	27	150	45	166
10	136	28	151	46	167
11	137	29	152	47	168
12	138	30	153	48	169
13	139	31	153	49	170
14	140	32	154	50	170
15	141	33	155		
16	142	34	156		
17	142	35	157		

Answer Explanations

(PRACTICE TEST 4)

Section 1:

1. B $2^x = 3$, $2^{3x} = (2^x)^3 = 3^3 = 27$. Quantity B is greater.

2. A $0.91 - 0.19 = 0.72$ or $0.80 - 0.08 = 0.72$. We have $A = 9$ or 8. Quantity A is greater.

3. C The sum of the measures of the interior angles of a 5-sides pentagon is $180^\circ \times (5-2) = 540^\circ$. They are equal.
$x = 540 - (150 + 150 + 90 + 90) = 60$.

4. C $3^{25} + 3^{25} + 3^{25} = 3(3^{25}) = 3^{26}$. They are equal.

5. D $a^3b^2 > 0$, we have $a^3 > 0$ and $a > 0$ (positive). Cannot be determined.
$a^2b^3 < 0$, we have $b^3 < 0$ and $b < 0$ (negative).
$a + b$ could be positive, negative, or 0.

6. B The average is: Quantity B is greater.
$$\frac{50(3)+60(13)+70(21)+80(27)+90(15)+100(7)}{86} = \frac{6610}{86} \approx 76.86$$
The median of the 86 scores is the average of the middle two scores when the 86 scores are listed in increasing order. The scores of the two middle scores located at 43 and 44 are both 80. The median is 80.

7. D There are $3+13+21+27+15+7 = 86$ scores. Cannot be determined.
The smallest possible average is:
$$\frac{45(3)+55(13)+65(21)+75(27)+85(15)+95(7)}{86} \approx 71.86$$
The greatest possible average is:
$$\frac{55(3)+65(13)+75(21)+85(27)+95(15)+100(7)}{86} \approx 81.45$$
The median is the average of the middle two scores located at 43 and 44. The histogram shows that the median is greater than 75 or less than 85.

8. D $a + b = -5$ Cannot be determined.
If $a = 0$, we have $b = -5$, $a > b$.
If $b = 0$, we have $a = -5$, $b > a$.

9. A Apply sequence: Quantity A is greater.

Sides	4	5	6	7	8	9
Diagonals	2	5	9	14	20	27

+3 +4 +5 +6 +7

10. A Volume $= \pi \cdot 10^2 \cdot h = 5000$, $h = 50/\pi$.

11. 0.215 $p(\text{outside the circles}) = \frac{(12\times6)-(\pi\cdot3^2)\times2}{12\times6} = \frac{72-18\pi}{72} = \frac{4-\pi}{4} \approx 0.215$.

12. A, B, and C Draw the graph:

F M F M
a b c Range

We have the system of equations: $a + b = 15$ and $b + c = 19$.
Except D, each of the statements on A ($c = 6$), B ($a = 2$), or C ($b = 13$). provides data to solve the system. We have: Range $= a + b + c = 21$.

Answer Explanations
(PRACTICE TEST 4)

Section 1:

13. C — The radius of the circle $x^2 + y^2 = 1^2$ is 1. Circumference $= 2\pi r = 2\pi(1) = 2\pi$.

14. $\frac{5}{18}$ (or any equivalent fraction) — $\frac{100}{360} = \frac{50}{180} = \frac{25}{90} = \frac{5}{18}$.

15. E — | 2 | x | x | x | x | — $5! = 5 \cdot 4 \cdot 3 \cdot 2 \cdot 1 = 120$, $120 \times 2 = 240$.

16. D

$$\sum_{n=1}^{30}\left(\frac{1}{n+1} - \frac{1}{n+3}\right) = \sum_{n=1}^{30}\left(\frac{1}{n+1}\right) - \sum_{n=1}^{30}\left(\frac{1}{n+3}\right)$$

$$= \left(\frac{1}{2} + \frac{1}{3} + \frac{1}{4} + \frac{1}{5} + \cdots + \frac{1}{30} + \frac{1}{31}\right) - \left(\frac{1}{4} + \frac{1}{5} + \cdots + \frac{1}{30} + \frac{1}{31} + \frac{1}{32} + \frac{1}{33}\right)$$

$$= \left(\frac{1}{2} + \frac{1}{3}\right) - \left(\frac{1}{32} + \frac{1}{33}\right)$$

17. B — The parabola $y^2 = x - 5$ has vertex $(5, 0)$. The x – intercept is 5.
The circle $x^2 + y^2 = 36$ has center (0, 0) and radius 6. The x – intercept is 6.
The vertex of the parabola locates within the radius of the circle.
They intercept at two points only.

18. 1 — $f(x) = 3x + 12$, $f(3y) = 3(3y) + 12 = 21$, $9y + 12 = 21$, $y = 1$.

19. E — The original solution has $25 \times 40\% = 10$ gallons pure salt.
Let x = water added, we have the equation:
Solve $\frac{10}{25 + x} = 0.16$ for x, $x = 37.5$ gallons of water added.

20. A — (See page 72)
The score is 2 standard deviation below the mean of 22. The standard deviation of all games is 3.5. We have the score $22 - 2(3.5) = 15$.

21. A, B, D, and E — To determine the area of a circle, we need the length of its radius.
A. Apply Pythagorean Theorem, $r^2 + r^2 = (\sqrt{2})^2$, $r = 1$.
B. The minor arc AB is $\frac{1}{4}$ of the circumference. $\frac{1}{2}\pi = \frac{1}{4}(2\pi \cdot r)$, $r = 1$.
C. Degree measure of arc provides shape of the circle, not the size.
D. The circumference is $2\pi\ r = 2\pi$, $r = 1$.
E. Area of the triangle $= \frac{1}{2}r^2 = \frac{1}{2}$, $r = 1$.

22. C — $\frac{\pi \times 12 \times 80}{\pi \times 8} = \frac{960}{8} = 120$ revolutions per minute.

23. D — $\frac{5080}{5740} \approx 0.8850$;
A. $\frac{5}{6} \approx 0.8333$
B. $\frac{6}{7} \approx 0.8571$
C. $\frac{7}{8} = 0.8750$
D. $\frac{8}{9} \approx 0.8889 \leftarrow$ closest to 0.8850
E. $\frac{9}{10} = 0.9$

24. E — $\frac{2400 - 1800}{1800 - 1600} = \frac{600}{200} = 3$.

25. B — $\frac{5740 - 5390}{2} = 175$, $5740 + 175 = 5915 \approx 5910$ students.

Answer Explanations
(PRACTICE TEST 4)

Section 2: 1. B Apply Pythagorean Theorem: Quantity B is greater.

$9^2+(4.3)^2 \approx 99.5 < 10^2$, $a < 90$.

We need two longer sides to have 10 on the third side.

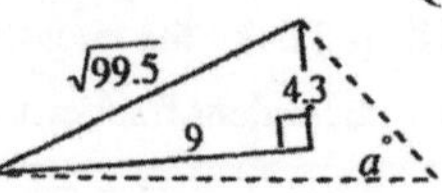

2. B Apply Pythagorean Theorem: Quantity B is greater.

The angle is larger than 90°.

Since a and b are shorter than the two corresponding sides of a right triangle.

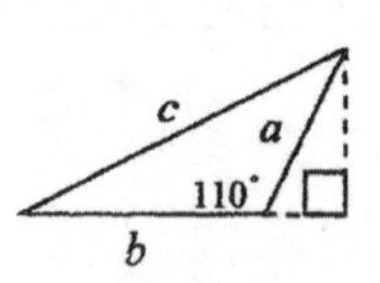

Hint: Apply trigonometry. $c^2 = a^2 + b^2 - 2ab\cos\theta$

If $0^\circ \le \theta < 90^\circ$, $\cos\theta$ is positive, $c^2 < a^2 + b^2$.

If $90^\circ < \theta \le 180^\circ$, $\cos\theta$ is negative, $c^2 > a^2 + b^2$.

3. A $\dfrac{1}{3+\dfrac{1}{4+\frac{1}{5}}} = \dfrac{1}{3+\dfrac{1}{21/5}} = \dfrac{1}{3+\frac{5}{21}} = \dfrac{1}{68/21} = \dfrac{21}{68} \approx 0.31$ Quantity A is greater.

4. C It is a 45-45-90 triangle. They are equal.

The length of the hypotenuse is $\sqrt{1^2+1^2}=\sqrt{2}$.

The perimeter of the triangle is $1+1+\sqrt{2}=2+\sqrt{2}$.

5. D If n is a positive integer, $n^2 \le n^3$. Cannot be determined.

If n is a negative integer, $n^2 \ge n^3$.

6. A The ratio of the volumes $= \dfrac{4\times4\times4}{3\times3\times3} = \dfrac{64}{27}$. Quantity A is greater.

The ratio of the surface areas $= \dfrac{6\times(4\times4)}{6\times(3\times3)} = \dfrac{16}{9} = \dfrac{48}{27}$.

7. A (See normal distribution on page 72.) Quantity A is greater.

The probability of an event having the value of 1 standard deviation above the mean is about 16 %.

8. C The point (a, b), the origin $(0, 0)$, and the y-axis form a 45-45-90 triangle having equal legs. They are equal.

9. D Based on the information given, we can form two different triangles: Cannot be determined.

50-50-80 triangle : $50 + 50 = 100$.

50-65-65 triangle : $65 + 65 = 130$.

10. 5 They are on a line having the same slope.

Slope $m = \dfrac{2-(-1)}{1-(-3)} = \dfrac{3}{4}$, solve $\dfrac{a-2}{5-1} = \dfrac{3}{4}$, $a = 5$.

11. B The units digits of 7^1, 7^2, 7^3, and 7^4 are 7, 9, 3, 1.

The units digits of 7^5, 7^6, 7^7, and 7^8 are 7, 9, 3, 1.

7th row: 7^{25}, 7^{26}, 7^{27}, and 7^{28} are 7, 9, 3, 1.

The pattern of 4 units digits 7, 9, 3, 1, repeat without end.

The exponent of the 1st. number of the 7th row is: $1+4(7-1)=25$.

Answer Explanations
(PRACTICE TEST 4)

Section 2: 12. A, C, and D

A. If $a>0$ and $b>0$, then $|a|+|b|=a+b$, $|a+b|=a+b$.

B. If $a>0$ and $b<0$, then $|3|+|-2|=3+2$, $|3-2|=1$.

C. If $a<0$ and $b<0$, then $|-3|+|-2|=3+2$, $|-3-2|=|-5|=5$.

D. If $a\times b$ are positive,
then a and b are both positive, or both negative.

13. E There are 26 English alphabetical letters.
$26\times25\times24=15{,}600$ codes.

14. C Let $x=$ the other number.
Solve $\frac{x+n}{2}=3m$ for x, we have $x=6m-n$.

15. 200 $x=\sqrt{(100\sqrt{2})^2+(100\sqrt{2})^2}=\sqrt{20000+20000}=\sqrt{40000}=200$ feet.

16. E $\frac{4}{5}\times\frac{3}{8}=\frac{1}{7}\times\frac{x}{10}$, $\frac{12}{40}=\frac{x}{70}$, $x=\frac{12\times70}{40}=21$.

17. C Arrange the data from least to greatest. The median is the middle number in the data. 40, 40, 50, **60**, 70, 70, 90
Hint: For the even number of items, the average of the two numbers in the middle is the median.

18. $\frac{2}{5}$ (or any equivalent fraction)

$(9^{3x+1})(3^{-x})=81$, $(3^2)^{3x+1}\cdot3^{-x}=3^4$, $3^{6x+2}\cdot3^{-x}=3^4$

$3^{6x+2-x}=3^4$, $3^{5x+2}=3^4$, $5x+2=4$, $5x=2$, $x=\frac{2}{5}$.

19. D $\frac{3x-1}{6}-\frac{x-2}{12}=\frac{x-2}{6}-\frac{3x+1}{12}$, $\frac{6x-2}{12}-\frac{x-2}{12}=\frac{2x-4}{12}-\frac{3x+1}{12}$,

$6x-2-x+2=2x-4-3x-1$

$5x=-x-5$, $6x=-5$, $x=-\frac{5}{6}$.

20. B ΔABE is a 30-60-90 triangle. $AE=4$, we have $BE=2$, $AB=2\sqrt{3}$.
Area of the square $=2\sqrt{3}\times2\sqrt{3}=12$.

21. A (See page 72) $170+2.4(12)=198.8$

$170-2.4(12)=141.2$

The value must be **more than** 2.4 standard deviations above the mean or below the mean. $x<141.2$ or $x>198.8$.

22. B and C $-1<x<0$ indicates x is a negative fraction between -1 and 0.

A. $(-\frac{1}{2})(-\frac{1}{2})(-\frac{1}{2})>-\frac{1}{2}$, $x^3>x$. $x^3<x$ is incorrect.

B. x^5+x^3 is negative. x^4+x^2 is positive. $x^5+x^3<x^4+x^2$ is correct.

C. If $(x^4-x^3)<x^2-x$,

$x^3(x-1)<x(x-1)$, divide each side by $x-1$ (a negative).

$x^3>x$, divide each side by x (a negative).

$x^2<1$ is correct. $(-\frac{1}{2})(-\frac{1}{2})<1$

Answer Explanations
(PRACTICE TEST 4)

Section 2: 23. E The slope of the line: $m = \frac{3-1}{4-0} = \frac{2}{4} = \frac{1}{2}$.

The equation of the line: $y = \frac{1}{2}x + 1$.

The equation of the translating line:

$$y' = \frac{1}{2}(x+5) + 1$$
$$y' = \frac{1}{2}x + \frac{5}{2} + 1$$
$$y' = \frac{1}{2}x + \frac{7}{2}\text{ , } y\text{-intercept is } \frac{7}{2}.$$

24. A Let $\frac{B}{G}$ = the ratio of boys to girls at the beginning of year.

At the end of the year, we have:

$$\frac{1.10B}{0.88G} = \frac{110}{88}\left(\frac{B}{G}\right) = \frac{5}{4}\left(\frac{B}{G}\right) = 1.25\left(\frac{B}{G}\right)$$

25. 28,800 $20 \times 30 \times 48 = 28{,}800$ choices.

Notes

Notes

PRACTICE TEST 5
(GMAT Test)

QUANTITATIVE SECTION
Time–75 minutes
37 Questions

Problem Solving Questions: Solve each problem and indicate the best of the answer choices given.
Numbers: All numbers used are real numbers.
Figures: A figure accompanying a problem solving question is intended to provide information useful in solving the problem. Figures are drawn as accurately as possible EXCEPT when it is stated in a specific problem that its figure is not drawn to scale. Lines that appear jagged can be assumed to be straight..
All figures lie in a plane unless otherwise indicated.

Data Sufficiency Questions: It consists of a question and two statements, labeled (1) and (2), in which certain data are given. You have to decide whether the data given in the statements are sufficient for answering the question.
Numbers: All numbers used are real numbers.
Figures: A figure accompanying a data sufficiency question will conform to the information given in the question but will not necessary conform to additional information given in statements (1) and (2). Lines that appear jagged can be assumed to be straight. All figures lie in a plane unless otherwise indicated.
The positions of points, angles, regions, etc. exist in the order shown and that angle measures are greater than zero.

To answer the question asked, you must indicate whether ~

A Statement (1) ALONE is sufficient, but statement (2) alone is not sufficient.
B Statement (2) ALONE is sufficient, but statement (1) alone is not sufficient.
C BOTH statements TOGETHER are sufficient, but NEITHER statement ALONE is sufficient.
D EACH statement ALONE is sufficient.
E Statements (1) and (2) TOGETHER are not sufficient.

1. a is what percent of b ?

(1) $a = 0.875\,b$
(2) $8 \times a = 7 \times b$

2. Roger rented a car. The car rental was $100 plus $0.45 per mile. If the total bill was $199, how many miles did he drive by using this car ?

(A) 200 miles
(B) 210 miles
(C) 220 miles
(D) 230 miles
(E) 240 miles

GO ON TO THE NEXT PAGE.

3. How many degree of arc are there in $\frac{3}{5}$ of a circle ?

(A) 210°
(B) 216°
(C) 220°
(D) 224°
(E) 228°

4. In the rectangular coordinate system, what is the $x-$intercept of the linear equation $5x+4y=18$?

(A) 4.5
(B) 4.2
(C) 3.9
(D) 3.6
(E) 2.4

5. If you deposited $8,500 in a saving account at an annual simple interest rate 4 %, how much will be in the account after two years ?

(A) $8,840
(B) $8,960
(C) $9,020
(D) $9,100
(E) $9,180

6. If you deposited $8,500 in a saving account at an annual interest rate 4 %, compounded yearly, how much will be in the account approximately after two years ?

(A) $9,194
(B) $9,246
(C) $9,298
(D) $9,320
(E) $9,384

A B C

7. In the figure above, what is the length of segment AB ?

(1) AB is one-third of AC.
(2) $AC=9$

8.

Group	Average Weight	No. of students
A	170 pounds	120
B	171 pounds	140

To compare two groups of data above, does the data in group A has more spread out about mean (average weight) than group B ?

(1) The standard deviation of data in group A is 12.
(2) Group A has greater standard deviation than that of Group B.

9. A group of students has an average weight 170 pounds and a standard deviation 12. Which of following values is more than 2.4 standard deviations from the mean (average weight) ?

(A) 190 pounds
(B) 172 pounds
(C) 165 pounds
(D) 150 pounds
(E) 140 pounds

GO ON TO THE NEXT PAGE.

10. If a, b, and c are positive integers, is $a < b < c$?

(1) $ab < bc$
(2) $ac < bc$

11. If a, b, and c are positive integers, is $a < b < c$?

(1) $ab < ac$
(2) $ac < bc$

12. What is the value of $x^2 - y^2$?

(1) $x+y=8$
(2) $x-y=2$

13. What is the value of $x^2 - y^2$?

(1) $x+y=0$
(2) $x-y=2$

14. What is the value of $x^2 + y^2$?

(1) $x+y=8$
(2) $x-y=2$

15. What is the value of $x^2 + y^2$?

(1) $x+2y=8$
(2) $2x+4y=16$

16. You saved $120 by purchasing a suitcase at a 20 percent discount on the list price. What is the list price of the suitcase ?

(A) $550
(B) $600
(C) $650
(D) $700
(E) $750

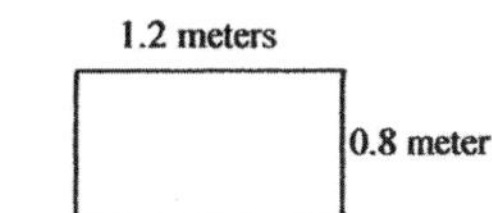

Note: Figure not drawn to scale.

17. The graph above shows the length and the width of a rectangular field on a map. If the actual length is 144 meters, what is the actual width ?

(A) 94 meters
(B) 96 meters
(C) 98 meters
(D) 100 meters
(E) 102 meters

18. Which of the following numbers is the median among these numbers ?

(A) 0.1×10^{-2}
(B) 0.001×10^{2}
(C) 0.01×10^{-3}
(D) 0.001×10^{-1}
(E) 1000×10^{-5}

GO ON TO THE NEXT PAGE.

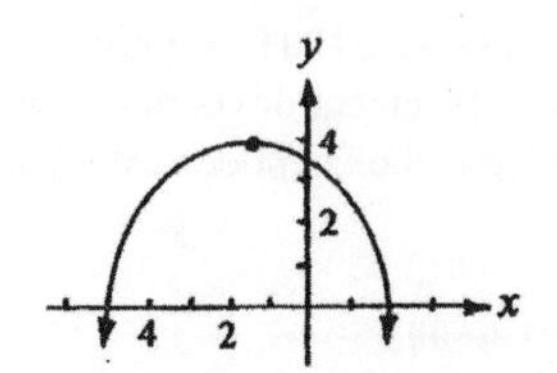

19. The figure above shows the graph of a quadratic function $y = f(x)$. The maximum value of the function is $f(a) = 4$. Which of the following could be the value of a ?

(A) 2
(B) 1.5
(C) 0
(D) −1.5
(E) −5

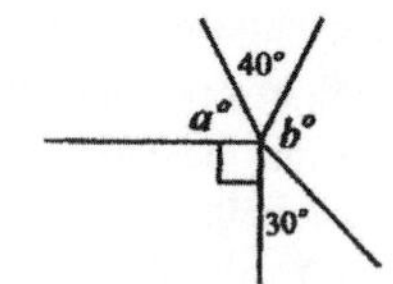

Note: Figure not drawn to scale.

20. In the figure above, what is the value of a in terms of b ?

(A) $260 - b$
(B) $240 - b$
(C) $220 - b$
(D) $200 - b$
(E) $180 - b$

21. What are all values of x for which $|x+5| > 9$?

(A) $x < 4$ or $x > 14$
(B) $-5 < x < 9$
(C) $5 < x < 9$
(D) $4 < x < 14$
(E) $x < -14$ or $x > 4$

22. If $a \neq -b$, What is the value of $a - b$?

(1) $a = b + \sqrt{40}$

(2) $\dfrac{a^2 - b^2}{a + b} = \sqrt{40}$

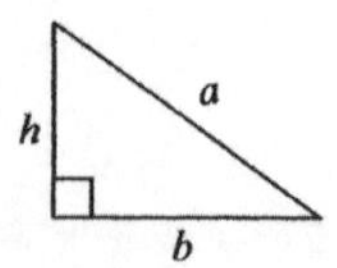

23. What is the area of the triangle above ?

(1) $a + b = 8$
(2) $b \times h = 12$

24. There are 20 boys and 15 girls in a class. A committee to be formed consisting of 3 boys and 3 girls. How many different committees are possible ?

(A) 518,700
(B) 429,500
(C) 330,000
(D) 332,000
(E) 234,000

GO ON TO THE NEXT PAGE.

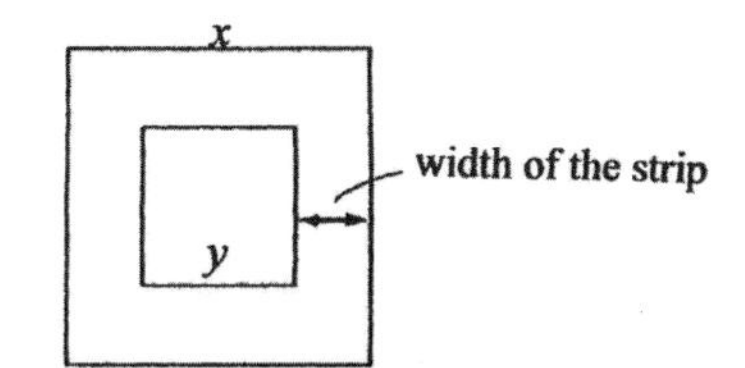

Note: Figure not drawn to scale.

25. In the figure above, A small square is placed in the center of a larger square. The small square has a side length y and the larger square has a side length x. The ratio of the area of the small square to the area of the larger square is 9 to 16. Which of the following could be the width of the strip around small square ?

(A) $\frac{1}{2}x$
(B) $\frac{1}{3}x$
(C) $\frac{1}{4}x$
(D) $\frac{1}{6}x$
(E) $\frac{1}{8}x$

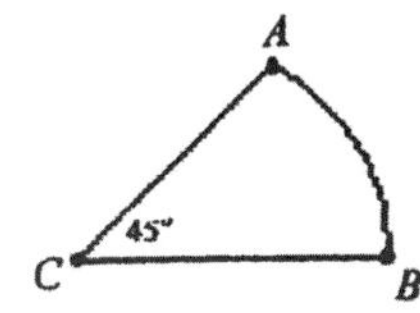

26. In the figure above, AB is the arc of a circle. If the length of the arc AB is 3π, what is the area of the sector ABC ?

(A) 14π
(B) 15π
(C) 16π
(D) 17π
(E) 18π

27. Is n an even integer ?

(1) n^2 is an even integer.
(2) $\sqrt{n}$ is an even integer.

28. N is an integer. Is N divisible by 20 ?

(1) N is divisible by 4.
(2) N is divisible by 5.

29. N is an integer. Is N divisible by 40 ?

(1) N is divisible by 4.
(2) N is divisible by 10.

30. N is an integer. Is N a multiple of 28 ?

(1) N is a multiple by 4.
(2) N is a multiple by 7.

31. N is an integer. Is N a factor of 24 ?

(1) N is a factor of 8.
(2) N is a factor of 18.

32. N is an integer. Is N a factor of 48 ?

(1) N is a factor of 8.
(2) N is a factor of 24.

GO ON TO THE NEXT PAGE.

Questions 33-34: refer to the following information.

A and B are two sets of students. A denotes the set of 31 students who are taking Algebra. B denotes the set of 38 students who are taking Geometry. Among them, there are 8 students who are taking both courses.

33. How many students are taking Algebra or Geometry only, but not both ?

(A) 53
(B) 52
(C) 51
(D) 50
(E) 49

34. There are 70 students in the class. How many students who are not taking either class ?

(A) 10
(B) 9
(C) 8
(D) 7
(E) 6

35. 100 salmons were caught and tagged by the scientists. Then, returned to the river. A few days later, 50 were caught again , of these 2 have tags. Estimate how many salmons are in the river ?

(A) 500
(B) 900
(C) 1,500
(D) 2,200
(E) 2,500

36. $(3\sqrt{6}-6\sqrt{3})\,(3\sqrt{6}+6\sqrt{3})=$

(A) −108
(B) −54
(C) −36
(D) 54
(E) 108

37. How many multiples of 6 are there between 24 and 1,000, inclusive ?

(A) 165
(B) 164
(C) 163
(D) 162
(E) 161

STOP

IF YOU FINISH BEFORE TIME IS CALLED, YOU MAY CHECK YOUR WORK ON THIS SECTION ONLY. DO NOT TURN TO ANY OTHER SECTION IN THE TEST.

PRACTICE TEST 5
(GMAT Test)

Quantitative Section Answer Key (Page 419 ~ 424)

1. D	**2.** C	**3.** B	**4.** D	**5.** E	**6.** A	**7.** C	**8.** B	**9.** E	**10.** E
11. C	**12.** C	**13.** A	**14.** C	**15.** E	**16.** B	**17.** B	**18.** A	**19.** D	**20.** D
21. E	**22.** D	**23.** B	**24.** A	**25.** E	**26.** E	**27.** D	**28.** C	**29.** E	**30.** C
31. A	**32.** D	**33.** A	**34.** B	**35.** E	**36.** B	**37.** C			

CALCULATE YOUR MATH SCORE

	Correct	Incorrect
Questions 1 ~ 37	( )	( )

Total Unrounded Raw Score ( ) – ( ) × 0.25 = ( )
Total Rounded Raw Score ( )

The actual GMAT multiple-choice test scores are determined by a complex procedure. The quantitative (math) and verbal scores are combined (37 + 41) to determine your total GMAT score on multiple-choice questions. It will vary from the score table presented here.
Using the table below to convert your raw score to scaled score. Without verbal score, you may double your math raw score and use the table to estimate your total GMAT scaled score on multiple-choice questions.

GMAT VERBAL AND MATH TESTS CONVERSION TABLE

Raw Score	Scaled Score	Raw Score	Scaled Score	Raw Score	Scaled Score
0 to 10	200 ~ 340	34	520	58	650
11	350	35	520	59	650
12	360	36	530	60	660
13	370	37	530	61	670
14	380	38	540	62	670
15	390	39	540	63	680
16	400	40	550	64	680
17	410	41	550	65	690
18	420	42	560	66	700
19	430	43	560	67	710
20	440	44	570	68	720
21	450	45	570	69	730
22	460	46	580	70	740
23	460	47	590	71	750
24	470	48	590	72	760
25	470	49	600	73	770
26	480	50	600	74 to 78	780 ~ 800
27	480	51	600		
28	490	52	610		
29	490	53	610		
30	500	54	620		
31	500	55	620		
32	510	56	630		
33	510	57	640		

Answer Explanations
(PRACTICE TEST 5)

1. D (1) $a = 0.875\,b = 87.5\,\%\,b$; Sufficient.
   (2) $8\,a = 7\,b$, $a = \frac{7}{8}\,b = 0.875\,b = 87.5\%\,b$; Sufficient.

2. C $100 + 0.45x = 199$, $x = \dfrac{199-100}{0.45} = \dfrac{99}{0.45} = 220$ miles.

3. B $360^\circ \times \frac{3}{5} = 216^\circ$.

4. D $5x + 4y = 18$. Let $y = 0$, $5x + 4(0) = 18$, $x = 3.6$.

5. E $8500 + (8500 \times 0.04) \times 2 = 8500 + 340 \times 2 = 8500 + 680 = 9180$.

6. A $8500 \times (1 + 0.04)^2 = 8500 \times (1.04)^2 = 8500 \times 1.0816 = 9193.60 \approx 9194$.

7. C (1) $AB = \frac{1}{3}(AC)$. No information about the length of AC; Not Sufficient.
   (2) $AC = 9$. No information about the lengths relating AB and AC; Not Sufficient.
   Together: $AB = \frac{1}{3}AC = \frac{1}{3}(9) = 3$; Sufficient.

8. B (1) No information about the standard deviation of data in group B; Not Sufficient.
   (2) Data with greater standard deviation are more spread (dispersion) out about the mean; Sufficient.

9. E (see page 72) $170 + 2.4(12) = 198.8$.
   $170 - 2.4(12) = 141.2$.
   The value x must be more than 2.4 standard deviations below or above the mean.
   $x < 141.2$ or $x > 198.8$.
   The only value outside the range of 141.2 and 198.8 is 140.

10. E (1) $ab < bc$, $a < c$. No information relating b to either a or c; Not Sufficient.
    (2) $ac < bc$, $a < b$. No information relating c to either a or b; Not Sufficient.
    Together: Still no information relating b and **c**; Not Sufficient.

11. C (1) $ab < ac$, $b < c$. No information relating a to either b or c; Not Sufficient.
    (2) $ac < bc$, $a < b$. No information relating c to either a or b; Not Sufficient.
    Together: The ordering of the three numbers can be determined; Sufficient.

12. C (1) $x + y = 8$. No information about the value of $(x - y)$; Not Sufficient.
    (2) $x - y = 2$. No information about the value of $(x + y)$; Not sufficient.
    Together: $x^2 - y^2 = (x + y)(x - y) = 8 \times 2 = 16$; Sufficient.

13. A (1) $x + y = 0$. $x^2 - y^2 = (x + y)(x - y) = 0 \times (x - y) = 0$; Sufficient.
    (2) $x - y = 2$. No information about the value of $(x + y)$; Not Sufficient.

14. C (1) $x + y = 8$. The values of the two variables are unknown; Not Sufficient.
    (2) $x - y = 2$. The values of the two variables are unknown; Not Sufficient.
    Together: Solve the system of two equations, we have $x = 5$ and $y = 3$; Sufficient.

15. E (1) $x + 2\,y = 8$ The values of the two variables are unknown; Not Sufficient.
    (2) $2\,x + 4\,y = 16$ The values of the two variables are unknown; Not Sufficient.
    Together: The two equations are consistent and dependent; Not Sufficient.

16. B Let x = list price, $x \times 0.2 = 120$, $x = \dfrac{120}{0.2} = 600$.

Answer Explanations
(PRACTICE TEST 5)

17. B $\frac{144}{1.2}\times 0.8 = 96$ meters.

18. A Arrange the numbers from the least to the greatest. The number in the middle is the median.
$0.1\times 10^{-2} = 0.001$, $0.001\times 10^{2} = 0.1$, $0.01\times 10^{-3} = 0.00001$, $0.001\times 10^{-1} = 0.0001$, $1000\times 10^{-5} = 0.01$.
0.00001, 0.0001, **0.001**, 0.01, 0.1

19. D Maximum of y locates at $f(-1.5)=4$, $a=-1.5$.

20. D $a+b+90+40+30=360$, $a+b+160=360$, $a=200-b$.

21. E $|x+5|>9$, we have $x+5<-9$ or $x+5>9$
$x<-14$. $x>4$.

22. D (1) $a=b+\sqrt{40}$, $a-b=\sqrt{40}$; Sufficient.
(2) $\frac{a^2-b^2}{a+b}=\sqrt{40}$, $\frac{(a+b)(a-b)}{a+b}=\sqrt{40}$, $a-b=\sqrt{40}$; Sufficient.

23. B (1) $a+b=8$. The values of the two lengths a and b are unknown ; Not Sufficient.
(2) $b\times h=12$. Area $=\frac{1}{2}\times b\times h=\frac{1}{2}\times 12=6$; Sufficient.

24. A ${}_{20}C_3\times{}_{15}C_3=\frac{20!}{17!\times 3!}\times\frac{15!}{12!\times 3!}=\frac{20\times 19\times 18}{6}\times\frac{15\times 14\times 13}{6}=1140\times 455=518{,}700$ committees.

25. E Given the ratio of the area of the small square to the area of the larger square is $\frac{9}{16}$.
Then, the ratio of the side length of the small square to the side length of the larger square is: $\frac{y}{x}=\sqrt{\frac{9}{16}}=\frac{3}{4}$.
The width of the strip is: $w=\frac{x-y}{2}=\frac{x-\frac{3}{4}x}{2}=\frac{\frac{1}{4}x}{2}=\frac{1}{8}x$.

26. E The circumference of the circle is: $c=3\pi\times\frac{360}{45}=24\pi=2\pi r$, radius $r=12$.
Area $=\pi\times 12^2\times\frac{45}{360}=144\pi\times\frac{1}{8}=18\pi$.
For any positive value of x, there is a width w. All choices are possible.

27. D (1) n^2 is even. The product of two or more even integers is always even ; Sufficient.
(2) $\sqrt{n}$ is an even integer. $\sqrt{n}\times\sqrt{n}=n$ is also even ; Sufficient.

28. C (1) Numbers which are divisible by 4 → 4, 8, 12, ····, **20**, ····, **40**, ····, **60**, ···; Not Sufficient.
(2) Numbers which are divisible by 5 → 5, 10, 15, **20**, ····, **40**, ····, **60**, ···; Not Sufficient.
Together: Numbers which are divisible by 4 and 5 → **20, 40, 60, 80**, ···; Sufficient.

29. E (1) Numbers which are divisible by 4 → 4, 8, 12, ····, **20**, ····, **40**, ····, **60**, ···; Not Sufficient.
(2) Numbers which are divisible by 10 → 10, **20**, 30, **40**, 50, **60**, ····, ···· ; Not Sufficient.
Together: Numbers which are divisible by 4 and 10 → 20, **40, 60, 80**, ···; Not Sufficient.

30. C (1) Numbers which are divisible by 4 → 4, 8, 12, ····, **28**, ·····, **56**, ·····, **74**, ···; Not Sufficient.
(2) Numbers which are divisible by 7 → 7, 14, 21, **28**, ·····, **56**, ·····, **63**, ···; Not Sufficient.
Together: Number which are divisible by 4 and 7 → **28, 56, 84**, ·····; Sufficient.

31. A (1) Numbers which are factors of **8 → 1, 2, 4, 8** ; Sufficient.
(2) Numbers which are factors of 18 → **1, 2, 3, 6**, 9, 18 ; Not sufficient.

Answer Explanations
(PRACTICE TEST 5)

32. D (1) Numbers which are factors of 8 $\rightarrow$ **1, 2, 4, 8** ; Sufficient.
(2) Numbers which are factors of 24 $\rightarrow$ **1, 2, 3, 4, 6, 8, 12, 24** ; Sufficient.

33. A Draw a Venn Diagram.
There are $31 - 8 = 23$ students taking Algebra only.
$38 - 8 = 30$ students taking Geometry only.
$23 + 30 = 53$ students taking Algebra or Geometry only, but not both.

34. B $70 - (23 + 30 + 8) = 70 - 61 = 9$ students. Or: $70 - (31 + 38 - 8) = 70 - 61$ students.

35. E $100 \times \frac{50}{2} = 2500$ salmons.

36. B $(3\sqrt{6} - 6\sqrt{3})(3\sqrt{6} + 6\sqrt{3}) = (3\sqrt{6})^2 - (6\sqrt{3})^2 = 54 - 108 = -54.$

37. C $6 \times 4 = 24,\ 6 \times 5 = 30,\ \cdots\cdots,\ 6 \times 166 = 996.$
There are $166 - 3 = 163$ multiples of 6 between 24 and 1,000.

PRACTICE TEST 6
(GMAT Test)

QUANTITATIVE SECTION
Time–75 minutes
37 Questions

Problem Solving Questions: Solve each problem and indicate the best of the answer choices given.

Numbers: All numbers used are real numbers.

Figures: A figure accompanying a problem solving question is intended to provide information useful in solving the problem. Figures are drawn as accurately as possible EXCEPT when it is stated in a specific problem that its figure is not drawn to scale. Lines that appear jagged can be assumed to be straight.. All figures lie in a plane unless otherwise indicated.

Data Sufficiency Questions: It consists of a question and two statements, labeled (1) and (2), in which certain data are given. You have to decide whether the data given in the statements are sufficient for answering the question.

Numbers: All numbers used are real numbers.

Figures: A figure accompanying a data sufficiency question will conform to the information given in the question but will not necessary conform to additional information given in statements (1) and (2). Lines that appear jagged can be assumed to be straight. All figures lie in a plane unless otherwise indicated.

The positions of points, angles, regions, etc. exist in the order shown and that angle measures are greater than zero.

To answer the question asked, you must indicate whether ~

A Statement (1) ALONE is sufficient, but statement (2) alone is not sufficient.
B Statement (2) ALONE is sufficient, but statement (1) alone is not sufficient.
C BOTH statements TOGETHER are sufficient, but NEITHER statement ALONE is sufficient.
D EACH statement ALONE is sufficient.
E Statements (1) and (2) TOGETHER are not sufficient.

1. $3\sqrt{18}-2\sqrt{8}+6\sqrt{50}=$

(A) $32\sqrt{2}$
(B) $33\sqrt{2}$
(C) $34\sqrt{2}$
(D) $35\sqrt{2}$
(E) $36\sqrt{2}$

2. If $x^{\otimes}$ is defined such that $x^{\otimes}=x^3-2x$, what is the value of $4^{\otimes}-2^{\otimes}$?

(A) 51
(B) 52
(C) 53
(D) 54
(E) 55

GO ON TO THE NEXT PAGE.

3. If $a > b$ and $c < 0$, which of the following must be true ?

I. $ac < bc$
II. $ac > bc$
III. $a + c > b + c$

(A) I only
(B) II only
(C) I and II only
(D) I and III only
(E) II and III only

4. If x and y are nonzero integers, what is the ratio of x to y ?

(1) $x^2 = y^2$
(2) $x = 1$

5. What is the value of a ?

(1) $a = -a$
(2) $a + b = 1$

6. In a 300 grams sugar-water solution, the ratio of sugar to water is 1 to 5. How many grams of sugar are there ?

(A) 65 grams
(B) 60 grams
(C) 50 grams
(D) 45 grams
(E) 40 grams

7. The side lengths of a triangle are a, b, and c. Is the triangle a right triangle ?

(1) $a = 6$
(2) $b^2 + c^2 = 36$

8. What is the simple annual interest rate if you earned interest \$18 on \$1,200 after 3 months ?

(A) 4.5 %
(B) 6 %
(C) 6.5 %
(D) 7 %
(E) 7.5 %

9. The average (arithmetic mean) of 40 integers is what percent of the sum of the 40 integers ?

(A) 25 %
(B) 18 %
(C) 4.5 %
(D) 3.5 %
(E) 2.5 %

10. What is the perimeter of rectangle $ABCD$?

(1) The rectangle is a square.
(2) $AB = 8$

11. In a math test, 60 percent of the all 30 girls passed the test, all of the boys passes the test. What was the total number of students who passed the test ?

(1) There are 40 students in the class.
(2) There are 70 percent of the students who passed the test

GO ON TO THE NEXT PAGE.

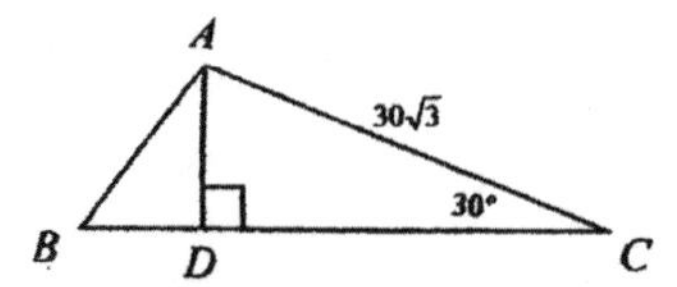

12. In the figure above, what is the measure of $\overline{CD}$?

(A) 42
(B) 43
(C) 44
(D) 45
(E) 46

13. If $\frac{x-6}{x}=\frac{11}{12}$, then x =

(A) 72
(B) 71
(C) 70
(D) 69
(E) 68

14. What is the sum of the measures of the interior angles in degrees of a polygon with 18 sides ?

(A) 2880
(B) 2790
(C) 2630
(D) 2500
(E) 1460

15. Is $x^3 < 0$?

(1) $x < 0$
(2) $-x > 0$

16. A bag contains 10 red balls and 8 white balls. 7 balls are drawn at random from the bag. How many of the balls left in the bag are red ?

(1) 3 of the first 5 balls drawn are white.
(2) Among the 7 balls drawn, the ratio of the number of red balls to the number of white balls is $4:3$.

17. Is $x = -9$?

(1) $x^2 - 81 = 0$
(2) $x^2 + 8x - 9 = 0$

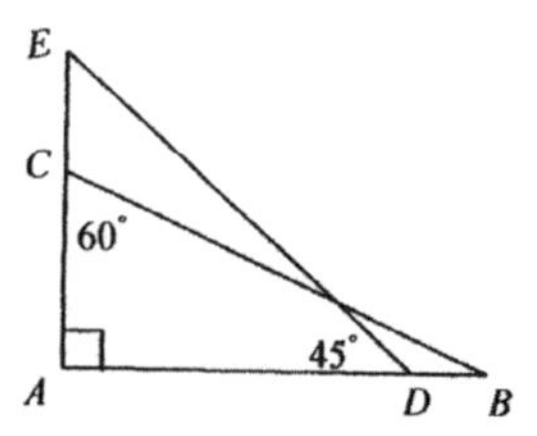

18. In the figure above, $\overline{BC} = \overline{DE}$, what is the value of $\overline{AB} - \overline{AD}$?

(1) $\overline{BC} = 12$
(2) $\overline{AD} = 6\sqrt{2}$

GO ON TO THE NEXT PAGE.

19. Among the senior students in a high school, there are $\frac{1}{10}$ of the students who are taking Calculus, $\frac{1}{3}$ of the students who are taking Pre-Calculus, and $\frac{2}{5}$ of the students who are taking Statistics. Each of the above students takes only one of the above math courses. The remaining 40 students who are not taking any of the above math courses. How many senior students are there ?

(A) 240
(B) 260
(C) 280
(D) 300
(E) 320

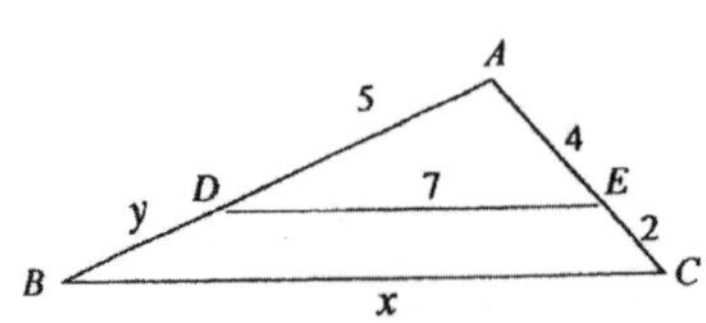

Note: Figure not drawn to scale.

20. In the figure above, $DE // BC$, what is the value of $x+y$?

(A) 16
(B) 15
(C) 14
(D) 13
(E) 12

21. If x is a prime number, what is the value of x ?

(1) $6 < x < 20$
(2) $(x-7)$ is a factor of 16.

22. If x is a prime number, what is the value of x ?

(1) $6 < x < 20$
(2) $(x-7)$ is a multiple of 5.

Party	Candidate 1	Candidate 2	Undecided
Rep.	280	160	200
Dem.	240	320	180
Ind.	180	240	200

23. The above table shows the result of the poll of 2,000 people to determine their voting preference by political party for each candidate in U.S. What percent of the people had decided their voting preferences on the candidates ?

(A) 29 %
(B) 46 %
(C) 54 %
(D) 62 %
(E) 71 %

24. If $2^{4n-5} = 128$, what is the value of n ?

(A) 1
(B) 2
(C) 3
(D) 4
(E) 5

25. If $ab > 0$, does $(a-1)(b-1) = 1$?

(1) $ab = 2$
(2) $ab = a + b$

26. x is a positive integer. Is $\sqrt{x}$ an integer ?

(1) $x = 36$
(2) $x = n^2$ and n is an integer.

GO ON TO THE NEXT PAGE.

27. What is the smallest integer having 4, 8, and 11 as factors ?

(A) 88
(B) 121
(C) 176
(D) 352
(E) 704

28. If $2^x + 2^x + 2^x + 2^x = 128$, what is the value of x ?

(A) 2
(B) 3
(C) 4
(D) 5
(E) 6

29. If n is a number among the set of numbers 24, 27, 32, 45, 66, 70, 83, what is the value of n ?

(1) n is odd.
(2) n is divisible by 9.

30. In a high school class, what is the ratio of the number of boys to the number of girls ?

(1) In the class, $\frac{3}{11}$ of the students are boys and $\frac{8}{11}$ of the students are girls.
(2) There are 12 boys and 32 girls in the class.

31. What are all values of x for which $|x+5| < 9$?

(A) $5 < x < 9$
(B) $4 < x < 9$
(C) $-4 < x < 4$
(D) $-9 < x < 4$
(E) $-14 < x < 4$

32. A company had a total sale for this year 3 times the average of the sales for the past 9 years. What is the ratio of the sale this year to the total sales for the whole 10 years ?

(A) $\frac{3}{5}$
(B) $\frac{2}{5}$
(C) $\frac{1}{5}$
(D) $\frac{1}{2}$
(E) $\frac{1}{4}$

33. Scientists estimate that the diameter of the solar system is about 118,000,000,000 miles. Which of the following is the correct way to write this number in scientific notation ?

(A) 11.8×10^{10}
(B) 118×10^{9}
(C) 1.18×10^{11}
(D) 0.118×10^{12}
(E) 1.18×10^{12}

GO ON TO THE NEXT PAGE.

34. A box contains 40 red balls and 40 white balls. Three balls are drawn at random with replacement. What is the probability that the three balls drawn at least one ball is red ?

(A) $\frac{1}{8}$
(B) $\frac{1}{4}$
(C) $\frac{3}{8}$
(D) $\frac{1}{2}$
(E) $\frac{5}{8}$

35. If $x \neq 1$ and -3, then $\dfrac{x^3+3x^2-x-3}{x^2+2x-3}=$

(A) $x+1$
(B) $x-1$
(C) x^2+x-3
(D) x^2+1
(E) x^2-1

36. One-third of all Matts are Patts. Half of all Patts are Natts. No Matt is a Natt. All Natts are Patts. There are 25 Natts and 45 Matts. How many Patts are neither Matts nor Natts ?

(A) 8
(B) 10
(C) 12
(D) 14
(E) 16

37. How many copies were sold of the Algebra books ?

(1) The copies of Algebra books sold were 5,000 copies more than the copies of Geometry books sold.
(2) The copies of Calculus books sold were 2,000 copies less than the copies of Geometry books sold.

STOP

IF YOU FINISH BEFORE TIME IS CALLED, YOU MAY CHECK YOUR WORK ON THIS SECTION ONLY. DO NOT TURN TO ANY OTHER SECTION IN THE TEST.

PRACTICE TEST 6
(GMAT Test)

Quantitative Section **Answer Key** **(Page 429 ~ 434)**

1. D **2.** B **3.** D **4.** E **5.** A **6.** C **7.** C **8.** B **9.** E **10.** C

11. D **12.** D **13.** A **14.** A **15.** D **16.** B **17.** C **18.** D **19.** A **20.** D

21. C **22.** E **23.** E **24.** C **25.** B **26.** D **27.** A **28.** D **29.** E **30.** D

31. E **32.** E **33.** C **34.** C **35.** A **36.** B **37.** E

CALCULATE YOUR MATH SCORE

	Correct	Incorrect
Questions 1 ~ 37	( )	( )

Total Unrounded Raw Score ( ) − ( ) × 0.25 = ( )
Total Rounded Raw Score ( )

The actual GMAT multiple-choice test scores are determined by a complex procedure. The quantitative (math) and verbal scores are combined (37 + 41) to determine your total GMAT score on multiple-choice questions. It will vary from the score table presented here.
Using the table below to convert your raw score to scaled score. Without verbal score, you may double your math raw score and use the table to estimate your total GMAT scaled score on multiple-choice questions.

GMAT VERBAL AND MATH TESTS CONVERSION TABLE

Raw Score	Scaled Score	Raw Score	Scaled Score	Raw Score	Scaled Score
0 to 10	200 ~ 340	34	520	58	650
11	350	35	520	59	650
12	360	36	530	60	660
13	370	37	530	61	670
14	380	38	540	62	670
15	390	39	540	63	680
16	400	40	550	64	680
17	410	41	550	65	690
18	420	42	560	66	700
19	430	43	560	67	710
20	440	44	570	68	720
21	450	45	570	69	730
22	460	46	580	70	740
23	460	47	590	71	750
24	470	48	590	72	760
25	470	49	600	73	770
26	480	50	600	74 to 78	780 ~ 800
27	480	51	600		
28	490	52	610		
29	490	53	610		
30	500	54	620		
31	500	55	620		
32	510	56	630		
33	510	57	640		

Answer Explanations
(PRACTICE TEST 6)

1. D $3\sqrt{18}-2\sqrt{8}+6\sqrt{50}=3\sqrt{(9)(2)}-2\sqrt{(4)(2)}+6\sqrt{(25)(2)}=9\sqrt{2}-4\sqrt{2}+30\sqrt{2}=35\sqrt{2}$.

2. B $x^{\otimes}=x^3-2x$
   $4^{\otimes}-2^{\otimes}=4^3-2(4)-\left(2^3-2(2)\right)=64-8-4=52$.

3. D $a>b$, $c<0$ indicates c is negative.
   I. $ac<bc$ is true. Inequality changes sign (< or >) if it is multiplied by a negative number.
   II. $ac>bc$ is not true.
   III. $a+c>b+c$ is true. Inequality will not change sign if it is added a same number on each side.

4. E (1) $x^2=y^2$, $\left(\frac{x}{y}\right)^2=1$, $\frac{x}{y}=\pm 1$.
   No single value of the ratio can be determined ; Not Sufficient.
   (2) $x=1$. No information about the value of y ; Not Sufficient.
   Together: The single value of ratio still cannot be determined ; Not sufficient.

5. A (1) $a=-a$, $2a=0$, $a=0$; Sufficient.
   (2) $a+b=1$. No information about the value of b ; Not sufficient.

6. C $300\times\frac{1}{6}=50$ grams.

7. C (1) $a=6$. No information about the side lengths of the other two sides; Not Sufficient.
   (2) $b^2+c^2=36$. No information about the side length of a ; Not Sufficient.
   Together: Apply Pythagorean Theorem, $b^2+c^2=a^2$; Sufficient.

8. B The simple monthly interest rate is: $r=\dfrac{18\div 3}{1200}=\dfrac{6}{1200}=0.005$.
   The simple annual interest rate is: $r'=0.005\times 12=0.06=6\%$.

9. E Let s = the sum of the 40 integers, $\frac{s}{40}$ is the average.
   We have the ratio $\dfrac{\frac{s}{40}}{s}=\dfrac{s}{40}\cdot\dfrac{1}{s}=\dfrac{1}{40}=0.025=2.5\%$.

10. C (1) The rectangle is a square. No information about the side length ; Not Sufficient.
   (2) $AB=8$. No information about the other three side-lengths ; Not Sufficient.
   Together: Perimeter of the square is: 8 + 8 + 8 +8 =32 ; Sufficient.

11. D (1) There are 18 girls and 10 boys passed the test . Total $18+10=28$; Sufficient.
   (2) Solve $0.70(30+b)=18+b$, $b=10$ (boys) ; Total $18+10=28$; Sufficient.

12. D (See page 229)
   $\overline{AD}=\frac{1}{2}(30\sqrt{3})=15\sqrt{3}$, $\overline{CD}=\sqrt{3}(15\sqrt{3})=45$.

13. A $\dfrac{x-6}{x}=\dfrac{11}{12}$, $12(x-6)=11x$, $12x-72=11x$, $x=72$.

14. A $180\times(18-2)=180\times 16=2{,}880$.

15. D (1) If $x<0$ (negative), then x^3 is negative ; Sufficient.
   (2) If $-a>0$, then $a<0$ (negative), x^3 is negative ; Sufficient.

Answer Explanations
(PRACTICE TEST 6)

16. B (1) The color of the last two balls drawn is unknown. Not sufficient.
(2) $7\times\frac{4}{7}=4$ red balls were drawn. $10-4=6$ red balls are left ; Sufficient.

17. C (1) $x^2=81$, $x=\pm 9$. No single value of x can be determined ; Not sufficient.
(2) $x^2+8x-9=0$, $(x+9)(x-1)=0$, $x=-9$ and 1. No single value of x ; Not Sufficient.
Together: $x=-9$ only ; Sufficient.

18. D They are 36-60-90 and 45-45-90 special right triangles in the figure.
(1) $\overline{BC}=\overline{DE}=12$ is given. The length of the other two sides can be determined ; Sufficient.
(2) $\overline{AD}=6\sqrt{2}$ and $\overline{BC}=\overline{DE}$ are given, $\overline{DE}$, $\overline{BC}$, and $\overline{AB}$ can be determined ; Sufficient.

19. A There are $\frac{1}{10}+\frac{1}{3}+\frac{2}{5}=\frac{3}{30}+\frac{10}{30}+\frac{12}{30}=\frac{25}{30}=\frac{5}{6}$ of the students who take only one course.
Let $x=$ total students; $x\times\frac{1}{6}=40$, $x=240$ students.

20. D $\frac{4}{6}=\frac{7}{x}$, $4x=42$, $x=10.5$. ; $\frac{4}{6}=\frac{5}{5+y}$, $20+4y=30$, $4y=10$, $y=2.5$.
$x+y=10.5+2.5=13$.

21. C (1) $6<x<20$. Prime numbers: 7, 11, 13, 17, and 19.
No single prime number can be determined ; Not Sufficient.
(1) $(x-7)$ is a factor of 16 when $x=11$ and 23.
No single prime number of x can be determined ; Not Sufficient.
Together: $x=11$ only ; Sufficient.

22. E (1) $6<x<20$, Prime numbers: 7, 11, 13, 17, and 19.
No single prime number of x can be determined ; Not Sufficient.
(2) $(x-7)$ is a multiples of 5 when $x=7, 12, 17, 22, 27, \cdots$.
No single prime number of x can be determined ; Not Sufficient.
Together: $x=7$ and 17. Still no single value of x can be determined ; Not Sufficient.

23. E $\frac{200+180+200}{2000}=\frac{580}{2000}=0.29=29\%$ (undecided), $1-29\%=71\%$ (decided).

24. C $2^{4n-5}=128$, $2^{4n-5}=2^7$, $4n-5=7$, $4n=12$, $n=3$.

25. B (1) $ab=2$, $(a-b)(b-1)=ab-(a+b)+1$. The value of $(a+b)$ is unknown ; Not Sufficient.
(2) $ab=a+b$, $(a-1)(b-1)=ab-(a+b)+1=ab-ab+1=1$; Sufficient.

26. D (1) $x=36$, $\sqrt{x}=\sqrt{36}=6$ is an integer ; Sufficient.
(2) $x=n^2$, $\sqrt{x}=\sqrt{n^2}=n$ is an integer ; Sufficient.

27. A Find the least common denominator (LCM) = $4\times1\times2\times11=88$.
$$\begin{array}{r|l} 4 & 4,\ 8,\ 11 \\ \hline & 1,\ 2,\ 11 \end{array}$$

28. D $2^x+2^x+2^x+2^x=128$
$4\times 2^x=2^7$
$2^2\times 2^x=2^7$, $2^{2+x}=2^7$, $2+x=7$, $x=5$.

Answer Explanations
(PRACTICE TEST 6)

29. E (1) 27, 45, 83; No single value of x can be determined ; Not Sufficient.
(2) 27 and 45 are divisible by 9; No single value of x can be determined ; Not Sufficient.
Together: $x = 27$ and 45; No single value of x can be determined ; Not Sufficient.

30. D (1) Ratio $= \dfrac{n \times \frac{3}{11}}{n \times \frac{8}{11}} = \dfrac{3}{8}$; Sufficient.
(2) Ratio $= \dfrac{12}{32} = \dfrac{3}{8}$; Sufficient.

31. E $|x+5| < 9$, we have: $-9 < x+5 < 9$
$-14 < x < 4$.

32. E Let s = the average of sales for the past 9 years.
We have: $\dfrac{3s}{9s+3s} = \dfrac{3s}{12s} = \dfrac{3}{12} = \dfrac{1}{4}$.

33. C $118{,}000{,}000{,}000 = 1.18 \times 10^{11}$.

34. C Red, Red, Red : $\frac{1}{2} \times \frac{1}{2} \times \frac{1}{2} = \frac{1}{8}$
Red, Red, White : $\frac{1}{2} \times \frac{1}{2} \times \frac{1}{2} = \frac{1}{8}$
Red, White, White : $\frac{1}{2} \times \frac{1}{2} \times \frac{1}{2} = \frac{1}{8}$
We have the probability that at least one ball is red: $\frac{1}{8} + \frac{1}{8} + \frac{1}{8} = \frac{3}{8}$.

35. A $\dfrac{x^3+3x^2-x-3}{x^2+2x-3} = \dfrac{x^2(x+3)-(x+3)}{x^2+2x-3} = \dfrac{\cancel{(x+3)}(x^2-1)}{\cancel{(x+3)}(x-1)} = \dfrac{(x+1)\cancel{(x-1)}}{\cancel{x-1}} = x+1$.

36. B Draw a Venn Diagram:

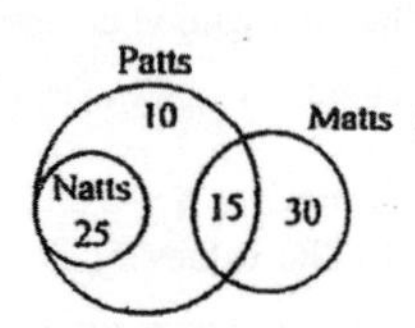

37. E Let A: copies of Algebra , G: copies of Geometry , C: copies of Calculus .
(1) $A = G + 5000$: One equation with 2 unknowns; Not Sufficient.
(2) $C = G - 2000$: One equation with 2 unknowns; Not Sufficient.
Together: The system of two equations has 3 unknown; Not Sufficient.

PRACTICE TEST 7
(GMAT Test)

QUANTITATIVE SECTION
Time–75 minutes
37 Questions

Problem Solving Questions: Solve each problem and indicate the best of the answer choices given.
Numbers: All numbers used are real numbers.
Figures: A figure accompanying a problem solving question is intended to provide information useful in solving the problem. Figures are drawn as accurately as possible EXCEPT when it is stated in a specific problem that its figure is not drawn to scale. Lines that appear jagged can be assumed to be straight..
All figures lie in a plane unless otherwise indicated.

Data Sufficiency Questions: It consists of a question and two statements, labeled (1) and (2), in which certain data are given. You have to decide whether the data given in the statements are sufficient for answering the question.
Numbers: All numbers used are real numbers.
Figures: A figure accompanying a data sufficiency question will conform to the information given in the question but will not necessary conform to additional information given in statements (1) and (2). Lines that appear jagged can be assumed to be straight. All figures lie in a plane unless otherwise indicated.
The positions of points, angles, regions, etc. exist in the order shown and that angle measures are greater than zero.

To answer the question asked, you must indicate whether ~

A Statement (1) ALONE is sufficient, but statement (2) alone is not sufficient.
B Statement (2) ALONE is sufficient, but statement (1) alone is not sufficient.
C BOTH statements TOGETHER are sufficient, but NEITHER statement ALONE is sufficient.
D EACH statement ALONE is sufficient.
E Statements (1) and (2) TOGETHER are not sufficient.

---

1. What is the value of x ?

(1) $x^2 - 2x - 3 = 0$
(2) $2x - 3 = 3x - 2$

2. The average (arithmetic mean) of Roger's scores for the 8 math tests is 76. What does the average score of the next 2 math tests have to be if the average of the entire 10 tests equals 80 ?

(A) 85
(B) 89
(C) 92
(D) 96
(E) 98

GO ON TO THE NEXT PAGE.

3. If three points $(1, -1)$, $(2, 1)$, and $(4, a-2)$ are on a straight line, what is the value of a ?

(A) 3
(B) 4
(C) 5
(D) 6
(E) 7

4. A plane intersects a rectangular solid. Which of of the following could be the intersection ?

I. A triangle
II. A rectangle
III. A parallelogram

(A) I only
(B) II only
(C) *I and II only*
(D) II and III only
(E) I, II, and III

5. If $n!$ be defined such that
$$n! = n(n-1)(n-2) \cdots\cdots 3 \cdot 2 \cdot 1.$$
What is the value of $10! \div 8!$?

(A) 90
(B) 85
(C) 82
(D) 78
(E) 72

6. How many boys are in the class ?

(1) 45 students are in the class.
(2) The ratio of boys to girls is 4 : 5.

7. What is the average age of all 45 students in a class ?

(1) The highest age is 18, and the lowest age is 15.
(2) The average age of 20 boys is 17.5, and the average age of 25 girls is 16.

8. How many of the integers between 1 and 250, inclusive, are multiples of 4 or 5 ?

(A) 100
(B) 112
(C) 124
(D) 132
(E) 140

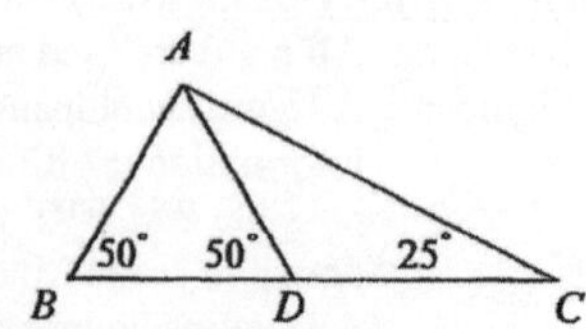

Note: Figure not drawn to scale.

9. In the triangle above, what is the length of side AB ?

(1) $AD = 7$
(2) $CD = 7$

10. In the flu season, 9 % of the students in a school were absent in November, and 11 % of the students were absent in December. If the number of the students was the same for these 2 months, how many students were absent in December ?

(1) The number of students was 3,200 in November.
(2) There were 64 more students in the school were absent in December than in November.

GO ON TO THE NEXT PAGE.

11. If x equals to two percent of y, then $200y=$

(A) $400x$
(B) $800x$
(C) $1{,}200x$
(D) $10{,}000x$
(E) $12{,}000x$

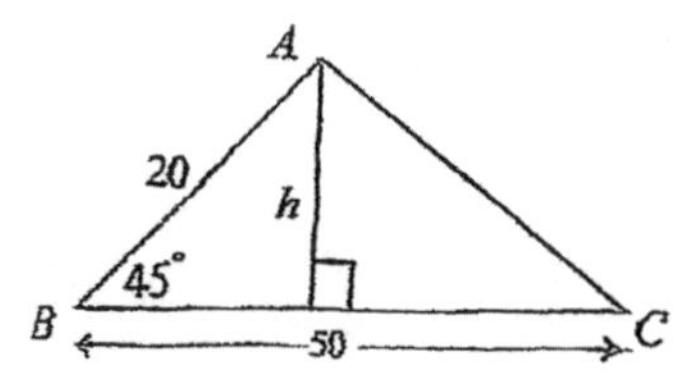

Note: Figure not drawn to scale.

12. In the figure above, what is the area of the triangle ABC ?

(A) $250\sqrt{2}$
(B) $200\sqrt{2}$
(C) $150\sqrt{2}$
(D) $100\sqrt{2}$
(E) $50\sqrt{2}$

13. A bottle of salt-water solution needs m liters of pure salt and $m \times n$ liters of water. The pure salt costs x dollars and water is considered no cost. The solution is sold for y dollars per liter. What is the profit of the solution ?

(A) $m(y+ny-x)$
(B) $m(n+y-x)$
(C) $m(y-x)$
(D) $m(ny-x)$
(E) $m(y-nx)$

14. If $a \neq 0$, what is the value of $\frac{a^x}{a^y}$?

(1) $x=y$
(2) $a=1$

15. What fraction of the total number of book sales does the Algebra book have ?

(1) The sales of Calculus book have one half as many sales as Geometry book.

(2) The sales of Geometry book have two-thirds as many sales as Algebra book.

16. The graph of a function is made after three transformations: reflect about the x-axis, shift up 4 units, and shift right 6 units, applied to the graph of $y=x^3$. Which of the following is the function ?

(A) $y=(x+6)^3-4$
(B) $y=-(x-6)^3+4$
(C) $y=-(x+6)^3-4$
(D) $y=(x-4)^3+6$
(E) $y=-(x-4)^3+6$

17. Are the integers a, b, c consecutive ?

(1) $a+c=2b$
(2) $a+b+c=3b$

GO ON TO THE NEXT PAGE.

18. Can n be written as the sum of two odd integers ?

(1) n is an integer greater than 1.
(2) n is even.

19. A triangle has the lengths of three sides 9, 19, and x, what are the possible values of x ?

(A) $10<x<28$
(B) $9<x<19$
(C) $10<x<19$
(D) $9<x<10$
(E) $6<x<9$

20. A triangle has an area of k. If all of the side lengths are increased by twice of the original lengths, what is the new area ?

(A) $2k$
(B) $4k$
(C) $6k$
(D) $8k$
(E) $9k$

21. A rectangular solid has a volume of k. If all of the side lengths are increased by twice of the original lengths, what is the new volume ?

(A) $2k$
(B) $3k$
(C) $6k$
(D) $8k$
(E) $9k$

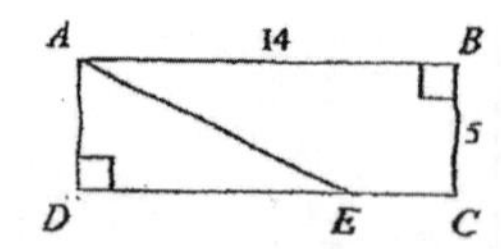

Note: Figure not drawn to scale.

22. In the figure above, $\overline{DE}$ is five-seventh the length of $\overline{DC}$. What is the area of quadrilateral $ABCE$?

(A) 43 (B) 44 (C) 45 (D) 46 (E) 47

23. There are 42 students in a class. 24 of them took Algebra class. 22 of them took Biology class. 2 of them took neither courses. How many students took both courses ?

(A) 8
(B) 7
(C) 6
(D) 5
(E) 4

24. How many different seat arrangements can be made for 5 students seated in a row if one of them does not want to sit in the first seat ?

(A) 120
(B) 96
(C) 84
(D) 72
(E) 58

25. A triangle has base b and height h. What is its area ?

(1) $h=3b+2$
(2) $2b=\frac{32}{h}$

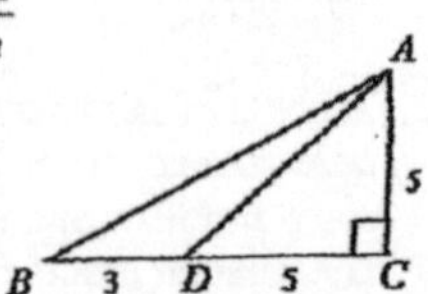

Note: Figure not drawn to scale.

26. In the figure above, what is the area of the triangle ABD ?

(A) 12.5
(B) 9.5
(C) 7.5
(D) 6.5
(E) 4.5

GO ON TO THE NEXT PAGE.

27. If $3^{4x+3} = 27^{2x-3}$, what is the value of x ?

(A) 7
(B) 6
(C) 5
(D) 4
(E) 3

28. How many lines can be drawn by 8 noncollinear points ?

(A) 25
(B) 26
(C) 27
(D) 28
(E) 29

29. David can complete a job in 6 hours. Maria can complete the same job in 12 hours. If they work together, how many hours will it take them to complete the job ?

(A) 3.0 hours
(B) 3.2 hours
(C) 3.5 hours
(D) 3.7 hours
(E) 4.0 hours

30. David can complete a job in 6 hours. How much less hours would it take to complete the job if David and Maria work together ?

(1) It takes Maria twice time as long as it takes David to complete the job.
(2) It takes David one and one-half time as long as it takes David and Maria together to complete the job.

31. Is the positive integer n a multiple of 36 ?

(1) n is a multiple of 9.
(2) n is a multiple of 4.

32. Is the positive integer n a multiple of 27 ?

(1) n is a multiple of 9.
(2) n is a multiple of 3.

33. Is n^3 an odd integer ?

(1) n is an odd integer.
(2) $\sqrt{n}$ is an odd integer.

34. Is x greater the 0 and less than 1 ?

(1) $x^2 < x$
(2) $x^2 > 0$

GO ON TO THE NEXT PAGE.

35. Lisa took a test that had two sections. Section 1 has 40 questions and section 2 has 50 questions. How many of the questions did she answered correctly ?

(1) The number of questions she answered correctly in Section 1 was 12 less than the number she answered correctly in Section 2.
(2) She answered 80 percent of the questions in Section 1 correctly and 90 percent of the questions correctly in Section 2.

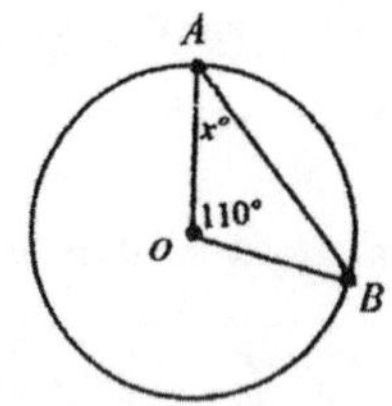

Note: Figure not drawn to scale.

36. In the figure above, o is the center of the circle. What is the value of x ?

(A) 25 (B) 30 (C) 35 (D) 40 (E) 45

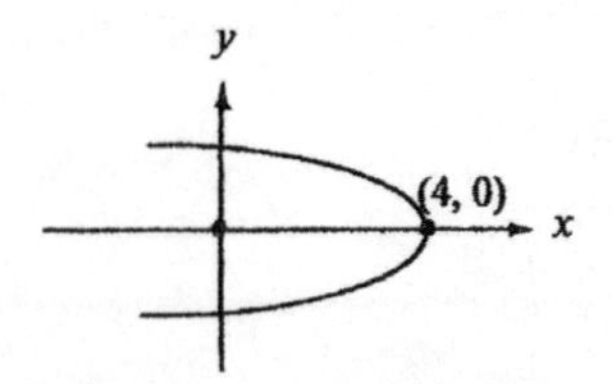

Note: Figure not drawn to scale.

37. The graph above is a parabola that is symmetric about the x-axis. Which of the following could be the equation of the parabola ?

(A) $y^2 = x - 4$
(B) $y^2 = -(x-4)$
(C) $y^2 = x + 4$
(D) $y^2 = -(x+4)$
(E) $y^2 = -x$

STOP

IF YOU FINISH BEFORE TIME IS CALLED, YOU MAY CHECK YOUR WORK ON THIS SECTION ONLY. DO NOT TURN TO ANY OTHER SECTION IN THE TEST.

PRACTICE TEST 7

(GMAT Test)

Quantitative Section Answer Key (Page 439 ~ 444)

1. B **2.** D **3.** E **4.** E **5.** A **6.** C **7.** B **8.** A **9.** D **10.** D

11. D **12.** A **13.** A **14.** D **15.** C **16.** B **17.** D **18.** B **19.** A **20.** B

21. D **22.** C **23.** C **24.** B **25.** B **26.** C **27.** B **28.** D **29.** E **30.** D

31. C **32.** E **33.** D **34.** A **35.** B **36.** C **37.** B

CACULATE YOUR MATH SCORE

	Correct	Incorrect
Questions 1 ~ 37	( )	( )

Total Unrounded Raw Score ( ) − ( ) × 0.25 = ( )
Total Rounded Raw Score ( )

The actual GMAT multiple-choice test scores are determined by a complex procedure. The quantitative (math) and verbal scores are combined (37 + 41) to determine your total GMAT score on multiple-choice questions. It will vary from the score table presented here.
Using the table below to convert your raw score to scaled score. Without verbal score, you may double your math raw score and use the table to estimate your total GMAT scaled score on multiple-choice questions.

GMAT VERBAL AND MATH TESTS CONVERSION TABLE

Raw Score	Scaled Score	Raw Score	Scaled Score	Raw Score	Scaled Score
0 to 10	200 ~ 340	34	520	58	650
11	350	35	520	59	650
12	360	36	530	60	660
13	370	37	530	61	670
14	380	38	540	62	670
15	390	39	540	63	680
16	400	40	550	64	680
17	410	41	550	65	690
18	420	42	560	66	700
19	430	43	560	67	710
20	440	44	570	68	720
21	450	45	570	69	730
22	460	46	580	70	740
23	460	47	590	71	750
24	470	48	590	72	760
25	470	49	600	73	770
26	480	50	600	74 to 78	780 ~ 800
27	480	51	600		
28	490	52	610		
29	490	53	610		
30	500	54	620		
31	500	55	620		
32	510	56	630		
33	510	57	640		

Answer Explanations
(PRACTICE TEST 7)

1. B (1) Solve $x^2-2x-3=0$, $(x-3)(x+1)=0$, $x=3$ and -1.
No Single x-value can be determined ; Not Sufficient.
(2) Solve $2x-3=3x-2$, $x=-1$; Sufficient.

2. D Let x = the average of the next 2 tests.
We have the equation and solve for x : $\frac{76\times 8+2x}{10}=80$, $x=96$.

3. E The slope of the line is: $\frac{1-(-1)}{2-1}=\frac{1+1}{1}=2$. The two line segments have the same slope. Solve the equation and solve for a : $\frac{(a-2)-1}{4-2}=2$, $a=7$.

4. E They are a rectangle, a parallelogram, and a triangle. If a plane slices through three adjacent faces at a corner of the rectangular solid, the intersection is a triangle.

5. A $\frac{10!}{8!}=\frac{10\cdot 9\cdot 8!}{8!}=10\cdot 9=90$.

6. C (1) No information to determine the numbers of boys or girls ; Not Sufficient.
(2) No information to determine the total number of the students ; Not Sufficient.
Together: $45\times\frac{4}{9}=20$ boys ; Sufficient.

7. B (1) No information to determine the total ages of 45 students ; Not Sufficient.
(2) Average age $=\frac{17.5\times 20+16\times 25}{45}\approx 16.7$; Sufficient.

8. A 1×4, 2×4, 3×4, ·····, $62\times 4=248$
1×5, 2×5, 3×5, ·····, $50\times 5=250$
1×20, 2×20, 3×20, ·····, $12\times 20=240$
We have $62+50-12=100$ multiples of 4 or 5.

9. D (1) ΔABD is isosceles. $AD=7$, $AB=AD=7$; Sufficient.
(2) $\angle ADC=180^\circ-50^\circ=130^\circ$ and $\angle CAD=180^\circ-130^\circ-25^\circ=25^\circ$.
ΔADC and ΔABD are isosceles. $AB=AD=CD=7$; Sufficient.

10. D (1) $3{,}200\times 0.11=352$ students were absent in December ; Sufficient.
(2) We can find the number of the students n :
$(0.11-0.09)\times n=64$, $n=3{,}200$.
$3{,}200\times 0.11=352$ students were absent in December ; Sufficient.

11. D $x=0.02y$, $y=\frac{x}{0.02}=50x$, $200y=200(50x)=10{,}000x$.

12. A Apply Pythagorean Theorem to find the height h :
$h^2+h^2=20^2$, $2h^2=400$, $h^2=200$, $h=\sqrt{200}=10\sqrt{2}$.
The area $=\frac{1}{2}(50)(10\sqrt{2})=250\sqrt{2}$.

13. A The sale price (P) of the solution is : $p=y(m+mn)$.
The cost (C) of the solution is : $c=mx$.
The profit $=y(m+mn)-mx=my+mny-mx=m(y+ny-x)$.

14. D (1) $x=y$, then $\frac{a^x}{a^y}=a^{x-y}=a^0=1$; Sufficient.
(2) $a=1$, then $\frac{a^x}{a^y}=a^{x-y}=1^{x-y}=1$; Sufficient.

Answer Explanations
(PRACTICE TEST)

15. C (1) Cal. = $\frac{1}{2}$ (Geo.). No information to determine the number of Algebra book. Not Sufficient.

(2) Geo.= $\frac{2}{3}$ (Alg.). No information to determine the number of Calculus book. Not Sufficient.

Together: Fraction (Algebra book sales)

$$=\frac{A}{C+G+A}=\frac{A}{\frac{1}{2}(\frac{2}{3}A)+\frac{2}{3}A+A}=\frac{1}{\frac{1}{3}+\frac{2}{3}+1}=\frac{1}{2};$$ Sufficient.

16. B $y=x^3$; Reflect about the x-axis : $y=-x^3$

Shift up 4 units: $y=-x^3+4$

Shift right 6 units: $y=-(x-6)^2+4$.

17. D The consecutive integers a, b, c are $n,\ n+1,\ n+2$.

(1) $a+c=n+n+2=2n+2=2(n+1)=2b$; Sufficient.

(2) $a+b+c=n+n+1+n+2=3n+3=3(n+1)=3b$; Sufficient.

18. B The sum (difference) of two odd integers is always even.

(1) If n is an odd integer, it can't. $3=1+2,\ 5=2+3,\ \cdots\cdots$; Not Sufficient.

(2) If n is an even integers, it can. $2=1+1,\ 4=1+3,\ \cdots\cdots$; Sufficient.

19. A The sum of the lengths of any two sides is greater than the length of the third side. The difference of the lengths of any two sides is less than the length of the third side.

20. B If two geometric figures are similar, the ratio of the areas (A) equals to the square of the ratio of the side lengths (S).

$$\left(\frac{A_1}{A_2}\right)=\left(\frac{S_1}{S_2}\right)^2=\left(\frac{2}{1}\right)^2=\frac{4}{1}.$$

21. D If two geometric figures are similar, the ratio of the volumes (V) equals to the cube of the ratio of the side lengths (S).

$$\left(\frac{V_1}{V_2}\right)=\left(\frac{S_1}{S_2}\right)^3=\left(\frac{2}{1}\right)^3=\frac{8}{1}.$$

22. C $\overline{DE}=14\times\frac{5}{7}=10$, $\overline{EC}=14-10=4$

The quadrilateral $ABCE$ is a trapezoid. The area $=\dfrac{(14+4)\times 5}{2}=45$.

23. C Draw a Venn Diagram.

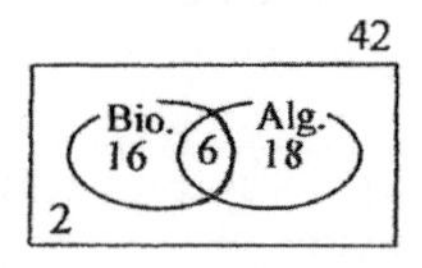

$42-2=40$
$24+22-40=6$

24. B

5	4	3	2	1

$5\times4\times3\times2\times1=120$ seat arrangements.

x	4	3	2	1

$4\times3\times2\times1=24$ seat arrangements for any one sitting in the first seat.

$120-24=96$ seat arrangements if one of them does not want to sit in the first seat.

Answer Explanations
(PRACTICE TEST 7)

25. B (1) $h=3b+2$, $bh=3b^2+2b$. Area $=\frac{1}{2}bh=\frac{1}{2}(3b^2+2b)$;
No information to determine the value of b; Not Sufficient.
(2) $2b=\frac{32}{h}$, $bh=16$. Area $=\frac{1}{2}bh=\frac{1}{2}(16)=8$; Sufficient.

26. C Area of ΔABD = Area of ΔABC − Area of $\Delta ADC=\frac{1}{2}(8)(5)-\frac{1}{2}(5)(5)=20-12.5=7.5$.

27. B $3^{4x+3}=27^{2x-3}$, $3^{4x+3}=(3^3)^{2x-3}$, $3^{4x+3}=3^{6x-9}$, $4x+3=6x-9$, $12=2x$, $x=6$.

28. D

Points	2,	3,	4,	5,	6,	7,	8
Lines	1	3	6	10	15	21	28
	+2	+3	+4	+5	+6	+7	

29. E In one hour, they can complete $\frac{1}{6}+\frac{1}{12}=\frac{2}{12}+\frac{1}{12}=\frac{3}{12}=\frac{1}{4}$ of the job.
To complete the job, they need $1\div\frac{1}{4}=1\times4=4$ hours.

30. D (1) Maria can complete the job in 12 hours.
To complete the job together, they can complete the job in
$1\div(\frac{1}{6}+\frac{1}{12})=4$ hours. $6-4=2$ less hours; Sufficient.
(2) They can complete the job in $6\div1.5=4$ hours. $6-4=2$ less hours; Sufficient.

31. C (1) Multiples of 9 : 9, 18, 27, **36**, 45, 54, 63, **72**, ····· .
Some are multiples of 36, some are not; Not Sufficient.
(2) Multiples of 4 : 4, 8, 12, 16, 20, ····· , **36**, 40, ···· , 68, **72** , ····· ;
Some are multiples of 36, some are not; Not Sufficient
Together, the multiples of 9 and 4 ; **36, 72, 108,** ····· . They are all multiples of 36. Sufficient.

32. E (1) Multiples of 9 : 9, 18, **27**, 36, 45, **54**, ····· ; Not Sufficient.
(2) Multiples of 3 : 3, 6, 9, ····· , **27**, 30, 33, ····· , **54**, ····· ; Not Sufficient.
Together, the multiples of 9 and 3 : 9, 18, **27**, 36, 45, **54**, ·····. Some are multiples of 27, some are not; Not sufficient.

33. D The product of two or more odd integers is always odd.
(1) Try some numbers: $3^3=27$, $5^3=125$, ····· ; Sufficient.
(2) Try some numbers: $\sqrt{9}=3$, $3^3=27$; $\sqrt{25}=5$, $5^3=125$, ··· ; Sufficient.

34. A (1) We choose any number from $0<x<1$, say $x=0.5$, $(0.5)^2<0.5$; Sufficient.
(2) Test any real number of x : $(-2)^2>0$, but $x<0$; Not Sufficient.

35. B (1) No information to determine the number of questions she answer correctly ; Not Sufficient.
(2) $40\times0.80+50\times0.90=77$ questions she answered correctly ; Sufficient.

36. C $OA=OB$, ΔOAB is isosceles. $m\angle A=m\angle B$, $x=\frac{180-110}{2}=35$.

37. B The standard form of parabola : $(y-k)^2=a(x-h)$.
The parabola has vertex $(h, k) = (4, 0)$ and open to the left ($a<0$).
The equation is : $(y-0)^2=-(x-4)$.

PRACTICE TEST 8
(GMAT Test)

QUANTITATIVE SECTION

Time–75 minutes
37 Questions

Problem Solving Questions: Solve each problem and indicate the best of the answer choices given.

Numbers: All numbers used are real numbers.

Figures: A figure accompanying a problem solving question is intended to provide information useful in solving the problem. Figures are drawn as accurately as possible EXCEPT when it is stated in a specific problem that its figure is not drawn to scale. Lines that appear jagged can be assumed to be straight.. All figures lie in a plane unless otherwise indicated.

Data Sufficiency Questions: It consists of a question and two statements, labeled (1) and (2), in which certain data are given. You have to decide whether the data given in the statements are sufficient for answering the question.

Numbers: All numbers used are real numbers.

Figures: A figure accompanying a data sufficiency question will conform to the information given in the question but will not necessary conform to additional information given in statements (1) and (2). Lines that appear jagged can be assumed to be straight. All figures lie in a plane unless otherwise indicated.

The positions of points, angles, regions, etc. exist in the order shown and that angle measures are greater than zero.

To answer the question asked, you must indicate whether ~

A Statement (1) ALONE is sufficient, but statement (2) alone is not sufficient.
B Statement (2) ALONE is sufficient, but statement (1) alone is not sufficient.
C BOTH statements TOGETHER are sufficient, but NEITHER statement ALONE is sufficient.
D EACH statement ALONE is sufficient.
E Statements (1) and (2) TOGETHER are not sufficient.

1. Which of the following is the reciprocal of $2\sqrt{3}$?

(A) $\frac{\sqrt{3}}{2}$ (B) $\frac{\sqrt{3}}{6}$ (C) $\frac{\sqrt{6}}{2}$
(D) $\frac{\sqrt{6}}{3}$ (E) $\frac{\sqrt{2}}{6}$

2. How many multiples of 6 are there between 24 and 5,000, inclusive ?

(A) 831 (B) 830 (C) 829
(D) 828 (E) 827

GO ON TO THE NEXT PAGE.

3. If $(9^{3x+1})(3^{-x}) = 81$, then $x =$

(A) $\frac{1}{2}$

(B) $\frac{1}{3}$

(C) $\frac{3}{2}$

(D) $\frac{2}{3}$

(E) $\frac{2}{5}$

4. If $|x-a| < b$, where a and b are positive numbers, then $x =$

(A) $a < x < b$
(B) $b < x < a$
(C) $a-b < x < a+b$
(D) $b-a < x < a+b$
(E) $x < a-b$ or $x > a+b$

5. If $x < -1$, which of the following must be true ?

I. $x^3 < x$
II. $x^5 + x^3 < x^4 + x^2$
III. $x^4 - x^3 < x^2 - x$

(A) I only
(B) II only
(C) I and II only
(D) II and III only
(E) I, II, and III

6. If $2^x + 2^x = \sqrt{8}$, then $x =$

(A) $\frac{3}{2}$

(B) 1

(C) $\frac{1}{2}$

(D) $\frac{1}{3}$

(E) $\frac{2}{3}$

7. The hourly wage of a teacher is twice the hourly wage of his assistant. They were paid a total of $266 for working on a project. The teacher worked 10 hours and his assistant worked 8 hours. What is the hourly wage in dollars of the assistant ?

(A) $ 9.50
(B) $10.50
(C) $11.50
(D) $12.50
(E) $13.00

8. There are 20 bulbs in a box and 5 are defective. You select 4 bulbs from the box. What is the probability (round to the nearest 0.01 %) that exactly 3 are defective ?

(A) 7.04 %
(B) 6.12 %
(C) 5.07 %
(D) 3.10 %
(E) 2.18 %

GO ON TO THE NEXT PAGE.

9. Roger picked up two-fifths of the books on the shelf. Maria picked up one-half of the remaining books. Jack picked up 6 books that were left. How many books were on the shelf to begin with ?

(A) 52
(B) 46
(C) 34
(D) 28
(E) 20

10. If $x^2 = 36$, $y^2 = 9$, $x > 0$ and $y > 0$, what is the value of $(x - y)^2$?

(A) 18
(B) 15
(C) 12
(D) 9
(E) 6

11. Line k is parallel to the line $3x - 4y = 12$. What is the slope of line k ?

(A) 0.34
(B) 0.75
(C) 1.25
(D) 1.82
(E) 2.36

12. Line p is perpendicular to the line $8x + y = 16$. What is the slope of line p ?

(A) 0.125
(B) 0.945
(C) 1.374
(D) 2.244
(E) 2.546

13. What is the sum of $1 + 2 + 3 + \cdots + 200$?

(A) 20,500
(B) 20,400
(C) 20,300
(D) 20,200
(E) 20,100

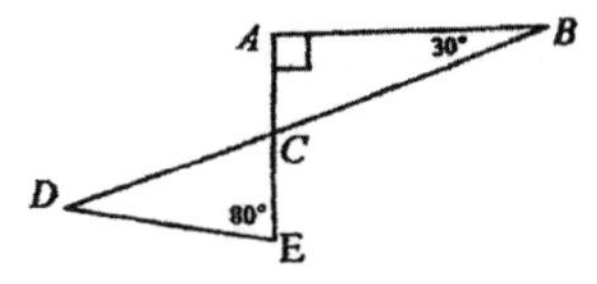

Note: Figure not drawn to scale.

14. In the figure above, what is the degree measure of $< D$?

(A) 30
(B) 35
(C) 40
(D) 45
(E) 50

15. What is the largest integer of the 4 consecutive even integers ?

(1) The average of the smallest integer and the largest integer is 7.
(2) The sum of the integers is 28.

GO ON TO THE NEXT PAGE.

16. David, John, and Mona are friends. Who is the oldest ?

(1) John is younger than one of them and older than the other one.
(2) John's age is closer to David's age than to Mona's age.

17. If the function $f(x) = x^2 + ax + 18$, where a is constant, what is the value of a ?

(1) $f(2) = 0$
(2) $f(9) = 0$

18. If $x^2 + ax + 18 = (x-9)(x-b)$, where a and b are constants, what is the value of a ?

(A) -18
(B) -11
(C) -9
(D) -7
(E) -2

19. A company makes 100,000 tires a year. The life of a tire is normally distributed with a mean of 32,000 miles and a standard deviation of 3,000 miles. Approximately how many tires will last between 35,000 and 38,000 miles ?

(A) 68,000 tires
(B) 48,000 tires
(C) 34,000 tires
(D) 20,000 tires
(E) 14,000 tires

20. Is $xy > 0$?

(1) $x + y = -5$
(2) $x < 0$ and $y < 0$

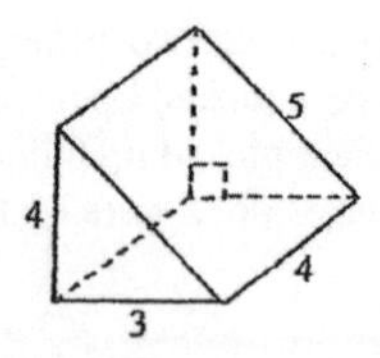

Note: Figure not drawn to scale.

21. In the figure above, what is the surface area of this right triangular prism ?

(A) 60
(B) 55
(C) 50
(D) 45
(E) 40

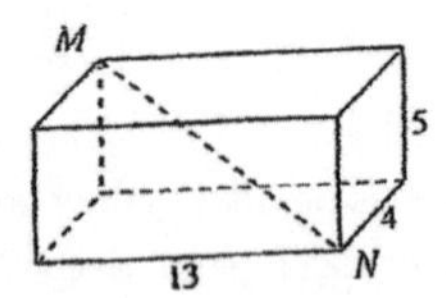

Note: Figure not drawn to scale.

22. In the figure above, what is the length of diagonal $\overline{MN}$ in the rectangular solid ?

(A) $\sqrt{180}$
(B) $\sqrt{200}$
(C) $\sqrt{210}$
(D) $\sqrt{220}$
(E) $\sqrt{230}$

GO ON TO THE NEXT PAGE.

23. If $|x+y| = |x| + |y|$, which of the following could be the value of x and y?

I. $x > 0$ and $y < 0$
II. $x < 0$ and $y < 0$
III. $x < 0$ and $y > 0$

(A) I only

(B) II only

(C) I and II only

(D) II and III only

(E) I, II, and III

24. what is the ratio of $a : b : c$?

(1) $ab = 2$ and $bc = 3$
(2) $ac = 6$ and $b = 1$

25. If ΔABC is a right triangle, what is the length of side AB?

(1) The length of side AC is 8.
(2) The length of side BC is 6.

26. In ΔABC, what is the length of side AB?

(1) The length of side AC is 8.
(2) The length of side BC is 6.

27. If ΔABC is a 30-60-90 triangle, AB is the hypotenuse, what is the area?

(1) The length of side AC is $6\sqrt{3}$.
(2) The length of side BC is 6.

28. In a high school, the number of boys increased 10 % and girls decreased by 12 % this year. The ratio of boys to girls at the end of this year was how many times the ratio at the beginning of this year?

(A) $\frac{9}{4}$

(B) $\frac{2}{1}$

(C) $\frac{7}{4}$

(D) $\frac{3}{2}$

(E) $\frac{5}{4}$

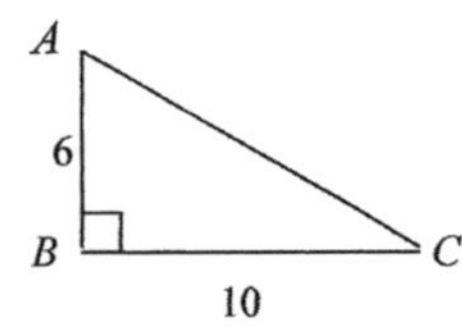

Note: Figure not drawn to scale.

29. The right triangle ABC above is revolved about side AB as an axis generating a right cone. What is the volume of the cone?

(A) 230 π
(B) 220 π
(C) 210 π
(D) 200 π
(E) 190 π

GO ON TO THE NEXT PAGE.

30. What is the value of x ?

(1) $2y+1=2(x-y)$
(2) $y-2x=4$

31. What is the value of x ?

(1) $2y+1=2(x+y)$
(2) $y-2x=4$

32. A wire of x feet in length is cut into exactly 10 pieces, each 3 feet 2 inches in length. What is the length of x ?

(A) $30\frac{4}{5}$ feet
(B) $31\frac{2}{3}$ feet
(C) $31\frac{2}{5}$ feet
(D) $32\frac{1}{4}$ feet
(E) $32\frac{2}{3}$ feet

33. What is the ratio of a to b ?

(1) The ratio of $0.4a$ to $5b$ is $\frac{4}{5}$.
(2) a is 2 more than 8 times b.

34. City A and City B each had residents increase. Which city had the greater number of residents increase ?

(1) The residents of City A increased 12 %.
(2) The residents of City B increased 15 %.

35. In a college, there are 6,000 students, one-third of them are female. What percent of the students are foreign students ?

(1) The number of males who are foreign students is twice the number of females who are foreign students.
(2) The number of males who are not foreign students is twice the number of females who are not foreign students.

36. In a college, there are 6,000 students, one-third of them are female. What percent of the students are foreign students ?

(1) The number of males who are foreign students is twice the number of females who are foreign students.
(2) The number of males who are not foreign students is 7 times the number of males who are foreign students ?

37. A 10-liter salt-water solution contains x liters of pure salt and y liters of water. How many liters of pure salt does the solution contain ?

(1) $y-x=6$
(2) $4x=y$

STOP

IF YOU FINISH BEFORE TIME IS CALLED, YOU MAY CHECK YOUR WORK ON THIS SECTION ONLY. DO NOT TURN TO ANY OTHER SECTION IN THE TEST.

PRACTICE TEST 8
(GMAT Test)

Quantitative Section Answer Key (Page 449 ~ 454)

(Page 449 ~ 454)

1. B	**2.** B	**3.** E	**4.** C	**5.** C	**6.** C	**7.** A	**8.** D	**9.** E	**10.** D
11. B	**12.** A	**13.** E	**14.** C	**15.** D	**16.** E	**17.** D	**18.** B	**19.** E	**20.** B
21. A	**22.** C	**23.** B	**24.** A	**25.** C	**26.** E	**27.** C	**28.** E	**29.** D	**30.** C
31. A	**32.** B	**33.** A	**34.** E	**35.** E	**36.** C	**37.** D			

CALCULATE YOUR MATH SCORE

	Correct	Incorrect
Questions 1 ~ 37	( )	( )

Total Unrounded Raw Score ( ) − ( ) × 0.25 = ( )
Total Rounded Raw Score ( )

The actual GMAT multiple-choice test scores are determined by a complex procedure. The quantitative (math) and verbal scores are combined (37 + 41) to determine your total GMAT score on multiple-choice questions. It will vary from the score table presented here.
Using the table below to convert your raw score to scaled score. Without verbal score, you may double your math raw score and use the table to estimate your total GMAT scaled score on multiple-choice questions.

GMAT VERBAL AND MATH TESTS CONVERSION TABLE

Raw Score	Scaled Score	Raw Score	Scaled Score	Raw Score	Scaled Score
0 to 10	200 ~ 340	34	520	58	650
11	350	35	520	59	650
12	360	36	530	60	660
13	370	37	530	61	670
14	380	38	540	62	670
15	390	39	540	63	680
16	400	40	550	64	680
17	410	41	550	65	690
18	420	42	560	66	700
19	430	43	560	67	710
20	440	44	570	68	720
21	450	45	570	69	730
22	460	46	580	70	740
23	460	47	590	71	750
24	470	48	590	72	760
25	470	49	600	73	770
26	480	50	600	74 to 78	780 ~ 800
27	480	51	600		
28	490	52	610		
29	490	53	610		
30	500	54	620		
31	500	55	620		
32	510	56	630		
33	510	57	640		

Answer Explanations
(PRACTICE TEST 8)

1. B $\frac{1}{2\sqrt{3}}=\frac{1}{2\sqrt{3}}\cdot\frac{\sqrt{3}}{\sqrt{3}}=\frac{\sqrt{3}}{2(3)}=\frac{\sqrt{3}}{6}$.

2. B $6\times4=24$, $6\times5=30$, $6\times6=36$, ⋯⋯, $6\times833=4998$.
There are $833-3=830$ multiples of 6.

3. E $(9^{3x+1})(3^{-x})=81$
$(3^2)^{3x+1}\cdot3^{-x}=3^4$, $3^{6x+2-x}=3^4$, $6x+2-x=4$, $5x=2$, $x=\frac{2}{5}$.

4. C $|x-a|<b$, we have $-b<x-a<b$
$a-b<x<a+b$.

5. C $x<-1$ indicates x is a negative number.
I. $(-1.1)(-1.1)(-1.1)<1.1$, $x^3<x$ is correct.
2. x^5+x^3 is negative. x^4+x^2 is positive. $x^5+x^3<x^4+x^2$ is correct.
3. If $x^4-x^3<x^2-x$,
$x^3(x-1)<x(x-1)$ → divide each side by $(x-1)$ (it is a negative number.).
$x^3>x$ → divide each side by x (it is a negative number.).
$x^2<1$ is incorrect. $(-1.1)(-1.1)>1$.

6. C $2^x+2^x=\sqrt{8}$, $2(2^x)=8^{\frac{1}{2}}$, $2^{1+x}=(2^3)^{\frac{1}{2}}$, $2^{1+x}=2^{\frac{3}{2}}$, $1+x=\frac{3}{2}$, $x=\frac{1}{2}$.

7. A Let x = the hourly wage of the assistant
$2x$ = the hourly wage of the teacher
We have the equation: $10(2x)+8x=266$
$28x=266$, $x=9.50$.

8. D $p=\frac{{}_5C_3\times{}_{15}C_1}{{}_{20}C_4}=\frac{10\times15}{4845}=\frac{150}{4845}\approx0.03096\approx3.10$ %.
Or: $p=\frac{15}{20}\times\frac{5}{19}\times\frac{4}{18}\times\frac{3}{17}+\frac{5}{20}\times\frac{15}{19}\times\frac{4}{18}\times\frac{3}{17}+\frac{5}{20}\times\frac{4}{19}\times\frac{15}{18}\times\frac{3}{17}+\frac{5}{20}\times\frac{4}{19}\times\frac{3}{18}\times\frac{15}{17}$
$=(\frac{15}{20}\times\frac{5}{19}\times\frac{4}{18}\times\frac{3}{17})\times4=\frac{900}{116280}\times4\approx0.03096\approx3.10$ %.

9. E Let n = number of books to begin with
We have the equation: $\frac{3}{5}n-\frac{1}{2}(\frac{3}{5}n)=6$
Solve for n: $\frac{3}{5}n-\frac{3}{10}n=6$, $\frac{6n-3n}{10}=6$, $3n=60$, $n=20$.

10. D $x^2=36$ and $x>0$, we have $x=6$.
$y^2=9$, and $y>0$, we have $y=3$. $(x-y)^2=(6-3)^3=3^2=9$.

11. B $3x-4y=12$, $y=\frac{3}{4}x-3$, the slope (m) of the line is: $m=\frac{3}{4}$.
The slope of the line k: $m'=\frac{3}{4}=0.75$.

12. A $8x+y=16$, $y=-8x+16$, the slope (m) of the line is: $m=-8$.
The slope of the line p: $m'=-\frac{1}{-8}=\frac{1}{8}=0.125$.

Answer Explanations
(PRACTICE TEST 8)

13. E There are 100 pairs of (200 + 1) : The sum $= (200+1)\times 100 = 20{,}100$.

14. C $\angle ACB = 180^\circ - 90^\circ - 30^\circ = 60^\circ$, $\angle DCE = \angle ACB = 60^\circ$, $\angle D = 180^\circ - 60^\circ - 80^\circ = 40^\circ$.

15. D The 4 consecutive even integers: n, $n+2$, $n+4$, $n+6$

(1) $\frac{n+n+6}{2} = 7$, $2n+6=14$, $n=4$. The largest integer is $n+6=10$; Sufficient.

(2) $n+(n+2)+(n+4)+(n+6)=28$, $4n+12=28$, $n=4$.
The largest integer is $n+6=10$; Sufficient.

16. E (1) David < John < Mona or Mona < John < David ; Not Sufficient.
(2) No information to determine the ages about David and Mona ; Not Sufficient.
Together: David John ——— Mona or Mona ——— John David
Still no information to determine who is the eldest, David or Mona. Not Sufficient.

17. D (1) $f(x) = x^2 + ax + 18$, $f(2) = 2^2 + 2a + 18 = 0$, $a = -11$; Sufficient.
(2) $f(x) = x^2 + ax + 18$, $f(9) = 9^2 + 2a + 18 = 0$, $a = -11$; Sufficient.

18. B
$$x^2 + ax + 18 = (x-9)(x-b)$$
$$x^2 + ax + 18 = x^2 - bx - 9x + 9b$$
$$x^2 + ax + 18 = x^2 - (b+9)x + 9b$$
Compare the coefficients, we have : $a = -(b+9)$ and $18 = 9b$.
Solve the system of two equations : $a = -11$, $b = 1$.

19. E (See normal distribution on page 72.)
The probability of an event having the value between 1 standard deviation from the mean (32,00 + 3,000) and 2 standard deviations from the mean (32,000 + 2×3,000) is about 13.6 %., $100{,}000 \times 0.136 = 13{,}600 \approx 14{,}000$ tires.

20. B (1) Try some numbers: $-2-3=-5, xy>0$, $-6+1=-5, xy<0$; Not Sufficient.
(2) Both x and y are negative, therefore $xy>0$; Sufficient.

21. A Surface area $= \frac{1}{2}(3\times 4)\times 2 + 3\times 4 + 4\times 4 + 5\times 4 = 12+12+16+20 = 60$.

22. C $\overline{MN}$ is the diagonal of the rectangular solid. Apply Pythagorean Theorem in three-dimensions. Length of diagonal : $d = \sqrt{13^2+4^2+5^2} = \sqrt{169+16+25} = \sqrt{210}$.

23. B Test some numbers : $|x+y| = |x|+|y|$

I. $|3-2| \neq |3|+|-2| \to 1 \neq 3+2$.

II. $|-3-2| = |-3|+|-2| \to 5 = 3+2$.

III. $|-2+3| \neq |-2|+|3| \to 1 \neq 2+3$.

24. A (1) $ab=2$ and $bc=3$.
We have $a = \frac{2}{b}$ and $c = \frac{3}{b}$. $a:b:c = \frac{2}{b}:b:\frac{3}{b} = 2:1:3$; Sufficient.
(2) $b=1$, $ac = 6 = 1\times 6 = 2\times 3 = 3\times 2 = 6\times 1$, a and c cannot be determined. Not Sufficient.

25. C (1) No information to determine the length of side BC; Not Sufficient.
(2) No information to determine the length of side AC; Not Sufficient.
Together: It is a right (90°) triangle. Apply Pythagorean Theorem, we can find the length of the third side AB; Sufficient.

Answer Explanations
(PRACTICE TEST 8)

26. E (1) No information to determine the degree of any angle and the length of BC ; Not Sufficient.
(2) No information to determine the degree of any angle and the length of AC ; Not Sufficient
Together: No information to determine the degree measure of any angle ; Not Sufficient

27. C (1) In a 30-60-90 triangle, the length of the hypotenuse is twice the length of the shorter leg. No information to determine whether AC is the shorter leg ; Not Sufficient.
(2) No information to determine whether BC is the shorter leg ; Not Sufficient.
Together: area $= \frac{1}{2} \times AC \times BC = \frac{1}{2} \times 6\sqrt{3} \times 6 = 18\sqrt{3}$; Sufficient.

28. E Let $\frac{B}{G} =$ the ratio of boys to girls at the beginning of the year.
At the end of the year: $\frac{1.10B}{0.88G} = \frac{110B}{88G} = \frac{5}{4}\left(\frac{B}{G}\right)$.

29. D Volume $= \frac{1}{3}\pi \cdot r^2 \cdot h = \frac{1}{3}\pi \cdot 10^2 \cdot 6 = 200\pi$.

30. C (1) $2y+1=2(x-y)$, $\cancel{2y}+1=2x-\cancel{2y}$, $4y-2x=-1$; Not Sufficient.
(2) $y-2x=4$; One equation with 2 variables ; Not Sufficient.
Together : we have a system of two equations with two variables ; Sufficient.

31. A (1) $2y+1=2(x+y)$, $2y+1=2x+2y$, $2x=1$, $x=\frac{1}{2}$; Sufficient.
(2) $y-2x=4$; One equation with 2 variables ; Not Sufficient.

32. B 1 feet = 12 inches, 3 feet 2 inches = 38 inches, 38 inches × 10 = 380 inches.
380 inches ÷ 12 = $31\frac{2}{3}$ feet.

33. A (1) $\frac{0.4a}{5b} = \frac{4}{5}$, $\frac{a}{b} = \frac{4 \times 5}{5 \times 0.4} = \frac{20}{2} = \frac{10}{1}$. ; Sufficient.
(2) $a = 8b+2$. $\frac{a}{b} = 8 + \frac{2}{b}$. No information to find the value of b or $\frac{a}{b}$; Not Sufficient.

34. E (1) No information about the original number of residents ; Not Sufficient.
(2) No information about the original number of residents ; Not Sufficient.
Together : Still no information about the original number of residents ; Not Sufficient.

35. E There are 2000 females and 4000 males.
Let b = male-foreigners, g = female-foreigners.
(1) $b = 2g$; One equation with 2 variables ; Not Sufficient.
(2) $4000 - b = 2(2000 - g)$ gives $b = 2g$; same as (1) Not Sufficient.
Together: Still one equation with two variables b and g ; Not Sufficient.

36. C There are 2000 females and 4000 males.
Let b = male-foreigner, g = female-foreigners.
(1) $b = 2g$; One equation with 2 variables ; Not Sufficient.
(2) $4000 - b = 7b$ gives $b = 500$; No information to find g ; Not Sufficient.
Together: $b = 500$, $g = 250$, $p = \frac{500+250}{6000} = 12.5\%$; Sufficient.

37. D (1) Solve the system of $x+y=10$ and $y-x=6$, we have $x=2$, $y=8$; Sufficient.
(2) Solve the system of $x+y=10$ and $4x=y$, we have $x=2$, $y=8$; Sufficient.

Challenging Examples

1. Suppose you throw a fair coin 4 times. What is the probability that you will have exactly 2 heads and 2 tails ?
Solution :
Method 1 : (Combinations, see page 74)

$${}_4C_0 = 1,\ {}_4C_1 = 4,\ {}_4C_2 = 6,\ {}_4C_3 = 4,\ {}_4C_4 = 1$$

$$p = \frac{6}{16} = \frac{3}{8}.$$

Method 2 : (Binomial Distribution, see page 70)

$$P = {}_4C_2\left(\frac{1}{2}\right)^2\left(\frac{1}{2}\right)^{4-2} = 6\left(\frac{1}{4}\right)\left(\frac{1}{4}\right) = \frac{6}{16} = \frac{3}{8}.$$

2. Suppose you throw a fair coin 4 times. What is the probability that you will have exactly 3 heads and 1 tail ?
Solution: (Binomial Distribution, see page 70)

$$P = {}_4C_3\left(\frac{1}{2}\right)^3\left(\frac{1}{2}\right)^{4-3} = 4\left(\frac{1}{8}\right)\left(\frac{1}{2}\right) = \frac{4}{16} = \frac{1}{4}.$$

3. A bookstore purchased a number of A-Plus math books at a total cost of \$4,000. On the first month, 80 percent of the books were sold, each copy at a price of 40 percent more than the purchased cost. The remaining books were sold, each copy at a price of 30 percent less than the price of the first month. What was the profit of the bookstore on these books ?
Solution:

$$\begin{aligned}\text{Total revenue} &= 0.80\ (\$4{,}000 \times 1.40) + 0.20\ (\$4{,}000 \times 1.40 \times 0.70)\\ &= 0.80\ (\$5{,}600) + 0.20\ (\$3{,}920)\\ &= \$4{,}480 + \$784\\ &= \$5{,}264\end{aligned}$$

The profit = \$5,264 − \$4,000 = \$1,264.

4. One-third of all Matts are Patts. Half of all Patts are Natts. No Matt is a Natt. All Natts are Patts. There are 25 Natts and 45 Matts. How many Patts are neither Matts nor Natts ?
Solution: (Venn Diagram, see page 399)

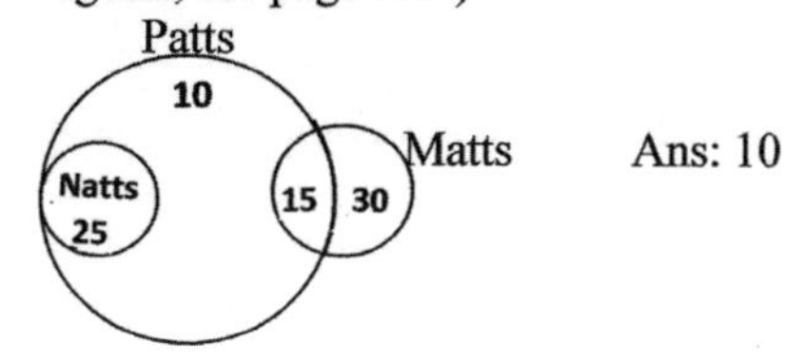

Ans: 10

Challenging Examples

5. The figure below shows the graph of a normal distribution with mean μ, and standard deviation σ. The graph also shows the approximate percents of the distribution corresponding to the regions shown. Suppose the heights of 2,000 students in a high school are approximately normally distributed with a mean 172 centimeters and a standard deviation of 4 centimeters.

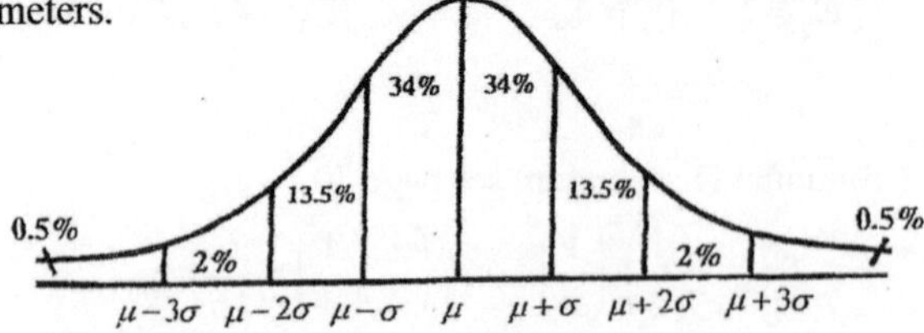

a. How many students are between 164 centimeter and 172 centimeters tall ?
b. If one student is chosen at random, what is the probability that the height of this student will be less than 160 centimeters ?
c. If one student is chosen at random, what is the probability that the height of this student will be greater than $176\,cm$ and less than $180\,cm$
d. If the height of a student is at the 15th percentile of the distribution, which of the heights $167\,cm$, $167.5\,cm$, $168\,cm$, or $168.5\,cm$ is the best estimate of the height of this student ?

Solution:

Let h = the random variable for height which is normally distributed with a mean of 172, and a standard deviation 4.

a. The heights between $164\,cm$ and $172\,cm$ are 2 standard deviations below the mean. $p(164 < h < 172) = 0.34 + 0.135 = 0.475$

$2{,}000 \times 0.475 = 950$ students.

b. The height $160\,cm$ is 3 standard deviations below the mean.

$p(h < 160) = 0.5\% = 0.005 \approx 0.01$.

c. The percentile between 176 cm and $180\,cm$ is 14 percent of every normal distribution between 1 and 2 standard deviations above the mean.

$p(176 < h < 180) = 0.14$.

d. The height $168\,cm$ is at 1 standard deviation below the mean of $172\,cm$. 16% of the normal distribution is less than $168\,cm$.

$p(h < 168) = 0.135 + 0.02 + 0.005 = 0.16$.

The 15th percentile of the distribution is at a value a little less than $168\,cm$. The best estimate of the height is $167.5\,cm$.

Notes

Notes

Notes

Notes

INDEX

A

B

C

D

E

F

G

H

I

J

L

M

N

O

P

Q

R

S

T

U

V

W , Z

The A-Plus Notes Math Series (Printed in USA)

Author: Rong Yang

1. **A-Plus Notes for Beginning Algebra (Pre-Algebra and Algebra 1)**
   ISBN: 978-0-9654352-2-2

2. **A-Plus Notes for Algebra (Algebra 2 and Pre-Calculus)**
   (A Graphing Calculator Approach)
   ISBN: 978-0-9654352-4-6

3. **A-Plus Notes for SAT Math (SAT/PSAT and Subject Tests)**
   ISBN: 978-0-9654352-6-0

4. **A-Plus Notes for GRE/GMAT Math (General and Subject Tests)**
   (GRE revised General Test)
   ISBN: 978-0-9654352-9-1

(To view the above math books, visit at www.amazon.com)

Notes

Notes

Notes